Comparative Vertebrate Lateralization

No longer viewed as a characteristic unique to humans, brain lateralization is considered a key property of most, if not all, vertebrates. This field of study provides a firm basis from which to examine a number of important issues in the study of brain and behaviour. In addition to neuroscience, it has implications in developmental biology, genetics, ethology, and comparative psychology.

This book takes a comparative and integrative approach to lateralization in a wide range of vertebrate species, including humans. It highlights model systems that have proved invaluable in elucidating the function, causes, development, and evolution of lateralization. The book is arranged in four parts, beginning with the evolution of lateralization, moving to its development, to its cognitive dimensions, and finally to its role in memory. Experts in lateralization in lower vertebrates, birds, non-primate mammals, and primates have contributed chapters in which they discuss their own research and consider its implications to humans. This is the first book to take a definitive look at lateralization in a truly comparative context.

Researchers, graduates, and advanced undergraduates in psychology, neuroscience, and the behavioural sciences will find this an important and authoritative text.

Lesley J. Rogers is Professor of Neuroscience and Animal Behaviour at the University of New England.

Richard J. Andrew is Professor Emeritus of Animal Behaviour at the University of Sussex.

COMPARATIVE VERTEBRATE LATERALIZATION

Edited by

LESLEY J. ROGERS
University of New England

RICHARD J. ANDREW
University of Sussex

CAMBRIDGE
UNIVERSITY PRESS

CAMBRIDGE UNIVERSITY PRESS
Cambridge, New York, Melbourne, Madrid, Cape Town, Singapore, São Paulo

Cambridge University Press
The Edinburgh Building, Cambridge CB2 8RU, UK

Published in the United States of America by Cambridge University Press, New York

www.cambridge.org
Information on this title: www.cambridge.org/9780521781619

First published 2002
Re-issued in this digitally printed version 2008

A catalogue record for this publication is available from the British Library

Library of Congress Cataloguing in Publication data

Comparative vertebrate lateralization / edited by Lesley J. Rogers, Richard J. Andrew.
    p. cm.
Includes bibliographical references and index.
ISBN 0-521-78161-2
1. Cerebral dominance. 2. Comparative neurobiology. I. Rogers, Lesley J., 1943– II. Andrew,
Richard John, 1932–
QP385.5.C65 2002
573.8′616–dc21   2001035239

ISBN 978-0-521-78161-9 hardback
ISBN 978-0-521-78700-0 paperback

Contents

Contributors

Richard J. Andrew
Sussex Centre for Neuroscience, School of Biological Sciences, University of Sussex, Brighton BN1 9QG, UK

Angelo Bisazza
Department of General Psychology, University of Padua, Padua, Italy

Samuel Fernández Carriba
Division of Psychobiology, Yerkes Regional Primate Research Center, Atlanta, GA 30322, USA, and Departmento de Psicología Biológica y de Salud, Facultad de Psicología, Universidad Autónoma de Madrid, 28049 Madrid, Spain

Patricia E. Cowell
Department of Human Communication Sciences, University of Sheffield, Sheffield, UK

Eric Damerose
Centre de Recherche en Neurosciences Cognitives, CRNC-CNRS, Marseille, France, and Station de Primatologie, CNRS, Rousset-sur-Arc, France

Victor H. Denenberg
Biobehavioral Sciences Graduate Degree Program, University of Connecticut, Storrs, CT, USA

Chao Deng
Division of Zoology, School of Biological Sciences, University of New England, Armidale, NSW 2351, Australia

Asif A. Ghazanfar
Primate Cognitive Neuroscience Laboratory, Department of Psychology, 33 Kirkland Street, Harvard University, Cambridge, MA 02138, USA

Onur Güntürkün
Biopsychologie, Fakultät für Psychologie, Ruhr-Universität, Bochum, 44780 Bochum, Germany

Marc D. Hauser

Primate Cognitive Neuroscience Laboratory, Department of Psychology, 33 Kirkland Street, Harvard University, Cambridge, MA 02138, USA

William D. Hopkins

Department of Psychology, Berry College, Mount Berry, GA 30149, USA, and *Division of Psychobiology, Yerkes Regional Primate Research Center, Atlanta, GA 30322, USA*

Amy N.B. Johnston

Department of Biology, The Open University, Walton Hall, Milton Keynes, MK7 6AA, UK

Cory T. Miller

Primate Cognitive Neuroscience Laboratory, Department of Psychology, 33 Kirkland Street, Harvard University, Cambridge, MA 02138, USA

Lucia Regolin

Department of General Psychology, University of Padua, Via Venezia 8, 35131 Padova, Italy

Lesley J. Rogers

Division of Zoology, School of Biological Sciences, University of New England, Armidale, NSW 2351, Australia

Steven P.R. Rose

Department of Biology, The Open University, Walton Hall, Milton Keynes, MK7 6AA, UK

Giorgio Vallortigara

Department of Psychology and B.R.A.I.N. Centre for Neuroscience, University of Trieste, 34123 Trieste, Italy

J. Vauclair

Centre de Recherche en Neurosciences Cognitives, CRNC-CNRS, Marseille, France, and *UFR de Psychologie, Université de Provence, Center for Research in Psychology of Cognition, Language and Emotion, Aix-en-Provence, France*

J.A.S. Watkins

Sussex Centre for Neuroscience, School of Biological Sciences, University of Sussex, Brighton BN1 9QG, UK

Daniel J. Weiss

Department of Brain and Cognitive Science, Rochester University, Meliora Hall, Rochester, NY 14627-0268, USA

Preface

The idea of this book was conceived at the time Richard Andrew retired from his professorial position at Sussex University. It is a celebration of his lifelong contribution to the study of ethology. In that sense, however, it is premature because Richard Andrew has far from retired from highly active research, shown most clearly by his contributions to this book. If anything, his 'retirement' has allowed him time to develop his ideas, especially on lateralization, and structure them in a way that unifies the field. For this reason, it was important for Richard and me to work together in editing this volume and for Richard to contribute key chapters that develop his recent lines of thought. Our aim was to collect together a volume of recent research on lateralization in a wide range of vertebrate species and to highlight model systems that have been valuable in elucidating the function of lateralization, its causes, its development and its evolution.

Although Richard Andrew's contribution to ethology has been much wider than in the field of lateralization, this book celebrates his important contribution to the study of lateralization. We hope that it demonstrates that knowledge of lateralization clarifies certain problems in the study of brain and behaviour, especially memory formation, and provides an excellent basis for future research with broad implications in the fields of development and genetics. The field of lateralization has 'come of age' largely because it provides a firm basis on which to examine a number of important issues in the study of brain and behaviour.

Some of the chapters have been written by students of Richard Andrew, others by his close colleagues in research and yet others by fellow researchers sharing an interest in comparative vertebrate lateralization. We are sincerely grateful for their contributions. Finally, I would like to acknowledge that I have had the privilege of being both a student and research colleague of Richard Andrew.

L.J.R.

Introduction

LESLEY J. ROGERS AND R. J. ANDREW

Interest in lateralization of function in a non-human vertebrate species was stimulated by Fernando Nottebohm's finding of differential effects on song production of sectioning the left and right tracheosyringeal nerves supplying the musculature of the avian syrinx (Nottebohm, 1971). He found that, in the canary, severing the left nerve impairs singing, whereas severing the right nerve has no effect on song. Within the next decade, such lateralization of song control had been traced to centres in the forebrain (Nottebohm, Stokes and Leonard, 1976), and Victor Denenberg and colleagues had discovered lateralization in the rat brain for control of activity and emotional responses (Denenberg et al., 1978; Denenberg, 1981). The latter had also been reported by Bianki, whose work received less recognition in the Western world largely because he was based in Russia (in translation Bianki, 1988). Also, in the next decade, lateralization for visual responding had been discovered in the domestic chicken brain, shown first by unilateral treatment of the forebrain hemispheres with either cycloheximide (Rogers and Anson, 1979) or glutamate (Howard, Rogers and Boura, 1980). Treatment of the left hemisphere led to a set of behavioural changes that differed from those resulting from the same treatment of the right hemisphere. Later it was found that the behavioural lateralization in the chick was matched by asymmetry in the visual projections from the thalamus to the Wulst region of the forebrain (Rogers and Sink, 1988) and that glutamate treatment of the Wulst reveals lateralization by stimulating neural transmission in the Wulst (Rogers and Hambley, 1982) and enhancing the growth of visual projections unilaterally (Khyentse and Rogers, 1997).

Knowledge of lateralization in these species became a basis for a series of in-depth studies. This development was greatly assisted by the recognition that lateralization in birds, and in other species with optic nerve fibres completely crossed at the chiasma, can be revealed by testing the animals

1

monocularly (Mench and Andrew, 1986). Not only did the technique of monocular testing to reveal lateralization simplify the procedure for revealing lateralization, but it also met ethical guidelines for research more adequately, and so laid the basis for investigating lateralization in species living in their natural environment. Once it was clear that chicks tested using the right eye performed very differently from those using the left eye (Andrew, 1983; Rogers, Zappia and Bullock, 1985), it was a logical step to assume that animals with eyes placed laterally on the sides of their head would respond differently to stimuli detected on their left and right sides (Andrew and Dharmaretnam, 1991) and, in other cases, choose to view particular stimuli with either their left or right eye. First demonstrated clearly in the chick (Andrew and Dharmaretnam, 1993), eye preferences for viewing can be measured without a great deal of difficulty and can now be applied to study of animals in the wild.

The study of lateralization has so expanded over the last two decades that it is no longer seen merely as an interesting idiosyncrasy of a few species (admittedly including our own), but instead as a key property of most or all vertebrates. Until recently, investigation of lateralization in species other than humans was neglected, largely because of the widespread notion that lateralization was unique to humans. Indeed, so long as lateralization was seen as intimately linked to tool use, consciousness and language, no other conclusion was possible. It is now obvious that this earlier assumption was incorrect. Some researchers, while recognizing the existence of lateralization in nonhuman species, then developed the idea that human uniqueness resided in the fact that humans are more lateralized than all other species (Corballis, 1991). This too is incorrect, as many examples in various chapters of this book show. Not only is lateralization common in vertebrates, but there also appears to be a common pattern of lateralization that evolved at least as early as fish, which has been retained amongst all of the major groups of vertebrates.

At a statistical level there are two kinds of lateralization. One is present in individuals within a population but is inconsistent between individuals, so that there is no overall bias in the population (or species) as a whole. Handedness in rats and mice is the best known example of this kind of lateralization: half of a population is left-handed for retrieving food and the other half is right-handed (Collins, 1985). The other kind of lateralization is present at the population level as well as in individuals. In this case, the majority of individuals are lateralized in the same direction so that a frequency histogram for the population is skewed to the right or left side of the no-preference value. Lateralization for control of activity or emotional responses in rats fits this kind of distribution. It also shows that, within one species (rats in this case), there may be lateralization of both the first and second kinds depending on the behaviour scored.

Different kinds of lateralization occur at different levels of neural organization. To use the example of rats again, there are cortical lateralizations, occurring at both the individual and population levels, and also lateralizations occurring at hypothalamic level (Nordeen and Yahr, 1982). It is likely that the presence of different types of lateralization at different levels of neural organization explains why individual-level and population-level lateralization can coexist in the same species. Learning and individual experience may also modulate lateralization at different levels of neural organization.

The book is divided into four parts. The first part considers the evolution of lateralization of the type present at the population level and, in so doing, traces it back to the primitive chordates. Remarkably, it seems the basic pattern of lateralization that first evolved has been retained by fish, amphibians, birds and mammals, even primates. The resilience of this basic pattern suggests that it confers an advantage to the individual and to the population. Possible advantages, as well as disadvantages, of being lateralized are, therefore, discussed.

The second part discusses development of lateralization and how shifts in which hemisphere is in control at different ages determine transitions in behaviour as development takes place. The effect of early visual experience before and after hatching on lateralization is discussed and so is maternal influence on the development of hand preferences in primates. The role of the corpus callosum is important in the lateralization of the mammalian brain, and one chapter discusses its development and the influence of sex hormones on its development.

Cognition and lateralization form the basis of Part Three. The basic pattern of lateralization, it seems, is common to the visual, auditory and olfactory senses. The eventual evolution of language and its lateralization deserves reconsideration on this basis. Moreover, lateralized responding of primates to vocalizations and asymmetry of facial muscle movement in primates when communicating by facial expressions have allowed the identification of precursors to lateralization in humans. In addition, this section presents evidence for unexpected cognitive abilities of young chicks and relates them to hemispheric specialization.

Finally, in Part Four, lateralization of memory processes, both biological and behavioural, is discussed. Using the young chick as a model, a large number of studies have revealed neurochemical asymmetries in different regions of the forebrain. These are related to specialized use of the hemispheres in memory formation and recall.

Many chapters mention lateralization in humans, in comparison to lateralization in animals. Thus, the evolution of hemispheric specialization in humans is considered in a number of contexts and placed within the broader scheme of comparative vertebrate lateralization.

References

Andrew, R.J. (1983). Lateralization of cognitive function in higher vertebrates, with special reference to the domestic chick. In *Advances in Vertebrate Neuroethology*, ed. J.-P. Ewert, R.R. Capranica & D. Ingle, pp. 477–509. New York: Plenum Press.

Andrew, R.J. & Dharmaretnam, M. (1991). A timetable of development. In *Neural and Behavioural Plasticity: The Use of the Domestic Chick as a Model*, ed. R.J. Andrew, pp. 166–173. Oxford: Oxford University Press.

Andrew, R.J. & Dharmaretnam, M. (1993). Lateralization and strategies of viewing in the domestic chick. In *Vision, Brain, and Behavior in Birds*, ed. H.P. Zeigler & H.-J. Bischof, pp. 319–332. Cambridge, MA: MIT Press.

Bianki, V.L. (1988). *The Right and Left Hemispheres: Cerebral Lateralization of Function. Monographs in Neuroscience*, Vol. 3. New York: Gordon and Breach.

Collins, R.L. (1985). On the inheritance of the direction and degree of asymmetry. In *Cerebral Lateralization in Nonhuman Species*, ed. S.D. Glick, pp. 41–71. New York: Academic Press.

Corballis, M.C. (1991). *The Lopside Ape: Evolution of the Generative Mind*. New York: Oxford University Press.

Denenberg, V.H. (1981). Hemispheric laterality in animals and the effects of early experience. *The Behavioral and Brain Sciences*, 4, 1–49.

Denenberg, V.H., Garbanati, J.A., Sherman, G., Yutzey, D.A. & Kaplan, R. (1978). Infantile stimulation induces brain lateralization in rats. *Science*, 201, 1150–1152.

Howard, K.J., Rogers, L.J. & Boura, A.L.A. (1980). Functional lateralization of the chicken forebrain revealed by use of intracranial glutamate. *Brain Research*, 188, 369–382.

Khyentse, M.D. & Rogers, L.J. (1997). Glutamate affects the development of the thalamofugal visual projections of the chick. *Neuroscience Letters*, 230, 65–68.

Mench, J. & Andrew, R.J. (1986). Lateralization of a food search task in the domestic chick. *Behavioral and Neural Biology*, 46, 107–114.

Nordeen, E.J. and Yahr, P. (1982). Hemispheric asymmetries in the behavioral and hormonal effects of sexually differentiating mammalian brain. *Science*, 218, 391–394.

Nottebohm, F. (1971). Neural lateralization of vocal control in a passerine bird. I. Song. *The Journal of Experimental Zoology*, 177, 229–261.

Nottebohm, F., Stokes, T.M. & Leonard, C.M. (1976). Central control of song in the canary, *Serinus canarius. Journal of Comparative Neurology*, 165, 457–486.

Rogers, L.J. & Anson, J.M. (1979). Lateralization of function in the chicken forebrain. *Pharmacology, Biochemistry and Behavior*, 10, 679–686.

Rogers, L.J. & Hambley, J.W. (1982). Specific and nonspecific effects of neuro-excitatory amino acids on learning and other behaviours in the chicken. *Behavioural Brain Research*, 4, 1–18.

Rogers, L.J. & Sink, H.S. (1988). Transient asymmetry in the projections of the rostral thalamus to the visual hyperstriatum of the chicken, and reversal of its direction by light exposure. *Experimental Brain Research*, 70, 378–384.

Rogers, L.J., Zappia, J.V. & Bullock, S.P. (1985). Testosterone and eye–brain asymmetry for copulation in chickens. *Experientia*, 41, 1447–1449.

Part One

Evolution of Lateralization

1

How Ancient is Brain Lateralization?

GIORGIO VALLORTIGARA AND
ANGELO BISAZZA

1.1. Introduction

This chapter is devoted to discussion of the evolution of lateralization. We
have limited ourselves to vertebrates, and concentrated in particular on the
so-called lower vertebrates (i.e. fishes, amphibians and reptiles). We think
that we should declare our position from the start: we believe that, in its basic
and fundamental form, lateralization among higher vertebrates (i.e. birds and
mammals) is a phenomenon of homology (i.e. that it has been inherited from
a common ancestor). We also believe that important clues to the evolution of
lateralization can be obtained by investigation of extant vertebrate forms,
particularly fish, which are likely to come closest to retaining the original
conditions under which lateralization probably first appeared in early chordates.

1.2. A Brief History of the Comparative Study of Lateralization

The discovery of functional brain lateralization in the human being is associated with the classical observations by Broca in 1861 and, more recently, its
study received renewed impetus from the work carried out on split brain
patients by Sperry and his associates in the 1960s.

Interestingly, for a very long time it was maintained that there were no
anatomical data that could be associated with functional asymmetries in
humans (von Bonin, 1962). The first contrary evidence came in 1968, when
Geschwind and Levitsky reported that the planum temporale, which is part
of Wernicke's area, is larger in the left than in the right hemisphere. Thus, in
humans, observations on functional asymmetries preceded evidence of structural asymmetries. Quite the opposite occurred for research on non-human
animals, though the story is less widely known.

The existence of structural asymmetries in the brain (particularly the diencephalon) of vertebrates was common knowledge among neuroanatomists at the beginning of the century (see Braitenberg and Kemali, 1970; Harris, Guglielmotti and Bentivoglio, 1996). For instance, Gierse (1904, quoted in Shanklin, 1935) reported the right habenular nucleus to be larger than the left one in the fish *Cyclothone acclinidens*; Johnston (1902) and Roethig (1923, quoted by Frontera, 1952) reported a vastly preponderant right habenulopeduncolar tract in the lamprey Petromyzon. All of this very early evidence of animal lateralization came from lower vertebrates.

Subsequently any mention of these anatomical asymmetries disappeared from textbooks (see Braitenberg and Kemali, 1970) and apparently nobody searched for evidence of functional asymmetries in fishes, amphibians and reptiles for a very long period of time.

In the 1970s and early 1980s, came the results of Fernando Nottebohm with songbirds (Nottebohm, 1971, 1977; Nottebohm et al., 1990) and those of Lesley Rogers (Rogers and Anson, 1979) and Richard Andrew (Andrew, Mench and Rainey, 1982) with chickens, demonstrating both structural and functional lateralization in the avian brain (see also Andrew, 1983, 1988, 1991). At the same time Victor Denenberg (Denenberg et al., 1978) and Stanley Glick (Glick and Ross, 1981) reported lateralization in rodents; Denenberg also provided the very first comprehensive review of the subject (Denenberg, 1981), which was instrumental in producing a cascade of data from a variety of avian and mammalian species, summarized by Bradshaw and Rogers (1993).

We return at this point to cold-blooded vertebrates and the associated evolutionary issue: did brain lateralization evolve independently in birds and mammals by sheer coincidence? It is perhaps possible that, in the course of evolutionary history, similar solutions have been independently provided to similar problems in phylogenetically disparate species. Note that lineages of the amniote groups separated about 300 million years ago from the ancestral 'stem reptiles'. Alternatively, it is possible that lateralization in birds and mammals was inherited from common ancestors.

If two species with a common phylogenetic history exhibit structurally similar traits, we call such traits homologous (Campbell, 1988); if two species lack a common phylogenetic history but exhibit structurally similar traits, we call such traits homoplasic (Hodos, 1988). Homoplasy results because even distantly related species may evolve the same solution, selected from a limited set of possible adaptive solutions to the same environmental problem. Different evolutionary forces can be responsible for generating homologous and homoplasic similarities.

In order to understand such a prominent biological character as brain lateralization, we must find out what these forces might have been. If cerebral asymmetries of birds and mammals are homologous, we should find widespread traces of lateralization among current living fishes, amphibians and reptiles. In order to make an argument for homology, therefore, it is crucial for lateralization among lower vertebrates to be a widespread phenomenon. We therefore start with a short review of the current evidence for lateralization among fishes, reptiles and amphibians. More detailed reviews are given by Bisazza, Rogers and Vallortigara (1998) and Vallortigara, Rogers and Bisazza (1999).

1.3. Evidence for Lateralization in Fish, Amphibian and Reptilian Species

Table 1.1 summarizes the current evidence for lateralization (both functional and structural) in lower vertebrates. This is discussed under three headings: anatomical, motor and sensory asymmetries. Such subdivisions and terminology are simply based on convenience and do not necessarily correspond with any important theoretical issue. We discuss first asymmetries that are present at the population level (see Section 1.5 for a discussion of asymmetries at the individual level).

1.3.1. Anatomical Asymmetries

Left–right asymmetries in brain anatomy are ubiquitous in lower vertebrates. The habenular nuclei, located in the anterior dorsal diencephalon, behind the epiphysis (pineal gland), on either side of the third ventricle, are markedly asymmetrical in size in cyclostomes (lampreys and hagfish), sharks, and some teleost fish and amphibians (reviewed in Walker, 1980; Bradshaw and Rogers, 1993). Braitenberg and Kemali (1970) reported marked differences between the right and left habenular nuclei in a frog (*Rana esculenta*), newt (*Triturus cristatus*) and eel (*Anguilla anguilla*). In all three species the left habenula was found to be more lobate than its right counterpart. In the frog, the asymmetry is particularly striking because the left habenula consists of two distinct nuclei, whereas the right habenula has a single nucleus only. An asymmetry similar to that described by Braitenberg and Kemali in *Rana esculenta* has been reported in *Rana temporaria* by Morgan, O'Donnell and Oliver (1973). These authors also reported that tadpoles and young frogs are asymmetrical as well. The left nucleus is partially divided by a vertical septum, lacks cells adjoining the third ventricle along part of its length and

Table 1.1. *Evidence for brain asymmetries in cold-blooded vertebrates. Reference to studies where laterality has been shown only at the individual level is underlined*

| Species | Type of asymmetry | | |
	Sensory	Motor	Anatomical
Chordata			
Amphioxus			Kappers et al. (1936); Young (1962)
Jawless fish			
Petromyzon			Braitenberg and Kemali (1970);
Myxine glutinosa			Braitenberg and Kemali (1970)
Cartilaginous fish			
Sharks			Kappers et al. (1936)
Bony fish			
Gambusia holbrooki	Bisazza and Vallortigara (1997); Bisazza, Pignatti, and Vallortigara (1997a, b); Bisazza and Vallortigara (1997); Bisazza et al. (1998); Bisazza, De Santi and Vallortigara (1999); Bisazza, Facchin, and Vallortigara, (2000); Sovrano et al. (1999); De Santi et al. (2001)	Bisazza and Vallortigara (1996); Bisazza, Pignatti, Vallortigara (1997a); Bisazza et al. (1998)	
Gambusia nicaraguensis	Bisazza, Pignatti, and Vallortigara (1997a)		
Girardinus falcatus	Bisazza, Pignatti, and Vallortigara (1997b); Bisazza et al. (1998); Facchin, Bisazza and Vallortigara (1999); Bisazza et al. (2000); Bisazza, Facchin and Vallortigara (2000)	Cantalupo, Bisazza and Vallortigara (1995); Bisazza et al. (1998).	

Table 1.1. *(cont.)*

Species	Type of asymmetry		
	Sensory	Motor	Anatomical
Brachyraphis roseni	Bisazza, Pignatti and Vallortigara (1997b)		
Poecilia reticulata	Aronson and Clark (1952); Bisazza, Pignatti and Vallortigara (1997b)		
Xiphophorus maculatus	Aronson and Clark (1952)		
Xiphophorus helleri	Aronson and Clark (1952)		
Jenynsia lineata		Bisazza, Cantalupo and Vallortigara (1997a)	
Xenotoca eiseni	Sovrano et al. (1999)		
Xenopoecilus sarasinorum	Sovrano et al. (1999)		
Oreochromys niloticus		Gonçalves and Hoshino (1990a, b)	
Oreochromys mossambicus		Nepomnyashchikh and Gremyatchikh (1993)	
Pterophyllum scalare	Sovrano et al. (1999); Bisazza et al. (2000)		
Betta splendens	Cantalupo, Bisazza and Vallortigara (1996)		
Trichogaster trichopterus	Bisazza et al. (2000); Sovrano et al. (1999)	Bisazza, Lippolis and Vallortigara (2001)	
Knipowitschia punctatissima	Bisazza et al. (2000)		
Padogobius martensi	Bisazza et al. (2000)		
Lepomis gibbosus	Bisazza et al. (2000)		
Ictalurus punctatus		Fine et al. (1996)	

Table 1.1. *(cont.)*

| Species | Type of asymmetry | | |
	Sensory	Motor	Anatomical
Ancistrus sp.,	Bisazza et al. (2000)		
Barbus conchonius	Bisazza et al. (2000)		
Brachydanio rerio	Miklósi, Andrew and Savage (1998); Bisazza et al. (2000); Miklósi and Andrew (1999)	Heuts (1999)	
Carassius auratus		Heuts (1999)	Moulton and Barron, (1967)
Coregonus nasus Coregonus clupeaformis		Reist et al. (1987)	
Corydoras aeneus	Bisazza et al. (2000)		
Phoxinus phoxinus	Sovrano et al. (1999)		
Anguilla anguilla	Westin (1998)		Braitenberg and Kemali (1970)
Anuran amphibia *Bufo bufo*	Vallortigara et al. (1998); Robins et al. (1998); Lippolis et al. (2001)	Bisazza et al. (1996b); Bisazza et al. (1997)	
Bufo marinus	Vallortigara et al. (1998); Robins et al. (1998); Lippolis et al. (2001)	Bisazza et al. (1996b); Bisazza et al. (1997)	
Bufo viridis	Vallortigara et al. (1998); Robins et al. (1998); Lippolis et al. (2001)	Bisazza et al. (1997)	
Rana pipiens Rana catesbeiana		Bauer (1993); Wassersug, Naitoh and Yamashita (1999)	

Table 1.1. *(cont.)*

Species	Type of asymmetry		
	Sensory	Motor	Anatomical
Rana esculenta			Braitenberg and Kemali (1970); Vota-Pinardi and Kemali (1990); Kemali (1977); Kemali and Guglielmotti (1977); Kemali, Guglielmotti and Fiorino (1990)
Hyla regilla		Dill (1977)	
Rana temporaria			Morgan, O'Donnell and Oliver (1973)
Microhyla ornata		Yamashita et al. (1999)	
Urodelan amphibia			
Triturus vulgaris		Green (1997)	
Triturus cristatus			Braitenberg and Kemali (1970)
Reptiles			
Anolis carolinensis		Deckel (1995, 1998); Deckel et al. (1998)	
Thamnophis sirtalis		Shine et al. (2000)	
Uta stansburiana			Engbretson, Reiner and Brecha (1981)

contains fewer free cell bodies in its lumen than the right nucleus. In the fully grown adult, then, two distinct habenular nuclei are present on the left and only one on the right. Analysis of simultaneous recordings from the left and right habenulae of *Rana esculenta* revealed that both spontaneous and light-evoked activity in the left habenula is less marked than in the right. Density of spikes in the left habenula is smaller than that in the right and spike amplitudes are larger in the right habenula (Vota-Pinardi and Kemali, 1990). Braitenberg and Kemali (1970) and Kemali (1977) suggested that such an asymmetry might have appeared early in evolution if two originally paired and bilaterally symmetrical organs, such as two parietal eyes, each connected to the corresponding half of the epithalamus, had shifted their positions and rotated around each other, so that they ended as one dorsal and one ventral. If in this process the connections with the epithalamus were preserved, but one of the two parietal organs lost its function or took over a different function, then asymmetries in their corresponding central projections would be expected.

The idea that habenular asymmetries may be associated with peripheral asymmetry in organs such as the pineal gland seems to be well supported in some reptiles. Engbretson, Reiner and Brecha (1981) examined the central connections of the parietal eye of the lizard (*Uta stansburiana*). They found that the left medial habenular nucleus could be subdivided into *pars dorsolateralis* and *pars ventromedialis*, whereas no such subdivision was apparent in the right medial habenular nucleus, which instead resembles the *pars ventromedialis* of the left nucleus. Centripetal fibres arising from ganglion cells of the parietal eye project only to the *pars dorsalis* of the left medial habenular nucleus. This unilateral projection from the parietal eye probably explains the habenular asymmetry in reptiles. Yet, the presumed anuran homologue of the parietal eye, the frontal organ, does not project to the habenular region at all (Eldred, Finger and Nolte, 1980). In anurans the habenular asymmetry is probably related to an asymmetrical input from the pineal gland. The functions of the habenula are uncertain. It is possible that the habenular nuclei in the frog play a role in reproduction (Braitenberg and Kemali, 1970). It has been observed that female frogs (*Rana esculenta*), and to a lesser degree also males, have longer and larger habenular nuclei in spring than in winter (Kemali, Guglielmotti and Fiorino, 1990).

Moreover, the medial subnucleus of the left dorsal habenula in the frog has been shown to contain endocellular 'crystal-like' inclusions, which are not present in either the lateral subnucleus of the left dorsal habenula or in the right dorsal habenula (Kemali and Guglielmotti, 1977). These inclusions have been interpreted as photosensitive organelles (Kemali, 1977). These findings suggest that the asymmetries in the habenulae may reflect a distinct function of one unpaired nucleus in the entrainment of endogenous biological rhythms to light cycles via its connections with the pineal complex. In support of this hypothesis is the observation that the habenulae are symmetrical in reptiles that lack a parietal organ (Harris et al., 1996).

Interestingly, habenular asymmetry also exists in the male domestic chicken (and in the female treated postnatally with testosterone), where the medial habenular nuclei were found to be significantly larger on the right side of the brain (Gurusinghe and Ehrlich, 1985a, 1985b). The asymmetry in the medial habenula is lost in mammals, in which the neural mechanisms underlying photoreception and the entrainment of endogenous biological rhythms have undergone substantial evolution. However, residual asymmetries have been identified in some eutherian mammals, such as albino rats (Wree, Zilles and Schleicher, 1981), mice (Zilles, Schleicher and Wingert, 1976) and the macrosomatic mole (Kemali, 1984).

Habenular asymmetries are likely to be very ancient: they have been reported in the most primitive living vertebrates, the jawless fish (Cyclostomes), e.g. the lamprey *Petromyzon* and *Myxine glutinosa* (see Braitenberg and Kemali, 1970).

A neural asymmetry in Mauthner (M) cells which could have important implications for behavioural asymmetries in fish (Section 1.3.2.1), has been reported by Moulton and Barron (1967). The M-cells are a pair of giant reticulospinal neurones that are present in most teleost fish and a variety of amphibians (Zottoli, 1978; Will, 1991). The axon of the M-cell decussates and synapses upon motor neurones innervating the contralateral body musculature (Fetcho, 1991, 1992). M-cell activity has been correlated with the C-start response (Zottoli, 1977; Eaton and Emberley, 1991). When confronted with an approaching predator, potential prey, physical obstacle or vibratory stimulus, fish contract their fast right- or left-sided trunk muscles in order to escape. The first stage of this startle response consists of a 'C-bend' of the trunk in a horizontal plane, either to the left or to the right, and is followed by a second phase, a 'tail flip', during which the fish gains considerable acceleration (Heuts, 1999). Moulton and Barron (1967) reported that in some goldfish (*Carassius auratus*) the left M-cell was approximately three times the size of the right M-cell; the right and left M-cells also appeared to be oriented somewhat differently. These asymmetrical neurones were found in different batches of fish from the same supplier, though only four individuals were examined. In a single specimen, it was found that the M-axon of the left cell alone seemed to be present (on the right side of the spinal cord). There was no indication of the corresponding axon of the right cell, which was either lacking or, more probably, reduced to the size of other ordinary neurones in the vicinity and thus not revealed by the staining technique used by the authors. Moulton and Barron (1967) and Moulton, Jurand and Fox (1968) also reported similar asymmetries in amphibians (in urodele and anuran tadpoles). Stefanelli (1951) also had previously reported that, in anurans, M-cells do not always atrophy at the same rate on the two sides.

Finally, an anatomical asymmetry has been recently reported in reptiles. Gartersnakes (*Thamnophis sirtalis*) exhibit an asymmetry in hemipenis size and usage (Shine et al., 2000). Male snakes possess paired reproductive systems. However, testes, kidneys and hemipenes on the right-hand side of the body are larger than those on the left. Data from mating in the field suggest that the asymmetry has implications for reproductive behaviour and possibly reproductive success. For instance, male gartersnakes preferentially used their larger (right) hemipenis when mating at high body temperatures; the authors suggest perhaps because they are more able to make subtle postural adjustments and thus select the best side under these conditions.

1.3.2. *Motor Asymmetries*

1.3.2.1. *Turning Biases*

Measures of turning and/or rotational preferences have been widely used to assess lateralization in mammals (reviewed in Bradshaw and Rogers, 1993). Rats tend to turn or rotate spontaneously at night (or after drug treatment in the daytime) with a consistent preferred direction at the individual level (reviewed in Glick and Shapiro, 1985). Unilateral lesions to the nigrostriatal motor system produce ipsiversive turning towards the side of the lesion and this effect can be potentiated by dopaminergic drugs. In normal rats there are asymmetries in the dopamine concentration in the two striata, and high doses of D-amphetamine increase dopamine asymmetry and induce daytime rotation in the direction contralateral to the side with higher dopamine levels (Glick and Shapiro, 1985).

Humans, too, exhibit natural (and perhaps drug-induced) turning biases (see Bradshaw, 1991). Asymmetries in the concentration of striatal dopamine in the basal ganglia correlate with turning biases, handedness and unilateral Parkinson's disease (Bracha et al., 1987; Glick and Shapiro, 1985; Kooistra and Heilman, 1988).

Dolphins show a marked tendency for counterclockwise swimming when placed in pools (Ridgway, 1986; Sobel, Supin and Myslobodsky, 1994). There is anecdotal account that people promenading around enclosed spaces also tend to progress counterclockwise (Bradshaw, 1991).

There are only two documented attempts to check for lateralization during spontaneous swimming in tanks by fish. Nepomnyashchikh and Gremyatchikh (1993) investigated the swimming activity of young *Oreochromis mossambicus* (Cichlidae) in Petri dishes lit uniformly. They found that most individuals preferred to move counterclockwise (though this was not significant at the population level), and individual fish appeared to be consistent in their direction of swimming over time. Spontaneous rotational preferences in swimming of mosquitofish (*Gambusia holbrooki*) have been investigated by Bisazza and Vallortigara (1996). They found that in circular, centrally lighted tanks, females, but not males, swam preferentially clockwise in the morning and counterclockwise in the afternoon. The rotational preferences of the females appeared to be related to a sun-compass orientation mechanism (Goodyear and Ferguson, 1969), since they disappeared under diffuse lighting conditions and when using naive females (i.e. individuals that had never been exposed to sunlight). Bisazza and Vallortigara (1996) also reported, however, that, following repeated testing, the population bias produced by the sun-compass mechanism tended to disappear, but females still showed rotational preferences that became stable and consistent at the individual level. Males continued to show consistent individual rotational preferences under repeated testing. It seemed, therefore, that the rotational bias in mosquitofish was due to two independent phenomena: a sun-compass navigation mechanism, which is related to the intensity of predation, and a behavioural lateralization at the individual level, which is likely to reflect neural asymmetries in motor or sensory systems.

Evidence for a bias at the population level was, in fact, obtained in males (irrespective of time of day) by placing a predator in the centre of the tank, when a significant bias to rotate counter-clockwise was observed. When using a group of females as a target in the centre of the tank, there was a non-significant trend for a bias in the same direction, whilst during spontaneous swimming in the absence of any target, no bias occurred (Bisazza and Vallortigara, 1997). It should be noted that the same species shows a striking lateralization at the population level in a detour test (see Section 1.4): when males were required to circle around an obstacle (a barrier with vertical bars) to approach a group of females or a dummy predator as a target, they showed a consistent bias to detour on the left side. It was unlikely that this reflected a motor bias, for the lateral asymmetry disappeared when males were tested using a group of males as a target or in the absence of any target (Bisazza, Pignatti and Vallortigara, 1997a, 1997b). It seems that these biases reflect preferences in eye use rather than motoric asymmetries, and as such they will be discussed in the next section.

Work done on the freshwater fish tilapia (*Oreochromis niloticus*) by Gonçalves and Hoshino (1990a) investigated the frequency with which one or other side of a T-maze was chosen by Tilapia fry. The fry showed a higher number of choices for the right side. The attacked side and the side of turning away made in order to avoid attacks by submissive Tilapias were significantly correlated with the side chosen in the T-maze, whereas spontaneous turning during exploratory swimming was not (Gonçalves and Hoshino, 1990b). This suggests that the same motor asymmetries were expressed in both the T-maze test and aggressive encounters with conspecifics. Similar conclusions arise from another test, in which the time required to correct natural posture after its forced loss induced by attaching a spherical lead weight to a ribbon lace fixed at the base of the first ray of the dorsal fin was determined (Gonçalves and Hoshino, 1990a). Postural correction data showed 80% accordance with the T-maze data.

Direction of turning in front of an opaque barrier has been studied in the poeciliid fish *Gambusia holbrooki* and *Girardinus falcatus* (Bisazza, Pignatti and Vallortigara, 1997a, 1997b). Fish moved along a swimway at the end of which was an opaque barrier. Both species showed a population-level bias to turn to the right when facing the barrier.

Heuts (1999) found a significant bias towards right-hand startle C-bends in vibration-stimulated zebrafish (*Brachydanio rerio*) and goldfish (*Carassius auratus*), but not in guppies (*Poecilia reticulata*) and in another four Cichlid species. In an undisturbed situation, the fast swimming turns of iso- lated goldfish and grouped zebrafish were also significantly right-biased; interestingly, however, slow turns were significantly left-biased (except for female zebrafish, which showed significant right-biased slow turns even dur- ing periods of non-attack by group mates). Heuts suggested that the differ- ence between right and left biases in fast and slow turning could be explained by an asymmetry at the muscular level. Recruitment of mainly white muscle occurs in fast swimming as opposed to the mainly red muscle recruitment in slow swimming (see Heuts, 1999, for details and references). In fact, the right side of the trunk has a larger white-muscle mass in the zebrafish (at least in the anal region), whilst the left side of the trunk has a larger red-muscle mass. Thus, it seems that behavioural asymmetries in turning direction in fish are associated with neural asymmetries in the Mauthner cells (previous section) and with morphological asymmetries in the muscles that are innervated by Mauthner cells. Heuts also explained the absence of fast-turn lateralization in four Cichlid and one Poeciliid species by the hypothesis that their habit of hiding on the substrate (which is associated with an avoidance of open water) made it less necessary for them to possess fast and lateralized swimming-turn responses in order to avoid predatory fish. Fish taxa, that are often exposed in their natural open-water habitat to predatory fish, are likely to have evolved stronger muscular and behavioural lateralization than fish that are more bound to the substrate.

Wassersug, Naitoh and Yamashita (1999) studied turning biases in tad- poles. After surfacing to breathe air, most tadpoles descend by turning shar- ply to the left or right. Bullfrog (*Rana catesbeiana*) tadpoles show an overall bias to turn left, whereas clawed frogs (*Xenopus laevis*) larvae show no bias. The authors suggested that the difference may relate to the fact that ranid tadpoles, including *R. catesbeiana*, are externally asymmetric, whereas pipid tadpoles, including *X. laevis*, are not. Ranid tadpoles have a single sinistral spiracle through which water that enters the mouth is expelled from the body. Pipid tadpoles, in contrast, have dual symmetrical spiracles, one on each side of their body. However, it is possible that the asymmetry in spiracle position and the turning bias are not functionally linked. Yamashita et al. (1999) have recently tested the turning bias in the startle response of *Microhyla ornata* tadpoles. *Microhyla* tadpoles are, like *Xenopus*, externally symmetrical but phylogenetically they are more closely related to Rana. *Microhyla* tadpoles were startled by a vibratory shock, revealing a left-handed turning bias while still in the earliest free-swimming stage and that bias persisted until forelimbs emerged, when it faded away. Since *Microhyla* larvae are externally symmetrical, yet preferentially turn to the left, the hypothesis that the turning biases in tadpoles are caused by external morphological asymmetry appears untenable.

1.3.2.2. Asymmetries in Limb Usage

In humans, the most striking form of motor asymmetry associated with cerebral lateralization is handedness. It has long been denied that any other animal species show differences in the use of the limbs in any way comparable to human handedness. 'Handedness' is here used to mean consistent preferential use of one limb in most individuals and across most tasks. However, avian species that use their feet to manipulate food and objects have shown significant 'footedness' present at the population level, with proportions similar to those of handedness in humans (Friedman and Davis, 1938; Rogers, 1980; Harris, 1989; Rogers and Workman, 1993; Snyder and Harris, 1997; Tommasi and Vallortigara, 1999). Even in rodents, previously believed to exhibit limb preferences only at the individual but not population level, a recent report, in which large samples of inbred mice were used, showed significant right pawedness in one test of lateral paw preference and left pawedness in another reaching test (Waters and Denenberg, 1994).

The situation in primates, however, still arouses controversy (see Hopkins, 1996; McGrew and Marchant, 1997). A reassessment of the data suggests that handedness is indeed present. The lower primates seem to be left handed for holding food, whereas their right hand is stronger and used for holding on to branches (Ward et al., 1993). MacNeilage et al. (1987) suggested that, as primates became less arboreal, the right hand became available for manipulation, with a shift to right handedness for fine motor acts. There is indeed some evidence in support of a right-hand bias in tool use by capuchin monkeys (Westergaard and Suomi, 1996) and in manipulation by chimpanzees (Hopkins, 1996), but handedness varies between species of primates. For example, orang-utans show a left-hand preference in manipulating parts of their own faces, when cleaning their teeth, eyes or ears (Rogers and Kaplan, 1996).

The criticism that the departure from an unlateralized 50:50 random distribution in these animal populations is typically small, compared to handedness in humans, appears to be questionable. Marchant et al. (1995) showed that, when a wide range of everyday behavioural patterns of hand use in preliterate cultures was examined, the right handedness appeared to be consistent but rather weak (about 45:55 for left:right). The notable exception is hand preference in tool using and manipulative activities, particularly when using a precision grip for tool use: here preference is markedly right-handed even in preliterate cultures. However, these specialized forms of hand use are not typical of the use of the limbs in most non-human species and there have been few studies of hand preferences in tool using by non-human animals. As already mentioned, however, parrots and cockatoos that use their feet to manipulate food and objects with a high degree of sophistication have significant 'footedness' present at the population level with proportions similar to those of handedness of precision-gripping tool use in humans.

Any investigation of the evolution of limb preferences would be incomplete without an examination of the earliest tetrapods, the amphibians. Bisazza et al. (1996, 1997) tested the common European toad *Bufo bufo*, using a task in which animals have to remove from the head an elastic balloon or a strip of paper. Results revealed a significant population preference for the use of the right forepaw to remove the balloon (59%) or the paper strip (55%). In another test, it was shown that the South American cane toad, *Bufo marinus*, uses the right forepaw preferentially (66%) to control rolling to an upright position after the body has been turned over and submerged in water. Naitoh and Wassersug (1996) observed that the ingestion of toxic material by several species of frogs provokes vomiting, and in some cases the stomach itself is regurgitated. Before the prolapsed stomach is reswallowed, remaining vomitus is wiped from the gastric lining with the forepaw. These researchers suggested that a right forepaw preference in wiping tests might, therefore, have developed because asymmetric mesentery attachment causes the prolapsed stomach to hang to the right. However, Robins et al. (1998) found that asymmetric use of the limbs also occurs for hindlimbs, which are not used in any wiping behaviour. Three species of toads (*B. marinus*, *B. viridis* and *B. bufo*) were overturned on a horizontal surface. In this condition the toad uses one of its hindlimbs actively to push against the substrate or throws a hindlimb across the body, thereby providing momentum for the righting response by displacing the body's centre of gravity. The other hindlimb assumes a more permissive role during this initial phase of the righting response. Here, rotation of the pelvis and the pectoral girdle, and hence the involvement of the forepaws, are all secondary to the use of the hindlimbs in the righting response. Both *B. marinus* and *B. bufo* toads showed preferential right hindlimb use, whereas *B. viridis* showed preferential left hindlimb use. Interestingly, this species also showed a slight preference for left forepaw use in wiping tests (Bisazza et al., 1996, 1997).

Handedness in toads is thus task- and species-specific; its relation to anatomical asymmetries is however still unclear. Greer and Mills (1997) suggested that the right forelimb bias in *Bufo* is linked to asymmetry in the toad pectoral girdle. In *Bufo*, a percentage variable from 65% to more than 95% (depending on population and species; see Martin, 1972) have the right epicoracoid cartilage superficial to the left. However, it is doubtful whether this could cause the behavioural asymmetry. The epicoracoid asymmetry is constant in direction in all of the genus *Bufo*, but *B. viridis* shows a preference for the left forelimb rather than for the right as occurs in *B. bufo* and *B. marinus* (Bisazza et al., 1997); also, *B. viridis* shows the same reverse pattern even with respect to hindlimb usage (Robins et al., 1998). Borkhvardt and Ivashintsova (1994, 1995) have suggested that, in fact, preferences in limb usage and/or order of limb eruption at metamorphosis may affect whether toads develop right- or left-handed pectoral girdles (see also Borkhvardt and Malashichev, 1997; Malashichev and Starikova, 2001).

Motoric lateralization seems to be present even in urodelan amphibians. It has been shown that the smooth newt (*Triturus vulgaris*) shows lateral asymmetries at the population level during sexual behaviour (Green, 1997). Following courtship, the female newt crawls parallel to and behind the male and touches his tail with her snout. The male then deposits his spermatophore and turns through 90 degrees so forming a barrier ahead of, and perpendicular to the female, with its tail folded along the flank facing the female. The female crawls further until the male blocks her path, and at this point she may pick up the spermatophore in her cloaca. Green (1997) has investigated the direction of turning of male newts, and has found a significant population bias for turning leftwards. Green (1997) argued that the asymmetrical turning of smooth newts is unlikely to be due to visual lateralization: the sexual behaviour of this species appears to be mainly guided by internal programming and perhaps tactile stimuli from the substrate. Lateralization of general use of the limbs is also unlikely because male newts do not use their limbs for manipulating objects (though females do use their hindpaws for manipulating their eggs) and limb movements do not appear to be under fine visual control during turning behaviour. The author thus suggested that turning asymmetry in smooth newts may reflect lateralized use of the tail rather than of the limbs. Males hold their tails to the side of the body throughout courtship, and after depositing a spermatophore they turn their body and fold their tails along the inner flank. The hypothesis that lateralized tail use predates the evolution of lateralized limb use is interesting. Striking lateralized tail use in the coiling of prehensive tails in reptiles has been reported (Heinrich and Klaassen, 1985), though the low number of individuals examined makes it unclear whether such asymmetry is present at the individual or population level.

A phenomenon that resembles handedness has been recently observed in a species of fish (Bisazza, Lippolis and Vallortigara, 2001), which would contrast with the idea that lateralization of tail use precedes lateralization of limb use. The blue gourami, *Trichogaster trichopterus* (family Belontiidae, order Perciformes), is a facultative air-breathing fish (Burggren, 1979). This fish possesses paired filamentous ventral fins in which the first ray is greatly elongated, with the remaining rays being vestigial. Sharrer et al. (1947) have observed the presence of both taste buds and free nerve endings in the ventral fins of the blue gourami. This suggests that these appendages serve both gustatory and tactile functions. The fins are used (rather like 'limbs') for tactile inspection of nearby objects. When exposed to a sequence of novel plastic objects, varying in shape and colour, the blue gourami showed preferential use of the left fin during initial contacts. Lateralization apparently depends on the nature of the stimulus: fish exposed to a randomized series of natural objects showed preferential use of the left fin with mineral objects, but no asymmetry was apparent with animal or vegetable objects. This would suggest that some form of 'handedness' may have appeared sporadically even prior to the evolution of tetrapods. However, measurements of fish monocular viewing revealed that fin use was strongly associated with preferential use of the ipsilateral eye before stimulus touching, thus suggesting that the asymmetry in fin usage may be related to a lateralization of the visual system. The blue gourami has in fact been proved to be lateralized in eye use in other tests (see Sovrano et al., 1999; Section 1.4). There are, however, reasons to believe that visual lateralization cannot entirely explain the lateral bias in use of the ventral fins. Apparently, the bias depended on the material employed (which would be consistent with a gustatory function of the fins): when animal or vegetable materials were used, no reliable asymmetries occurred, and there was also a certain tendency for a shift to right fin use with some of these stimuli. Moreover, it is not so obvious that vision at a distance would guide subsequent touching: at least with organic material it is likely that diffusion of molecules occurred at some distance in the water. Thus, the possibility that a slight turning to favour approaching with the right or the left fin might be the basic phenomenon; it cannot be ruled out. Furthermore, the concordance between eye and fin use decreased with distance, which would also be compatible with response to diffusing substances. Thus, at this stage, it cannot be established whether the primary, causal, mechanism in generating the asymmetry is vision or chemical senses (or indeed some other sense).

1.3.2.3. Control of Vocalization

Control of vocalization might also be ascribed to motor asymmetries, though this is far from certain. Anurans have several types of vocalization, some of which are used to communicate with conspecifics. When a male mounts (clasps) another male or a non-gravid female, the mounted animal emits a series of calls that cause its release. If the mounted animal is a gravid female, no release calls are emitted, and clasping is maintained until the eggs have been laid and fertilized (Aronson and Noble, 1945; Rand, 1988; Kelley and Tobias, 1989). It seems that the clasping vocalization allows male frogs to obtain the information necessary to discriminate between sexes or between gravid and non-gravid females. Neurones of the pretrigeminal area (PTA), located just anterior to the border of the tectum and cerebellum, seem to be the major generators of clasping (as well as mating) calls (Schmidt, 1973, 1976; Wetzel, Haerter and Kelley, 1985; Schneider, 1988; Kelley and Tobias, 1989).

Bauer (1993) induced vocalization in a group of frogs (*Rana pipiens*) by clasping the animals behind the forelimbs for 30 s in the morning and 30 s in the afternoon for 6 weeks. Then, on the basis of the number of baseline vocalizations, he divided the animals in four matched groups: non-operated animals, sham-operated animals and animals with right or left lesioning of the vocalization generators. He found that animals receiving knife cuts on the left side just posterior to the PTA areas showed, in the subsequent 6-week period of recovery, a much more pronounced reduction in clasping vocalizations than those receiving the same lesions on the right side. Anurans are the earliest vertebrates with true vocal cords (Vial, 1973). Apparently, they show an asymmetric brain control of vocalization similar to that observed in birds, rodents, primates and humans (see Section 1.4 for a review).

The first ray of the pectoral fin of catfish is a bilaterally symmetrical spinous structure that is minimally important for movement and is used to produce stridulatory sounds in several families of catfish (Tavolga, 1962; Ladich and Fine, 1994). In the channel catfish (*Ictalurus punctatus*) the stridulatory apparatus consists of an expanded process on the first pectoral spine, which is rubbed against a complementary surface on a groove in the pectoral girdle. The details of this mechanism have been described recently by Fine et al. (1996). The ventral aspect of the process has a slightly convex surface with a series of ridges that radiate outward from the centre. The matching ventral surface of the groove is slightly concave and has a rough surface punctuated with lacunae. Depression of the ridges against this rough surface produces a series of pulses per fin sweep, and the durations of the sound in a sweep ranges from 56 to 177 ms. Many of the sounds are produced during individual sweeps, but others that occur in a rapid series utilize both fins.

Fine et al. (1996) videorecorded 20 catfish to allow measurement of pectoral fin sweeps for handedness analysis (many of the sonic movements are too fast to be resolved using normal vision). It appeared that, although all fish were capable of using both fins, half of the individuals had a marked preference for one fin over the other; of these, nine were right-finned and only one favoured its left fin. Nothing is known about the neural control of this behaviour: fish have a region, which is equivalent to the striatum of mammals, that is responsible for the control of movement (Northcutt and Davis, 1983) as well as an anterior commissure, which may be worth investigating for a possible role in lateralization using lesioning techniques.

1.3.2.4. Predator-escape Responses

Indirect evidence for functional lateralization can arise from the observation of asymmetrical distribution of external scars resulting from unsuccessful attempts of predation. In the whitefish (*Coregonus nasus* and *C. clupeaformis*) scars appeared to be located far more commonly on the left side of the fish and below the lateral line (Reist et al., 1987). Scars belonged to three types: small round scars probably caused by the marine parasitic copepod *Coregonicola* or by Arctic lampreys (*Lampetra japonica*), larger rounded scars probably caused by either attacks by lampreys or by previous gill net capture, and slash scars, the causation of which is uncertain (the authors claimed that parasite and lamprey attacks are not satisfactory explanations for these scars: they probably result from predation attempts by bears, birds or piscivorous fish). In principle the asymmetry could be attributed to an asymmetry of the predators rather than of the whitefish. However, the asymmetry in the location of the scars was the same for all three types of scars. The preponderance of slash scars on the left side of the fish is particularly intriguing: the authors suggested 'handedness' in evasive actions by the fish as a possible explanation. Interestingly, Donnelly and Reynolds (1994) found that copepod parasites infect more frequently the left rather than the right side of the lateral line of male wrasse (*Crenilabrus melops*: Labridae) and attribute this to asymmetries of fish involving curling the body preferentially to the left side [see Buchmann, 1988; McCarthy and Rita, 1991, for similar asymmetries in the branchial arches of eels (*Anguilla anguilla*)].

More direct evidence for functional lateralization has been collected. Cantalupo, Bisazza and Vallortigara (1995) used a species of poeciliid fish, *Girardinus falcatus*, and, as a behavioural measure, the direction of turning during escape response evoked by a simulated predator (see also Bisazza, Cantalupo and Vallortigara, 1996). Immature fish were repeatedly faced with the simulated predator and the results revealed a significant bias to escape rightwards in the initial session (each session comprised 65 stimulus presentations), after which the bias was reduced progressively, until, in the fifth session, it reversed, yielding a preference to escape leftwards. Similar trends were present in both males and females. In adult *G. falcatus* only a slight population bias to turn right was observed in the first session, but a strong bias to turn left developed after repeated sessions.

It seems likely that such a shift in bias could be related to the asymmetries of fast and slow swimming described by Heuts (1999; see Section 1.3.2.1). However, asymmetries in other brain structures are also possible. Although the motor commands for the C-start reaction are ballistic, and, therefore, do not rely on sensory information from the stimulus once the movement begins, sensory information is necessary to coordinate the C-start successfully before its initiation. The fish need to know what and where the stimulus is, when to begin the escape sequence and where to go (Eaton and Emberley, 1991). An excitatory visual input is known to project from the optic tectum to the M-cell's ventral dendrite (Zottoli et al., 1987). Thus, the behavioural asymmetry may be generated at the level of the tecta, in relation to the analysis of the visual stimuli that precedes the C-start-initiated escape sequence. Visual lateralization due to tectal asymmetrical inhibitory interactions has been demonstrated in the pigeon (Güntürkün and Bohringer, 1987) and intertectal commissural connections appear to suppress lateralization in the chick, probably because of removal of tonic intertectal inhibition (Parsons and Rogers, 1993).

Evidence for lateralization in anti-predatory responses in anurans has been reported by Dill (1977). He measured the jumping directions of Pacific tree frogs (*Hyla regilla*) in response to a model predator. Frogs were presented with a model predator consisting of a rubber ball, suspended by a thread attached above the centre of a pedestal. The frog was placed on the pedestal and the 'predator' was released from a position 50 cm in front of the frog, moving down an arc to cross the centre of the pedestal. The angle of the direction of jumps was recorded, revealing a bias towards leftward jumps. Interestingly, most frogs had longer right than left hindlimbs, but there was no correlation between the proportion of left jumps by an individual and the amount of asymmetry in the hindlimb bones.

Lateralization of responses to presentation of a simulated predator has been investigated also in three species of toads, *B. bufo*, *B. viridis* and *B. marinus* (Lippolis, 1998; Lippolis et al., 2001). A simulated snake was presented moving rapidly towards the toad in the frontal field of vision and the toad's escape jumps to the right and to the left were recorded. No bias appeared in this test. In another test the simulated snake was presented in the left or right lateral field of vision. Escape and defensive responses were elicited more strongly when the stimulus was on the toad's left side compared to its right side. This bias was present in all three species. The asymmetry might be related to a dominance of right-side brain structures in emotional responding.

1.4. Sensory Asymmetries

In the last few years, striking evidence for visual lateralization in fish has appeared. Bisazza, Pignatti and Vallortigara (1997a) devised a simple behavioural test, originally developed for use in birds (see Vallortigara, and Regolin, 1996; Vallortigara, Rogers and Bisazza, 1999), which revealed functional lateralization in a variety of species of fishes. Male mosquitofish (*Gambusia holbrooki*), faced with an obstacle (a vertical-bar barrier) behind which a group of females was visible, preferentially circled around the obstacle leftwise (Bisazza, Pignatti and Vallortigara, 1997a, 1997b). The same bias was observed using a simulated predator as a target (predator-inspection behaviour is common among most prey fish species). The lateral asymmetry was task- and stimulus-dependent. Thus it disappeared when the task was changed by forcing the fish to lose visual contact with the goal (by using a U-shaped barrier) or using less attractive targets (i.e., a group of males or an empty environment).

Bisazza et al. (1998) compared detour responses of mosquitofish (*Gambusia holbrooki* and *G. falcatus*) faced with a vertical-bar barrier through which conspecifics of different sex or a dummy predator were visible. Both species showed a consistent bias to turn leftward when faced with the predator. Sexual stimuli elicited a leftward bias in females that had been deprived of the presence of males for 2 months, whilst no bias was apparent in non-deprived females.

Further interspecies comparisons have been made. Right–left direction of detour was measured in males of five species of poeciliid fishes (Bisazza, Pignatti and Vallortigara, 1997b). When motivational factors were assessed properly, all five species showed lateralization in the same direction. These results suggest that the direction of lateral asymmetries in detour behaviour tends to be strikingly similar in closely related species, at least in the family Poeciliidae (but see below for a discussion of this issue).

The dependence of the direction of lateralization on the stimulus used and on the difficulty of the task suggests that asymmetries at the motor level are implausible as an explanation. It has been suggested, therefore, that lateralization of detour behaviour may arise from asymmetries of eye use. Miklósi, Andrew and Savage (1998) provided independent evidence for asymmetric eye use during inspection of different stimuli in the zebrafish: strange objects were viewed on first exposure with the right eye and with the left eye thereafter; a familiar stimulus tended to be viewed with the left eye. Facchin et al. (1999) performed viewing tests on *G. falcatus*. They found that fish that tended to detour the barrier on the left side used the right eye to scrutinize a dummy predator and the left eye to scrutinize a neutral stimulus, whereas fish that tended to detour the barrier on the right side showed the reverse pattern of eye use. Fish that did not show any consistent bias in the detour test did not reveal any significant preference in the viewing test (see Section 1.7).

Asymmetries in the use of the eyes seem to be ubiquitous among animals with laterally placed eyes (see Vallortigara, 2000, for a review of evidence). It may appear somewhat counterintuitive that cerebral lateralization can impose differential use of the two eyes. Biologically relevant stimuli would occur equally often on either side and, therefore, there should be selective pressures maintaining the right and left eye equally capable of performing visual processing tasks. This may be true for the initial detection of stimuli. However, it should be noted that sustained viewing in animals with eyes placed laterally is commonly monocular and, therefore, after the initial detection and recognition, the choice of the right or left eye for viewing seems to be affected by lateralization of hemispheric function. The choice of right or left eye viewing would determine the type of visual analysis that follows. In fact, birds and fish may be able to bring into action the hemisphere most appropriate to particular conditions and to particular stimuli by using lateral fixation with the contralateral eye (see Dharmaretnam and Andrew, 1994; Vallortigara et al., 1996; Vallortigara, Rogers and Bisazza, 1999; Tommasi, Andrew and Vallortigara, 2000; Vallortigara, 2000).

Some mammals may show differential eye (and hemisphere) use too. White laboratory rats, for example, similarly to birds, have eyes that are placed laterally and 95% of the optic fibres decussate (see Bradshaw and Rogers, 1993, for a review). Thus direct monocular input is primarily to the opposite hemisphere. Consistent with this, rats tested monocularly can use spatial information to navigate in a maze provided that they are using the left eye (right hemisphere) but not when they are using the right eye (Cowell et al., 1997). Those mammals, including humans, with frontally placed eyes, which therefore fixate stimuli of interest binocularly (with obligate conjugate eye movements), do not show any directly comparable phenomena. However, a lateralized mechanism somewhat similar to that observed in birds may be available to humans as well. The engagement of one or other hemisphere in verbal or spatial tasks is revealed by eye movements to the right or the left, provided that there are no external factors affecting gaze (Gur and Gur, 1977). Also, direction of gaze to the left tends to promote analysis by right hemisphere strategies, while gaze to the right brings left hemisphere strategies into play (Gross, Franko and Lewin, 1978). Adjustment of head position and eye movements may thus play a similar role in mammals with frontal vision as does the choice for right or left lateral visual fields in birds.

It should also be noted that, even in mammals with incomplete decussation of optic fibres at the chiasma, there are differences in the inputs from one eye to each of the two hemispheres. In fact, although each eye relays inputs (via the geniculate) to both the right and left hemispheres, the fibres from the medial half of the retina, which cross to the contralateral hemisphere, are larger than those that arise from the lateral half of the retina and go to the ipsilateral hemisphere (Bishop, Jeremy and Lance, 1953). Fibres that cross the midline and go to the contralateral hemisphere, therefore, conduct neural signals faster than the uncrossed fibres and they dominate the uncrossed fibers during binocular stimulation (Proudfoot, 1983; Walls, 1953). Consistent with this, eye preferences for viewing in both humans (Adam, Szilagyi and Lang, 1992) and non-human primates (Hook-Costigan and Rogers, 1995; Rogers, Ward and Stafford, 1994) have been observed.

A most notable demonstration of complementary eye use in a lower vertebrate species was obtained with toads (see Vallortigara et al., 1998), which have a large binocular overlap. Prey-catching behaviour was studied in three species of toads (the European green toad *Bufo viridis*, the European common toad *Bufo bufo*, and the cane toad *Bufo marinus*) using a modification of the classic procedure known as the 'worm-test' (Ewert, 1980). A preferred prey was attached to a thread and suspended from a wire support that moved it mechanically in a horizontal plane around the toad, entering first either its right or its left monocular visual field, depending on the direction of rotation. When the prey was moving clockwise, and thus entered first the left and then the binocular field of vision, almost all of the tongue strikes occurred in the right half of the binocular field. When the prey was moving anticlockwise, and thus entered first the right and then the binocular field of vision, a more symmetrical distribution of strikes in the left and right halves of the binocular fields occurred. Thus, it seems to be necessary for prey to enter the right half of the binocular visual field in order to evoke predatory behaviour. Initial detection in the left visual field does not allow the toad to show prey catching until the prey has moved into the right half of the binocular visual field. In contrast, initial detection of the prey in the right visual field allows the toad to orient towards and follow the prey and strike at it anywhere in the binocular field. Toads were also tested for agonistic behaviours in the form of tongue strikes at competitors during feeding. Both *B. marinus* and *B. bufo* toads showed a population bias to strike with the tongue at other toads when these were occupying their left visual field (Robins et al., 1998; Vallortigara et al., 1998). A toad is more likely to attack prey on its right side (and ignore them on its left side) and to attack conspecifics on its left side (and ignore them on its right side).

The lizards of the genus *Anolis* (Reptilia, Lacertilia, Iguanidae), a common species indigenous to the Southern United States, present clearly identifiable aggressive behaviours. These include headbobbing, dewlapping (i.e. extension of the coloured throat fan in the male), changes in body posture and threatened bites. The eyes are set posteriolaterally on the snout and thus convergence on the same visual target is prevented because the frontal visual fields are physically obstructed by the snout (Deckel, 1995). This forces lizards to scan the environment with either the left or the right eye. Eye use was scored by Deckel (1995) in adult lizards (*Anolis carolinensis* and *Anolis sagrei*) during social interactions. This was done simply by assessing on a video recording which side of the body or head was oriented towards the other lizard. Eye use during aggressive, escape and assertion responses was recorded for each animal during a 20-min session. In addition, since aggressive responses in this species are accompanied by a change in colouration, the colour of each animal was recorded during each interaction and graded according to a standard scale from 'maximally light' to 'maximum dark' (see Deckel, 1995, for details). A predominant left-eye use was found during aggressive responses, whereas movements of a less obvious aggressive nature, such as 'assertion displays', showed no consistent bias. During aggressive displays, the animals not only used their left eye significantly more often than their right eye, but also were of a lighter colour when using the left eye. These results could not be due to a generic left-eye preference, as the non-aggressive animals watched their opponent, ambulated and fled the aggressive partner using either eye with similar frequency (there was also a trend for using the right eye in non-aggressive animals during motionless observation of an aggressive conspecific). Moreover, non-aggressive responses occurred when animals were darkened, but, unlike aggressive responses, there was no linkage between the degree of dark coloration and eye preference. Clearly, left-eye preference was confined to the most aggressive of responses (see also Deckel, 1998; Deckel et al., 1998, for further evidences).

In *Anolis* the vast majority of the retinal fibres project contralaterally to the opticus tegmenti, with only a vestigial number of these fibres projecting to the ipsilateral thalamus (Butler and Northcutt, 1971). Moreover, although some transfer of information between the hemispheres is possible through the hippocampal and anterior commissure (Armstrong et al., 1953; Greenberg, 1982), *Anolis* lacks the corpus callosum, which, in mammals, connects one hemisphere to the other with 'point-to-point' connections. It seems that the small interhemispheric projections present in the lizard brain do not allow a functional integration of information stored in the two hemispheres. Greenberg, Maclean and Ferguson (1979) and Greenberg, Scott and Crews (1984) reported that *Anolis* with lesions of the amygdala fail to respond to social stimulation when the contralateral, but not ipsilateral, eye is patched; moreover, in intact animals occlusion of either eye has no effect on responses. This indicates that brain areas ipsilateral to the occluded eye do not have access to the visual input available to the contralateral brain areas. All this suggests that, in *Anolis*, aggressive responses are mainly activated by the right hemisphere.

Virtually nothing is known about behavioural lateralization in other, non-visual, sensory modalities in lower vertebrates, although there is striking evidence for lateralization in the olfactory and auditory domains in higher vertebrates (e.g. Vallortigara and Andrew, 1994a, 1994b; Miklòsi et al., 1996; Rogers, Andrew and Burne, 1999). However, Westin (1998) recently reported a fascinating result: anosmic eels (*Anguilla anguilla*), with both nostrils blocked, and eels with only the left nostril blocked differed from natural eels in speed, direction and hibernation during migration; in contrast, eels with only the right nostril blocked behaved like controls. This suggests that olfaction is essential for orientation in eels, and that, since olfactory projections are mainly to ipsilateral structures, structures located to the left side of the brain are crucial for orientation.

1.5. Is the Direction of the Asymmetries Species-invariant?

Similarities in the direction of lateralization in different tasks among different species may be seen as evidence for possible homologies, but the direction of the lateralization is probably not as important as the fact that there is different functional specialization of the two sides of the brain. Similarity in direction is only moderately informative, considering that there are only two possible directions for an asymmetry to occur, either favouring the right or the left side, and that the direction of lateralization could be influenced by embryological and other environmental factors (see Rogers, 1997, for chicks; and Güntürkün, 1993, for pigeons). Nonetheless, there are quite impressive similarities in the direction of lateralization among vertebrates. The first one concerns the selective involvement of the right side of the encephalon in spatial tasks (see De Renzi, 1982, for humans). This has been largely documented in birds (i.e. chicks: Rashid and Andrew, 1989; Vallortigara et al., 1996; Vallortigara, Regolin and Pagni, 1999; Tommasi, Andrew and Vallortigara, 2000; *Parus*: Clayton and Krebs, 1994; but see Ulrich et al., 1999, and Gagliardo et al., 2001, for a discussion of the taxonomy of different 'spatial tasks') and in mammals (Crowne et al., 1992; Cowell, Waters and Denenberg 1997; King and Corwin, 1992). Unfortunately, there is at present no evidence (nor has any study been carried out as far as we know) to check whether vertebrates other than birds and mammals show a similar right hemisphere dominance in spatial tasks.

The other cases of species-invariance, on the other hand, comprise virtually all vertebrate classes. A striking example concerns the association between the right hemisphere and some species-specific behaviours, such as escape, sexual behaviour and attack. Aggressive behaviour towards conspecifics, for instance, is more likely to be evoked when the left eye is in use in the baboon (Casperd and Dunbar, 1996), chick and adult fowl (Rogers, 1991; Rogers, Zappia and Bullock, 1999; MacKenzie, Andrew and Jones, 1998), lizard (Deckel, 1995) and toad (Robins et al., 1998; Vallortigara et al., 1998). In humans there is a clear parallel in the more intense emotions, which are usually evoked when stimuli are seen with the left lateral visual field (Dimond, Farrington and Johnson, 1976; Gur and Gur, 1977).

Lateralization in the control of attack and copulation is also revealed by injecting cycloheximide or glutamate into the left or right hemisphere of the chick on day 2 posthatching (Rogers, 1980; Howard, Rogers and Boura, 1980). Following injection of the left (but not the right) hemisphere, attack and copulation levels were elevated. Similarly, when chicks were treated with testosterone (which also elevates attack and copulation) and tested using the left eye, attack and copulation levels were elevated, as expected following testosterone treatment, but when they were tested using the right eye these behaviours remained at the low levels characteristic of untreated controls (Rogers, Zappia and Bullock, 1985). These data suggest that in the chick the left hemisphere suppresses and the right hemisphere activates attack and copulation. Similar specialization of the right hemisphere of the rat's brain has been deduced from the control of mouse killing (muricide), a spontaneous, species-specific behaviour of rats. The rats with only an intact right hemisphere had scores of muricide that were higher than those of the intact, whole-brain controls, as well as those of rats with only an intact left hemisphere (Garbanati et al., 1983). Again, this suggests that muricide is activated by the right hemisphere and inhibited by the left. The inhibition seems to be mediated by the corpus callosum because its sectioning removes the inhibition and increases the level of muricide behaviour (Denenberg et al., 1986).

Recently, Rogers (see Cameron and Rogers, 1999; Rogers, 2000) has argued for a general vertebrate specialization of the right hemisphere for fear and negative emotional states, and of the left hemisphere for approach and positive emotional states (also citing similar evidence in humans; see Davidson, 1992, for evidence). In non-human primates, there is indeed evidence that the left and right hemispheres might be specialized for approach and withdrawal (or fear) responses, respectively (see Hauser, 1993; Casperd and Dunbar, 1996; Hook-Costigan and Rogers, 1998; Ifune, Vermeire and Hamilton, 1984). Evidence for right hemisphere involvement in aggressive responses in lizards (Deckel, 1995) and toads (Robins et al., 1998; Vallortigara et al., 1998), mentioned above, could be considered to be consistent with the hypothesis, but further research on more species seems to be necessary.

Specialization of the right hemisphere for face recognition in humans (Sergent and Signoret, 1992) might be an elaboration of similar processes found in social recognition in non-human species. Split-brain monkeys show similar specialization of the right hemisphere to discriminate faces (Hamilton and Vermeire, 1988; Morris and Hopkins, 1993; Vermeire et al., 1998) and in birds the left eye seems to be involved in recognition of individual conspecifics (Vallortigara, 1992; Vallortigara and Andrew, 1991, 1994b; Vallortigara et al., 2001). Evidence for a left-eye bias during scrutiny of conspecifics in several species of fish has been reported recently (Bisazza, De Santi and Vallortigara, 1999; Sovrano et al., 1999; Sovrano, Bisazza and Vallortigara, 2001).

A further example of species-invariance is relative to the control of species-specific vocalization: a left hemisphere dominance has been demonstrated in several different species of passerine birds (reviews in Nottebohm, 1980; Williams, 1990) and in monkeys (Heffner and Heffner, 1986; Petersen et al., 1978, 1984), mice (Ehret, 1987), frogs (Bauer, 1993) and catfish (Fine et al., 1996). However, since lateralization for language in humans is striking particularly with respect to phonology, it is unclear to which aspects the species-invariance could be referred. One possibility is that control by left hemisphere occurs for all stimuli having species-specific communicative relevance. However, a right-ear advantage has been reported in male rats for discriminating two- and three-tone sequences (but not single tones) by O'Connor, Roitblat and Bever (1992). Fitch et al. (1993) also reported that male rats showed significantly better discrimination of tone sequences with the right ear than with the left ear. These stimuli have no communicative relevance. A crucial finding seems to be that Japanese macaques rely on temporal information (frequency peak position) in making the 'coo' vocalization that shows a right-ear advantage (May, Moody and Stebbins, 1989). This suggests a left-hemisphere specialization for the processing of temporal acoustic information (see also Hauser et al., 1998), which could represent evolutionary precursors to lateralized speech perception and language processing in humans.

An interesting attempt to 'homologize' lateralization of the tetrapod type as common to all vertebrates has been recently pursued by Richard Andrew. As mentioned before, he found that past experience of a stimulus determined which eye is predominantly used during approach in the zebrafish. The right eye was used in the first trial on which a particular stimulus was seen and the left eye on the second (see Miklósi et al., 1998). Mosquitofish also tend to use the right eye in fixation of biologically relevant stimuli (model predator, female conspecifics for sexually motivated males; see Bisazza, Pignatti and Vallortigara 1997a, 1997b; Bisazza et al., 1998). Miklósi and Andrew (1999) suggested that the sharpness of the shift between first and second presentations in zebrafish could be due to the fact that right eye use is associated with the need to take a decision as to what is the appropriate way to treat the novel stimulus, whilst inhibiting escape during careful approach. In the following presentations, the shift to left eye use would probably be associated with visual analysis directed to confirming that the stimulus is identical to one that has been seen before. Obviously, dependence on motivational state could also account for the phenomena described in zebrafish. Nonetheless, the hypothesis fits in well with a large variety of phenomena. For instance, the same association of the right eye use with the decision to respond holds for biting at small targets in zebrafish, whilst the approach to investigate a stimulus, which is not then bitten, shows no bias in eye use (Miklósi and

Andrew, 1999). Association of right eye use with control of feeding pecks has been demonstrated in chicks (Rogers and Anson, 1979; Mench and Andrew, 1986), pigeons (Güntürkün and Kesch, 1987; Güntürkün, 1985, 1997a, 1997b) and zebrafinch (Alonso, 1998); toads are more likely to take food seen in the right hemifield (Vallortigara et al., 1998). Why should, however, other types of responses be associated with the opposite (right) hemisphere? The answer, suggested Andrew, may be the difference in the role played by inhibition of response whilst a decision is taken. Feeding is an activity in which repeated decisions, taken on the basis of previous experience, are crucial. Species-specific behaviours like copulation, attack and escape are, by contrast, behaviours under continuous inhibitory control until this is removed at their onset, after which the course of behaviour is governed by automatic pre-programmed response (a striking example being provided by escape via Mauthner cells in fish).

A partially different hypothesis would suggest a motoric origin for lateralization in vertebrates, with asymmetries in eye use arising from initial asymmetries for the execution of fast responses such as the C-start reaction. Specialization of the muscles and increase of soma size and axon size of the Mauthner cell controlling the right side of the body could have produced subsequent postural asymmetries forming the basis for specialization of the left-eye system in sustained fixation of stimuli of interest (Vallortigara, Rogers and Bisazza, 1999; Vallortigara, 2000). However, with respect to the subdivision of labour between the two sides of the brain, this hypothesis makes basically the same predictions: right hemisphere for spatial analysis and evocation of attack, copulation and escape; left hemisphere for decision to be taken as to subsequent course of action and stimulus categorization. Such a subdivision would arise from the idea of specializations for rapid versus considered responding: a left hemisphere tendency to make considered decisions before responding and to process information on the basis of categories, whereas the right hemisphere would attend to novelty and the immediate, unsorted features of a stimulus, so controlling rapid responses. The origins of such a dichotomy could perhaps be traced back to the specializations for the fast and slow swimming shown by fish associated with M-cells, extending (admittedly a far-reaching step) to humans as well. For instance, Warrington (1982) showed that, when viewing objects, humans use the right hemisphere to process the purely perceptual properties of a stimulus, whereas they use the left hemisphere to consider the purposes for which the stimulus might be used.

Clearly, with such wide and long-reaching dichotomies, it is difficult to think of experiments which could really test and falsify them. There is, however, an issue on which some evidence is available, namely whether different species of the same class actually show an invariant pattern of lateralization. Bisazza et al. (2000) investigated turning responses in 16 species of fish faced with a vertical-bar barrier through which a learned dummy predator was visible. Ten of these species showed a consistent lateral bias to turn preferentially to the right or to the left. Species belonging to the same family showed similar directions of lateral biases, but species phylogenetically less related showed different directions of turning bias. How can this be explained? Obviously, ecological factors might well determine whether a species can show lateralization in a certain task. For instance, Heuts (1999) noticed that fish taxa that in their natural open-water environment are often exposed to predatory fish might have evolved stronger muscular and behavioural lateralizations than fish that are more substrate-bound.

But what about changes in the direction of lateralization in the same task in different species? Motivational factors could be likely candidates for an explanation: a test is not necessarily the same 'task' for all species or for individuals within a species which differ motivationally. Some species might be more fearful, and this could affect the importance of inhibition of Mauthner cells mediating escape during approach to inspect the predator (see also Miklósi, Andrew and Savage, 1998). The speed of swimming of fish and thus the point at which decisions are taken could also be important. A species that dashes round the barrier might decide at a distance, whereas a slowly moving fish might decide close to the barrier, and this might affect eye use. It is quite intriguing that the data presented by Bisazza et al. (2000) suggest that closely related species tend to manifest similarities in the direction of the asymmetries. The fish families for which more than one species was available (i.e. Gobiidae, Poeciliidae and Cyprinidae) showed an identical direction of lateralization within the family and different directions for different families. Results obtained for three families are probably not sufficient to make a conclusion, but they certainly constitute an interesting starting point. Obviously, similarity in direction among more closely related fish is compatible with both the hypothesis of reversed organization of lateralization between certain (less related) species and with the hypothesis of extraneous factors affecting detour performance of different species (for instance, it is likely that closely related species would be more similar in emotional/ motivational responses). To clarify these issues, work is needed that is complementary to that which has been described here. It is necessary to study species in a variety of tasks to check whether reversal of the direction of lateralization occurs invariably in different species or is merely task-dependent.

Evidence that the nature of the task could be crucial has been recently provided by another study that examined eye use during mirror image inspection in six species of fish, all showing a consistent left-eye preference (Sovrano et al., 1999). The species-invariance is impressive when contrasted with the variability observed with regards to anti-predatory responses (Bisazza et al., 2000). In fact, some of the species that had shown different patterns of eye-use during predator-inspection responses (i.e. *Xenotoca eiseni*, *Pterophyllum scalare* and *Trichogaster trichopterus*; see Bisazza et al., 2000) showed identical direction of lateralization during mirror-image viewing. We hypothesize that this may reflect the different costs and benefits associated with directional asymmetries in relation to anti-predatory and social behaviours. One problem with directional asymmetries is that they can be exploited by other species. A predator could in principle learn that its prey tends to escape in a particular direction as a consequence of a behavioural lateralization (e.g. a preference in eye use). Thus, it could be adaptive for single individuals of a species to develop a different direction of lateralization and, assuming that selective pressures exist for the alignment of these asymmetries at the population level (e.g. in order to coordinate schooling behaviour; see Section 1.6), different species can develop different directions of lateralization. In contrast, no need for different direction of lateralization with respect to other species should be expected with regards to social interactions within a species (e.g. with respect to courtship behaviour and agonistic interactions). Therefore, it could be predicted that these behaviours would be more likely to manifest an invariant pattern of directional asymmetries even in unrelated species (providing that they have been inherited from some common ancestor), whereas for behaviours involving inter-species interactions and competitions there could have been advantages for species-specific variations in the direction of lateral asymmetries.

These changes in the direction of lateralization can be expected even if lateralization is genetically determined (Section 1.7). Evidence obtained in birds has shown clearly that the alignment of brain and behavioural asymmetries at the population level could easily be obtained (and modified) through an interplay of embryologic and environmental factors. For example, in chicks (Rogers, 1991) and in pigeons (Güntürkün, 1993), the position of the eye of the embryo just prior of hatching, and the consequent different amount of light stimulation to the two eyes, is responsible of at least some of the asymmetries documented both at the anatomical and behavioural level (Rogers, 1995, 1996). Individuals incubated in the dark are still lateralized, but with equiprobable distribution within the population (Rogers, 1982; and see also Chapter 6 by Deng and Rogers). There is presently no evidence of similar phenomena in fish, but the issue is worth testing.

1.6. Population Lateralization versus Individual Lateralization

Some authors maintain that, if the population is not lateralized even though individual animals are, the evolutionary significance is debatable, and therefore they prefer to concentrate only on lateralization at the population level (Denenberg, 1981). On the other hand, it has been stressed that if a brain needs to be lateralized to function efficiently (whatever that would mean), it may be irrelevant which side is used to conduct one set of functions versus the other, the important thing being that, at the level of the individual, lateralization is present in one direction or the other (Rogers, 1989). Thus, in principle, lateralization can be considered even though the asymmetries are balanced in the population, provided that enough individuals show an asymmetric bias (see Collins, 1981). The problem with this line of reasoning, however, is that asymmetries at the individual level can arise as the result of fortuitous factors, such as an inability of the organism to develop full symmetry; this phenomenon has been studied extensively as part of work on fluctuating asymmetry (Leary and Allendorf, 1989; Palmer and Strobeck, 1986). This hypothesis has been applied with regard to lateralization at the individual level in mice (see McManus, 1992; Collins et al., 1993).

A well-documented example of functional lateralization at the individual level in fish is asymmetry of sexual behaviour of certain poeciliids: this is related to morphological asymmetries in the male gonopodium and in the female genital opening. Aronson and Clark (1952) studied gonopodium movements during swinging, thrusting and copulatory actions in platyfish (*Xiphophorus maculatus*), swordtail (*Xiphophorus helleri*), and guppies (*Poecilia reticulata*) where the gonopodium is structurally symmetrical. They found that individual males showed no lateral bias in swinging behaviour (in which the gonopodium is brought forward and to one side in conjunction with a forward movement of the homolateral pelvis), whereas they showed significant right or left preferences in thrusting behaviour (in which the forward and sideward movements are similar to those of swinging but, in addition, the gonopodium is directed toward the genital region of the female and may actually contact her genital orifice) and in copulatory behaviour (in which the contact thrust is prolonged and during which sperm transfer is often realized). Thrusts (and copulations), in contrast to swinging behaviour, are always performed in relation to the female, which can therefore influence the direction of thrusts. The authors observed that frequently the female remained stationary close to a wall of the aquarium while permitting the male to thrust or to copulate. For at least some of the platyfish males, however, further analysis led to the conclusion that they exhibit a consistent and significant bias, which was independent of the female used (see Aronson and Clark, 1952). Why consistent lateralization occurred during thrusting and copulation but not swinging is unclear, given the similarity of the actions. The crucial factor could be that thrusting and copulation involve more precise and accurate neuromuscular mechanisms, or, alternatively, that different degrees of arousal are associated with these behaviours.

Evidence for lateralization at the individual level has been found when testing another species of fish, *Jenynsia lineata*, for lateralization of escape responses using the apparatus employed with *Girardinus falcatus* (Bisazza, Cantalupo and Vallortigara, 1997). No significant right–left bias was found at the population level; however, a high number of significantly lateralized individuals (15 out of 32 animals tested) was observed. Moreover, the frequency distribution of the percentages of rightward and leftward escapes was found to differ from normality and uniformity, fitting instead a bimodal distribution. Interestingly, the results appeared to be consistent at retest 1 month later. A high number of individuals were in fact still significantly lateralized and there was a highly significant positive correlation between the two tests, meaning that the same individuals, which were lateralized in the first test, tended to be lateralized in the second test too.

Evidence for lateralization at the individual level has been obtained in yet another species of fish, the Siamese fighting fish *Betta splendens*, tested for preferential eye use during courtship and aggressive behaviour. Cantalupo, Bisazza and Vallortigara (1996) placed male *B. splendens* individually into a circular plastic tank with the bottom consisting of a mirror. The fish could turn its body parallel to its mirror image in order to look at it with its right or left eye while performing a lateral display to its own image. The number of left and right lateral displays and their duration were recorded for 10 min in two successive days. Results did not reveal any group bias in the number or duration of right and left displays. However, a highly significant positive correlation between the data obtained in the two daily sessions was observed, meaning that the individual animals were consistent in their right or left eye use. Moreover, there was a significant positive correlation between the total number of lateral displays and their mean duration, suggesting that, if one side was preferred by an individual for exhibiting the lateral displays, the duration of the displays on that side was longer than the duration of the displays on the other side.

Twenty-nine adult male animals were used in a second experiment, and ten of them were the same ones used in the previous experiment. This time there was no mirror at the bottom but a small glass tube placed in the middle containing a female conspecific. The tube was inserted into a box, the walls of which were composed of unidirectional screens. The experimental subject was segregated in the peripheral portion of the tank by means of a transparent circular plexiglas cylinder, and its right- or left-sided courtship displays were recorded in two successive daily sessions. Again, no significant right–left population bias was found, but there was a significant positive correlation between the data obtained in the two daily sessions. Moreover, the frequency distribution of the duration of the lateral displays was found to differ from normality and uniformity. Interestingly, data for the ten subjects tested previously with the mirror showed a significant positive correlation between the two tests, indicating that, if one eye was preferred by an individual in fixating its mirror image in the bottom of the tank, the same eye tended to be preferred 2 months later in a quite different situation, that is in fixating a female.

Although some authors (see above) have stressed that lateralization in a population may be affirmed regardless of the population distribution of the directions of the asymmetries, lateralization at the individual level may reflect basic asymmetries in the morphology of individuals. Fluctuating asymmetries, consisting of random deviations from bilateral symmetry in individuals, have been described for a number of different species, including fish (Downhower et al., 1990; and references therein). They are associated with environmental stress or with reduced heterozygosis and are believed to be due to the incapacity of individuals to undergo identical development on both sides of the body (e.g. Leary and Allendorf, 1989). Fluctuating asymmetries can, in principle, affect behaviour, and so could account for lateralization at the individual level. The question whether behavioural asymmetries at the individual level in fish could be accounted for in terms of fluctuating asymmetries has recently been tested by measuring some basic morphologic asymmetries in *Jenynsia lineata* and *Betta splendens*, and checking whether any correlation exists between the degree of morphological and of behavioural asymmetry (Bisazza, Cantalupo and Vallortigara, 1996, 1997). Four bilateral meristic characters (i.e. morphological characters with discrete integer values) were examined for morphological left–right asymmetry. No correlation was observed between the degree of asymmetry of each character and the degree of lateralization in escape behaviour. Further, a general index of fluctuating asymmetry was calculated for each fish as the mean $|R - L/R + L|$ of the four morphological characters. Even in this case no correlation with the behavioural measures was observed. Even the gonopodial asymmetry did not appear to be related to the behavioural asymmetry.

Three bilateral meristic characters (number of rays of the pectoral fin before the first bifurcation, number of preocular pore openings, number of preopercular pore openings) were examined also in *B. splendens*. The results were comparable to those obtained with *J. lineata* (see Bisazza, Cantalupo and Vallortigara, 1996). Even in this case no correlation was observed between the degree of behavioural asymmetry and the absolute degree of asymmetry in each character or the general index of the fluctuating asymmetries. The logic underlying the reasoning about fluctuating asymmetries should be made explicit to understand our point. If different individuals vary in their capacity to fulfil a symmetric project of development, then those individuals unable – for either genetic or environmental reasons – to develop symmetrically should manifest fluctuating asymmetries in a variety of different characters. Thus, if behavioural/neural lateralization at the individual level merely represents an instance of fluctuating asymmetry, then the individuals more lateralized should be expected to be, on average, more asymmetric even in other characters (including those characters not directly related to behavioural/neural lateralization). The evidence discussed above, however, suggests that lateralization at the individual level is not (at least for the examples considered here) an instance of fluctuating asymmetry, i.e. of mere 'biological noise'. All this points to the idea that lateralization at the individual level could be a genuine phenomenon, i.e. that there could be advantages in making a nervous system asymmetric at the individual level. In fact, virtually all the advantages that have been traditionally associated to brain lateralization (e.g. saving space for neural computation, evolving separate mechanisms for incompatible purposes; see Vallortigara, Rogers and Bisazza 1999, for a review) do not require any alignment of asymmetries at the population level. Why then does lateralization at the population level exist? And why do certain species possess it and others apparently do not?

1.7. Possible Origins of Lateralization at the Population Level

One possibility is to argue that the direction of asymmetry is genetically fixed in certain species. That is, there are genes coding for left- and right-handed individuals (see Section 1.8). However, this does not solve the problem, for the mere existence of phenomena of lateralization at the individual level provides evidence that, whatever the advantages conferred by lateralization, they can be obtained at the individual level. Moreover, direct encoding of directional asymmetries at the population level is costly, and there should therefore be specific advantages for its evolution. In the case of predation, for example, a population bias for motor responses may be a disadvantage because predators could learn this bias and use it to their own advantage. In fact, organisms with asymmetries at the individual but not population level might be favoured because they possess the computational advantages of having a lateralized brain but lack the disadvantages associated with predictable lateral biases (actually, the behaviour of an individual would be predictable under repeated trials, but not at first encounter on the basis of a general sampling of the individuals of the same species in the population). We believe that the solution of the conundrum probably lies in the fact that, in most circumstances, the best for an individual depends on what is best for the majority of other individuals. In other words, sociality and gregarious behaviour (in the very simple sense of interacting with other conspecifics in meaningful ways) could have provided the selective pressures to align the direction of the asymmetries in most individuals of a population (see also Rogers, 1989).

We recently investigated turning responses in sixteen species of fish faced with a vertical-bar barrier through which a learned dummy predator was visible (see Bisazza et al., 2000). Ten of these species showed a consistent lateral bias to turn preferentially to the right or to the left. We performed an independent test of shoaling tendency and found that all gregarious species showed population lateralization, whereas only 40% of the non-gregarious species did. The results provide some support to the hypothesis that population lateralization might have been developed in relation to the need to maintain coordination among individuals in behaviours associated with social life.

In schooling fish, a population bias for turning in the same direction may be an advantage because it would keep the school together, and this might outweigh the disadvantage of being predictable by predators. All of these considerations have interesting consequences, namely that lateralization, far from being a sort of laboratory artefact, should deeply affect animals' everyday behaviour.

It is common for pairs of fish to leave their shoal in order to approach and inspect a potential predator (Magurran and Pitcher, 1987; Magurran and Seghers, 1990). In such cases, both fish share the risk of being preyed upon, but not if one of the fish remains at a distance. The fish are thus believed to face a classical 'prisoner's dilemma' in this situation, and predator-inspection behaviour has been used as a model to analyse the evolution of mutual cooperation among unrelated individuals. Milinski (1987) found that sticklebacks are more likely to approach a predator when a mirror is placed parallel to the tank so that the image appears to swim along with the fish (simulating a cooperative partner) than when the mirror is angled so that the image appears to swim away from the fish (simulating a non-cooperative partner). Recently, Bisazza, De Santi and Vallortigara (1999) duplicated Milinsky's original procedure in order to check for the effects of positioning a mirror on either the left or the right side of the fish. They found that cooperative predator inspection is more likely to occur when the mirror image is visible on the left rather than on the right side of mosquitofish *Gambusia holbrooki*. (The effect cannot be due to asymmetries of the predator because control experiments with right–left inverted video images confirmed the asymmetry.)

De Santi et al. (2001) recently tried to unravel whether such a peculiar pattern of results arises because of a preference in the use of the right eye in fixation of the predator or of the left eye in fixation of the mirror image, or some combination of both eye preferences. They found that mosquitofish do preferentially use the left eye during sustained scrutiny of their mirror image when tested in the absence of any predator. On the other hand, when tested in a swimway for predator inspection in the absence of any mirror image (or other social stimuli), mosquitofish revealed a tendency to explore the environment using the left eye when at a distance and a preferential use of the right eye when coming closer to the predator. It is likely that use of the right eye only when in proximity of the predator is associated with higher ability of the left hemisphere to inhibit escape responses to allow predator-inspection responses (see Andrew, Tommasi and Ford, 2000). In fish, therefore, the visual scenes seen on the right and left sides may evoke different types of social behaviour, as a result of differing modes of analysis of perceptual information carried out by the left and right sides of the brain. Experiments on predator inspection, when in the presence of social fellows, provide for the first time a striking demonstration of how perceptual (and consequently motor) asymmetries play a crucial role in everyday behaviour of biological organisms and, at the same time, a rationale for the emergence of such population-level asymmetries.

Population biases may also be important for other aspects of social behaviour. Groups of young chicks with a population bias for visual lateralization (exposed to light before hatching) form more stable social hierarchies than groups of chicks without a population bias (for some behaviours) since they had been incubated in the dark [Rogers and Workman, 1989; see also Chapters 5 by (Andrew) and 6 by (Deng and Rogers)].

Environmental demands for interactive exploration versus more cautious observation might also affect population biases for some forms of lateralization. Right-handed primates exhibit more interactive exploration than left-handed ones (chimpanzees: Hopkins and Bennett, 1994; marmosets: Cameron and Rogers, 1999). This relationship between hand preference and exploration may reflect a population bias for hemispheric specialization coupled with dominance of the hemisphere opposite to the preferred hand, at least in some contexts.

1.8. Individual Variation in Lateralization and its Genetic Basis

In humans, it is usually maintained that there is some form of polymorphism in cerebral asymmetries at the population level (Bradshaw and Nettleton, 1981; Hellige, 1990). The great majority of humans are right handed for object manipulation, but around 10–13% are left handed. (It is also worth noting that humans vary somewhat in the alignment of asymmetries, e.g. language and handedness are not always consistent.) Does evidence for lateralization in animals also suggest some form of polymorphism? This is usually taken for granted, but we are not aware of any direct evidence. Behavioural experiments reveal, statistically, lateralization at the group level, which means that not all the animals are necessarily lateralized. This could be ascribed to either measurement errors and random individual variations, or to the existence of a minority of individuals with a somewhat different pattern of lateralization (like left-handers in humans).

The genetic bases of cerebral lateralization are, however, largely unsettled. Studies with animals have been mostly confined to the mouse model of pawedness, and have led to different conclusions about inheritance of direction of lateralization in humans and mice. Within an inbred strain, approximately half the mice are left-pawed and half are right-pawed. Among different strains, the left–right direction of paw preference seems to be genetically neutral (Collins, 1977, 1985). In contrast, the degree of lateralization of paw preference is genetically determined as demonstrated by the fact that both highly lateralized and weakly lateralized strains can be selected starting from a heterogeneous stock (Collins, 1985; Collins, Sargent and Neumann, 1993). In apparent distinction to mice studies, human handedness has generally shown significant, though low, parent–offspring correlations for direction of handedness (Collins, 1977), whereas it is uncertain whether there is heritability for degree of handedness (Annett, 1998; Collins, 1977; McManus, 1984, 1991). McManus (1991, 1992) argued that genetic variation in handedness is fundamentally different in humans and in mice (but see in contrast

Collins et al., 1993), suggesting the existence of a locus in humans at which the dominant allele confers right-handedness and that 50% of homozygotes for the recessive allele are left-handed. Similar arguments have been put forward by Corballis (1997), who argued for a human-specific D (dextral) allele that would underlie both right-handedness and left cerebral control of speech.

However, the issue of the genetic basis of handedness in mice is not clear-cut. In fact, it has been claimed that, in some strains of mice, the direction of paw usage may not be a genetically neutral trait, and that the direction of paw preference may also be genetically determined (Biddle and Eales, 1996; Biddle et al., 1993).

Genetic studies with non-human animals other than the mouse are rare (Neville, 1976, 1978). Hopkins et al. (1994) reported heritability in direction of hand preference in chimpanzees. They found that, for two separate mea-sures of handedness, offspring had the same hand preference as their biolo-gical parent significantly more often than would have been predicted by chance. Half-siblings also showed hand preferences that deviated signifi-cantly from chance, whereas no effects of the rearing environment were observed. Westergaard and Suomi (1997) investigated lateral biases for look-ing, reaching, and turning in capuchin monkeys. They found evidence for maternal and paternal contributions to the direction of looking bias, but no evidence for contribution of either parent to the direction of reaching bias. They also found evidence of differential maternal and paternal contribution to the direction of offspring turning bias. No evidence of parental contribu-tion to the strength of lateral bias for any of such measures was observed.

Recently, we began to study intraspecific variability and its causation in the teleost fish *Girardinus falcatus*. For this analysis we employed the detour test as described above, using a dummy predator as target. Although at the population level fish turn approximately 65% of times left and 35% right, there is a wide individual variation and analyses of large samples indicate that extreme values are significantly over-represented (Figure 1.1).

The first step in our analysis was to determine whether scores at the extremes of the distribution correspond to individuals with different latera-lization or whether they simply represent fluctuations in the measurement of otherwise identical individuals. We tested fish scoring 90% or more detour to the left (LD) and fish scoring 90% or more detour to the right (RD) in an independent test of lateralization, the viewing test (Miklósi et al., 1998). Singly housed fish were exposed in their home cage to two novel stimuli, a dummy predator (different from the one used in the detour test) or a neutral stimulus, represented by a red ball, and we measured eye preference while

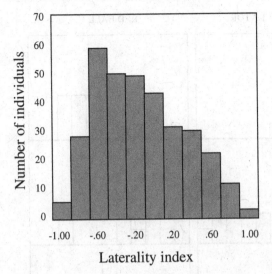

Figure 1.1.   Frequency distribution of percentage of right turning in a large sample of *Girardinus falcatus* (N = 341) in the detour test with a dummy predator as target. Extremes are over-represented (Kolmogorov-Smirnov test $z = 2.47$, $p < 0.001$). [Pooled from published (Bisazza et al., 1998; Facchin, Bisazza and Vallortigara, 1999) and unpublished data.]

fixating the novel objects. Fish that tended to detour the barrier on the left side used the right eye to scrutinize the dummy predator and the left eye to scrutinize the neutral stimulus while the fish that tended to detour the barrier on the right side showed the reverse tendency in eye use (Facchin, Bisazza and Vallortigara, 1999). Differences among groups were clear-cut and there was almost no overlap in viewing preferences between RD and LD fish (Figure 1.2).

Different directions for detouring the barrier are thus related to differences in viewing preferences. In addition these results show a complementarity in eye use that is probably related to the different motivational states evoked by the two stimuli, and indicate the possible existence of individual differences in the localization of cognitive functions that goes beyond a simple motor tendency to turn right or left in order to detour an obstacle.

These data do not establish whether the central portion of the distribution of Figure 1.1 is produced by superimposing the tails of two distributions of RD and LD fish or whether it represents fish with more balanced use of right and left eye. In a study (Facchin, Bisazza and Vallortigara, 1999), we thus examined in the viewing test an additional group of fish, chosen among those animals detouring 50% in each direction. The majority of the fish that showed no preferences in the detour test, alternated the two eyes while view-

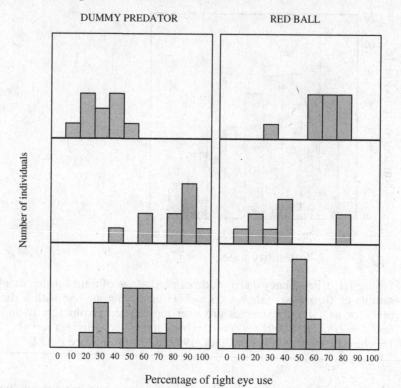

DUMMY PREDATOR   RED BALL

Figure 1.2.   Frequency distribution of eye use during viewing of two different stimuli in individuals of *Girardinus falcatus* selected for different lateralization in the detour test. (Data from Facchin, Bisazza and Vallortigara, 1999.)

ing the novel stimuli, suggesting that there is at least a third type of fish in the population, characterized by reduced visual lateralization (Figure 1.2). In a subsequent study (Bisazza, Facchin and Vallortigara, 2000) we investigated possible sources of the individual differences in detour direction. Males and females with the same score at the detour test were paired and their progeny were tested with the same behavioural procedure. There was a strong correlation between scores of parents and that of offspring (Figure 1.3). Heritability exceeded 0.5, a notably high value in view of the nature of the character that was measured. In this first experiment progeny were raised with their parents. In order to exclude non-genetic influences (e.g. learning), the experiment was repeated separating the progeny from their parents at birth, but the results were identical (Figure 1.3). These data thus strongly suggest the existence of genes controlling the direction of brain lateralization,

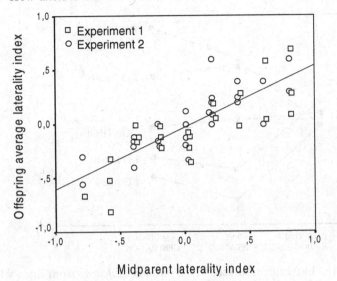

Figure 1.3. Regression of offspring on mid-parent for lateralization score in the detour test. In Experiment 1 offspring were raised with their parents, whilst in Experiment 2 they were separated at birth and raised in a different tank.

although maternal influences in these viviparous fish cannot completely be excluded.

Finding for the first time in vertebrates a clear indication for the existence of genes for direction of lateralization suggested that it would be rewarding to select artificial lines differing for lateralization. We have started to select two lines of fish that turn right in the detour test, two lines of fish that turn left and one unselected control line. Figure 1.4 shows the results of the first two generations of selection. In all four lines, there is a significant departure from the value of parental population and the response to selection is very close to the response predicted from heritability. There is no significant gender difference in response to selection and, as far as one can tell from the first few generations, the response is similar for both directions of selection.

Subsequent analysis was directed to assess whether fish belonging to the two selected lines differed for a single trait controlling only the behaviour in the detour test or had more profound differences in the organization and localization of multiple behavioural and cognitive functions. Fish from the LD and RD lines were compared using several different tests of lateralization. To date, comparison has been completed for five tests, two of which are mainly of a motor nature and the remaining three more related to visual lateralization.

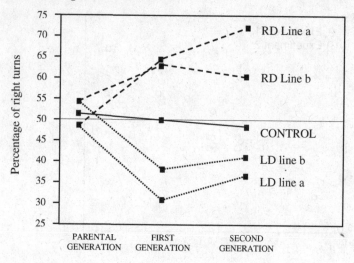

Figure 1.4.   Lateralization score in *Girardinus falcatus* from lines selected for right or left turning in the detour test.

Rotational preferences have been used as a way to measure lateralization in several species of mammals and more recently in the teleost fishes (Section 1.3.2.1). The procedure is the same used for testing *Gambusia holbrooki* (Bisazza and Vallortigara, 1996, 1997), and consists of measuring the proportion of clockwise and anticlockwise direction of rotation of individuals introduced singly into a circular arena. LD and RD lines of *G. falcatus* differed significantly in the preferred direction of rotation. LD fish rotated more often clockwise (59% of time; Figure 1.5) while RD fish showed the opposite tendency (37% of time clockwise) with no significant difference between sexes.

Turning direction was measured in a T-maze, in which fish were tested in ten consecutive trials. There was a clear preference for RD fish to turn right and for LD fish to turn left. Percentages of right turn are respectively 57% and 31% with no gender difference. As shown in Figure 1.5, both distributions are very skewed towards the extremes.

The eye preferred for viewing a companion fish, with which the tested animal was swimming, was measured using the mirror test (Sovrano et al., 1999), in which the companion is provided by the mirror image of the subject itself. Only females were used in this test (since males show little affiliative tendencies in this species). RD fish tended to swim keeping their mirror image on the left side, while LD fish did the reverse. Right eye use was 39% and 67% respectively (Figure 1.5). When we measured the time spent by fish more or less facing the mirror, we still found a statistically significant difference in

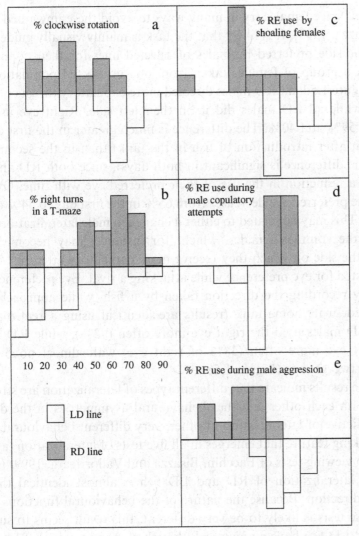

Figure 1.5. Frequency distribution of lateralization scores in five behavioural tests done on *Girardinus falcatus* from lines selected for opposite lateralization tendencies in the detour test.

eye preference but this time in the opposite direction (RD fish 54% of right eye use, LD fish 44%).

Male *G. falcatus*, like most poeciliids, perform continuous attempts to fertilize females (Bisazza and Pilastro, 1997). The male approaches the female from behind trying to reach her side and insert its copulatory organ, a behaviour normally referred to as copulatory thrust. This task requires precise coordination, since females try in many ways to avoid these unwanted copulations (Bisazza, 1993). It is likely that the task is mainly visually guided. We measured the side preferred by males of selected lines for their copulatory thrusts with a group of females taken from one unselected population. As shown in Figure 1.5d, all RD males directed more copulatory thrusts on their right side, while all LD males did it on their left side. Right eye use was respectively 59% and 40%. The difference is much greater in the first day of testing (soon after introduction of fish in the tank) than in the second one (although the difference is significant in both days), since both RD and LD males show a reduction in the use of their preferred eye with time. Overall, the use of the preferred eye decreases from 65% in the first day to 54% on the second day. This may be related to either a change in male lateralization or to changes in the stimulus females, which (for instance) may become more vigilant on the side on which they receive more copulatory attempts. Males were also tested for eye preference while attacking a rival. Eye preference was measured by recording the direction taken by a fish, while approaching a mirror placed in its home tank (results are identical using a real male as stimulus). LD males used the right eye more often (62%), while RD males used the left eye more often (38% of right eye) with almost no overlap between the two groups (Figure 1.5e).

Overall the results indicate that different types of lateralization are strongly correlated with each other and that behavioural asymmetries in the detour test are predictive of lateralization in other, very different behavioural tests. Another striking feature that emerges in all five tests of lateralization (as well as in the two viewing tests of Facchin, Bisazza and Vallortigara, 1999) is that the average lateralization of RD and LD fish is almost identical though opposite in direction. Because the nature of the behavioural functions measured in these tests is likely to be very different, this result seems to suggest that RD and LD fish have an identical, though reversed pattern of localization of cognitive/behavioural functions.

The inquiry into the individual variation in lateralization of G. falcatus suggests a close resemblance with the conditions found in our own species. The fish population appears to be polymorphic and, at least in part, individual differences appear to be genetically determined. Different manifestations of behavioural lateralization tend to be associated and highly correlated (apparently, even more strikingly than in humans, but in G. falcatus fish this is likely to be an artefact due to artificial selection focused on the individuals at the extreme ends of the distribution).

The population bias appears to be determined by the prevalence of one direction over the other (in the natural population, about 65% of fish turn leftwise when tested in the detour test with the dummy predator). The reason for the initial prevalence of a particular direction is unknown (in fish as well as in humans), but this is an issue we shall return to at the end of the next section.

1.9. Conclusion: Advantages of a Lateralized Brain

When exactly did lateralization first evolve? Fossil records show that the most primitive chordates (calcichordates) were asymmetrical in the head region (Jefferies, 1979), and chordate ancestors seem to have lain on the right side, thereby receiving different sensory inputs through structures on the left and right sides of the body (Jefferies and Lewis, 1978). Andrew, Tommasi and Ford (2000) have suggested that visual detection of potential prey to allow inhibition of avoidance may have been the first step in the evolution of cerebral lateralization in these ancestors of modern vertebrates. In Amphioxus the mouth is innervated by the left side of the brain, as a result of its left-side position in the larva. Evolution of predation would require control of mouth reflexes, which include rejection responses to anterior input. This control would be predominantly a matter for the left brain, and with the evolution of eyes capable of detecting prey, preparations for ingestion would be likely to be added as a further responsibility of visual mechanisms of the left side of the brain [see Andrew, Tommasi and Ford, 2000; and Chapter 2 by (Andrew)].

It is clear that lateralization of perceptual functions is ancient, first occurring probably just after the brain duplicated itself into two halves. But what are the continuing advantages of lateralization? We have argued that two types of advantages are conferred by lateralization: one at the individual level, the other at the population level. Advantages at the individual level include enhanced skill in performance and faster responses in lateralized compared to non-lateralized individuals. There is now clear evidence at the behavioural level for such advantages: Rogers (2000) reported that lateralized chicks detect a stimulus resembling a predator with shorter latency than non-lateralized chicks (see Chapter 4 by Rogers). She suggests that having a lateralized brain allows dual attention to the tasks of feeding (right eye and left hemisphere) and vigilance for predators (left eye and right hemisphere). It has been shown that, in cats, reach durations are shorter for the preferred paw than for the non-preferred paw; moreover, cats that are lateralized in this action tend to prefer their left paws and have shorter movement times than cats that do not exhibit lateralized reaching behaviour (Fabre-Thorpe et al., 1993). McGrew and Marchant (1999) reported that wild chimpanzees showing stronger and more complete handedness are also more efficient at fishing for termites than those with incomplete handedness. Finally, Güntürkün et al. (2000) showed that an increase of visual asymmetry enhances success in visually guided foraging of pigeons. It seems to us, therefore, that there are reasons to believe that the early origins of brain lateralization could have been related to simple computational advantages associated with the possession of asymmetric brains by single individuals (and thus with an equiprobable distribution of 'left' and 'right' phenotypes within the population). Such very simple advantages could have been related originally to the increase in the speed of predator-evasion responses of creatures with (roughly) bilaterally symmetrical bodies that possessed neural circuits such as those of the Mauthner cells. The intrinsic asymmetry needed for performing this sort of escape-response sets a requirement for an asymmetry at the neuronal, muscular and behavioural levels. This in turn could have produced other asymmetries at the sensory and perceptual level.

Obviously, there are other possible advantages associated with the appearance of asymmetries in the nervous system, which have been largely discussed in the literature. For example, the need to avoid useless duplication of function in relation to the saving of neural space (Nottebohm, 1977), or the particular logical demands associated with phenomena of functional incompatibility within neural mechanisms originally devoted to the solution of a particular and circumscribed problem, when they have to be coopted for novel tasks (Vallortigara, Rogers and Bisazza, 1999). It is likely that there are several advantages associated with the possession of asymmetries in the brain. The crucial point is, however, that all these advantages could be gained and maintained within individually asymmetric brains, without any need to align the direction of the asymmetries at the population level. To understand directional asymmetries, i.e. asymmetries in which the direction is the same among the majority of individuals in a population, we need a further simple logical step. Once natural selection built up behaviourally asymmetric organisms, these asymmetric organisms had to interact with each other. As a result, disadvantages were likely to arise for an individual if the direction of its behavioural asymmetries was different from that of the majority of the other asymmetric individuals of the population. Hence, evolution of population lateralization is therefore an example of what Maynard-Smith (1976) has called an 'evolutionarily stable strategy'.

It seems to us that such an evolutionary approach could prove useful in understanding problems in neuropsychology, which have long been intractable, such as the functions of left-handedness. What we have said here apparently would argue against the possibility of left-handedness as an adaptive strategy. Let us consider a hypothetical school of primitive fish. If it is advantageous for all the individuals to be similar in their lateral responses and, for instance, to escape to the right as a group, then a 'left-handed' fish escaping to the left would be an isolated fish and probably a dead fish. That is true but it is probably not the entire story. 'Left-handers' are likely to have their own advantages, and these advantages could, paradoxically, be associated with the fact that they are only a few individuals within a large 'right-handed' population. This is what evolutionary biologists call 'frequency-dependent selection'. Suppose that a predator could learn about and thus exploit the consistent escape tendencies of our hypothetical school of fish, for instance, paying more attention to a particular side or by developing a motor bias. As a result of such learning by a predator, a 'left-handed' fish would have a selective advantage. However, and this is the key point, such an advantage is frequency-dependent, since it holds only if left-handed fish remain a minority within a large group of fish that are right-handed. Hori (1993) has provided evidence of left–right morphological asymmetries associated with frequency-dependent natural selection in scale-eating Cichlid fish.

Work by Raymond et al. (1996) has provided evidence for frequency-dependent maintenance of left-handedness in humans. They proposed that left-handers have a frequency-dependent advantage in fights and for that reason a fitness advantage. Consistent with this hypothesis, they found a higher proportion of left-handed individuals in interactive sports (reflecting some fighting elements) but not in non-interactive sports. In fish, advantages associated with the minority form of lateralization are likely to be associated with predator-evasion responses and survival of cannibalistic attacks by conspecifics. The evidence for a preferred side for copulatory attempts suggests another possible advantage (at least for Poeciliid fish): given that females actively keep watch for unwanted copulatory attempts, if for the majority of males copulatory attempts occur on a particular side, then males of the less frequent form of lateralization, which attempt to copulate on the other side, might have a selective advantage.

It could also be that 'choice' of a particular direction of lateralization has been associated, from the start, with small physiological differences between the two possible phenotypes. Variation in fitness could arise by a small variation in immune response, and there is evidence in mammals (including humans) that right- and left-biased individuals differ in the pattern of immune response (e.g. Geschwind and Behan, 1982; Neveu, 1996). Small differences in fitness could have produced an initial imbalance in the frequencies of the right and left phenotype and this could have been subsequently amplified by other factors, such as the advantages associated with standardization of lateralization for social purposes mentioned above. Obviously, the advantages associated with social life are likely to be (today) very different in different species. They may comprise such different phenomena as the need to coordinate swimming in schooling of fish and the advantages in communication and cultural transmission of manual abilities in humans. Nevertheless, comparative investigation is worth trying in order to establish firmly the study of brain lateralization in an ethological and evolutionary framework.

References

Adam, G.Y., Szilagyi, N. & Lang, E. (1992). The effect of monocular viewing on hemispheric functions. *International Journal of Psychology*, 27, 401.

Alonso, Y. (1998). Lateralization of visual guided behaviour during feeding in zebra finches (*Taeniopygia guttata*). *Behavioural Processes*, 43, 257–263.

Andrew, R.J. (1983). Lateralization of emotional and cognitive function in higher vertebrates, with special reference to the domestic chick. In *Advances in Vertebrate Neuroethology*, ed. J.-P. Ewert, R.R. Capranica & D.J. Ingle, pp. 477–505. New York: Plenum Press.

Andrew, R.J. (1988). The development of visual lateralization in the domestic chick. *Behavioural Brain Research*, 29, 201–209.

Andrew, R.J. (1991). The nature of behavioural lateralization in the chick. In *Neural and behavioural Plasticity. The Use of the Chick as a Model*, ed. R.J. Andrew, pp. 536–554. Oxford: Oxford University Press.

Andrew, R.J., Mench, J. & Rainey, C. (1982). Right–left asymmetry of response to visual stimuli in the domestic chick. In *Analysis of Visual Behaviour*, ed. D.J. Ingle, M.A. Goodale & R.J. Mansfield, pp. 197–209. Cambridge, MA: MIT Press.

Andrew, R.J., Tommasi, L. & Ford, N. (2000). Motor control by vision and the evolution of cerebral lateralization. *Brain and Language*, 73, 220–235.

Annett, M. (1998). Handedness and cerebral dominance: The right shift theory. *Journal of Neuropsychiatry*, 4, 459–469.

Armstrong, J.A., Gamble, H.J. & Goldby, F. (1953). Observations on the olfactory apparatus and the telencephalon of *Anolis*, a microsomatic lizard. *Anatomy*, 8, 288–307.

Aronson, L.R. & Clark, E. (1952). Evidence of ambidexterity and laterality in the sexual behavior of certain poeciliid fish. *The American Naturalist*, 828, 161–171.

Aronson, L.R. & Noble, G.K. (1945). The sexual behavior of anura: 2. Neural mechanisms controlling mating in the male leopard frog, *Rana pipiens*. *Bulletin of the American Museum of Natural History*, 86, 83–139.

Bauer, R.H. (1993). Lateralization of neural control for vocalization by the frog (*Rana pipiens*). *Psychobiology*, 21, 243–248.

Biddle, F.G. & Eales, B.A. (1996). The degree of lateralization of paw usage (handedness) in the mouse is defined by three major phenotypes. *Behavioral Genetics*, 26, 391–406.

Biddle, F.G., Coffaro, C.M., Ziehr, J.E. & Eales, B.A. (1993). Genetic variation in paw preference (handedness) in the mouse. *Genome*, 36, 935–943.

Bisazza, A. (1993). Male competition, female choice and sexual size dimorphism in poeciliid fishes. In *The Behavioural Ecology of Fishes*, ed. F. Huntingford & P. Torricelli, pp. 257–286. Chur, Switzerland: Harwood Academic Publishers.

Bisazza, A. & Pilastro, A. (1997). Small mating advantage and the reverse dimorphism in poeciliid fishes. *Journal of Fish Biology*, 50, 397–406.

Bisazza, A. & Vallortigara, G. (1996). Rotational bias in mosquitofish (*Gambusia hoolbrooki*): The role of lateralization and sun-compass navigation. *Laterality*, 1, 161–175.

Bisazza, A. & Vallortigara, G. (1997). Rotational swimming preferences in mosquitofish (*Gambusia holbrooki*): Evidence for brain lateralization? *Physiology and Behaviour*, 62, 1405–1407.

Bisazza, A., Cantalupo, C. & Vallortigara, G. (1996). Lateralization of functions in the brain and behaviour of lower vertebrates: New evidences. *Atti e Memorie dell'Accademia Patavina di Scienze, Lettere ed Arti. Classe di Scienze Matematiche e Naturali*, 108, 93–138.

Bisazza, A., Cantalupo, C. & Vallortigara, G. (1997). Lateral asymmetries during escape behaviour in a species of teleost fish (*Jenynsia lineata*). *Physiology and Behaviour*, 61, 31–35.

Bisazza, A., De Santi, A. & Vallortigara, G. (1999). Laterality and cooperation: Mosquitofish move closer to a predator when the companion is on their left side. *Animal Behaviour*, 57, 1145–1149.

Bisazza, A., Facchin, L. & Vallortigara, G. (2000). Heritability of lateralization in fish: Concordance of right-left asymmetry between parents and offspring. *Neuropsychologia*, 38, 907–912.

Bisazza, A., Lippolis, G. & Vallortigara, G. (2001). Lateralization of ventral fins use during object exploration in the blue gourami (*Trichogaster trichopterus*). *Physiology and Behaviour*, 72, 575–578.

Bisazza, A., Pignatti, R. & Vallortigara, G. (1997a). Detour tests reveal task- and stimulus-specific neural lateralization in mosquitofish (*Gambusia holbrooki*). *Behavioural Brain Research*, 89, 237–242.

Bisazza, A., Pignatti, R. & Vallortigara, G. (1997b). Laterality in detour behaviour: Interspecific variation in poeciliid fishes. *Animal Behaviour*, 54, 1273–1281.

Bisazza, A., Rogers, L.J. & Vallortigara, G. (1998). The origins of cerebral asymmetry: A review of evidence of behavioural and brain lateralization in fishes, amphibians, and reptiles. *Neuroscience and Biobehavioral Reviews*, 22, 411–426.

Bisazza, A., Cantalupo, C., Capocchiano, M. & Vallortigara, G. (2000). Population lateralization and social behaviour: A study with sixteen species of fish. *Laterality*, 5, 269–284.

Bisazza, A., Cantalupo, C., Robins, A., Rogers, L. & Vallortigara, G. (1996). Right-pawedness in toads. *Nature*, 379, 408.

Bisazza, A., Cantalupo, C., Robins, A., Rogers, L. & Vallortigara, G. (1997). Pawedness and motor asymmetries in toads. *Laterality*, 2, 49–64.

Bisazza, A., Facchin, L., Pignatti, R. & Vallortigara, G. (1998). Lateralization of detour behaviour in poeciliid fishes: The effect of species, gender and sexual motivation. *Behavioural Brain Research*, 91, 157–164.

Bishop, P.O., Jeremy, D. & Lance, J.W. (1953). The optic nerve: Properties of a central tract. *Journal of Physiology*, 121, 415–432.

Borkhvardt, V.G. & Ivashintsova, E.B. (1994). On the position of the epicoracoids in amphibian arciferal pectoral girdles. *Russian Journal of Herpetology*, 1, 114–116.

Borkhvardt, V.G. & Ivashintsova, E.B. (1995). Arciferal pectoral girdle of amphibians – an instrument for recognizing right- and left-handedness? *Russian Journal of Herpetology*, 2, 34–35.

Borkhvardt, V.G. & Malashichev, Y.B. (1997). Position of the epicoracoids in arciferal pectoral girdles of the fire-bellies *Bombina* (Amphibia: Discoglossidae). *Russian Journal of Herpetology*, 4, 28–30.

Bracha, H.S., Seitz, D.J., Otemaa, D.J. & Glick, S.D. (1987). Rotational movement (circling) in normal humans: Sex difference and relationship to hand, foot, and eye preference. *Brain Research*, 411, 231–235.

Bradshaw, J.L. (1991). Animal asymmetry and human heredity: Dextrality, tool use and language in evolution – 10 years after Walker (1980). *British Journal of Psychology*, 82, 39–59.

Bradshaw, J.L. & Nettleton, N.C. (1981). The nature of hemispheric specialization in man. *Behavioural and Brain Sciences*, 4, 51–91.

Bradshaw, J.L. & Rogers, L.J. (1993). *The Evolution of Lateral Asymmetries, Language, Tool Use, and Intellect*. New York: Academic Press.

Braitenberg, V. & Kemali, M. (1970). Exceptions to bilateral symmetry in the epithalamus of lower vertebrates. *Journal of Comparative Neurology*, 138, 137–146.

Broca, P. (1861). Perte de la parole. Rammollissement cronique et partielle du lobe anterieur gauche de cerveau. *Bullettin de la Societé d'Anthropologie*, 2, 235–237.

Buchmann, K. (1988). Spatial distribution of *Pseudodactylogyrus anguillae* and *P. bini* (Monogenea) on the gills of the European eel *Anguilla anguilla*. *Journal of Fish Biology*, 32, 801–802.

Burggren, W.W. (1979). Bimodal gas exchange during variation in environmental oxygen and carbon dioxide in the air breathing fish *Trichogaster trichopterus*. *Journal of Experimental Biology*, 82, 197–213.

Butler, A.B. & Northcutt, R.G. (1971). Retinal projections in *Iguana iguana* and *Anolis carolinensis*. *Brain Research*, 26, 1–13.

Cameron, R. & Rogers, L.J. (1999). Hand preference of the common marmoset (*Callithrix jacchus*): Problem solving and responses in a novel setting. *Journal of Comparative Psychology*, 113, 149–157.

Campbell, C.B.G. (1988). Homology. In *Comparative Neuroscience and Neurobiology: Readings from the Encyclopedia of Neuroscience*, ed. L.N. Irwin, pp. 44–45. Boston: Birkhäuser.

Cantalupo, C., Bisazza, A. & Vallortigara, G. (1995). Lateralization of predator-evasion response in a teleost fish (*Girardinus falcatus:* Poeciliidae). *Neuropsychologia*, 33, 1637–1646.

Cantalupo, C., Bisazza, A. & Vallortigara, G. (1996). Lateralization of displays during aggressive and courtship behaviour in the Siamese-fighting fish (*Betta splendens*). *Physiology and Behaviour*, 60, 249–252.

Casperd, L.M. & Dunbar, R.I.M. (1996). Asymmetries in the visual processing of emotional cues during agonistic interactions by gelada baboons. *Behavioural Processes*, 37, 57–65.

Clayton, N.S. & Krebs, J.R. (1994). Memory for spatial and object-specific cues in food-storing and non-storing birds. *Journal of Comparative Physiology A*, 174, 371–379.

Collins, R.L. (1977). Origins of the sense of asymmetry: Mendelian and non-Mendelian models of inheritance. *Annals of the New York Academy of Sciences*, 299, 283–305.

Collins, R.L. (1981). On asymmetries exhibiting a near-equiprobable distribution of directions. *Behavioural Brain Sciences*, 4, 23–24.

Collins, R.L. (1985). On the inheritance of direction and degree of asymmetry. In *Cerebral Lateralization in Nonhuman Species*, ed. S.D. Glick, pp. 41–71. New York: Academic Press.

Collins, R.L., Sargent, E.E. & Neumann, P.E. (1993). Genetic and behavioral tests of the McManus hypothesis relating response to selection for lateralization of handedness in mice to degree of heterozygosity. *Behavioural Genetics*, 23, 413–421.

Corballis, M.C. (1997). The genetics and evolution of handedness. *Psychological Review*, 104, 714–727.

Cowell, P.E., Waters, N.S. & Denenberg, V.H. (1997). The effects of early environment on the development of functional laterality in Morris maze performance. *Laterality*, 2, 221–232.

Crowne, D.P., Novotny, M.F., Mier, S.E. & Vitols, R.W. (1992). Spatial deficits and their lateralization following unilateral parietal cortex lesions in the rat. *Behavioural Neuroscience*, 106, 808–819.

Davidson, R.J. (1992). Emotion and affective style: Hemispheric substrates. *Psychological Science*, 3, 39–43.

Deckel, A.W. (1995). Laterality of aggressive responses in *Anolis*. *Journal of Experimental Zoology*, 272, 194–200.

Deckel, A.W. (1998). Hemispheric control of territorial aggression in *Anolis carolinensis*: Effects of mild stress. *Brain, Behaviour and Evolution*, 51, 33–39.

Deckel, A.W., Lillaney, R., Ronan, P.J. & Summers, C.H. (1998). Lateralized effects of ethanol on aggression and serotonergic systems in *Anolis carolinensis*. *Brain Research*, 807, 38–46.

Denenberg, V.H. (1981). Hemispheric laterality in animals and the effects of early experience. *Behavioural and Brain Sciences*, 4, 1–49.

Denenberg, V.H., Gall, J.S., Berrebi, A. & Yutzey, D.A. (1986). Callosal mediation of cortical inhibition in the lateralized rat brain. *Brain Research*, 397, 327–332.

Denenberg, V.H., Garbanati, J.A., Sherman, G., Yutzey, D.A. & Kaplan, R. (1978). Infantile stimulation induces brain lateralization in rats. *Science*, 201, 1150–1152.

De Renzi, E. (1982). *Disorders of Space, Exploration and Cognition*. New York: Wiley.

De Santi, A., Sovrano, V.A., Bisazza, A. & Vallortigara, G. (2001). Mosquitofish display differential left- and right-eye use during mirror-image scrutiny and predator-inspection responses. *Animal Behaviour*, 61, 305–310.

Dharmaretnam, M. & Andrew, R.J. (1994). Age- and stimulus-specific use of right and left eyes by the domestic chick. *Animal Behaviour*, 48, 1395–1406.

Dill, L.M. (1977). 'Handedness' in the Pacific tree frog (*Hyla regilla*). *Canadian Journal of Zoology*, 55, 1926–1929.

Dimond, S.J., Farrington, L. & Johnson, P. (1976). Differing emotional response from right and left hemispheres. *Nature*, 261, 690–692.

Donnelly, R.E. & Reynolds, J.D. (1994). Occurrence and distribution of the parasitic copepod *Leposphilus labrei* on Corkwing Wrasse (*Crenilabrus melops*) from Mulroy Bay, Ireland. *Journal of Parasitology*, 80, 331–332.

Downhower, J.F., Blumer, L.S., Lejeune, P., Gaudin, P., Marconato, A. & Bisazza, A. (1990). Otolith asymmetry in *Cottus bairdi* and *C. gobio*. *Polish Archives of Hydrobiology*, 37, 209–220.

Eaton, R.C. & Emberley, D.S. (1991). How stimulus direction determines the trajectory of the Mauthner-initiated escape response in a Teleost fish. *Journal of Experimental Biology*, 161, 469–487.

Ehret, G. (1987). Left hemisphere advantage in the mouse brain for recognizing ultrasonic communication calls. *Nature*, 325, 249–251.

Eldred, W.D., Finger, T.E. & Nolte, J. (1980). Central projections of the frontal organs of Rana pipiens, as demonstrated by the anterograde transport of horseradish peroxidase. *Cell Tissue Research*, 211, 215–222.

Engbretson, G.A., Reiner, A. & Brecha, N. (1981). Habenular asymmetry and the central connections of the parietal eye of the lizard. *Journal of Comparative Neurology*, 198, 155–165.

Ewert, J.-P. (1980). *Neuro-Ethology*. Berlin: Springer.

Fabre-Thorpe, M., Fagot, J., Lorincz, E., Levesque, F. & Vaucliar, J. (1993). Laterality in cats: paw preference and performance in a visuomotor activity. *Cortex*, 29, 15–24.

Facchin, L., Bisazza, A. & Vallortigara, G. (1999). What causes lateralization of detour behaviour in fish? Evidence for asymmetries in eye use. *Behavioural Brain Research*, 103, 229–234.

Fetcho, J.R. (1991). Spinal network of the Mauthner cell. *Brain, Behavior and Evolution*, 37, 298–316.

Fetcho, J.R. (1992). Excitation of the motoneurons by the mauthner axon in goldfish: complexities in a 'simple' reticulospinal pathway. *Journal of Neurophysiology*, 67, 1574–1586.

Fine, M.L., McElroy, D., Rafi, J., King, C.B., Loesser, K.E. & Newton, S. (1996). Lateralization of pectoral stridulation sound production in the channel catfish. *Physiology and Behavior*, 60, 753–757.

Fitch, R.H., Brown, C.P., O'Connor, K. & Tallal, P. (1993). Functional lateralization for auditory temporal processing in male and female rats. *Behavioural Neuroscience*, 107, 844–850.

Friedman, H. & Davis, M. (1938). 'Left-handedness' in parrots. *Auk*, 80, 478–480.

Frontera, J.G. (1952). A study of the anuran diencephalon. *Journal of Comparative Neurology*, 96, 1–69.

Gagliardo, A., Ioale, P., Odetti, F., Bingham, V.P. & Vallortigara, G. (2001). Homing pigeons: Differential role of left and right hippocampal formation in the acquisition of the navigational maps. *European Journal of Neuroscience*, 13, 1617–1624.

Garbanati, J.A., Sherman, G.F., Rosen, G.D., Hofmann, M., Yutzey, D.A. & Denenberg, V.H. (1983). Handling in infancy, brain laterality, and muricide in rats. *Behavioural Brain Research*, 71, 351–359.

Geschwind, N. & Behan, P.O. (1982). Left-handedness: Association with immune disease, migraine and developmental learning disorders. *Proceedings of the National Academy of Sciences USA*, 79, 5097–5100.

Geschwind, N. & Levitski, W. (1968). Left-right asymmetry in the temporal speech region. *Science*, 161, 186–187.

Glick, S.D. & Ross, D.A. (1981). Right-sided population bias and lateralization of activity in normal rats. *Brain Research*, 205, 222–225.

Glick, S.D. & Shapiro, R.M. (1985). Functional and neurochemical mechanisms of cerebral lateralization in rats. In *Cerebral Lateralization in Nonhuman Species*, ed. S.D. Glick, pp. 158–184. New York: Academic Press.

Gonçalves, E. & Hoshino, K. (1990a). Behavioral lateralization in the freshwater fish *Oreochromis niloticus*. *Abstract translated from the Annals of the 5th Annaula Meeting of the Federation of Brazilian Societies for Experimental Biology (FESBE)*, Caxambu, M.G., p. 25.

Gonçalves, E. & Hoshino, K. (1990b). Lateralized behavior of Nile Tilapia in natural conditions. *Abstract translated from the Annals of the 5th Annaula Meeting of the Federation of Brazilian Societies for Experimental Biology (FESBE)*, Caxambu, M.G., p. 423.

Goodyear, C.P. & Ferguson, D.E. (1969). Sun-compass orientation in the mosquitofish, *Gambusia affinis*. *Animal Behaviour*, 17, 636–640.

Green, A.J. (1997). Asymmetrical turning during spermatophore transfer in the male smooth newt. *Animal Behaviour*, 54, 343–348.

Greenberg, N. (1982). A forebrain atlas and stereotaxic technique for the lizard *Anolis carolinensis* (Reptilia, Lacertilia, Iguanidae). *Journal of Morphology*, 174, 217–236.

Greenberg, N., MacLean, P.D. & Ferguson, J.L. (1979). Role of the paleostriatum in species-typical display behavior of the lizard (*Anolis carolinensis*). *Brain Research*, 172, 229–241.

Greenberg, N., Scott, M. & Crews, D. (1984). Role of the amygdala in the reproductive and aggressive behavior of the lizard, *Anolis carolinensis*. *Physiology and Behaviour*, 32, 147–151.

Greer, A.E. & Mills, A.C. (1997). Directional asymmetry in the amphibian pectoral girdle: additional data and a brief overview. *Journal of Herpetology*, 31, 594–596.

Gross, Y., Franko, R. & Lewin, I. (1978). Effects of voluntary eye movements on hemispheric activity and choice of cognitive mode. *Neuropsychologia*, 17, 653–657.

Güntürkün, O. (1985). Lateralization of visually controlled behavior in pigeons. *Physiology and Behaviour*, 34, 575–577.

Güntürkün, O. (1993). The ontogeny of visual lateralization in pigeons. *German Journal of Psychology*, 17, 276–287.

Güntürkün, O. (1997a). Avian visual lateralization: A review. *NeuroReport*, 8, 3–11.

Güntürkün, O. (1997b). Visual lateralization in birds: From neurotrophins to cognition? *European Journal of Morphology*, 35, 290–302.

Güntürkün, O. & Bohringer, P.G. (1987). Reversal of visual lateralization after midbrain commisurotomy in pigeons. *Brain Research*, 408, 1–5.

Güntürkün, O. & Kesch, S. (1987) Visual lateralization during feeding in pigeons. *Behavioral Neuroscience*, 101, 433–435.

Güntürkün, O., Diekamp, B., Manns, M., Nottelmann, F., Prior, H., Schwarz, A. & Skiba, M. (2000). Asymmetry pays: Visual lateralization improves discrimination success in pigeons. *Current Biology*, 10, 1079–1081.

Gur, R. & Gur, R. (1977) Correlates of conjugate lateral eye movements in man. In *Lateralization in the Nervous System*, ed. S. Harnad, R.W. Doty, L. Goldstein, J. Jaynes & G. Krauthamer, pp. 261–281. New York: Academic Press.

Gurusinghe, C.J. & Ehrlich, D. (1985a). Sex dependent structural asymmetry of the medial habenular nucleus of the chicken brain. *Cellular and Tissue Research*, 240, 149–152.

Gurusinghe, C.J. & Ehrlich, D. (1985b). Age, sex, and hormonal effects on structural asymmetry of the medial habenular nucleus of the chicken brain. *Neuroscience Letters*, 19, S67.

Hamilton, C.R. & Vermeire, B.A. (1988). Complementary hemispheric specialization in monkeys. *Science*, 242, 1691–1694.

Harris, L.J. (1989). Footedness in parrots: Three centuries of research, theory, and mere surmise. *Canadian Journal of Psychology*, 43, 369–396.

Harris, J.A., Guglielmotti, V. & Bentivoglio, M. (1996) Diencephalic asymmetries. *Neuroscience and Biobehavioral Reviews*, 20, 637–643.

Hauser, M.D. (1993). Right hemisphere dominance for the production of facial expression in monkeys. *Science*, 261, 475–477.

Hauser, M.D., Agnetta, B. & Perez, C. (1998). Orienting asymmetries in rhesus monkeys: the effect of time-domain changes on acoustic perception. *Animal Behaviour*, 56, 41–47.

Heffner, H.E. & Heffner, R.S. (1986). Effects of unilateral and bilateral auditory cortex lesions on the discrimination of vocalizations by Japanese macaques. *Journal of Neurophysiology*, 56, 683–701.

Heinrich, M.L. & Klaassen, H.D. (1985). Side dominance in constricting snakes. *Journal of Herpetology*, 19, 531–533.

Hellige, J.B. (1990). Hemispheric asymmetry. *Annual Review of Psychology*, 41, 55–80.

Heuts, B.A. (1999). Lateralization of trunk muscle volume, and lateralization of swimming turns of fish responding to external stimuli. *Behavioural Processes*, 47, 113–124.

Hodos, W. (1988). Homoplasy. In *Comparative Neuroscience and Neurobiology: Readings from the Encyclopedia of Neuroscience*, ed. L.N. Irwin, pp. 47. Boston: Birkhäuser.

Hook-Costigan, M.A. & Rogers L.J. (1995). Hand, mouth and eye preferences in the common marmoset (*Callithrix jacchus*). *Folia Primatologica*, 64, 180–191.

Hook-Costigan, M.A. & Rogers, L.J. (1998). Lateralized use of the mouth in production and vocalizations by marmosets. *Neuropsychologia*, 36, 1265–1273.

Hopkins, W.D. (1996) Chimpanzee handedness revisited: 55 years since Finch (1941). *Psychonomic Bulletin and Reviews*, 3, 449–457.

Hopkins, W.D. & Bennett, A.J. (1994). Handedness and approach-avoidance behavior in chimpanzees (Pan). *Journal of Experimental Psychology: Animal Behavior Processes*, 20, 413–418.

Hopkins, W.D., Bales, S.A. & Bennett, A.J. (1994). Heritability of hand preference in chimpanzees (*Pan*). *International Journal of Neuroscience*, 74, 17–26.

Hori, M. (1993). Frequency-dependent natural selection in the handedness of scale-eating Cichlid fish. *Science*, 260, 216–219.

Howard, K.J., Rogers, L.J. & Boura, A.L. (1980). Functional lateralization of the chicken forebrain revealed by use of intracranial glutamate. *Brain Research*, 188, 369–382.

Ifune, C.K., Vermeire, B. & Hamilton, C.R. (1984). Hemispheric differences in split-brain monkeys viewing and responding to videotape recordings. *Behavioural and Neural Biology*, 41, 231–235.

Jefferies, R.P.S. (1979). Calcichordates. In *Encyclopedia of Earth Sciences, Vol. VII: The Encyclopedia of Paleontology*, ed. R.W. Fairbridge & D. Jablonski, pp. 161–167. Strondsburg, PA: Dowden, Hutchinson & Ross Inc.

Jefferies, R.P.S. & Lewis, D.N. (1978). The English Silurian Fossil *Polacocystites forebesianus* and the ancestors of the vertebrates. *Philosophical Transactions of the Royal Society of London B*, 282, 204–223.

Johnston, J.B. (1902). The brain of *Petromyzon*. *Journal of Comparative Neurology*, 12, 1–86.

Kelley, D.B. & Tobias, L.M. (1989). The genesis of courtship song: Cellular and molecular control of sexually differentiated behavior. In *Prospectives in Neural Systems and Behavior*, ed. T.J. Carew & D.B. Kelley, pp. 175–194. New York: Wiley.

Kemali, M. (1977). Morphological relationship established through the habenulo-interpeduncolar system between the right and left portions of the frog brain. In *Structure and Function of the Cerebral Commissures*, ed. I. Steele-Russell, M.W. van Hof & G. Berlucchi, pp. 13–33. Baltimore: University Park Press.

Kemali, M. (1984). Morphological asymmetry of the habenulae of a macrosomatic mammal, the mole. *Zeitschrift für Mikroskopie-anatomie Forschung*, 98, 951–954.

Kemali, M. & Guglielmotti, V. (1977). An electron microscope observation of the right and the two left portions of the habenular nuclei of the frog. *Journal of Comparative Neurology*, 176, 133–148.

Kemali, M., Guglielmotti, V. & Fiorino, L. (1990). The asymmetry of the habenular nuclei of female and male frogs in spring and winter. *Brain Research*, 517, 251–255.

King, V.R. & Corwin, J.V. (1992). Spatial deficits and hemispheric asymmetries in the rat following unilateral and bilateral lesions of posterior parietal or medial agranular cortex. *Behavioural Brain Research*, 50, 53–68.

Kooistra, C.A. & Heilman, K.M. (1988). Motor dominance and lateral asymmetry of the globus pallidus. *Neurology*, 38, 388–390.

Ladich, F. & Fine, M.L. (1994). Localization of swimbladder and pectoral motoneurons involved in sound production in pimelodid catfish. *Brain, Behaviour and Evolution*, 44, 86–100.

Leary, R.F. & Allendorf, F.W. (1989). Fluctuating asymmetry as an indicator of stress: implication for conservation biology. *Trends in Ecology and Evolution*, 4, 214–217.

Lippolis, G. (1998). Lateralizzazione nell'uso degli arti posteriori e in alcuni comportamenti guidati dalla vista in due specie di rospo (*Bufo bufo* e *Bufo viridis*). (Lateralization of hindlimbs use and in visual-guided behaviours in two species of toads (*B. bufo* and *B. viridis*). Tesi di Laurea, Università di Padova, Graduation thesis not published, University of Padua, Italy.

Lippolis, G., Bisazza, A., Rogers, L.J. & Vallortigara, G. (2001). Lateralization of predator avoidance responses in three species of toads. *Laterality*, in press.

MacKenzie, R., Andrew, R.J. & Jones, R.B. (1998). Lateralization in chicks and hens: New evidence for control of response by the right eye system. *Neuropsychologia*, 36, 51–58.

MacNeilage, P.F., Studdert-Kennedy, M.G. & Lindblom, B. (1987). Primate handedness reconsidered. *Behavioural and Brain Sciences*, 10, 247–303.

Magurran, A.E. & Pitcher, T.J. (1987). Provenance, shoal size and the sociobiology of predator-evasion behavior in minnow shoals. *Proceedings of the Royal Society of London B*, 229, 439–465.

Magurran, A.E. & Seghers, B.H. (1990). Population differences in predator recognition and attack cone avoidance in the guppy *Poecilia reticulata*. *Animal Behaviour*, 40, 443–452.

Malashichev, Y.B. & Starikova, N. (2001). Preferential limb use and epicoracoid overlap in toads. *Laterality*, in press.

Marchant, L.F., McGrew, W.C. & Eibl-Eibesfeldt, I. (1995). Is human handedness universal? Ethological analyses from three traditional cultures. *Ethology*, 101, 239–258.

Martin, R.F. (1972). Arciferal dextrality and sinistrality in anuran pectoral girdles. *Copeia*, 376–381.

May, B., Moody, D.B. & Stebbins, W.C. (1989). Categorical perception of conspecific communication sounds by Japanese macaques, *Macaca fuscata*. *Journal of Acoustical Society of America*, 85, 837–847.

Maynard-Smith, J. (1976). Evolution and the theory of games. *American Scientist*, 64, 41–45.

McCarthy, T.K. & Rita, S.D. (1991). The occurrence of the monogean *Pseudodactylogyrus anguillae* (Yin and Sproston) on migrating silver eels in western Ireland. *Irish Naturalists' Journal*, 23, 473–477.

McGrew, W.C. & Marchant, L.F. (1997). On the other hand: Current issues in a meta-analysis of the behavioral laterality of hand function in nonhuman primates. *Yearbook of Physical Anthropology*, 40, 201–232.

McGrew, W.C. & Marchant, L.F. (1999). Laterality of hand use pays off in foraging success for wild chimpanzees. *Primates*, 40, 509–513.

McManus, I.C. (1984). The inheritance of asymmetries in man and flatfish. *Behavioral and Brain Sciences*, 7, 731–733.

McManus, I.C. (1991). The inheritance of left-handedness. In *Biological Asymmetry and Handedness*. Ciba Foundation Symposium No. 162, pp. 251–267. Chichester: John Wiley and Sons.

McManus, I.C. (1992). Are paw preference differences in HI and Lo mice the result of specific genes or of heterosis and fluctuating asymmetry? *Behavior Genetics*, 22, 435–451.

Mench, J.A. & Andrew, R.J. (1986). Lateralization of a food search task in the domestic chick. *Behavioral and Neural Biology*, 46, 107–114.

Miklósi, A. & Andrew, R.J. (1999). Right eye use associated with decision to bite in zebrafish. *Behavioural Brain Research*, 105, 199–205.

Miklósi, A., Andrew, R.J. & Dharmaretnam, M. (1996). Auditory lateralization: shifts in ear use during attachment in the domestic chick. *Laterality*, 1, 215–224.

Miklósi, A., Andrew, R.J. & Savage, H. (1998). Behavioural lateralization of the tetrapod type in the zebrafish (*Brachydanio rerio*), as revealed by viewing patterns. *Physiology and Behavior*, 63, 127–135.

Milinski, M. (1987). Tit for tat in sticklebacks and the evolution of cooperation. *Nature*, 325, 433–435.

Morgan, M.J., O'Donnell, J.M. & Oliver, R.F. (1973). Development of left-right asymmetry in the habenular nuclei of *Rana temporaria*. *Journal of Comparative Neurology*, 149, 203–214.

Morris, R.D. & Hopkins, W.D. (1993). Perception of human chimeric faces by chimpanzees: Evidence for a right hemipshere advantage. *Brain and Cognition*, 21, 111–122.

Moulton, J.M. & Barron, S.E. (1967). Asymmetry in the Mauthner cells of the goldfish brain. *Copeia*, 836–837.

Moulton, J.M., Jurand, A. & Fox, H. (1968). A cytological study of Mauthner's cells in *Xenopus laevis* and *Rana temporaria* during metamorphosis. *Journal of Embryology and Experimental Morphology*, 19, 415–431.

Naitoh, T. & Wassersug, R. (1996). Why are toads right-handed? *Nature*, 380, 353.

Nepomnyashchikh, V.A. & Gremyatchikh, V.A. (1993). The relation between structure of trajectory and handedness of direction of locomotion on the *Oreochromys mossambicus* Peters (Cichlidae). *Journal of General Biology*, 5, 618–626 [in Russian].

Neveu, P.J. (1996). Lateralization and stress response in mice: interindividual differences in the association of brain, neuroendocrine, and immune responses. *Behavior Genetics*, 26, 373–377.

Neville, A.C. (1976). *Animal Asymmetry*. The Institute of Biology's Studies on Biology, No. 67. London: Edward Arnold.

Neville, A.C. (1978). On the general problem of asymmetry. *Behavioral Brain Sciences*, 2, 308–309.

Northcutt, R.G. & Davis, R.E. (1983). Telencephalic organization in ray-finned fishes. In *Fish Neurobiology*. Vol. 2: Higher Brain Areas and Functions, ed. R.E. Davis & R.G. Northcutt, pp. 203–236. Ann Arbor: University of Michigan Press.

Nottebohm, F. (1971). Neural lateralization of vocal control in a Passerine bird. I. Song. *Journal of Experimental Zoology*, 177, 229–261.

Nottebohm, F. (1977). Asymmetries in neural control of vocalization in the canary. In *Lateralization of the Nervous System*, ed. S. Harnad, R.W. Doty, L. Goldstein, J. Jaynes & G. Krauthamer, pp. 23–44. New York: Academic Press.

Nottebohm, F. (1980). Brain pathways for vocal learning in birds: A review of the first 10 years. In *Progress in Psychobiology and Physiological Psychology*, ed. J.M. Sprague & A.N. Epstein, pp. 85–124. New York: Academic Press.

Nottebohm, F., Alvarez-Buylla, A., Cynx, J., Chang-Ying, L., Nottebohm, M., Suter, R., Tolles, A. & Williams, H. (1990). Song learning in birds: The relation between perception and production. *Philosophical Transactions of the Royal Society of London*, 329, 115–124.

O'Connor, K.N., Roitblat, H.L. & Bever, T.G. (1992). Auditory sequence complexity and hemispheric asymmetry of function in rats. In *Language and Communication: Comparative Perspectives*, ed. H.L. Roitblat, L.M. Herman & P.E. Nachtigall. Hillsdale, NJ: Erlbaum.

Palmer, R.A. & Strobeck, C. (1986). Fluctuating asymmetry: Measurement, analysis, patterns. *Annual Review of Ecology and Systematics*, 17, 391–421.

Parsons, C.H. & Rogers, L.J. (1993). Role of the tectal and posterior commissures in lateralization of the avian brain. *Behavioral Brain Research*, 54, 153–164.

Petersen, M.R., Beecher, M.D., Zoloth, S.R., Moody, D.B. & Stebbins, W.C. (1978). Neural lateralization of species-specific vocalizations by Japanese macaques (*Macaca fuscata*). *Science*, 202, 324–327.

Petersen, M.R., Beecher, M.D., Zoloth, S.R., Green, S., Marler, P.R., Moody, D.B. & Stebbins, W.C. (1984). Neural lateralization of vocalizations by Japanese macaques: Communicative significance is more important than acoustic structure. *Behavioural Neuroscience*, 98, 779–790.

Proudfoot, R.E. (1983). Hemiretinal differences in face recognition: Accuracy versus reaction time. *Brain and Cognition*, 2, 25–31.

Rand, A.S. (1988). An overview of anuran acoustic communication. In *The Evolution of the Amphibian Auditory System*, ed. B. Fritzsch, J.J. Ryan, W. Wilczynski & T.E. Hethington, pp. 415–431. New York: Wiley.

Rashid, N.Y. & Andrew, R.J. (1989). Right hemisphere advantage for topographical orientation in the domestic chick. *Neuropsychologia*, 27, 937–948.

Raymond, M., Pontier, D., Dufour, A.-B. & Pape Møller, A. (1996). Frequency-dependent maintenance of left handedness in humans. *Proceedings of the Royal Society of London B*, 263, 1627–1633.

Reist, J.D., Bodaly, R.A., Fudge, K.J., Cash, K.J. & Stevens, T.V. (1987). External scarring of whitefish, *Coregonus nasus* and *C. clupeaformis* complex, from the western Northwest Territories, Canada. *Canadian Journal of Zoology*, 65, 1230–1239.

Ridgway, S.H. (1986). Physiological observations on dolphin brains. In *Dolphin Cognition and Behavior: A Comparative Approach*, ed. R.J. Schusterman, J.A. Thomas & F.G. Wood, pp. 31–60. Hillsdale, NJ: Erlbaum.

Robins, A., Lippolis, G., Bisazza, A., Vallortigara, G. & Rogers, L.J. (1998). Lateralization of agonistic responses and hind-limb use in toads. *Animal Behaviour*, 56, 875–881.

Rogers, L.J. (1980). Lateralization in the avian brain. *Bird Behavior*, 2, 1–12.

Rogers, L.J. (1982). Light experience and asymmetry of brain function in chickens. *Nature*, 297, 223–225.

Rogers, L.J. (1989). Laterality in animals. *International Journal of Comparative Psychology*, 3, 5–25.

Rogers, L.J. (1991). Development of lateralization. In *Neural and Behavioural Plasticity: The Use of the Domestic Chick as a Model*, ed. R.J. Andrew, pp. 507–535. Oxford: Oxford University Press.

Rogers, L.J. (1995). *The Development of Brain and Behaviour in the Chicken*. Wallingford: CAB International.

Rogers, L.J. (1996). Behavioral, structural and neurochemical asymmetries in the avian brain: A model system for studying visual development and processing. *Neuroscience and Biobehavioral Reviews*, 20, 487–503.

Rogers, L.J. (1997). Early experimental effects on laterality: Research on chicks has relevance to other species. *Laterality*, 2, 199–219.

Rogers, L.J. (2000). Evolution of hemispheric specialization: advantages and disadvantages. *Brain and Language*, 73, 236–253.

Rogers, L.J. & Anson, J.M. (1979). Lateralization of function in the chicken forebrain. *Pharmacology, Biochemistry and Behavior*, 10, 679–686.

Rogers, L.J. and Kaplan, G. (1996). Hand preferences and other lateral biases in rehabilitated orang-utans (*Pongo pygmaeus pygmaeus*). *Animal Behaviour*, 51, 13–25.

Rogers, L.J. & Workman, L. (1989). Light exposure during incubation affects competitive behaviour in domestic chicks. *Applied Animal Behaviour Science*, 23, 187–198.

Rogers, L.J. & Workman, L. (1993). Footedness in birds. *Animal Behaviour*, 45, 409–411.

Rogers, L.J., Andrew, R.J. & Burne, T.H.J. (1999). Light exposure of the embryo and development of behavioural lateralization in chicks: 1. Olfactory responses. *Behavioural Brain Research*, 97, 195–200.

Rogers, L.J., Ward, J.P. & Stafford, D. (1994). Eye dominance in the small-eared bushbaby, *Otolemur garnettii*. *Neuropsychologia*, 32, 257–264.

Rogers, L.J., Zappia, J.V. & Bullock, S.P. (1985). Testosterone and eye–brain asymmetry for copulation in chickens. *Experientia*, 1, 1447–1449.

Schmidt, R.S. (1973). Central mechanisms of frog calling. *American Zoologist*, 13, 1169–1177.

Schmidt, R.S. (1976). Neural correlates of frog calling: Isolated brainstem. *Journal of Comparative Physiology*, 154, 847–853.

Schneider, H. (1988). Peripheral and central mechanisms of vocalization. In *The Evolution of the Amphibian Auditory System*, ed. B. Fritzsch, J.J. Ryan, W. Wilczynski & T.F. Hethington, pp. 537–558. New York: Wiley.

Sergent, J. & Signoret, J.-L. (1992). Functional and anatomical decomposition of face processing: Evidence from prosopagnosia and PET study of normal subjects. *Philosophical Transactions of the Royal Society of London B*, 335, 55–62.

Shanklin, W.M. (1935). On diencephalic and mesencephalic nuclei and fibre paths in the brains of three deep sea fish. *Philosophical Transactions of the Royal Society of London B*, 516, 224–361.

Sharrer, E., Smith, S. & Palay, S.L. (1947). Chemical sense and taste in the fishes *Prionotus* and *Trichogaster*. *Journal of Comparate Neurology*, 86, 183–198.

Shine, R., Olsson, M.M., LeMaster, M.P., Moore, I.T. & Mason, R.T. (2000). Are snakes right-handed? Asymmetry in hemipenis size and usage in gartersnakes. *Behavioral Ecology*, 11, 411–415.

Snyder, P.J. & Harris, L.J. (1997). Lexicon size and its relation to foot preference in the African Grey parrot (*Psittacus erithacus*). *Neuropsychologia* 35, 919–926.

Sobel, N., Supin, A. Ya. & Myslobodsky, M.S. (1994). Rotational swimming tendencies in the dolphin (*Tursiops truncatus*). *Behavioural Brain Research*, 65, 41–45.

Sovrano, V.A., Bisazza, A. & Vallortigara, G. (2001). Lateralization of response to social stimuli in fishes: A comparison between different methods and species. *Physiology and Behavior*, in press.

Sovrano, V., Rainoldi, C., Bisazza, A. & Vallortigara, G. (1999). Roots of brain specializations: Preferential left-eye use during mirror-image inspection in six species of teleost fish. *Behavioural Brain Research*, 106, 175–180.

Stefanelli, A. (1951). The Mauthnerian apparatus in the Ichthyopsida; its nature and function and correlated problems of neurohistogenesis. *Quarterly Review of Biology*, 26, 17–34.

Tavolga, W.N. (1962). Mechanisms of sound production in the ariid catfishes Galeichthys and *Bagre*. *Bulletin of the American Museum of Natural History*, 124, 1–30.

Tommasi, L. & Vallortigara, G. (1999). Footedness in binocular and monocular chicks. *Laterality*, 4, 89–95.

Tommasi, L., Andrew, R.J. & Vallortigara, G. (2000). Eye use in search is determined by the nature of task in the domestic chick (*Gallus gallus*). *Behavioural Brain Research*, 112, 119–126.

Ulrich, C., Prior, H., Duka, T., Leschchins'ka, I., Valenti, P., Güntürkün, O. & Lipp, H-P. (1999). Left-hemispheric superiority for visuospatial orientation in homing pigeons. *Behavioural Brain Research*, 104, 169–178.

Vallortigara, G. (1992). Right hemisphere advantage for social recognition in the chick. *Neuropsychologia*, 30, 761–768.

Vallortigara, G. (2000). Comparative neuropsychology of the dual brain: A stroll through left and right animals' perceptual worlds. *Brain and Language*, 73, 189–219.

Vallortigara, G. & Andrew, R.J. (1991). Lateralization of response by chicks to change in a model partner. *Animal Behaviour*, 4, 187–194.

Vallortigara, G. & Andrew, R.J. (1994a). Olfactory lateralization in the chick. *Neuropsychologia*, 32, 417–423.

Vallortigara, G. & Andrew, R.J. (1994b). Differential involvement of right and left cerebral hemisphere in individual recognition in the domestic chick. *Behavioural Processes*, 33, 41–58.

Vallortigara, G. & Regolin, L. (1996). Detour behaviour in the domestic chick: Cognition and lateralization. *Abstracts of the Avian Brain and Behaviour Meeting*, 25–28 August 1996, Tihany, Hungary.

Vallortigara, G., Regolin, L. & Pagni, P. (1999). Detour behaviour, imprinting, and visual lateralization in the domestic chick. *Cognitive Brain Research*, 7, 307–320.

Vallortigara, G., Rogers, L.J. & Bisazza, A. (1999). Possible evolutionary origins of cognitive brain lateralization. *Brain Research Reviews*, 30, 164–175.

Vallortigara, G., Cozzutti, C., Tommasi, L. & Rogers, L.J. (2001). How birds use their eyes: Opposite left–right specialisation for the lateral and frontal visual hemifield in the domestic chick. *Current Biology*, 11, 29–33.

Vallortigara, G., Regolin, L., Bortolomiol, G. & Tommasi, L. (1996). Lateral asymmetries due to preferences in eye use during visual discrimination learning in chicks. *Behavioural Brain Research*, 74, 135–143.

Vallortigara, G., Rogers, L.J., Bisazza, A., Lippolis, G. & Robins, A. (1998). Complementary right and left hemifield use for predatory and agonistic behaviour in toads. *NeuroReport*, 9, 3341–3344.

Vermeire, B.A., Hamilton, C.R. & Erdmann, A.L. (1998). Right-hemispheric superiority in split-brain monkeys for learning and remembering facial discriminations. *Behavioral Neuroscience*, 112, 1048–1061.

Vial, J.L. (1973). *Evolutionary Biology of Anurans*. Columbia: University of Missouri Press.

von Bonin, G. (1962). Anatomical asymmetries of the cerebral hemispheres. In *Interhemispheric Relations and Cerebral Dominance*, ed. V.B. Mountcastle, pp. 483–488. Baltimore: Johns Hopkins University Press.

Vota-Pinardi, U. & Kemali, M. (1990). Neuroelectrophysiology of the morphologically asymmetric habenulae of the frog. *Comparative Biochemistry and Physiology*, 96A, 421–424.

Walls, G.L. (1953). The lateral geniculae nucleus and visual histophysiology. *University of California Publications in Physiology*, 9, 1–100.

Walker, S.F. (1980). Lateralization of functions in the vertebrate brain: A review. *British Journal of Psychology*, 71, 329–367.

Ward, J.P., Milliken, G.W. & Stafford, D. K. (1993). Patterns of lateralized behavior in prosimians. In *Primate Laterality: Current Behavioral Evidence of Primate Asymmetries*, ed. J.P. Ward & W.D. Hopkins, pp. 43–74. Springer-Verlag: New York.

Warrington, E.K. (1982). Neuropsychological studies of object recognition. *Philosophical Transactions of the Royal Society of London B*, 298, 15–53.

Wassersug, R.J., Naitoh, T. & Yamashita, M. (1999). Turning bias in tadpoles. *Journal of Herpetology*, 33, 543–548.

Waters, N.S. & Denenberg, V.H. (1994). Analysis of two measures of paw preference in a large population of inbred mice. *Behavioural Brain Research*, 63, 195–204.

Westergaard, G.C. & Suomi, S.J. (1996). Hand preference for stone artefact production and tool-use by monkeys: possible implications for the evolution of right-handedness in hominids. *Journal of Human Evolution*, 30, 291–298.

Westergaard, G.C. & Suomi S.J. (1997). Lateral bias in capuchin monkeys (*Cebus apella*): Concordance between parents and offspring. *Developmental Psychobiology*, 31, 143–147.

Westin, L. (1998). The spawning migration of European silver eel (*Anguilla anguilla* L.) with particular reference to stocked eel in the Baltic. *Fisheries Research*, 38, 257–270.

Wetzel, D.M., Haerter, U.L. & Kelley, D.B. (1985). A proposed neural pathway for vocalization in South African clawed frogs, *Xenopus laevis*. *Journal of Comparative Physiology*, 15, 749–761.

Will, U. (1991). Amphibian Mauthner cells. *Brain, Behavior and Evolution*, 37, 317–332.

Williams, H. (1990). Bird song. In *Neurobiology of Comparative Cognition*, ed. R.P. Kesner & D.S. Olton, pp. 77–126. Hillsdale, NJ: Erlbaum.

Wree, A., Zilles, K. & Schleicher, A. (1981). Growth of fresh volumes and spontaneous cell death in the nuclei habenulae of albino rats during ontogenesis. *Anatomical Embryology*, 161, 419–431.

Yamashita, M., Naitoh, T. & Wassersug, R.J. (1999). Startle response and turning bias in Microhyla tadpoles. *Zoological Sciences*, 17, 185–189.

Zilles, K., Schleicher, A. & Wingert, F. (1976). Quantitative analyse des wachstums der frischvolumina limberscherkerngebiete im diencephalon und mesencephalon einer ontogenetischen reihe von albinomäusen. 1. Nucleus habenulare. *Journal für Hirnforschung*, 17, 1–10.

Zottoli, S.J. (1977). Correlation of the startle reflex and Mauthner cell auditory responses in unrestrained goldfish. *Journal of Experimental Biology*, 66, 243–254.

Zottoli, S.J. (1978). Comparative morphology of the Mauthner cell in fish and amphibians. In *Neurobiology of the Mauthner Cell*, ed. D. Faber & H. Korn, pp. 13–45. New York: Raven Press.

Zottoli, S.J., Hordes, A.R. & Faber, D.S. (1987). Localization of optic tectal input to the ventral dendrite of the goldfish Mauthner cell. *Brain Research*, 401, 113–121.

2

The Earliest Origins and Subsequent Evolution of Lateralization

RICHARD J. ANDREW

2.1. Lateralization and predation

Recent studies of cerebral lateralization in teleost fish (Cantalupo, Bisazza and Vallortigara, 1995; Bisazza, Pignatti and Vallortigara, 1997a, 1997b; Miklósi, Andrew and Savage, 1998; Bisazza, Rogers and Vallortigara, 1998; Miklósi and Andrew, 1999; Chapter 1 by Vallortigara and Bisazza) have provided a sketch of the basic properties of early vertebrate lateralization. As a result, it is now possible to think concretely about the even earlier steps in evolution, by which this condition may have originated. Cerebral lateralization in extant fish is likely to have changed less during evolution than in tetrapods because the selection pressures imposed by life in water have remained (to some degree) the same.

The features that seem most likely to have been present very early in vertebrate evolution are the use of the right eye (RE) to fixate objects, (1) which the fish intends to bite or (2) which it is investigating, whilst at the same time inhibiting escape. This is accompanied by use of the left eye (LE) to fixate (3) familiar objects or (4) conspecifics that evoke sexual or social (schooling) behaviour. These features have been found in a number of species from different major taxa: all four are present in *Brachydanio rerio* (Cypriniformes, Ostariophysii (Miklósi, Andrew and Savage, 1998; Miklósi and Andrew, 1999; Sovrano, pers. comm.), whilst 2 and 4 have been described (Bisazza, De Santi and Vallortigara, 1999; Sovrano et al., 2000) for a number of Poeciliids and other families of the Cyprinodontiformes, which belong to the Acanthopterygii, rather than the Ostariophysii. Whilst a wider range of systematic groups needs to be examined in teleosts (as in other vertebrates), the most economical hypothesis at present is that cerebral lateralization of basically the type that has just been outlined was present in the common ancestor of the teleosts.

The presence of the same basic properties of cerebral lateralization in birds, as well as teleosts, argues for a much earlier origin of lateralization. Comparison of teleosts and birds is more straightforward than comparisons that involve mammals. Although birds must have changed in response to the new needs of terrestrial life, they retain independent eye movements and have, therefore, not undergone the reorganization of visual systems that conjugate eye movements have brought in mammals.

A key finding for the present argument is that birds, as well as teleosts, use the RE to guide the mouth to a visible target. In the chick, it is clear that readiness to peck (e.g. during rapid approach to a food dish within which the chick knows there is food that it is eager to eat) is not sufficient to bring about RE use (Chapter 3 by Andrew and Rogers; Andrew, Tommasi and Ford, 2000). Fixation with the RE is sustained throughout approach during which the chick can see a target that it must manipulate or grasp with the bill on arrival. Such fixation, once it appeared in evolution, would provide a way by which appropriate settings of motor reflexes could be established on the basis of visual information and then be sustained until contact with the target occurred.

In order to trace the evolutionary story further back in time, it is necessary to turn to evidence from comparative anatomy and the fossil record. I argue here that the origin of lateralization came about as part of the first evolution of visually controlled predation. Webb (1969) proposed, on the basis of evidence from the feeding of larval Amphioxus (*Branchiostoma*), that a key step in the origin of vertebrates was the appearance of predation in a free-swimming filter feeder. Mallatt (1985) extended this thesis by arguing that this resulted in the evolution of vision.

The extreme asymmetry of the common ancestor of chordates and echinoderms is crucial to understanding the probable early evolution of cerebral lateralization. This asymmetry seems to have appeared in an animal evolved from sessile forms, which gathered food by ciliary currents set up on branching arms. Pharyngeal gill slits allowed the exit of the water currents that entered the mouth during feeding. A descendant of such forms became mobile, using the tail-like appendage that had been its attaching stalk. It lay on its right side, and had viscera better developed on the side, and structures like gill slits and mouth confined to the left side that was exposed to the open water (Jefferies and Lewis, 1978). Much of subsequent evolution in early chordates involved the (partial) restoration of secondary bilateral symmetry.

If, as is here argued, cerebral lateralization arose in the descendants of such animals, the common assumption that cerebral asymmetries have appeared progressively, or only belatedly, during the evolution of the vertebrates, will have to be reversed. Instead the brain, like the body, may have been at its most asymmetrical early in evolution.

Amphioxus provides evidence for the sort of anatomical asymmetries that are likely to have been present in early chordates. In particular, the mouth of Amphioxus (and presumably therefore the mouth of comparable free-swimming ancestral chordates) is a structure of the left side. In adult Amphioxus, the mouth moves to become medial, as part of the development of secondary symmetry, the most important aspect of which is the development of gill slits on the right side as well as the left. In the larva gill slits are present only on the left. Visceral asymmetry (e.g. the presence of a gut diverticulum only on the right) persists in the adult (as it does in vertebrates).

Interestingly, the asymmetric development of the Amphioxus mouth area depends on left–right asymmetry of expression of the Amphioxus hedgehog gene (Shimeld, 1999). There is evidence that hedgehog genes continue to play a part in determining left–right asymmetries in chick (Pagán-Westphal and Tabin, 1998) and zebrafish (Chen et al., 1997).

The presence of gill slits on both sides in adult Amphioxus help to explain the absence of obvious asymmetries in the other group of primitive chordates: the tunicates. It is clearly adaptive to have a full complement of gill slits in an animal that relies exclusively on the ingestion of fine particulate food. This is true of adult tunicates. The larva is here a dispersal phase only, so that it is not unexpected that asymmetry should be deleted from development. (It is, of course, quite possible that studies of gene expression during development will show residual asymmetries in the tunicates as well.)

The Amphioxus larva is also important in that it shows how an animal retaining the original extreme asymmetry can make a living as a free-swimming predator. Webb (1969, 1975) recorded the presence of copepods and chain diatoms in the pharynx and gut of pelagic larvae of Amphioxus, which were about the same size as the mouth of the larva. He argued that the larval gut musculature was capable of moving such items by peristalsis and that they clearly formed an important part of the larval diet, in addition to the tiny particles, which were taken by filter feeding. Indeed, even in apparently strict microphages like the pluteus larva of Ophiuroid Echinoderms, objects somewhat larger than the mouth are regularly taken (Hendler, 1991), implying mouth dilation and perhaps some active swallowing. Echinoderm larvae often have very active muscular peristalsis in the oesophagus (Pearse and Cameron, 1991), which would allow them to cope with small active organisms.

In the adult Amphioxus, large items are expelled by atrial contraction: the 'cough reflex' (Guthrie, 1975). The contraction of pharyngeal muscles in the larva also constricts the pharynx by a similar reflex (Webb, 1969). Since this is here accompanied by mouth closure, it seems likely that here the function is to put pressure on the prey, rather than to expel it. The evolution of hardenings within the pharynx would allow more effective crushing of the prey.

The recent reconstructions (Aldridge et al., 1993; Aldridge and Purnell, 1996; Donoghue, 1998; Donoghue, Forey and Aldridge, 2000) of conodont anatomy suggest that this did indeed happen in early vertebrate evolution. The conodonts are now accepted as key early vertebrates, whose possession of tooth-like structures that are made of calcium phosphate, appears to put them crownward of hagfish and lampreys (Donoghue, Forey and Aldridge, 2000). Their small size, and the character of the deposits in which they were found, suggest that they were commonly free-swimming predators on small planktonic prey (Nicoll, 1987). Much at least of the conodont apparatus lay within the pharynx, and crushed or sheared food. It is thus exactly what might have evolved in an ancestral form with an anatomy and a way of life like those of an Amphioxus larva.

The fact that lampreys and hagfish have predatory organs, which normally lie internal to the mouth (e.g. toothed 'tongue' of lampreys), suggests that conodonts, lampreys and hagfish had a common ancestor with a non-mineralized intrapharyngeal feeding apparatus. A possible example of such an animal has been described for the early Cambrian. *Yunnanozoon lividum* is argued (Chen et al., 1995; but see Shu, Conway Morris and Zhang, 1996) to be a close relative of Amphioxus, but one which possesses a 'denticular' structure within the pharynx.

The second important feature of the conodonts is the presence of large anterior paired eyes, which almost certainly had as a major function the detection of prey. It is worth noting that modern heteropod molluscs provide an example of planktonic predators that use vision to guide predation. Such animals have paired eyes that scan the environment; resolution is good within a visual field that is restricted by the possession of a strip retina (Land, 1982). Heteropods provide an instructive example of a relatively recent evolution of pelagic predators whose adaptations parallel in some ways those which must have occurred during the origin of the vertebrates.

2.2. Evolution of Visual Control of Response

2.2.1. Control of the Mouth

The first improvement of the ability to take relatively large prey is likely to have been the evolution of the ability to inhibit expulsion reflexes and avoidance, which might be evoked by vigorous tactile stimulation of the mouth. How this may have occurred is best understood by turning again to the way of life of Amphioxus larvae. Such larvae are free swimming and pelagic. The mouth lies entirely on the left; in one mode of feeding, the larva sinks passively with its left side turned downwards, so that the mouth meets potential food items (Figure 2.1). These include prey like copepods, which are large in relation to the mouth (Webb, 1969). The left side of the body behind the mouth is made sticky by secretions of a (left-sided) structure, the 'club-

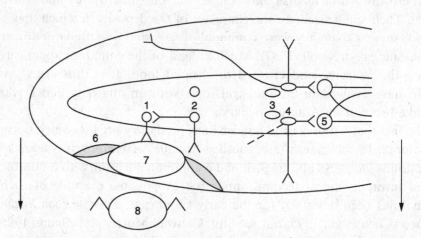

Figure 2.1.   Stylized representation of a stage in evolution corresponding to an Amphioxus larva. The animal feeds whilst sinking passively with left side and mouth (7) downwards. Encounter with a large and active prey (8) is about to occur, and tactile stimulation of mouth and flank will tend to initiate avoidance through a giant motor cell (Rohde cell: 5). A pair of such cells is shown here, with bilateral outflow, corresponding to the condition in larval Amphioxus; these cells appear to be lost in adult Amphioxus (see text). It is assumed that, in the hypothetical ancestral form, mouth dilation could occur through appropriate sensory input to the mouth via a sensory neuron (1), and a motor neuron (2), controlling dilator muscles (6). It is possible that the same reflex is present in larval Amphioxus (text).

Strong tactile stimulation of the rostrum or side of the body can initiate avoidance by inputs to what may be homologues (text) of vertebrate tectal cells (3 and 4). Note that such avoidance is likely to have been undirected and involved only rapid swimming forward or back (see text). The circuit for stimulation of the flank is shown in full for the right side only. It is assumed that a similar circuit existed for the left side, but this is shown only partially for the sake of clarity. It is assumed that this circuit would sometimes be activated by stimulation from the prey; it is likely that mechanisms existed, which allowed this to be inhibited when mouth dilation had begun. These are not shown here but a modification of existing circuits of this kind is assumed to have given rise to the condition shown in Figure 2.2.

shaped gland', which may help to secure potential prey (Webb, 1969). Tactile input from prey would thus be likely to be from receptors of the left side of the body, and so enter the left side of the central nervous system (CNS). The other key structure for predation is the mouth. The inside of the mouth, which would be strongly stimulated by large or struggling prey, is innervated from the left side of the anterior CNS via an extensive 'plexus of Fusari' (Drach, 1948). Further, the larval mouth has muscles in front and behind it, which are lost in the adult (Drach, 1948). They may allow dilation of the mouth to allow the entry of prey in the larva.

In an animal that was beginning to take larger and more difficult prey, it would be necessary to be able to reject and avoid prey that were potentially damaging; equally, an increased ability to inhibit rejection, when ingestion was possible, would be important (Figure 2.1).

The sensory input, which might have to be used to initiate either rejection or preparations for ingestion (i.e. inhibition of rejection and perhaps dilation of the mouth), is thus likely to have been routed to the left side of the anterior end of the CNS. Motor control of the mouth is likely to be similarly situated. Such a condition would provide a basis from which the control of prey taking might evolve to become the responsibility of structures on the left side of the vertebrate brain.

The argument from this point is presented in two parts: first, a hypothetical sequence of changes, which might have allowed visual control of predation to evolve; and, secondly, a consideration of evidence of existing asymmetric visual control of motor response in Amphioxus larvae (Lacalli, 1996).

It would be advantageous if likely encounters with suitable prey could be predicted before contact so that rejection and avoidance could be inhibited to allow ready entry of the prey. In forms feeding diurnally, visual detection of prey would be ideal. The homologue of the paired eyes of vertebrates has been identified in Amphioxus as the unpaired pigment spot, together with associated cells with sensory cilia, at the anteriormost tip of the CNS (Lacalli, Holland and West, 1994; Lacalli, 1996). Eye cups are present throughout the length of the CNS in adult Amphioxus (Drach, 1948). Thus, appropriate starting points for the evolution of visual control of predation were available very early in evolution. Reflex inhibition of avoidance during predation that is somewhat comparable to that postulated here for the vertebrate ancestor is present in modern teleosts: avoidance (mediated by Mauthner cells) of a variety of startling inputs, including visual inputs, is inhibited during visually guided approaches to food (goldfish: Canfield and Rose, 1993).

It is proposed that the first stage in the evolution of visual lateralization was the detection of potential prey by a simple eye, which provided input to the anterior left side of the CNS (Figure 2.2). Detection was followed by inhibition of avoidance due to tactile stimulation of the area around the mouth (such as might be produced by contact with the prey), and by preparations for ingestion such as mouth dilation. This would have occurred before pursuit or even effective directed swimming was possible. Note that larval Amphioxus (like adults) rotate whilst swimming (Lacalli, 1996); it is likely that any chordate ancestor, which was in the process of shifting from filter feeding to predation, would be inefficient in fast-directed locomotion.

Such limitations on locomotion do not exclude avoidance. It is important here to distinguish 'undirected' avoidance, a term used here for avoidance that does not involve locomotion whose direction is determined by the position of the stimulus that causes avoidance. The latter is termed here 'directed avoidance'. It is argued here that the ability to turn right or left appropriately was evolved first to allow such directed avoidance and only subsequently became available in prey catching.

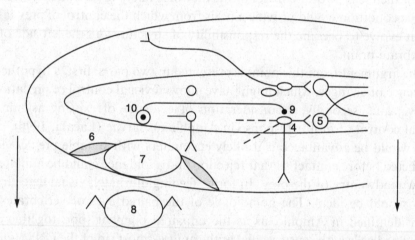

Figure 2.2.   The animal from Figure 2.1 is shown here, with the addition of a circuit that allowed photic input to be used to inhibit avoidance to tactile inputs from prey that was about to be taken by mouth dilation. The eye (10) is shown as looking out through the mouth, and is supposed to be like the eye cups that are present throughout the length of the CNS of adult Amphioxus (see text). It provides an inhibitory input (9) to the interneuron (4) that would otherwise initiate avoidance. It is likely that appropriate input to the eye could also promote mouth dilation, but this is not shown. An alternative route of evolution is also discussed in the text, involving the single frontal eye of Amphioxus. There is evidence suggesting that input to this eye may drive a lateral leap in the larva, but it is neither clear how this is mediated nor how a laterally placed (or asymmetric) eye might have evolved from the frontal eye. Other symbols are as for Figure 2.1.

The first effect of visual input on avoidance to evolve is suggested here to be inhibition of undirected avoidance, which might otherwise have been evoked by stimulation from the prey (Figure 2.3).

Adult Amphioxus swim quickly forwards or backwards in response to vigorous tactile input (Guthrie, 1975). The accompanying stiffening of the notochord during vigorous swimming (Flood, 1975) makes bending unlikely and it is perhaps the case that bending is not used in avoidance by the adult.

Further constraints are imposed by body form in extant fish. Here studies of avoidance have concentrated on the role of the Mauthner cells. In advanced teleosts, with short highly manoeuvrable bodies, avoidance turns initiated by these cells are directed away from a detected source of danger, whatever the position and posture of the body (Eaton and Emberley, 1991). However, this is not true in teleosts with elongated bodies (e.g. spiny eels: Currie, 1991): here the direction of escape tends to be that allowing the fastest exit from the posture in which the fish finds itself. In lampreys (also long-bodied), the Mauthner cells do not show reciprocal inhibition (Currie, 1991) as is usual in teleosts. Instead, they provide bilateral activation of the spinal

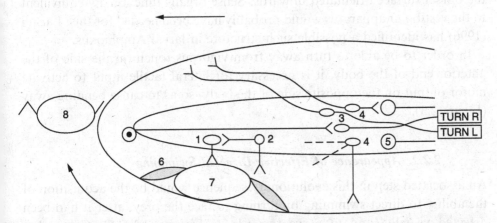

Figure 2.3.   Efficient forward swimming has evolved at this point, in which the animal maintains a standard body posture, with the dorsal surface uppermost. The animal now swims forwards in a straight line. In association with this, the ability to turn actively away from lateral inputs has evolved. This is shown as mediated by another of the 'tectal' interneurons (3). The circuit is shown only for the right side but a corresponding circuit is assumed to have been present on the left. Note that the motor outflow does not involve the paired giant motor cells, which would have continued to mediate undirected avoidance.

The single eye that controls feeding behaviour now looks forward so that prey can be detected in the line of advance. However, feeding remains inefficient. There is no steering of the body to bring potential prey into line with the mouth, which is still laterally placed.

cord, initiating swimming, presumably without controlling its direction. The same is probably true of the homologues of the Mauthner cells in adult Amphioxus, where the largest cell of all has a median descending fibre, rather than an ipsilateral one, and electrical coupling between cells is likely (Guthrie, 1975).

However, it must be stressed that recent studies in teleosts have shown that there are many reticulospinal interneurones, which are also concerned with control of spinal motor mechanisms; these have been argued (Fetcho, 1991, 1992) to be responsible for modulating the exact form of muscle contractions and so of movement. A better model for present purposes may be given by the observation that ammocoete larvae of lampreys, despite having Mauthner cells that give bilateral spinal activation, do turn before showing rapid escape swimming after an aversive withdrawal (Currie, 1991).

It seems almost certain that the ability to turn away from lateral input would have been present soon after the evolution of effective swimming, if not before. Note that 'effective swimming' in animals derived from an Amphioxus-like ancestor means the ability to swim in a direct line, keeping the dorsal surface orientated upwards. Sense organs functionally equivalent to the vestibular apparatus would probably have been needed for this; Lacalli (1996) has identified a possible such structure in larval Amphioxus.

In order to be able to turn away from vigorous touch on the side of the anterior end of the body, it is necessary for lateral tactile input to activate motor output on the opposite side of the body, so as to cause bending away (Figure 2.3).

2.2.2. *Appearance of Effective Directed Swimming*

An associated step in this evolutionary sequence would be the acquisition of the ability to direct swimming by turning to face the prey, after it had been detected visually, and to move towards it (Figures 2.3–2.5). Given that crossed motor control, which allowed turning in avoidance, was already well established, the simplest way of allowing optic input to initiate turning towards laterally placed prey would have been to use the existing motor control systems. Note that the fact that the mouth was still placed on the left increases the need for the ability to turn. So long as the mouth remained on the left, prey lying immediately ahead would be likely to strike the tip of the body, rather than the mouth, unless there was visually initiated turning (Figure 2.4).

It is suggested that the ability to turn, in order to bring the laterally placed mouth in line with prey lying in the line of forward swimming, evolved as

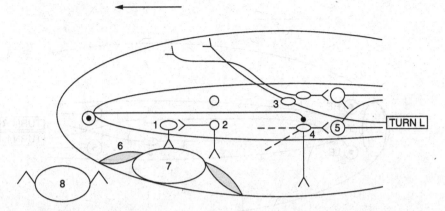

Figure 2.4.   The animal can now turn to bring prey that has been detected as lying directly ahead into contact with the lateral mouth. There would have been intense selection for this ability. It is shown as derived from a modification of part of the circuit that already allowed turning away in avoidance. The circuit allowing turning to prey would have existed only on the left side of the CNS. Avoidance circuits are assumed to have continued to be similar on both sides, but only that mediating turning left is shown.

shown in Figure 2.4. The motor interneurones that Lacalli (1996) has identified as homologues of the vertebrate tectum were already responsible for turning to avoid stimuli. They already receive input from the frontal eye in larval Amphioxus (below), although how this is used is still quite unknown. Figure 2.4 shows one such tectal neurone to have acquired the input from the eye that is necessary to allow turning to the right when prey is detected directly ahead. As a result, the prey would strike the mouth directly.

At this point in evolution, there would be a need for two eyes to allow scanning on both sides and turning to prey, wherever detected (Figure 2.5). Paired eyes may in any case have been evolved as part of the establishment of secondary symmetry. Once this had happened, both eyes had crossed input to the brain (i.e. a true optic chiasma had appeared) and the animal could turn towards prey on either side. It is likely that the main function of the frontal eyes remained control of predation and that avoidance continued to be mediated by other sense organs (including other eyes, see below). It is assumed that, although both eyes could provide inputs to the CNS that would initiate turning towards prey, only the RE input was used (by structures of the left side of the CNS) to control mouth reflexes. The RE had of course originally been a structure of the left side.

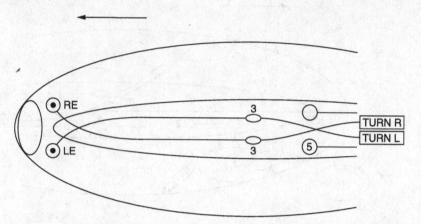

Figure 2.5. A second eye has been acquired as part of appearance of secondary bilateral symmetry during development. The original eye is now the right eye (RE) and the new one the left eye (LE); the optic chiasma is present as a result. The RE ended up on the right side because it maintained its position relative to the mouth (see text) as the latter moved to become terminal. Although this is not shown, the RE continues to supply input to the CNS, which controls mouth reflexes. There is no such circuit fed by the LE. Both eyes do provide inputs that allow the animal to turn to target prey on either side of the body. Mechanisms for turning in avoidance are still present but are not shown.

Some such sequence in evolution seems necessary to explain the control of targeting the mouth on to prey in vertebrates. Optic input is crossed in the optic chiasma so that supply is to the contralateral optic tectum, which mediates the targeting response via motor outflow through a second decussation in the floor of the tegmentum (Ingle and Hoff, 1990; cf. Figure 2.5).

Modern heteropod molluscs (see above) are capable of visually identifying preferred prey at a distance in the plankton (Hamner et al., 1975), suggesting that selection for improved visual discrimination is likely to have been strong from this point in evolution onwards.

2.2.2.1. Brain of Amphioxus larva

Support for the earlier part of this hypothetical sequence has been provided by a recent study (Lacalli, 1996) of the anatomy of the brain of the Amphioxus larva. This shows, firstly, that there is visual control of motor response. The relevant motor structures are two giant cells and two pairs of somewhat smaller cells, which together form the 'primary motor centre' (PMC). Bone (1960) notes that the former (at least) almost certainly disappear at metamorphosis, so that the effects of vision on motor response, which the PMC allows, are probably lost in the adult.

It should be noted at this point that evidence from Hox gene expression (Holland et al., 1992) shows that the anterior CNS of Amphioxus corresponds to structures of the vertebrate brain lying anterior to the hindbrain. Lacalli (1996) argues that the PMC corresponds to motor structures of the anterior part of the vertebrate hindbrain, and that a series of multimodal cells that supply the PMC (and provide one route for visual input to it) are homologues of the tectum. The lamellar body (pineal/parapineal, see below) and the infundibulum define the roof and floor of the diencephalon, respectively, whilst the frontal eye at the anterior tip of the brain is the vertebrate retina.

The left, but not the right, giant cell sends an axon forward to the level of the frontal eye, where the axon receives direct input from visual neurones with wide connections within the eye. Both right and left giant cells receive visual input indirectly, via synapses from eye cells on descending fibres in the ventrolateral tracts. The right giant cell also differs from its partner in that, together with the other two right cells of the PMC, it receives direct input from the tectal cells of the right side, which is absent on the left. The tectal cells receive inputs from the rostral nerves (perhaps a tactile input effective in initiating startle) and from dorsal sensory nerves, as well as from the eyes. The whole right PMC is thus clearly driven multimodally, whilst the giant cell on the left has a direct and effective visual drive, together with a direct input from the rostral nerves. It is striking (but perhaps coincidental) that the only two asymmetries in the chick brain, which have so far been described in visual structures, are in ascending thalamotelencephalic tracts (Chapter 6 by Deng and Rogers) and in the optic tectum (Chapter 7 by Güntürkün).

It is thus likely that visual control of motor response has been a special concern of the left side of the brain since before the evolution of paired eyes or of the ability to swim sustainedly towards or away from particular objects.

It is tantalizing that we do not know exactly what motor responses are affected by input to the frontal eye in Amphioxus larva. The anatomical evidence strongly suggests that the cells of the PMC, including the left giant cell, function to initiate or modulate muscle contraction. It is surprising, however, to find the photic input to be derived from the single frontal eye, which is in many ways little altered from the apical organ of the small planktonic larvae of Echinoderms and Hemichordates (Lacalli, Holland and West, 1994). In particular, it shows no obvious asymmetry (Lacalli, 1996), suggesting that the asymmetry of the PMC evolved in a different context. One possibility is that it appeared in animals ancestral to the chordates, which lay on their right side on or in the surface of the substrate (see above). Photoreceptors would be likely to be confined to the right side and could have provided input initiating escape (e.g. by burying the body by tail movements). These photoreceptors may not have been homologous with the later frontal eye: the CNS of Amphioxus has other photoreceptors throughout much of its length. In addition to the lamellar body, there are a number of eye cups in the adult CNS (Drach, 1948). The important point is that asymmetry in the organization of the PMC might have resulted.

However, this may not be the end of the story. Lacalli (1996) discusses a 'startle response' in the Amphioxus larva, in which body muscles contract almost but not quite simultaneously, so that 'the body arcs first to one side, then straightens out'. As a result, the larva moves suddenly to one side, where it may remain motionless or begin swimming. It is not certain whether photic stimulation of the frontal eye can affect startle, although Lacalli thinks this likely. An effect from the frontal eye on ciliary beat of epidermal cells is also likely, although the route by which control might occur is quite unknown; it is, of course, very much what might be expected of a structure derived from an apical organ. Stokes and Holland (1995) describe 'ciliary hovering' in the larva, in which the body is held at about 30° to the vertical by ciliary currents; the body is slowly turned (taking about 20 min for full adjustment) so that the frontal eye is maximally shielded from light. Output from the eye is probably used to guide this, although other photosensitive structures (e.g. lamellar body) could be responsible. Note that, if the leap is indeed sometimes initiated by photic stimulation of the frontal eye, it is likely to be then biased to one side, since it would depend on the firing of the left giant cell.

This, together with the fact that the hovering posture is presumably accompanied by filter feeding, raises an interesting possibility. Given that the larva also takes larger prey when sinking at night, it is possible that it also does this whilst hovering, when opportunity offers. If so, the 'startle' leap might have another function: that of bringing the left side of the body against a potential prey item. If this were to be the case, then firing the left giant cell to visual input from the prey would have to cause movement to the left. It is thus just possible that visual control of taking prey by structures in the left side of the brain is already present in Amphioxus. If so, the steps associated with the evolution of directed swimming and turning may have begun with this condition in a very early ancestor of the chordates.

2.2.3. Orientation to Distant Visual Cues

Once evolved, visual control of directed swimming would have brought the possibility of visually evoked response to conspecifics. Aggregation for gamete shedding must have occurred, before paired frontal eyes were acquired: presumably chemical cues were important (see below). Once larger prey began to be eaten, it would be advantageous, when in such aggregations, to avoid attempts at prey catching by other conspecifics. The readiest solution would be to evolve reflexes that promoted visual fixation of conspecifics with the left eye, during spawning, leaving avoidance to tactile input available (rather than inhibited), if the animal were endangered.

There may have been a second reason for the left eye to take on this role, feeding as it does the right side of the anterior CNS. Gamete shedding in response to chemical signals may have been especially the concern of the right side of the CNS. Extant pelagic relatives of Amphioxus (*Asymmetron*: Drach, 1948) have gonads only on the right, a condition likely to have been retained from the highly asymmetric ancestor of the chordates. The gonads are innervated from spinal nerves (deep visceral branches: Drach, 1948), so that a possible route for their control would be for chemical input during breeding to initiate shedding via the right side of the CNS.

In at least two groups of primitive chordates, 'epidemic spawning' (i.e. spawning initiated by the spawning of nearby conspecifics) has been described (Hemichordata: Hadfield, 1975; Tunicata: Berrill, 1975), strongly suggesting the promotion of spawning by chemical signals. In addition, in the sister group of the Chordata, the Echinoderms, neural control of gamete shedding by mechanisms that include pattern generators (Okada, Iwata and Yanagihara, 1984) is well attested and apparently widely distributed. Aggregations for breeding, both in pairs made up of a male and a female and in larger groups, are also common, and appear to depend on response to pheromones (Asteroidea: Hendler, 1991), which may include chemicals that also induce gamete release. Aggregation often includes responses more complex than simple proximity, such as amplexus-like behaviour (Holland, 1991; Hendler, 1991; Smiley et al., 1991).

In forms ancestral to the vertebrates, with gonads developed only on the right, neural systems controlling aggregation for breeding and breeding behaviour (central to which would be gamete shedding) might well, therefore, have been largely or entirely confined to the right side of the CNS.

2.2.4. *Pineal Eye*

The evolution of the pineal–parapineal complex provides another source of evidence bearing on early specialization of the right side of the brain. The originally medial photosensitive structure (lamellar body: Lacalli, 1996), from which the complex evolved, became paired, like the frontal eyes, but without achieving bilateral symmetry. The most probable reason for this is a pre-existing asymmetry of functioning of the brain structures that it supplies.

The lampreys are the most primitive forms, possessing a pineal organ for which we have appropriate anatomical information. The pineal–parapineal complex supplies the habenulae (as in all vertebrates), whose main outflow to the interpeduncular region (the habenulo-peduncular tracts) is strikingly asymmetric in *Petromyzon* (Frontera, 1952), the right tract being much larger than the left. A clue as to what functions might be involved is given by the fact that the pineal photoreceptors of the ammocoete larva of *Petromyzon* are very sensitive to sudden shadows (Pu and Dowling, 1981). This suggests involvement of the pineal outflow in escape; indeed, it is known for the larvae of the toad *Xenopus* that the pineal is responsible for the initiation of escape to shadows (Roberts, 1978).

It is thus possible that at least one cause of asymmetry of function in the pineal–parapineal complex and its connections is special involvement of the right side of the brain in escape early in evolution. This is of course consistent with later association of such behaviour with right hemisphere control (Chapter 3 by Andrew and Rogers), but it is not clear how this association might have first evolved. The special pattern of control of the right giant cell by tectal cells in the Amphioxus larva (see above) is consistent with a very early allocation of the initiation of avoidance by multimodal input to the right side of the brain. One sense that may have been important in the original bottom-living filter feeder is the perception of substrate vibration. In an animal lying on its right side, this is likely to have been routed to the right side of the CNS.

It has already been suggested that there may have been shifts of function from one photoreceptor to another during early evolution. Structures derived from the lamellar body of the Amphioxus larva would have been better placed to detect a shadow from above, once controlled swimming with the dorsal side always uppermost had evolved. The resulting input would then have been predominantly a matter for the right side of the brain, because it required the initiation of avoidance.

In summary, the teleost condition with left eye fixation of conspecifics is here assumed to have first appeared very early in vertebrate evolution, as a specialization that was complementary to right eye use in predation. A further likely step is the evolution on the right side of the CNS of the ability to recognize visual signals identifying conspecifics of the opposite sex, which were ready to breed. Visually mediated avoidance may have become a responsibility of the right side of the brain because avoidance (elicited via other senses) was already mediated by that side.

There is one ambiguous but intriguing piece of evidence for the existence of lateralized vertebrate predators in the Cambrian that were capable of attacking large prey. Healed scars are markedly and significantly commoner on the right side of the body of Cambrian trilobites than on the left (Babcock and Robison, 1989). The predator suggested was *Anomalocaris*, a large arthropod. However, there is little evidence for asymmetries of CNS function or behaviour in extant arthropods. More importantly (since such evidence may be absent because it has not been sought), the scars are most consistent with bites from jaws like those of vertebrates, being typically 'arcuate' in form. On balance, then, the asymmetry is more likely to arise from the behaviour of a vertebrate predator than that of an arthropod predator (or of the prey itself). Zebrafish, when faced with symmetrical paired targets, are more likely to choose the target in their right hemifield (Miklósi, Andrew and Gasparini, 2001). If predation attempts were made by a vertebrate predator with similar lateralization to the zebrafish and from a consistent orientation relative to the prey (e.g. from behind), the distribution of trilobite scars would be explained.

The presence of paired eyes in hagfish, lampreys and conodonts requires most or all of the above changes to have occurred before the origin of jawed vertebrates. A possible first test of the hypothesis is thus whether visual lateralization of the basic teleost type can be found in lampreys. Visual response to bright objects has been described in lampreys (Fontaine, 1958) and interpreted as the evocation of predation, so that the question can perhaps eventually be answered.

2.3. Use of Visual Topographical Cues

The point at which vision began to be used to control movement relative to topographical features (which in later vertebrates is especially the responsibility of the right side of the brain) is quite unknown. The best that can be done at present is to consider the possible needs of a pelagic conodont-like form. Although there can be no constant landmarks in the plankton, there are often many conspicuous local features (Hamner et al., 1975), whose relative positions could provide guidance in the short term. It seems likely that this would have been of use only after the acquisition of visual discrimination good enough to allow the identification of potential food patches, or other resources like spawning aggregations, sufficiently far away as to make necessary an approach over a considerable distance. The initial evolution of left eye functions is thus likely to have been dominated by needs other than the use of landmarks.

Schooling is very common amongst modern open-water fish and has evolved in a very similar form in squid. It is thus not unreasonable to suppose it to have appeared early in vertebrate evolution. Schooling requires that an animal should maintain its position relative to a number of other conspecifics. This would most readily be derived, on the arguments advanced here, from visual mechanisms evolved for use in breeding aggregations. Proximity rather than direct contact would be needed, given that simple shedding of gametes was involved, rather than amplexus. This would be achieved under visual control by holding bearings relative to particular conspecifics, rather than by steering directly at them. Note that the left eye is used preferentially by at least six species of teleosts to fixate conspecifics (Sovrano et al., 2000).

In schooling the need to hold position relative to several conspecifics calls for attention to multiple targets, with some specification of their separate properties. Such an ability could have come to be used to monitor the position of landmarks by one (or both) of two routes. Firstly, once visually evoked escape evolved, there would be great advantage, even for pelagic animals, in directing that escape, so as to avoid obstacles and sources of danger that had already been detected. In teleosts, such information is used to impose standing biases on escape direction, without increasing the latency of escape initiated by Mauthner cells (Eaton and Emberley, 1991). A comparable mechanism has been described for Anuran amphibia (Ingle and Hoff, 1990), which hold a standing record of the positions around them of obstacles. This is updated at each self-initiated body movement and is used to guide startle-induced escape leaps. Secondly, once life on the sea floor was adopted, with fixed rather than continuously changing features, brain structures fed by the LE would have been likely to acquire the further function of learning to use landmarks to allow return to important resources. Here the use of long-term memory to record environmental layout, rather than a continuously updated temporary record, would be useful for the first time.

2.4. Later Evolution of Functions of the Right Side of the Brain

The best available models for the next steps in evolution (i.e. from the condition exemplified by teleosts to that shown by non-mammalian tetrapods) are probably the domestic chick and other birds. There is extensive evidence for lateralization from the behaviour of birds with one or other eye covered, as well as from their spontaneous choice of eye with which to view. It will be convenient hereafter to use the terms right and left eye systems (RES and LES) for the structures fed by right and left eye, and responsible for asymmetries of performance. Although the structures making up RES and LES are likely to be mainly contralateral to the eye in use, there is often little direct evidence for this.

The LES has advantage in the use of topographical information to locate sites (chick: Rashid and Andrew, 1989; Tommasi, Vallortigara and Zanforlin, 1997; marsh tit: Clayton and Krebs, 1994). It has already been argued that the first steps in the evolution of this assignment of abilities may have been maintaining orientation within groups of conspecifics.

Once long-term memory of environmental layout became possible, the LES could have been used to examine familiar scenes to check whether any change had occurred. Zebrafish typically use the LE when fixating scenes or stimuli that they have seen once before (Miklósi, Andrew and Savage, 1998). Chicks also are better able to establish whether there has been change or not in a stimulus when using the LES (Vallortigara and Andrew, 1993).

Detection of change is an excellent general signal of significance, including possible danger, so that such use of the left eye would have increased the involvement of the LES in initiating avoidance. There is, in fact, evidence for teleosts that unexpected startling stimuli cause escape, which is biased towards the right, that is away from the left visual field (Cantalupo, Bisazza and Vallortigara, 1995).

Finally, in many vertebrates (Chapter 3 by Andrew and Rogers), attacks on competing conspecifics are more likely when the LES is in use. This is true of toads (Vallortigara et al., 1998), lizards (Deckel, 1995) and chicks (Rogers, 1982). In the last case (McKenzie, Andrew and Jones, 1998), the difference between right and left eye use arises from the fact that right eye use allows the inhibition of pecks at familiar conspecifics.

Commonly (e.g. toad, chick) the attack response resembles the main prey-catching response of the species. A possible way by which attack might have evolved is by the acquisition of the ability to initiate prey-catching behaviour by right hemisphere perceptual mechanisms, which had already evolved to respond to cues presented by conspecifics during breeding and schooling.

In humans, right hemisphere control gives much more negative response to potentially disturbing inputs (e.g. Dimond, Farrington and Johnson, 1976; Hugdahl, 1995). Evidence is given in Chapter 3 (Andrew and Rogers) that more intense positive as well as negative emotion is associated with right hemisphere control. The arguments advanced in the present chapter suggest that this is an ancient and general property of vertebrate brains. It appears to have evolved alongside and as a complement to the greater ability of the left hemisphere to inhibit behaviour such as fear responses, whilst a decision is taken as to what response to make.

2.5. Later Evolution of Functions of the Left Side of the Brain

Nearly all discussion of the lateralization of motor control has been confined to choice between use of right or left hand (or foot). However, once visual control is taken into account, the use of mouth or bill can be seen also to be likely to be under lateralized control. In advanced forms like teleosts or birds, visual control of mouth use may be divided into a number of aspects.

Firstly, in the approach phase, attention has to be sustained on a particular target, and a particular bearing or route of approach has to be specified and sustained. Both require focused attention and resistance to distraction. This in turn can be seen as a wide extension of the inhibition during prey catching of movements of rejection and avoidance, with which, it is argued here, the evolution of RES specializations began.

Secondly, the form of the final motor response may be determined by visual information and then be held, pre-programmed, ready to be performed. In pigeons, it is known that gape size is adjusted to target size on the basis of what can be seen (Zeigler, Bermejo and Bout, 1994). In the domestic chick, it has recently been shown that, when a manipulandum that has to be grasped by the bill can be seen at a distance, the chick sustains fixation with the right eye during approach (Andrew, Tommasi and Ford, 2000). This suggests that such fixation helps to maintain a motor plan. In contrast, the left eye is used in approach when no manipulandum can be seen.

Thirdly, there is evidence from birds that the RES records consequences of response in a way that the LES does not. This is to be expected, given that the RES is responsible for the visual control of response. It is the RES that will be better able to relate information about the evocation and performance of the response to information about its consequences. In chicks, the following findings are relevant: firstly, chicks using the right eye are quicker to learn to take familiar food rather than inedible, novel distracting targets (Rogers and Ehrlich, 1983). In addition, 'devaluation' of one type of food by prior consumption can be used to guide the choice of a site, where a different type of food is known to be available, only when the right eye is in use (Cozzutti and Vallortigara, 2001). It is also relevant that a strategy of shift in foraging is consistently applied only when the left hemisphere is controlling (Chapter 15 by Andrew).

Finally, when recall is after 24 h, items that were hoarded with both eyes in use can be retrieved by marsh tits, only when they are able to use the right eye (Clayton, 1993). This suggests readier access via the RES to the consequences of response than via the LES.

2.6. Handedness – Senses Other than Vision

In primates, choice of hand with which to perform a task is affected by factors, whose effects are sufficiently diverse as to have led to the widespread belief that only humans and perhaps great apes are consistently lateralized. This was corrected and some order was brought into studies of handedness by a seminal paper by MacNeilage, Studdert-Kennedy and Lindblom (1987). A key point was the distinction of support and manipulation, which of necessity compete in quadrupedal animals during hand use.

Support turns out also to be affected by lateralization in birds. Tommasi and Vallortigara (1998) have shown that when the controlling eye system is manipulated by covering one or other eye, then chicks support themselves (during scratching at the floor) with the foot that is contralateral to the eye that is in use. In other words, the chicks stand on the foot under the control of the side of the brain that is active because it can see. Tommasi and Vallortigara argue persuasively that the motor and perceptual needs of balancing on one foot are likely to be more demanding than those required in order to perform the relatively automatic and stereotyped scratch.

Clearly effects of this sort considerably complicate choice of hand or limb with which to perform a response. Task conditions that tend to put one cerebral hemisphere in charge could result either in the hand controlled by that hemisphere being used to grasp the target or being used to support. Presumably a relatively automatic task (e.g. a ballistic strike that is not under control by visual feedback) would result in the needs of balance and support being given priority. Thus the well-documented tendency for prosimians to use the left hand to catch moving prey (Ward and Hopkins, 1993) might reflect either right hemisphere control, and so left hand use, because of the need for competence in judging position within both right and left hemifields. Alternately, it might be due to left hemisphere control, mediating support by the right forelimb (leaving the left free to strike).

The field becomes even more difficult when more complex tasks are considered. Hemispheric specializations for perceptual processing become increasingly important in determining hand use in particular tasks (Chapter 3 by Andrew and Rogers).

Little attention has been paid in this chapter to the evolution of lateralization of senses other than vision. In Chapter 10 (Andrew and Watkins), evidence is reviewed, which suggests that, in birds and mammals, hearing at least shows features of lateralization that resemble those present for vision. It is argued that these include special involvement of left hemisphere mechanisms in controlling response to sounds, and of right hemisphere mechanisms in detection of details of their structure.

If the arguments advanced here are correct, this is what would be expected because the functional differences between the two sides of the brain are here assumed from the start to be related directly to the kind of behaviour that each controls. All of the senses would have been shaped, as a result, in similar ways.

References

Aldridge, R.J., Briggs, D.E.G., Smith, M.P., Clarkson, E.N.K. & Clark, N.D.L. (1993). The anatomy of conodonts. *Philosophical Transactions of the Royal Society of London B*, 340, 405–421.

Aldridge, R.J. & Purnell, M.A. (1996). The conodont controversies. *Trends in Ecology and Evolution*, 11, 463–468.

Andrew, R.J., Tommasi, L. & Ford, N. (2000). Motor control by vision and the evolution of cerebral lateralization. *Brain & Language*, 73, 220–235.

Babcock, L.E. & Robison, R.A. (1989). Preferences of Palaeozoic predators. *Nature*, 337, 695–696.

Berrill, N.J. (1975). Tunicata. In *Reproduction of Marine Invertebrates*, Vol. II, ed. A.C. Giese & J.S. Pearse, pp. 241–282. New York: Academic Press.

Bisazza, A., De Santi, A. & Vallortigara, G. (1999). Laterality and cooperation: Mosquitofish move closer to a predator when the companion is on their left side. *Animal Behaviour*, 57, 1145–1149.

Bisazza, A., Pignatti, R. & Vallortigara, G. (1997a). Laterality in detour behaviour: Interspecific variation in poeciliid fish. *Animal Behaviour*, 54, 1273–1281.

Bisazza, A., Pignatti, R. & Vallortigara, G. (1997b). Detour tasks reveal task- and stimulus-specific neural lateralization in mosquitofish (*Gambusia holbrooki*). *Behavioural Brain Research*, 89, 237–242.

Bisazza, A., Rogers, L.J. & Vallortigara, G. (1998). The origins of cerebral asymmetry: A review of evidence of behavioural and brain lateralization in fishes, reptiles and amphibians. *Neuroscience and Biobehavioral Reviews*, 22, 411–426.

Bone, Q. (1960). The CNS in Amphioxus. *Journal of Comparative Neurology*, 115, 27–64.

Canfield, J.G. & Rose, G.J. (1993). Activation of Mauthner cells during prey capture. *Journal of Comparative Physiology A*, 172, 611–618.

Cantalupo, C., Bisazza, A. & Vallortigara, G. (1995). Lateralization of predator-evasion response in a teleost fish (*Girardinus falcatus*). *Neuropsychologia*, 33, 1637–1646.

Chen, J.-N., Eeden, F.J.M. van, Warren, K.S., Chin, A., Nüsslein-Volhard, C., Haffter, P. & Fishman, M.C. (1997). Left-right pattern of cardiac BMP4 may drive asymmetry of the heart in zebrafish. *Development*, 124, 4373–4382.

Chen, J.-Y., Dzik, J., Edgecombe, G.D., Ramsköld, L. & Zhou, G.-Q. (1995). A possible Early Cambrian chordate. *Nature*, 377, 720–722.

Clayton, N. (1993). Lateralization and unilateral transfer of memory in marsh tits. *Journal of Comparative Physiology A*, 171, 799–806.

Clayton, N.S. & Krebs, J.R. (1994). Memory for spatial and object-specific cues in food-storing and non-storing birds. *Journal of Comparative Physiology A*, 174, 371–379.

Cozzutti, C. & Vallortigara, G. (2001). Hemispheric memories for the content and position of food caches in the domestic chick. *Behavioral Neuroscience*, 115, 305–313.

Currie, S.N. (1991). Vibration-evoked startle behaviour in larval lampreys. *Brain, Behaviour and Evolution*, 37, 260–271.

Deckel, A.W. (1995). Laterality of aggressive responses in *Anolis*. *Journal of Experimental Zoology*, 272, 194–200.

Dimond, S.J., Farrington, L. & Johnson, P. (1976). Differing emotional response from right and left hemispheres. *Nature*, 261, 690–692.

Donoghue, P.C.J. (1998). Growth and patterning in the conodont skeleton. *Philosophical Transactions of the Royal Society of London B*, 353, 633–666.

Donoghue, P.C.J., Forey, P.L. & Aldridge, R.J. (2000). Conodont affinity and chordate phylogeny. *Biological Reviews*, 75, 191–251.

Drach, P. (1948). Embranchement des Céphalocordés. In *Traité de Zoologie*, Vol. 11, ed. P.-P. Grassé, pp. 931–1037. Paris: Masson et Cie.

Eaton, R.C. & Emberley, D.S. (1991). How stimulus direction determines the trajectory of the Mauthner-initiated escape response in a teleost fish. *Journal of Experimental Biology*, 161, 469–487.

Fetcho, J.R. (1991). Spinal network of the Mauthner cell. *Brain, Behavior and Evolution*, 37, 298–316.

Fetcho, J.R. (1992). Excitation of motoneurons by the Mauthner axon in goldfish: complexities in a 'simple' reticulospinal pathway. *Journal of Neurophysiology*, 67, 1574–1586.

Flood, P.R. (1975). Fine structure of the notochord of amphioxus. *Symposia of the Zoological Society of London*, 36, 81–104.

Fontaine, M. (1958). Classe des cyclostomes. Formes actuelles. In *Traité de Zoologie*, vol. 13, part 1, ed. P.-P. Grassé, pp.13–106, Paris: Masson et Cie.

Frontera, J.G. (1952). A study of the anuran diencephalon. *Journal of Comparative Neurology*, 96, 1–69.

Guthrie, D.M. (1975). The physiology and structure of the nervous system of Amphioxus (the Lancelet) (*Branchiostoma lanceolatum* Pallas). In *Protochordates*, ed. E.J.W Barrington & R.P.S. Jefferies, pp. 43–80. London: Academic Press.

Hadfield, M.G. (1975). Hemichordates. In *Reproduction of Marine Invertebrates*, Vol. 2, ed. A.C. Giese & J.S. Pearse, pp. 185–240. New York: Academic Press.

Hamner, W.P., Madin, L.P., Alldredge, A.L. & Hamner, R.P. (1975). Underwater observations of gelatinous zooplankton: Sampling problems, feeding biology and behaviour. *Limnology & Oceanography*, 20, 907–917.

Hendler, G. (1991). Echinodermata: Ophiuroidea. In *Reproduction of Marine Invertebrates*, Vol. 6, ed. A.C. Giese, J.S. Pearce & V.B. Pearce, pp. 355–511. Pacific Grove, CA: Boxwood Press.

Holland, N.D. (1991). Echinodermata: Crinoidea. In *Reproduction of Marine Invertebrates*, Vol. 6, ed. A.C. Giese, J.S. Pearce & V.B. Pearce, pp. 247–299. Pacific Grove, CA: Boxwood Press.

Holland, P.W.H., Holland, L.Z., Williams, N.A. & Holland, N.D. (1992). An amphioxus homeobox gene: Sequence conservation, spatial expression during development and insights into vertebrate evolution. *Development*, 116, 653–661.

Hugdahl, K. (1995). Classical conditioning and implicit learning: the right hemisphere hypothesis. In *Brain Asymmetry*, ed. R.J. Davidson & K. Hugdahl, pp. 235–267. Cambridge, MA: MIT Press.

Ingle, D.J. & Hoff, K. vS. (1990). Visually evoked evasive behaviour in frogs. *BioScience* 40, 284–291.

Jefferies, R.P.S. & Lewis, D.N. (1978). The English Silurian fossil *Placocystites forbesianus* and the ancestry of the vertebrates. *Philosophical Transactions of the Royal Society of London B*, 282, 207–321.

Lacalli, T.C. (1996). Frontal eye circuitry, rostral sensory pathways and brain organisation in amphioxus larvae; evidence from 3D reconstructions. *Philosophical Transactions of the Royal Society of London B*, 351, 243–263.

Lacalli, T.C., Holland, N.D. & West, J.E. (1994). Landmarks in the anterior central nervous system of amphioxus larvae. *Philosophical Transactions of the Royal Society of London B*, 344, 165–185.

Land, M.F. (1982). Scanning eye movements in a heteropod mollusc. *Journal of Experimental Biology*, 96, 427–430.

MacNeilage, P.F., Studdert-Kennedy, M.G. & Lindblom, B. (1987). Primate handedness reconsidered. *Behavioural and Brain Sciences*, 10, 247–303.

Mallatt, J. (1985). Reconstructing the life cycle and the feeding of ancestral vertebrates. In *The Evolutionary Biology of Primitive Fishes*, ed. R.E. Foreman, A. Gorbman, J.M. Dodd & R. Olsson, pp. 59–68. New York: Plenum Press.

McKenzie, R., Andrew, R.J. & Jones, R.B. (1998). Lateralization in chicks and hens; new evidence for control of response by the right eye system. *Neuropsychologia*, 36, 51–58.

Miklósi, A. & Andrew, R.J. (1999). Right eye use associated with decision to bite in zebrafish. *Behavioural Brain Research*, 105, 199–205.

Miklósi, A., Andrew, R.J. & Gasparini, S. (2001). Role of right hemifield in visual control of approach to target in zebrafish. *Behavioural Brain Research*, 106, 175–180.

Miklósi, A., Andrew, R.J. & Savage, H. (1998). Behavioural lateralization of the tetrapod type in the zebrafish (*Brachydanio rerio*). *Physiology and Behavior*, 63, 127–135.

Nicoll, R.S. (1987). Form and function of the Pa element in the conodont animal. In *Palaeobiology of Conodonts*, ed. R.J. Aldridge, pp. 77–90. Chichester: Ellis Horwood Ltd.

Okada, Y., Iwata, K.S. & Yanagihara, M. (1994). Synchronised rhythmic contractions among five gonadal lobes in the shedding sea urchin: Coordinating functions of the oboral nerve ring. *Biological Bulletin*, 166, 228–236.

Pagán-Westphal, S.M. & Tabin, C.J. (1998). The transfer of left-right positional information during chick embryogenesis. *Cell*, 93, 25–35.

Pearse, J.S. & Cameron, R.A. (1991). Echinodermata: Echinoidea. In *Reproduction of Marine Invertebrates*, Vol. 6, ed. A.C. Giese, J.S. Pearse & V.B. Pearse, pp. 513–662. Pacific Grove, CA: Boxwood Press.

Pu, G.A. & Dowling, J.E. (1981). Pineal photoreceptor cells in larval lamprey. *Journal of Neurophysiology*, 46, 1018–1038.

Rashid, N.Y. & Andrew, R.J. (1989). Right hemisphere advantage for topographical orientation in the domestc chick. *Neuropsychologia*, 7, 937–948.

Roberts, A. (1978). Pineal eye and behaviour in *Xenopus* tadpoles. *Nature*, 273, 774–775.

Rogers, L.J. (1982). Light experience and asymmetry of brain function in chickens. *Nature*, 297, 223–225.

Rogers, L.J. & Ehrlich, D. (1983). Asymmetry in the chicken forebrain during development and possible involvement of the supraoptic decussation. *Neuroscience Letters*, 37, 123–127.

Shimeld, S.M. (1999). The evolution of the hedgehog gene family in chordates: Insights from amphioxus hedgehog. *Development, Genes and Evolution*, 209, 40–47.

Shu, D.-G., Conway Morris, S. & Zhang, X.-L. (1996). A *Pikaia*-like chordate from the Lower Cambrian of China. *Nature*, 384, 157–158.

Smiley, S., McEuen, F.S., Chaffee, C. & Krishnan, S. (1991). Echinodermata: Holothuroidea. In *Reproduction of Marine Invertebrates*, Vol. 66, ed. A.C. Giese, J.S. Pearse & V.B. Pearse, pp. 663–750. Pacific Grove, CA: Boxwood Press.

Sovrano, V.A., Rainoldi, C., Bisazza, A. & Vallortigara, G. (2000). Roots of brain specialisations: Preferential left-eye use during mirror-image inspection in six species of teleost fish. *Behavioural Brain Research*, 106, 175–180.

Stokes, M.D. & Holland, N.D. (1995). Ciliary hovering in larval lancelets (Amphioxus). *Biological Bulletin*, 188, 231–233.

Tommasi, L. & Vallortigara, G. (1998). Footedness in binocular and monocular chicks. *Laterality*, 4, 89–94.

Tommasi, L., Vallortigara, G & Zanforlin, M. (1997). Young chickens learn to localise the centre of a spatial environment. *Journal of Comparative Physiology A*, 180, 567–572.

Vallortigara, G. & Andrew, R.J. (1993). Lateralization of response by chicks to change in a model partner. *Animal Behaviour*, 41, 187–194.

Vallortigara, G., Rogers, L.J., Bisazza, A., Lippolis, G. & Robins, A. (1998). Complementary right and left hemifield use for predatory and agonistic behaviour in toads. *NeuroReport*, 9, 3341–3344.

Ward, J.P. & Hopkins, W.D. (1993). *Primate Laterality: Current Behavioural Evidence of Primate Asymmetries*. New York: Springer-Verlag.

Webb, J.E. (1969). On the feeding and behaviour of the larva of *Branchiostoma lanceolatum*. *Marine Biology*, 3, 58–72.

Webb, J.E. (1975). The distribution of amphioxus. *Symposia of the Zoological Society of London*, 36, 179–217.

Zeigler, H.P., Bermejo, R. & Bout, R. (1994). Ingestive behaviour and the sensorimotor control of the jaw. In *Perception and Motor Control in Birds*, ed. M.N.O. Davies & P.R. Green, pp. 182–200. Berlin: Springer-Verlag.

3

The Nature of Lateralization in Tetrapods

RICHARD J. ANDREW AND LESLEY J. ROGERS

3.1. Introduction

The hypothesis that all vertebrate groups inherit a common basic pattern of lateralization from a common chordate ancestor was advanced in Chapter 2. Here we examine the evidence available from tetrapods to see how far this hypothesis can be sustained. We consider mainly tetrapods other than primates, as primates are examined in other chapters. Nevertheless, since so much more is known of human lateralization than that of any other vertebrate, a final test of each aspect of the hypothesis is how far it is consistent with evidence of lateralization in humans.

The evidence available so far supports our view that there is a common basic pattern of lateralization in all vertebrates. It is, of course, possible that there has been loss or reorganization of the basic pattern of lateralization in some species, but there is no certain example of this as yet. The absence of asymmetry in one or a few tests on a particular species is not strong evidence of its absence, as is clearly exemplified by the work of Hamilton and Vermeire (1983, 1988), who, after providing a large body of negative findings for the rhesus monkey, went on to produce an impressive body of evidence for brain lateralization in that species. Their findings are now corroborated by other studies (see Chapter 12 by Hopkins and Carriba).

In this context, and before beginning a comparative review of evidence for lateralization in tetrapods, some comment is needed on lateralization in rats and mice. The focus of much work on the choice of paw, and on the direction of rotational bias in free locomotion, has led to a common (but far from universal) belief that, in rats and mice, lateralization is clear only at the individual level, and is expressed chiefly or entirely in individual motor biases. In other words, since half of the individuals use the right paw or circle clockwise and the other half use the left paw or circle anticlockwise, there is no bias in the population (Glick and Cox, 1978; Collins, 1985). Collins (1985) even extended this view as applying to animals in general. It thus should be stressed that the same researchers who have supplied most of the evidence on which this position is based have also made it clear that, on other criteria, rodents are very consistently lateralized at the population level.

A few examples will suffice to demonstrate lateralization at the population level in rodents (for more examples, see Bradshaw and Rogers, 1993; Denenberg, 1981). The degree of consistency of these results in demonstrating a population bias is shown by the fact that significant population bias is commonly revealed with group sizes of 10–20 subjects. The examples are as follows:

1. Ablation of the right hemisphere (RHem) of rats elevates their activity levels in open field tests to a greater extent than does ablation of the left hemisphere (LHem) (Denenberg et al., 1978).
2. Greater activity in the left prefrontal region of the cortex is shown by uptake of 2-deoxyglucose (Ross and Glick, 1981) and this is accompanied by larger size of the left prefrontal region, both during periods of development and in adults, coupled with higher dopamine levels (van Eden, Uylings and van Pelt, 1984).
3. Cortical lesions in the RHem, especially anteriorly, lead to a drop in norepinephrine levels on both sides of the brain, and to a rise in open field activity, whilst corresponding LHem lesions have no such effects (Robinson, 1985).
4. Rats, trained to find a hidden escape platform in the Morris water maze and then tested monocularly are able to find the platform, using spatial information, when they use the left eye (RHem) but not when they use the right eye (LHem) (Cowell, Waters and Denenberg, 1997).

The significance of these findings is considered further below; they are mentioned at this point to show that rodents do not contradict the thesis that consistent lateralization at the population level is a general property of vertebrates.

The basic pattern of lateralization in tetrapods may be summarized as follows, and is dealt with more fully later.

1. *Attention, perceptual processing and control of motor response.* All of these aspects of lateralization are closely linked. An example of this can be seen, in a simple form, when an animal approaches a target, which it can see and for which it has a planned response (see also Chapter 2 by Andrew). Both fish and birds (discussed below) use the right eye (RE) to fixate a manipulandum as they approach in order to grasp it with the mouth or

bill. Such use of the RE implies use of the LHem, since the main projection of visual input from each eye is to its contralateral hemisphere (discussed further in Chapter 6 by Deng and Rogers). The use of the RE can be shown to be associated with maintenance of the readiness to manipulate: attention is clearly locked onto the target during such approach. In an approach under identical conditions, but with no target of response visible until the food dish is reached, the left eye (LE) is used.

A complementary set of specializations is shown when the RHem and the LE are in use. Here there is much better use of topographical cues to identify position in space, an ability that requires the animal to use diffuse attention. Spatial context and the detailed properties of objects are attended to, and recorded, more fully than when the RE is in use. This facilitates the detection of novelty and establishment of identity.

The resemblance to human dichotomies of hemispheric function is obvious. The RHem shows diffuse or global attention, spatial analysis and no special involvement in control of response. The LHem shows focused attention, recording of local cues and control of response.

2. *Emotion and inhibition of response*. The best example of a simple behavioural situation, which calls for inhibition of response, is examination of a strange and strongly motivating object from a distance. When examining a potentially dangerous object, escape must be inhibited whilst the object is assessed. In fact, the need for such inhibition is not confined to situations where escape is likely. When a naive chick is deciding whether to approach an attractive object or sound, on which it might imprint, and which it now encounters for the first time (i.e. an attractive but unknown stimulus), approach has to be inhibited while a decision to approach or not is made. At such times the RE or right ear is turned towards the stimulus, indicating LHem control (Miklósi, Andrew and Dharmaretnam, 1996; McKenzie, Andrew and Jones, 1998).

Conversely, there is extensive evidence (discussed below) that responses such as escape, attack and sexual behaviour are evoked more vigorously when the RHem is controlling. This is clearly what would be expected when LHem inhibition is absent. Although direct evidence is lacking, it would be advantageous for the topographical abilities of the RHem to be freely available during the uninhibited performance of such responses. It is necessary for obstacles to be avoided during escape or pursuit, and the position of refuges needs to be used to guide locomotion, despite the dominance of brainstem motivational mechanisms. It is interesting in this context that in goldfish, sudden escape, which is mediated by Mauthner cells, has directional bias determined by visual (and no doubt other perceptual) information about the layout of the environment (Eaton and Emberley, 1991). In toads, the position of escape routes is updated at each bodily movement and guides startle-induced escape leaps (Ingle and Hoff, 1990). As we will discuss later, toads display lateralization of escape leaping and are more responsive to a simulated predator seen on their left side (Lippolis et al., 2001).

Here we note the existence of comparable evidence (discussed later) in humans for association of intense emotion with RHem control, and inhibited emotion with LHem control. In fact, there is also clear evidence for two species of primate (rhesus monkey and common marmoset) that the RHem is used for the expression of intense emotion: this is manifested as greater movement of the mouth and other facial features on the left side of the face (controlled by the RHem) in expressing fear (Hauser, 1993; Hook-Costigan and Rogers, 1998; discussed also in Chapter 13 by Weiss et al.).

3.2. Left Hemisphere: Attention, Perceptual Processing and Control of Motor Response

The use of the RE to control placement of, and manipulation by, the mouth has been recorded in fish, amphibia and birds. In the zebrafish, the RE is used to fixate a target that the fish intends to bite, but not when biting does not follow, even if the target is identical (Miklósi and Andrew, 1999). Toads (*Bufo bufo* and *Bufo viridis)* are more likely to strike at prey items when these are seen in the right visual hemifield (Vallortigara et al., 1998; see Figure 3.1). The same study showed that toads (*Bufo marinus*) were more likely to deliver aggressive tongue-strikes at conspecifics in their left visual hemifield (discussed in more detail below). Here again control using the RE is specific to taking a target into the mouth, whilst an 'emotional' response of similar form, but not involving seizing a target, is not.

Use of the RE to control approach to grasp a manipulandum has been demonstrated in the domestic chick. The study used three different manipulations, all requiring the displacement of a light paper lid to allow access to food in a dish (Andrew, Tommasi and Ford, 2000). In one test the chick displaced the (square) lid by grasping either its left or right protruding corner; in another the bill had to be inserted into a notch placed at the front of the lid, which was circular like the dish and difficult to grasp in any other way; in the last test, a short string protruding vertically up from the centre of the lid had to be pulled. In all cases, irrespective of whether the manipulandum was medially placed or not, RE fixation was used during approach with striking consistency. In the same apparatus, but in the absence of the lid, the LE was used with equal consistency during approach to the food dish.

A: Attack

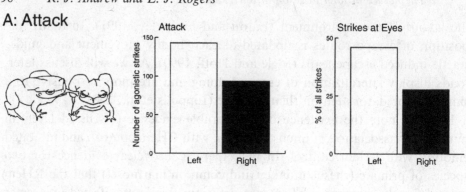

B: Feeding

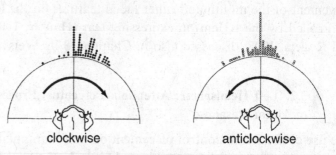

Figure 3.1.   Summary of the complementary specialization of the left and right visual hemifields of toads for attack and feeding responses. (A) Attack: *Bufo marinus* toads were tested in a group and were competing for prey (crickets). Significantly more of their agonistic strikes at each other were directed to the left (open bar) than to the right (black bar) visual hemifield. In addition, striking at the eye of a conspecific was avoided to a greater extent in the right hemifield than in the left hemifield. (B) Feeding: strikes at at prey were also lateralized but here the preference was to strike at prey on the right side. *Bufo viridis* toads were tested by placing them, one at a time, inside a glass cylinder and then rotating a prey (insect larva) around the toad and outside the cylinder. Strikes at the prey were recorded by videotaping. Each dot represents a strike. Note that when the prey was rotated clockwise (from the toad's left to right) strikes (black dots) were directed to it when it was in the right visual hemifield. When the prey was rotated anticlockwise, the strikes (open circles) were distributed more evenly around the midline but there was still a tendency for more strikes in the right hemifield. Lateralization is shown by the fact that the pattern of strikes is not simply reversed for the clockwise versus anticlockwise presentations. Data from Robins et al. (1998) and Vallortigara et al. (1998).

It should be noted that RE use is also shown when a target food item can be seen. During free search over an arena floor, chicks showed a preference to take food grains that had been located with the RE, although both eyes were used freely in searching (Andrew, Tommasi and Ford, 2000). The key factor determining eye use in the chick is thus visual guidance of the bill to the target that is to be grasped. When no target can be seen during approach to the dish, the LE is used, despite the fact that feeding is certainly the reason for approach, just as in the lid condition.

Hunt (2000) has provided evidence that the tropical crow *Corvus moneduloides* cuts its tools (hooked strips of Pandanus leaves used to obtain food) from the left edge of these large leaves. Hunt shows that the method of working along the edge away from the trunk means that such removal will normally require the bird to turn its head to view its work with the RE, and argues that this reflects 'specialization of the right-eye system for object-related tasks'. This fascinating finding is clearly consistent with the RE control of manipulation described above for the chick.

Control of response by the RE is likely to explain special involvement of the LHem in learning based on the association of the consequences of response with the stimulus to which the response was directed. At its simplest, in animals with independent eye movements, it will sometimes be only the RE that sees what is to be recorded. The recording of selected cues associated with the (successful) outcome of response is suggested by the more rapid acquisition by right-eyed chicks than by left-eyed chicks of a discrimination between familiar food grains and unfamiliar inedible distractors (Zappia and Rogers, 1987). The chicks tested monocularly had to search for grains of chick-mash scattered on a floor to which small pebbles had been adhered. The distractors (pebbles) roughly matched the food grains in size and hue, so that it was necessary to select appropriate cues to make a successful discrimination. The experiments showed that young chicks could inhibit pecking at pebbles and choose grain, provided that they used the RE in monocular tests, and provided that the LHem was fully functional; treatment of the LHem with cycloheximide or glutamate impaired the chick's ability to choose grain over pebbles, but treatment of the RHem had no such effect (Rogers and Anson, 1979; Howard, Rogers and Boura, 1980). Recent experiments using localized placement of glutamate injections in various regions of the hemispheres have shown that it is only the visual Wulst region of the LHem that controls the shift from pecking randomly at grains and pebbles to pecking predominantly at grain (Deng and Rogers, 1997; see also Chapter 6 by Deng and Rogers).

The use of shift strategies to choose potential feeding sites occurs when the LHem, but not the RHem, is controlling (see Chapter 15 by Andrew); again this result suggests LHem involvement in recording the outcome of response (here the consumption of food at the previous visit). Vallortigara and Regolin (Chapter 11) present evidence showing that the chick can shift the food type chosen following a devaluation procedure (with training and devaluation carried out with both eyes in use); such shifts take place only when the RE is in use during testing (Cozzutti and Vallortigara, 2001).

Very much the same pattern of lateralization is shown by marsh tits during recovery of hoarded food items, when the birds are tested with one or other eye covered (Clayton and Krebs, 1994). When the RE (LHem) is in use, retrieval just after hoarding makes use of the local cues that were associated with the hole into which the tit had previously inserted the food item. When the LE (RHem) is in use, spatial position relative to the test room is used instead. Recording local cues suggests once again that the LHem is more likely to record consequences of the response (successful hoarding) that is associated with the perception of these cues.

Any consideration of mammalian evidence must start with the consensus that the human LHem typically controls purposive movements (see review in Kimura, 1982). This is strikingly true of multiple movements, whether of hand or mouth (Kimura, 1982). However, it also holds for the selection of one out of a number of (arbitrary) motor responses, when this learned response has to be the one appropriate to the presented stimulus (Rushworth et al., 1998). The resemblance of this result to the bird condition is close enough to suggest that, underlying the complications of LHem control of spoken sequences, and of skilled manipulation by the right hand in human beings, is the basic vertebrate pattern assigning the LHem to control a planned response to a perceived stimulus.

This specialization of the LHem may also underlie its apparent greater activation in the marmoset during the emission of close-range, social contact calls (twitters) that signal intent to approach a conspecific (or human). These calls are accompanied by more vigorous contraction of facial muscles on the right side, thereby opening the mouth wider on the right side (Hook-Costigan and Rogers, 1998). This contrasts with the role of the RHem in producing, and responding to, mobbing and fear calls that require more diffuse attention and unplanned outcomes. Such calls are accompanied by greater muscular contraction on the left side of the face (Hauser, 1993; Hook-Costigan and Rogers, 1998). It is argued in Chapter 10 by Andrew and Watkins that the lateralization of control of species-specific vocalizations, the discovery of which by Nottebohm (1970) was largely responsible for the return of interest in cerebral lateralization in vertebrates in general, can also be understood as originating from control of response to a stimulus (here, a conspecific call). The responses of importance include approach (or avoidance), as well as reply by calling.

The next question is how far can a comparable condition be shown to hold for mammals other than primates. The rodents provide an excellent test case because they have been separate from the primate line since the Cretaceous (Archibald, 1996); resemblances between them and primates would thus probably be derived from the common ancestor of Eutherian mammals. Such evidence has been provided by Bianki and his collaborators in studies using rats, which began in the 1960s, but which were initially little quoted, no doubt because they were published in Russian. The full corpus of work is summarized and discussed by Bianki (1988) in English.

The usual approach was to use spreading depression of the cortex achieved by the application of potassium chloride to inactivate one or other hemisphere. A striking variety of studies using this technique show the same shift in hemispheric control during the elaboration of conditioned reflexes. Early in acquisition, RHem inactivation disturbs the conditioned response (CR) more than LHem inactivation but, when the CR is fully elaborated, it is LHem inactivation that disturbs performance. In this second phase of elaboration, not only does LHem control allow more correct responses but also they are emitted significantly more quickly. The studies used both active avoidance and food reinforcement, and so effects of hemispheric involvement in behaviour like escape are unlikely to be responsible for the shift. Instead, the involvement of the rat LHem in performance of an established CR is better compared with the human use of the LHem to select the learned response appropriate to the stimulus presented (as discussed previously). The RHem dominance early in acquisition can be understood as appropriate to a phase when the rat is not quite clear what is the signal, nor when to respond. As a result, diffuse general attention and attempts to process all detected stimuli would be appropriate.

A second finding by Bianki is consistent with special LHem involvement in recording the consequences of response, such as appears to be present in birds (see above). When the probability of food reinforcement associated with different stimuli was varied, LHem control allowed emission of responses with likelihoods proportional to the frequency with which each stimulus had been rewarded, whereas RHem control was associated with equal probabilities for all stimuli.

LHem recording of the outcome of recent responses may also explain the finding that LHem control was necessary to allow efficient rapid visiting of all the arms in a radial maze, without returns to arms that had already been visited (Bianki, 1988). In other words, rats using their LHem adopted an efficient, sequential searching strategy in the radial-arm maze, whereas rats using the RHem were unable to do so. A comparable result for the effects of unilateral removal of right whiskers on performance in a radial maze is discussed in Chapter 10 by Andrew and Watkins.

Finally, a very marked asymmetry was found for timing of responses. When the time approached at which a CS was expected to occur, normal rats, and rats in which the LHem was controlling, showed a substantial rise in responses, whilst RHem rats distributed responses with little reference to time. Somewhat comparable results were obtained by Mittleman, Whishaw and Robbins (1988), who tested rats with intact hemispheres in a task using stimuli presented sequentially. The rats were trained to hold their snouts in a central hole of a test chamber after a light had been illuminated inside the hole. A cue light was then presented to the left or right side of the hole. This signalled the rat to move its head to the left or right and, in doing so, to intercept the beam of a photocell that would lead to the delivery of a reward. The response was more rapid when the rat moved to its right side than to its left side. In other words, the LHem was better able to respond quickly to the relevant stimulus than was the RHem.

Bianki (1988) argued that the abilities of the rat LHem for sequential analysis resemble the LHem advantage seen also in humans for the analysis of sequences in time. It may also be interpreted as due to LHem control of response, with the LHem again being especially concerned with recording the circumstances under which a successful response was performed, and then using the record to control future responses.

Bianki (1982) also suggested that the LHem of the rat analyses abstract characteristics of stimuli, compared to the RHem, which processes and records concrete or absolute characteristics. One of the examples, on which he based this conclusion, was the rat's ability to discriminate between the areas of unfamiliar geometrical shapes. The rat had to respond, in an operant task, to the key that displayed shapes with a larger total area. When using the LHem only (RHem inactivated), rats could perform this task well, but they were less able to perform the task when using the RHem only. Bianki (1983a) noted that the same separate specializations of the hemispheres as seen in the rat (summarized in Bradshaw and Rogers, 1993) are found also in humans, and this is, perhaps, best exemplified by the LHem's specialization for abstraction.

Such an ability to analyse the abstract characteristics of stimuli would allow the LHem to categorize stimuli and, indeed, this ability of the LHem is present in both the chick and the pigeon. Chicks using the LHem are able to categorize grain from pebbles (discussed earlier; see also Figure 3.2) and also to respond to 'chicks' as a category. The latter was demonstrated by Vallortigara and Andrew (1991) by testing chicks with a choice of a familiar cage-mate and a stranger (a chick that the test chick had not seen previously). When tested monocularly using the LE, the test chick approached its cage-

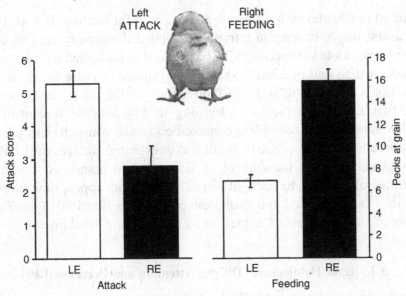

Figure 3.2. Complementary specialization of the left and right eyes of the chick for attack and feeding responses. The white bars represent means (with standard errors indicated) of groups of chicks (approximately 10 in a group) tested using the left eye (LE) and the black bars represent the same for chicks tested using the right eye (RE). Attack scores were obtained by testing testosterone-treated chicks using a standard hand-thrust test, in which the tester's hand is used to simulate an attacking chick. The responses of the chick are ranked according to intensity of attack. Attack is elevated when the chick uses its left eye, but not when it uses its right eye. Feeding scores were obtained by testing chicks on a search task of grain scattered on a background of small pebbles. The chicks could avoid pecking at pebbles when they were tested using their right eye but not when using their left eye. The scores plotted are number of pecks at grain in a block of 20 pecks, pecks 41–60 after the first peck given in the test. Adapted from Rogers, Zappia & Bullock (1985) and Rogers (1997).

mate in preference to the stranger but, when using its RE, the chick approached either, or both of the chicks. It seems, therefore, that the LHem attends to a chick as a category matched by all, or most, chicks, whereas the RHem records more specific and detailed information.

The ability of the pigeon's LHem to categorize stimuli was demonstrated clearly by von Fersen and Güntürkün (1990) by training pigeons in an operant situation to discriminate several abstract visual shapes from a large collection of similar shapes; the pigeon could do so only when using its RE.

We can conclude for these findings that rats, birds and humans all use the LHem for sequential analysis, and for abstracting characteristics of the stimuli to which planned responses will be directed. Comparable arguments are advanced in Chapter 10 by Andrew and Watkins for hearing. It is likely that the earliest use of hearing in tetrapods included the performance of appropriate responses to conspecific calls. In the most complicated use of vocalizations other than that by humans, namely that shown by passerine birds, it has been shown (Cynx, Williams and Nottebohm, 1992), for at least one songbird, that LHem use facilitates learning to associate reinforcement with familiar categories of (complex) conspecific sounds, whilst RHem use facilitates detection and response to small and unexpected changes of detail. The LHem specializations for control of response and associated erection of categories of stimuli, to each of which there is an appropriate response, may thus have affected the evolution of the lateralization of hearing as well as of vision (see also Chapter 10 by Andrew and Watkins).

3.3. Right Hemisphere: Diffuse Attention and Perceptual Processing

There is clear evidence for RHem advantage in the use of environmental layout to guide locomotion to a target site (i.e. to use spatial or topographical information). In the chick, Rashid and Andrew (1989) showed that, after binocular training, LE use allowed chicks to use both distant and local features to guide locomotion, whereas RE use led to almost complete failure to use distant features. Tommasi, Vallortigara and Zanforlin (1997) showed that, once chicks had learned to search in the centre of an arena, they went to the geometrical centre of distorted versions of the arena when using their LE. However, chicks used a much simpler rule when using the RE. They searched in a strip around the arena that was at about the distance from the wall that the centre of the training arena had been; in other words, they followed a rule that allowed them to find the centre in the unmodified arena (see Chapter 11 by Vallortigara and Regolin; also see Tommasi and Vallortigara, 2001). When a local landmark, which had identified the centre during training, was moved, LE birds used position in the arena whilst RE birds followed the local cue. As has already been noted, exactly the same pattern (position in space when the LE is used, local cues when the RE is used) is shown by marsh tits when retrieving a recently hoarded food item.

In the chick, RHem advantage is also shown in other tasks that involve spatial patterns but are not topographical. LE chicks respond to changes in the spatial context of a stimulus that are ignored in the RE condition. This is illustrated by the finding that chicks using the RE retain habituation of pecking at a coloured bead mounted on the end of a rod, even though the angle at which it is introduced to their home-cage is changed. In contrast, chicks using the LE respond to a change in the angle of introduction (spatial cue) by loss of habituation (Andrew, 1983, 1991). Further, LE chicks respond to moderate transformations in the appearance of model social partners (Vallortigara and Andrew, 1991) or partner chicks (Vallortigara and Andrew, 1994), which are ignored by RE chicks.

The same specialization of the RHem and LHem is shown by spontaneous patterns of eye use in chicks. Preferential use of one eye over the other can be measured by observing from overhead which lateral visual field is used by the chick to view the stimulus (i.e. by scoring head turning and visual fixation). Once chicks have attached to a model social partner, they use the LE when approaching it, whereas when deciding whether to approach at the first encounter they use the RE (McKenzie et al., 1998). Note that novelty detection or establishment of identity requires the use of detailed records of past appearance, which (it is here argued) is the responsibility of the RHem. The use of the LE when faced with a familiar stimulus or environment will usually confirm identity and so is the appropriate default condition; equally, unexpected novelty will be identified.

Once again, the human evidence is in agreement. Indeed, there appears to be no dissent from the proposition that the human RHem has advantage for spatial analysis, and that it shows global attention rather than focused (e.g. Posner and Petersen, 1990). Naturally enough, analysis has been pushed much further in work on humans; thus Kosslyn et al. (1992) have shown, using simple dot stimuli, that the RHem excels in judgements of position as measured by X-Y coordinates, whilst the LHem is as good or better at 'categorical' judgements, like above or below a reference position. They also show that the differences could be generated if field sizes of visual units were larger in visual analysis by the RHem.

Some evidence suggests that the RHem is specialized for processing spatial information in primates too. For example, capuchins (*Cebus apella*) show a stronger left-hand preference on haptic and haptic-visual tasks than they do on simple reaching tasks (Lacreuse and Fragaszy, 1999). As the researchers suggest, this may result because the right hemisphere is specialized to integrate the spatial processing and sensorimotor components of the actions demanded by the haptic tasks. Also, gorillas and baboons have been found to display left-hand preferences on spatial tasks, requiring them to align transparent doors in order to obtain food (Fagot and Vauclair, 1988a, 1988b). Other experiments have shown RHem advantage in primates for global processing, an ability that may be associated with spatial processing by that hemisphere: Deruelle and Fagot (1997) tested baboons (*Papio papio*) in an operant task in which they had to respond to a letter of the alphabet made up of smaller letters, which could be the same as, or different from, the larger letter. Attention to the larger letter was interpreted as 'global prece-dence' and this was associated with use of the RHem. Attention to the smaller letters indicated local attention and was associated with use of the LHem. Also, chimpanzees have a RHem advantage for locating a short line contained within a geometric figure, which suggests specialization of the RHem for spatial processing (Hopkins and Morris, 1989).

Once again, the rodent evidence is crucial in deciding whether mammalian lateralization resembles that of other vertebrates. Adelstein and Crowne (1991) found right, but not left, parietal lesions to impair the use of allo-centric cues, during navigation in a water maze. King and Corwin (1992) present similar findings and, as mentioned previously, rats are able to rely on spatial memory to find the escape platform in a Morris swim maze when they use the LE (RHem), but not when they use the RE (Cowell et al., 1997). Bianki (1988) presents evidence from a number of tests that show marked resemblance between the RHem advantage for spatial analysis, present in both rats and humans. There is RHem advantage for discriminations based on texture and on dot location. In discriminations based on arrays of three stimuli, RHem performance is better when matching has to be on arrangement in space (ordered in a linear array), whilst LHem performance is better when it is necessary to ignore spatial order and match on stimulus properties. This latter finding is true, both when the training stimuli are presented in a different order, and when only one of them is presented. LHem analysis thus is of separate stimuli, and does not stress arrangement or even the presence of the full set. RHem analysis includes the spatial arrangement of the full array, as well as the properties of individual stimuli.

Similar differences between the hemispheres probably underlie the results obtained in tasks testing generalization to transformations of the training stimuli (Bianki, 1988). In tests requiring the rat to choose between the origi-nal and a transformed stimulus, RHem analysis produces matching on abso-lute size, whereas LHem analysis tends to use the relative difference present in training (e.g. take the larger). However, this property of LHem analysis does not mean that absolute size cannot be used by the LHem. When only one stimulus is presented, so that the decision is whether to respond or not to a transformed stimulus, it is the LHem which tends to withhold responses after large changes in size (especially reduction) but no change in shape. As we have already mentioned, it is likely that categorization by the LHem determines the outcome, setting limits to the tolerable degree of transforma-tion of the key stimulus dimensions. The RHem apparently continues to assess the transformed stimulus as being to some degree similar to the posi-tive pattern.

Finally, when the positive pattern has a complex outline, and changes involve progressive stylization (i.e. loss of detail in the outline), it is the LHem that accepts greater change of the original, training stimulus (Bianki, 1983b). Here, it seems, the presence of some change in almost all features results in an estimate of very large change by the RHem, whereas the LHem relies on assessment of selected properties, in particular overall match in general outline. Again, this may depend on the ability of the LHem to attend to features defining a category, and so to accept major changes in the stimulus configuration.

This impressive body of work goes beyond establishing resemblance between rodent and human: it begins to extend our understanding of the general principles of vertebrate lateralization. In the rat, free from the complications introduced by LHem verbal abilities, it is possible to see more clearly that the LHem functions by categorizing objects using selected stimulus dimensions, such as size and general outline. The RHem tends to analyse in terms of all the properties of an object.

Recent studies in the chick (C. Jones, pers. comm.) have shown that when generalization along a single dimension (degree of rotation of a bar, in relation to the vertical) is studied, use of the LHem results in clear boundary values at which choice is concentrated, which shift progressively with experience, but remain sharp. It is likely that these are the current values by which the LHem defines a category. By contrast, use of the RHem results in choice, showing generalization that gradually decreases as the degree of transformation is increased.

3.4. Lateralized Control of Emotional Behaviour

3.4.1. *Intense Emotion and RHem*

Once again the chick makes a convenient starting point for this discussion. The use of the LE (and so of the RHem) of the chick facilitates attack, copulation and fear responses.

Following treatment of the young chick with testosterone, levels of attack and copulation are elevated, provided that the chick is tested either binocularly or monocularly using its LE. No effect of the testosterone treatment is evident when the same chicks are tested using the RE only (Rogers, Zappia and Bullock, 1985; Figure 3.2). Normal levels of these responses in chicks treated with testosterone are depressed by RE use. Thus, the LHem suppresses the attack and copulation response. Tests of agonistic behaviour in adult, untreated hens indicated the same role of the LHem in suppressing attack (Rogers, 1991).

RHem facilitation of fear behaviour is shown by the fact that lesioning of the right archistriatum reduces distress calling in an unfamiliar environment far more than corresponding left lesions (Phillips and Youngren, 1986). The archistriatum contains the homologue of the amygdala, the involvement of which in both fear and fear conditioning is well established for mammals (Maren, 1999). Adamec and Morgan (1994) have shown that kindling (i.e. chronic activation) of the right amygdala facilitates fear in the rat in a way which LHem kindling does not.

In humans, RHem control has been variously associated with intense emotions, or negative emotions or withdrawal (Davidson, 1995). Thus Dimond, Farrington and Johnson (1976) showed that the use of the left visual field to view film material resulted in a much more negative assessment than did right visual field viewing. Frontal and anterior regions of the RHem are selectively activated in withdrawn emotional states involving fear and disgust and, consistent with this, PET scans have revealed elevated activation of the RHem, during resting in panic-prone subjects (Davidson, 1995). Also, schizotypy with social and emotional withdrawal is associated with RHem dominance, seen in terms of scoring better memory of faces than words and poverty of speech (Gruzelier and Doig, 1996). Damage to the frontal region of the LHem, presumably forcing the equivalent region of the RHem to take control, leads to decreased interaction with other people and difficulty in initiating voluntary action (Davidson, 1995). Moreover, patients with injury to the left hemisphere resulting from stroke are significantly more depressed than those with equivalent injury to the right hemisphere (Robinson and Price, 1982; Robinson et al., 1984). RHem involvement in human emotion would, therefore, seem to be associated with negative and intense states, as well as with social withdrawal.

In contrast, the LHem has been assigned the opposite associations (i.e. positive emotions: Ahern and Schwartz, 1979, 1985; Davidson, 1992, 1995). Schizotypy with positive mood valence and eccentricity is associated with better memory of words than faces, indicating LHem dominance (Gruzelier and Doig, 1996). The inhibition of emotions by the LHem is sometimes suggested (e.g. Nestor and Safer, 1990). Clearly, this last position has much in common with the general vertebrate condition for which we argue here: that the LHem tends to inhibit responses like escape, attack and sexual behaviour. In fact, there is some evidence indicating that in humans the LHem also inhibits aggressive behaviour: subjects with epileptic seizures focused in the left temporal lobe (and so impaired LHem function) have higher than average levels of hostile feelings (Devinsky et al., 1994). Also, reduced activity in posterior regions of the LHem has been associated with suicidal and aggressive behaviour (Graae et al., 1996).

It is recognized in the human literature that there is an unresolved issue here: are all intense emotions, or only negative ones, associated with RHem control? The chick evidence suggests one way of resolving the issue: does RHem control go with facilitation of sexual behaviour in humans, as it does in the chick? Sexual behaviour is unlikely to be classified as involving 'negative' emotion in humans. There is extensive evidence that this is so in humans (review: Tucker and Frederick, 1989). Disturbed and exaggerated sexual behaviour is associated with RHem-related mania. RHem stroke is more likely to depress sexual function than LHem stroke. Sexual arousal is accompanied by central or posterior greater electroencephalogram (EEG) desynchronization in the RHem than in the LHem. Flor-Henry (1980) cites other evidence (e.g. RHem involvement in rapid eye movement sleep, and the typical accompaniment of such sleep by penile erection).

Evidence from other tetrapods is also consistent with RHem involvement in all intense (i.e uninhibited) emotional behaviour, rather than solely in negative emotional behaviour. For example, in *Bufo marinus* toads, aggressive tongue-strikes at conspecifics are more likely to occur when the conspecific is seen in the left visual hemifield than when it is seen with the right visual hemifield, indicating less inhibition, or activation, of aggressive responses by the RHem (Robins et al., 1998; Figure 3.1). In addition, those attack strikes that do occur to conspecifics on the toad's right side are aimed to avoid the recipient's eye (see Figure 3.1; more strikes at the eye occur to the left compared to the right). This bias may result from the LHem's ability to inhibit strikes directed at eyes on the right side, whereas the RHem is not able to do this.

In lizards, attack is again more likely when the conspecific is seen with the LE (Deckel, 1995). Since reptiles have much the same organization of the visual projections from the eyes to the brain as do birds, use of the LE means use of the RHem. Thus, in *Anolis,* aggression is initiated preferentially by the LE/RHem, but this aggression can be inhibited by the LHem in conditions that are mildly disturbing, and so call for such inhibition (Deckel, 1998).

The LE/RHem control of attack responses in the domestic fowl (both soon after hatching and in adulthood), which has already been described, is also revealed by differences between responses to conspecifics seen in the left or right monocular visual fields. Recently, Vallortigara et al. (2001) have shown that, when a young chick is temporarily paired with another that it has not seen before, in most cases the chick views the stranger using its left lateral field before pecking.

The preferential involvement of the right hemisphere in aggressive responses appears to have been conserved in a broad range of species, including primates. Gelada baboons (*Theropithecus gelada*) direct more agonistic responses to conspecifics on their left side than on the right (Casperd and Dunbar, 1996). In line with this finding, at least in terms of the presence of lateralization if not its direction, Drews (1996) found that carcasses of wild baboons (*Papio cynocephalus*) are marked by more injuries on the right side of the head region than on the left side. This finding is reminiscent of Jarman's (1972) earlier report of more scars on the right side of the pelts of impalas than on the left. Of course, lateralized occurrence of scars may depend on lateralized responding of the attacker or the one attacked but, comparing the results from these widely divergent species, the finding is suggestive of lateralization of aggressive responses.

The greater role of the RHem in initiating fear responses to novel stimuli has also been demonstrated in toads. Lippolis et al. (2001) tested fear responses of three species of toad (*Bufo bufo*, *Bufo viridis* and *Bufo marinus*) by introducing a simulated predator (a snake model) into the left or right lateral field. The toads were more reactive when the stimulus entered the left field than when it entered the right.

In rats, Robinson (1979, 1985) has shown that RHem lesions (infarcts, cortical undercuts and direct depression of noradrenergic activity) elevate activity in the open field, whereas corresponding LHem lesions have no effect. A rise in locomotion in the open field is likely to represent reduced immobility (freezing); a more general disinhibition is suggested by the fact that running wheel activity is also elevated. Robinson and Downhill (1995) compare these effects of RHem infarcts in rats, with effects of RHem insult in humans, such as general anxiety, without depression and secondary mania.

Finally, the association of RHem control with behaviour such as fear, aggression and sexual behaviour is paralleled by the fact that the RHem sympathetic outflow is the more effective, whereas parasympathetic outflow is under LHem control (Wittling, 1997). In humans (Hugdahl, 1995), whereas LHem controls the parasympathetic (vagal) outflow to the sinoatrial node (the heart 'pacemaker'), there is greater effectiveness of the sympathetic outflow from the RHem to the heart (via the stellate ganglia). In fact, the latter has been described for dogs, cats and humans (Lane and Jennings, 1995; Wittling, 1995). The stress hormone system (i.e. the hypothalamic–pituitary–adrenocortical axis) is, it appears, also under greater control by the RHem than the LHem (Wittling, 1997). Clearly, it is functionally appropriate for feed-forward preparations for exertion to accompany disinhibition of intense response.

It will be obvious that this preponderant involvement of the RHem in descending sympathetic outflow is paralleled by its own greater input of noradrenergic fibres. This is provided by neurones in the *locus coeruleus*, which in the rat are activated by startling and painful stimuli, but by little else (Aston-Jones et al., 1986). The RHem thus is affected more than the LHem by startling stimuli, as well as by the cognitive detection of novelty.

3.4.2. Inhibition of Emotional Behaviour by the LHem

It is argued here that LHem mechanisms provide inhibitory influences on behaviour such as fear and attack, and that the LHem is used when there is a need to assess the situation before taking a decision. Inhibition by the LHem of particular responses should be distinguished from states of general inhibition associated with fear and depression. Depression in humans, and immobility in fear (freezing) in the rat, are both actively organized (and probably comparable) conditions. It is known, in the rat, that they have direct dependence on a specific midbrain system, the ventrolateral periaqueductal gray (Bandler and Shipley, 1994). The inhibition of intense emotion can thus result in the removal of inhibition (e.g. of locomotion) at a lower level, as has already been noted for the effects of RHem lesions on locomotion in the rat.

However, such complications in the interpretation of findings obtained from studies of rats were avoided by a study of Adamec and Morgan (1994), in which unilateral kindling of the amygdala resulted in appropriate shifts in the amount of locomotion in open and covered parts of an elevated maze. Anxiety is known to increase the relative time spent under cover: RHem activation had this effect (suggesting that such activation increased fear), whereas LHem activation produced the opposite shift from control levels, indicating reduced fear.

Further evidence of inhibition of intense response by the LHem in the rat is provided by mouse killing, which is substantially elevated by LHem lesions, but not RHem lesions (Denenberg, 1984). Note that the tested rats were not experienced killers, and so they were faced with a stimulus that both presented strong releasers for the behaviour and was sufficiently unusual as to call for careful assessment with concomitant inhibition of response.

Asymmetries of facial expression provide further relevant evidence. Hauser (1993) reports more vigorous development of facial expressions on the left side of the face in rhesus monkeys, as mentioned earlier (see also Chapter 13 by Weiss et al.). All three of the expressions that showed significant asymmetry were related to fear or threat. It is possible that the left side of the face shows greater intensity of expression only when emotion is intense. Hook-Costigan and Rogers (1998) found that in marmosets there was greater intensity on the left for two fear expressions but on the right for a social contact call (the twitter). This appears to be the first demonstration in a primate other than humans of greater LHem control of production of a vocalization. Since the call is an affiliative social signal, it is likely that the marmoset at the same time inhibits the expression of intense emotion (e.g. withdrawal) in order to make social contact. Control by the LHem is, there-fore, entirely consistent with our previous argument for its role in inhibition of intense emotions.

The human evidence is, at least partially, inconsistent with this general pattern. Borod, Koff and Caron (1983) found greater intensity on the left side of the face for expressions that include greeting and clowning, as well as horror, grief and disgust. However, the evidence of association of affiliative behaviour with LHem control is explicable by the hypothesis that is advanced here. Much of the difficulty of comparing human and animal evidence arises from the existence in humans of states of amusement. The discussion may conveniently begin with Gainotti (1972, 1989), who summarized evidence that the diagnosis of LHem insult often evoked (appropriately enough) a 'depressive catastrophic' reaction in the patient: in other words the response was great disturbance and depression. A comparable diagnosis involving RHem lesion tended to be accompanied by 'denial of illness' and joking (Gainotti, 1979). The interpretation of these findings depends on whether it can be safely assumed that the dominant effect of the brain damage was to increase control by the intact hemisphere. This is made almost certain by evidence (see below) from normal subjects, using behavioural or brain acti-vation measures.

The association between LHem control and behaviour accompanied by laughter (humour, social interaction) is strengthened by the fact that patho-logical activation of the LHem, either by epileptic seizures or by RHem damage, is likely to be accompanied by uncontrolled laughter, whereas cor-responding RHem activation is more likely to be accompanied by crying (Sackeim et al., 1982). Evidence from normal subjects is in good agreement: Ahern and Schwartz (1979) found that positive emotional content to ques-tions brought about LHem involvement (as shown by rightward eye move-ments), whereas negative content involved the RHem. A specific connection of LHem functioning with laughter was revealed by Ahern and Schwartz (1985): when subjects thought about laughing there was LHem EEG activation frontally, whereas thought about fear produced such activation in the RHem.

A remarkable finding is that by repeatedly performing movements of the right or left side of the mouth, or the right or left hand, emotional state can be affected (Schiff and Lamon, 1989, 1994). Left side movement produces sadness (and even weeping), whereas right side produces emotion described as 'sarcastic, cocky, good, smug'.

Finally, there are studies in which stimuli are assessed according to their pleasantness. Painful or near-painful stimuli are judged more unpleasant when applied to structures of the left side (Schiff and Gagliese, 1994); this is true for both chronic (shoulder pain) and acute (hand in ice water) conditions. Ehrlichman (1986) found that odours presented to the right nostril (RHem input) are rated as more unpleasant than when they are presented to the left. This may also be the case in chicks: a chick will shake its head in a disgust response when it detects a noxious odour with its right nostril (left nostril occluded) but not when they use the left nostril (Rogers, Andrew and Burne, 1998). It should be noted here that the neural inputs from the olfactory epithelium of each nostril project to their ipsilateral hemisphere and do not cross over the midline to the contralateral hemisphere, as in the case of other sensory inputs. Thus, use of the right nostril reflects processing of olfactory information in the right RHem.

The subjective sensations described by humans during LHem control are beyond examination in animals. However, a humorous, joking or sarcastic approach allows humans to examine and evaluate stimuli and situations, which under other circumstances they would find too disturbing for rational treatment. Two components of 'humorous' states should be distinguished. Firstly, there is reduced likelihood of terminating examination whilst the state lasts. Secondly, the experience is not remembered as one to be avoided in the future; indeed, it may be subsequently sought after rather than avoided. Both are made likely in humans by positive affect; neither is likely in the absence of the special conditions of amusement.

The first component (i.e. reduced likelihood of terminating examination) is clearly present when animals persistently examine frightening and potentially dangerous objects (when RE and LHem are usually involved). The second (i.e. remembering the experience as being positive) may well also hold, since such viewing is likely to be adaptive; if it were accompanied by negative reinforcement, the animal would presumably learn not to do it again.

Association of laughter with emotional states, arising when a potentially disturbing experience is assessed as amusing or unimportant, is no doubt one mechanism for minimizing social disruption in humans by means of the effect of laughter on others. Behaviour like laughter may provide a way of examining this aspect of LHem inhibition of intense response in animals. It may turn out to be the case that the association of the social, twitter call with right side intensity of expression in marmosets (LHem control) is a first example of a general vertebrate condition. It certainly seems likely that LHem inhibition of behaviour like attack, sexual behaviour and escape will be usual during normal relaxed social behaviour, and in particular in greeting or reconciliation.

A final point is that it is likely that humour involves hemispheric interaction: there are indications that excessive activation of the LHem (e.g. in left temporal epilepsy: Tucker and Frederick, 1989) goes with humourless and obsessional thought. It may be that humour depends on RHem activation, followed by its control by the LHem, just as, when an animal assesses a genuine danger, escape is indeed activated, but for the moment is inhibited.

3.5. Handedness

The topic of handedness is discussed in other chapters (e.g. Chapter 4 by Rogers, and Chapter 12 by Hopkins and Carriba). Here we are concerned to see how evidence from the use of hands may be related to the evolution of lateralization in general in the tetrapods. The arguments advanced by MacNeilage, Studdert-Kennedy and Lindblom (1987) set the study of primate handedness on a firm basis for the first time and developed what has become known as the 'Postural Origins Theory'. In summary, the 1987 paper argued that consistent handedness at a population level could be demonstrated in primates in general, if the task for which a hand was being used was carefully taken into account. Left-hand use for strikes at prey appeared in prosimians over 50 million years ago because they involve the 'use of ballistic reaching movements'. MacNeilage (1998) ascribed this RHem control of strikes to a (presumably more ancient) RHem 'specialization for apprehension of the world for basic survival purposes'. This is specifically a 'visuospatiomotor' specialization.

The evidence already reviewed is at first sight discrepant with this hypothesis. In amphibians (toads: Vallortigara et al., 1998) and birds (Andrew et al., 2000) there is a bias to take food objects that are in the right visual field, not the left. This might have been expected to put the right hand in control of seizing food. The solution is probably in the way that the food is taken. Toads and birds use a medial effector, the tongue or beak, and therefore take up frontal fixation immediately before striking, although the decision to strike is made, most often, when the stimulus is in the right lateral visual field.

The main source of bias to control by one or other hemisphere is LHem control of a planned motor response.

Once a hand is used, it becomes more important to bring the hand to bear on the prey or food object rather than to point the head continuously at the prey. As a result, striking at a moving prey may occur with the prey in either visual hemifield (rather than in a predictable and constant position, frontal to the head). Under these circumstances, the RHem advantage in attention to both visual hemifields (as compared to the LHem focusing attention within the right hemifield) would be likely to put the RHem in control of striking and so lead to use of the left hand. We note also that cats show a left-paw preference when reaching for a moving target (Lorincz and Fabre-Thorpe, 1996).

This effect seems to operate throughout the primates. When striking with the hand at a moving (and difficult) target, namely a swimming goldfish, squirrel monkeys use their left hand, even though there is no such clear bias in other tasks (King and Landau, 1993). Dextral humans, with marked right-hand advantage when throwing a ball, show no such advantage when catching it (Watson and Kimura, 1989). Humans produce ballistic movements faster with their left hand (Guiard, Diaz and Beaubaton, 1983). Similarly, the prosimian, *Galago moholi*, performs its ballistic method of reaching for live prey faster when using the left hand than when using the right (Ward, 1999). This response is particularly relevant because some 80 per cent of the diet of galagos consists of insects and the reaching used has been referred to as 'smash and grab' (Bishop, 1964) with the ballistic motor action planned before its execution (Ward, 1999). The decision to grab and control of the action is, therefore, very likely to be controlled by the RHem.

The second major aspect of MacNeilage's theory is that specialization of the two hands for different purposes was present in prosimians, because their habitual posture required one hand to grasp a support, whilst the other was used to strike at prey. The consequent use of the right hand to grasp predisposed it in later evolution towards greater strength and, according to the Postural Origins Theory, towards a capacity for fine somatic sensorimotor control. Both factors may indeed have operated, but it seems likely that the pre-existing specialization of the LHem for the visual control of planned movements was more important in promoting manipulation by the right hand. Once visually controlled manipulation, starting with a carefully shaped grasp, became the main mode of hand use, such LHem control would have been strongly promoted.

Foot use in birds raises some of the same issues. Here too postural control appears to be important. In chicks, the main asymmetrical use of a foot is in scratching. This is a relatively automatic behaviour that is used when the substrate is covered with material that may be hiding food. It is unlikely that visual control (beyond selection of the area in which the chick is standing) is important. Nevertheless, asymmetry is clear in the choice of foot with which to begin scratching (Tommasi and Vallortigara, 1998). When both eyes are in normal use, 10–15-day-old chicks initiate a bout of scratching with the right foot (Rogers and Workman, 1993). Tommasi and Vallortigara (1998) found the same right foot preference to start scratching in 16-day-old chicks. When one or other eye was covered, it was the foot contralateral to the eye in use that took the lead in scratching. In other words, the 'seeing hemisphere' (i.e. the one receiving most of the monocular visual input) uses the foot directly under its control for support.

In the monocular condition, it is very likely that the controlling hemisphere is that contralateral to the eye in use. Therefore, if there were any visual control of the scratching itself, this would have to be from the hemisphere contralateral to the eye in use. Instead scratching is with the foot ipsilateral to the eye in use. The authors therefore argue that the difficult part of the task is not the foot stroke in scratching, but the postural demands of balancing on one foot (compounded by the forces generated by scratching). If visual cues are used to stabilize the animal at such times, which is likely in view of the importance of visual cues in stabilizing the head in adult fowl during loco-motion (Davies and Green, 1988), then there is a direct reason why the seeing hemisphere should determine which foot supports.

Why then did binocular chicks lead with the right foot and balance on the left? The answer is probably that the chicks had to find and then search in the centre of the test arena. Since the floor was covered with sawdust, the most challenging part of the task was to judge the position of the centre, and this resulted in RHem control.

Parrots are of particular interest in that they hold food and other objects with one foot whilst manipulating it with the bill. The majority of species of parrots examined so far have been found to hold food objects in the left foot and balance on the right foot (Harris, 1989; Rogers, 1981, 1996). This means that they use the opposite foot for balance compared to chicks. In the case of feeding in parrots, however, holding and manipulating the food with the foot and relating this to use of the beak requires constant visual guidance. This is shown directly by the way in which parrots continually tilt the head to view the food object with the eye ipsilateral to the foot used for holding (i.e. the LE and left foot). In most species of parrot, therefore, the RHem is used both for viewing the food, and for the motor acts required in both holding and manipulating it. This is typically the case in Australian parrots that feed on large acacia seed heads, each containing several seeds that need to be extracted. It is likely that there is a need to compare information from the grasping with visual information obtained with the ipsilateral eye and tactile information from the beak as it probes the seed head. A similar process occurs in manipulation by humans: the left hand typically holds and turns, whilst the right pokes and pulls. The parrots may thus be using the RHem because the task is a difficult spatial task requiring intermodal comparison of information.

The rosellas (*Platycercus* spp.) are exceptions to this rule: they show preferential use of the right foot to hold food objects (Cannon, 1983; Rogers, 1980). The exceptional condition shown by these species may depend on the different type of food that they eat (small seed heads of grass) or the relative instability of their perch (most often, grass stalks). The LHem is likely to be dominant when a series of small targets are being grasped with the bill in rapid succession.

This may therefore lead to RHem control of posture and the use of the left foot to grasp the perch. Further work on foot use in parrots is likely to be very rewarding.

Finally, something needs to be said about handedness and preferred direction of rotation in rats and mice. This topic was put to one side at the beginning of the chapter because these asymmetries clearly vary independently of the consistent lateralization of the type standard for tetrapods, and which is also present in rats (see above). The asymmetries in rotation are known to depend upon asymmetries of function of the nigrostriatal dopaminergic system, with rotation tending to occur towards the side of the brain on which that system is less active (Glick, Jerussi and Zimmerberg, 1977). Rotation is also usually, but not invariably, towards the side of the preferred paw (e.g. that preferred in a two-lever Skinner box).

Initially (Glick, Jerussi and Zimmerberg, 1977), it was speculated that the presence of bias had a function quite independent of its direction (which was thought to be functionless). The function that was suggested was odd, namely that it allowed proprioceptive feedback that enabled the animal to distinguish left from right. The specificity of neural connections from the periphery to the central nervous system, and the presence of entire systems like the optic tectum (superior colliculus) devoted to targeting of laterally placed stimuli seem entirely adequate to distinguish left from right for both reception and response.

However, it soon became clear that the independence of variation between individuals in nigrostriatal and frontal asymmetries of function generates different behavioural phenotypes (Glick and Ross, 1981). Lateral preferences are said to be clearer when frontal and nigrostriatal biases are in the same direction; since the frontal bias is consistent across populations, such animals must predominantly have higher nigrostriatal activity in the LHem. Glick and Ross (1981) found that right rotators do, indeed, differ from left rotators in that they show much higher levels of spontaneous activity. This is what the findings of Robinson (see above) would predict, with LHem control resulting in chronic inhibition of freezing. Right rotators also show a stronger bias to their preferred direction of rotation.

Other relationships between the degree of side preference and performance in tests are likely also to be determined (at least in part) by the degree of LHem control. Glick and Cox (1976) found that consistent side preferences (i.e. a greater bias to LHem control) are associated with greater ability to withhold response in DRL schedules. There was also evidence in such rats of better ability to time the required delay, which Glick and Ross associate with performance of stereotypies performed by the rat whilst it is waiting to respond. Similar features of LHem control (e.g. ability to withhold response for the correct interval in delayed response schedules) were described by Bianki (see above). Glick and Ross also showed that consistent side preferences went with better performance on fixed ratio schedules, where this required high rates of responding, rather than inhibiting response whilst waiting.

There is little doubt that differences of these kinds between individuals will be subject to natural selection, as well as learning. Why they continue to be generated at strikingly high frequencies is considered in Chapter 5 by Andrew. Chapter 5 also considers sex differences, the existence of which further confirms that interindividual differences are functional. Here it will be enough to note that the evidence for the existence, in rats, of the standard vertebrate pattern of lateralization suggests that different behavioural phenotypes differ in the way in which lateralized abilities are brought to bear in a range of different situations. Since lateralization is intimately involved in every aspect of behaviour, its expression will be subject to intense selection pressures, often no doubt discordant and divergent, and changing with age and environmental conditions. Given the wide range of selection pressures affecting lateralization, the fact that all vertebrates have the same basic pattern of lateralization may be seen as a problem. How do the different species in their different habitats retain the same basic pattern of lateralization? One solution, at least in tetrapods, may have been to use modulation of the development of lateralization by the environment to make adjustments to conditions that hold only for the life of an individual and so do not require the genetic basis of lateralization to change.

A whole set of unresolved issues is raised by the existence of variations in the patterns of lateralization, however generated. There is, clearly, varying bias as to which hemisphere tends to take charge of behaviour. One straightforward consequence of this is that the animal will attend to different aspects of the environment and of other animals according to which hemisphere is in control. The age-dependent shifts in bias in the chick appear to allow learning about the environment and social fellows at different and appropriate times (Dharmaretnam and Andrew, 1994; Chapter 5 by Andrew).

However, it is not clear how a long-term shift to control by the RHem might be accommodated since control of response is the basic specialization of the LHem. This control by the LHem includes categorizing stimuli by selected cues in order to determine what is the appropriate response, and inhibition of behaviour while this is being done and a decision is being made. Perhaps bias to RHem control is accompanied by changes in the interaction between hemispheres so that emission of an appropriate response continues to involve LHem mechanisms. It will be of great interest to study various kinds of reversal of lateralized functions, as they become available, to see exactly how the functioning of lateralized mechanisms is adjusted.

References

Adamec, R.E. & Morgan, H.D. (1994). The effect of kindling of different nuclei in the left and right amygdala in the rat. *Physiology and Behavior*, 55, 1–12.

Adelstein, A. & Crowne, D.P. (1991). Visuospatial asymmetries and interocular transfer in the split-brain rat. *Behavioural Neuroscience*, 105, 459–469.

Ahern, G.L. & Schwartz, G.E. (1979). Differential lateralization for positive versus negative emotion. *Neuropsychologia*, 17, 693–698.

Ahern, G.L. & Schwartz, G.E. (1985). Differential lateralization for positive and negative emotion in the human brain: EEG spectral analysis. *Neuropsychologia*, 23, 745–755.

Andrew, R.J. (1983). Lateralization of emotional and cognitive function in higher vertebrates, with special reference to the domestic chick. In *Advances in Vertebrate Neuroethology*, ed. J.P. Ewert, R.R. Capronica & D. Ingle, pp. 477–509. New York: Plenum Press.

Andrew, R.J. (1991). The nature of behavioural lateralization in the chick. In *Neural and Behavioural Plasticity: The Use of the Domestic Chick as a Model*, ed. R.J. Andrew, pp. 536–554. Oxford: Oxford University Press.

Andrew, R.J., Tommasi, L. & Ford, N. (2000) Motor control by vision and the evolution of cerebral lateralization. *Brain and Language*, 73, 220–235.

Archibald, J.D. (1996). Fossil evidence for a late Cretaceous origin of 'hooved' mammals. *Science*, 272, 1150–1152.

Aston-Jones, G., Ennis, M., Pieribone, V.A., Nickell, W.T. & Shipley, M.T. (1986). The brain nucleus *locus coeruleus*: restricted afferent control of a broad efferent network. *Science*, 234, 734–737.

Bandler, R. & Shipley, M.T. (1994). Columnar organisation in the midbrain periaqueductal gray: modules for emotional expression? *Trends in Neuroscience*, 17, 379–389.

Bianki, V.L. (1982). Lateralization of concrete and abstract characteristics analysis in the animal brain. *International Journal of Neuroscience*, 17, 233–241.

Bianki, V.L. (1983a). Hemisphere specialization in the animal brain for information processing principles. *International Journal of Neuroscience*, 20, 75–90.

Bianki, V.L. (1983b). Hemisphere lateralization of inductive and deductive processes. *International Journal of Neuroscience*, 20, 59–74.

Bianki, V.L. (1988). *The Right and Left Hemispheres of the Animal Brain: Cerebral Lateralization of Function*. Monographs in Neuroscience, Vol. 3. New York: Gordon and Breach.

Bishop, A. (1964). Use of the hand in lower primates. In *Evolutionary and Genetic Biology of Primates*, Vol. 37, ed. J. Buettner, pp. 133–225. New York: Academic Press.

Borod, J.C., Koff, E. & Caron, H.S. (1983). Right hemisphere specialisation for the expression and appreciation of emotion: a focus on the face. In *Cognitive Processing in the Right Hemisphere*, ed. E. Perecman, pp. 83–110. Orlando: Academic Press.

Bradshaw, J.L. & Rogers, L.J. (1993). *The Evolution of Lateral Asymmetries, Language, Tool Use, and Intellect*. San Diego: Academic Press.

Cannon, C.E. (1983). Descriptions of foraging behaviour of eastern and pale-headed rosellas. *Bird Behaviour*, 4, 63–70.

Casperd, J.M. & Dunbar, R.I.M. (1996). Asymmetries in the visual processing of emotional cues during agonistic interactions in gelada baboons. *Behavioral Processes*, 37, 57–65.

Collins, R.L. (1985). On the inheritance of direction and degree of asymmetry. *Cerebral Lateralization in Nonhuman Species*, ed. S.D. Glick, pp. 41–71. Orlando: Academic Press.

Clayton, N.S. & Krebs, J.R. (1994). Memory for spatial and object-specific cues in food-storing and non-storing birds. *Journal of Comparative Physiology A*, 174, 371–379.

Cowell, P.E., Waters, N.S. & Denenberg, V.H. (1997). Effects of early environment on the development of functional laterality in Morris maze performance. *Laterality*, 2, 221–232.

Cozzutti, C. & Vallortigara, G. (2001). Hemispheric memories for the content and position of food caches in the domestic chick. *Behavioral Neuroscience*, 115, 305–313.

Cynx, J., Williams, H. & Nottebohm, F. (1992). Hemispheric differences in avian song discrimination. *Proceedings of the National Academy of Sciences USA*, 89, 1372–1375.

Davidson, R.J. (1992). Emotion and effective style. *Psychological Science*, 3, 39–43.

Davidson, R.J. (1995). Cerebral asymmetry, emotion, and affective style. In *Brain Asymmetry*, ed. R.J. Davidson & K. Hugdahl, pp. 361–387. Cambridge, MA: MIT Press.

Davies, M.N.O. & Green, P.R. (1988). Head-bobbing during walking, running and flying: Relative motion perception in the pigeon. *Journal of Experimental Biology*, 138, 71–91.

Deckel, A.W. (1995). Laterality of aggressive responses in *Anolis*. *Journal of Experimental Zoology*, 272, 194–200.

Deckel, A.W. (1998). Hemispheric control of territorial aggression in *Anolis carolinensis*: effects of mild stress. *Brain, Behavior and Evolution*, 51, 33–39.

Deng, C. & Rogers, L.J. (1997). Differential contributions of the two visual pathways to functional lateralization in chicks. *Behavioural Brain Research*, 87, 173–182.

Denenberg, V.H. (1981). Hemispheric laterality in animals and the effects of early experience. *The Behavioral and Brain Sciences*, 4, 1–49.

Denenberg, V.H. (1984). Effects of right hemisphere lesions in rats. In *The Right Hemisphere: Neurology and Neuropsychology*, ed. A. Ardila & F. Ostrovsky-Solis, pp. 241–262. New York: Gordon and Breach.

Denenberg, V.H., Garbanati, J.A., Sherman, G.F., Yutzey, D.A. & Kaplan, R. (1978). Infantile stimulation induces brain lateralization in rats. *Science*, 201, 1150–1152.

Deruelle, C. & Fagot, J. (1997). Hemispheric lateralization and global precedence effects in the processing of visual stimuli by humans and baboons (*Papio papio*). *Laterality*, 2, 2233–2246.

Devinsky, O., Ronsaville, D., Cox, C., Witt, E., Fedio, P. & Theodore, W.H. (1994). Interictal aggression in epilepsy: The Buss–Durkee hostility inventory. *Epilepsia*, 35, 585–590.

Dharmaretnam, M. & Andrew, R.J. (1994). Age- and stimulus-specific use of right and left eyes by the domestic chick. *Animal Behaviour*, 48, 1395–1406.

Dimond, S.J., Farrington, L. & Johnson, P. (1976). Differing emotional response from right and left hemispheres. *Nature*, 261, 690–692.

Drews, C. (1996). Context and patterns of injuries in free-ranging male baboons (*Papio cynocephalus*). *Behaviour*, 133, 443–474.

Eaton, R.C. & Emberley, D.S. (1991). How stimulus direction determines the trajectory of the Mauthner initiated escape response in a teleost fish. *Journal of Experimental Biology*, 161, 469–487.

Eden, C.G. van, Uylings, H.B.M. & Pelt, J. van. (1984). Sex-difference and left–right asymmetry in the prefrontal cortex during postnatal development in the rat. *Developmental Brain Research*, 12, 146–153.

Ehrlichman, H. (1986). Hemispheric asymmetry and positive-negative affect. In *Duality and Unity of the Brain*, ed. D. Ottoson, pp. 194–206. Dordrecht, The Netherlands: Kluwer.

Fagot, J. & Vauclair, J. (1988a). Handedness and bimanual coordination in the lowland gorilla. *Brain, Behavior and Evolution*, 32, 89–95.

Fagot, J. & Vauclair, J. (1988b). Handedness and manual specialization in the baboon. *Neuropsychologia*, 26, 795–804.

Fersen, L. von & Güntürkün, O. (1990). Visual memory lateralization in pigeons. *Neuropsychologia*, 28, 1–7.

Flor-Henry, P. (1980). Cerebral aspects of the orgasmic response: normal and deviational. In *Medical Sexology*, ed. R. Forlec & W. Pasini, pp. 256–262. Elsevier: Amsterdam.

Gainotti, G. (1972). Emotional behaviour and hemispheric side of the lesion. *Cortex*, 8, 41–55.

Gainotti, G. (1979). The relationship between emotion and cerebral dominance: A review of clinical and experimental evidence. In *Hemisphere Asymmetries of*

Function in Psychopathology, ed. P. Flor-Henry, pp. 21–34. Elsevier: Amsterdam.

Gainotti, G. (1989). Disorders of emotions and affect in patients with unilateral brain damage. In *Handbook of Neuropsychology*, Vol. 3, ed. F. Boller & J. Grafman, pp. 345–381. Elsevier: Amsterdam.

Glick, S.D. & Cox, R.D. (1976). Differential effects of unilateral and bilateral caudate lesions on side preferences and timing behaviour in rats. *Journal of Comparative and Physiological Psychology*, 90, 528–535.

Glick, S.D. & Cox, R.D. (1978). Nocturnal rotation in normal rats: Correlation with amphetamine-induced rotation and effects of nigrostriatal lesions. *Brain Research*, 150, 149–161.

Glick, S.D. & Ross, D.A. (1981). Right-sided population bias and lateralization of activity in normal rats. *Brain Research*, 205, 222–225.

Glick, S.D., Jerussi, T.P. & Zimmerberg, B. (1977). Behavioural and neuropharmacological correlates of nigrostriatal asymmetry in rats. In *Lateralization in the Nervous System*, ed. S. Harnad, R.W. Doty, L. Goldstein, J. Jaynes & G. Krauthamer, pp. 213–249. Orlando: Academic Press.

Graae, F., Tenke, C., Bruder, G., Rotheram, M.-J., Piacentini, D. C.-B., Leite, P. & Towey, J. (1996). Abnormality of EEG alpha asymmetry in female adolescent suicide attempters. *Biological Psychiatry*, 40, 706–713.

Gruzelier, J.H. & Doig, A. (1996). The factorial structure of schizotypy: Part II. Cognitive asymmetry, arousal, handedness, and sex. *Schizophrenia Bulletin*, 22, 621–634.

Guiard, Y., Diaz, G. & Beaubaton, D. (1983). Left hand advantage in right handers for spatial constant error: Preliminary evidence in a manual ballistic aimed movement. *Neuropsychologia*, 21, 111–115.

Hamilton, C.R. & Vermeire, B.A. (1983). Discrimination of monkey faces by split-brain monkeys. *Behavioural Brain Research*, 9, 263–275.

Hamilton, C.R. & Vermeire, B.A. (1988). Complementary hemispheric specialisation in monkeys. *Science*, 242, 1694–1696.

Harris, L.J. (1989). Footedness in parrots: Three centuries of research, theory, and mere surmise. *Canadian Journal of Psychology*, 43, 369–396.

Hauser, M.C. (1993). Right hemisphere dominance in the production of facial expression in monkeys. *Science*, 261, 475–477.

Hook-Costigan, M.A. & Rogers, L.J. (1998). Lateralized use of the mouth in production of vocalisations by marmosets. *Neuropsychologia*, 36, 1265–1273.

Hopkins, W.D. & Morris, R. (1989). Laterality for visual-spatial processing in two language-trained chimpanzees (*Pan troglodytes*). *Neuropsychologia*, 35, 343–348.

Howard, K.J., Rogers, L.J. & Boura, A.L.A. (1980). Functional lateralization of the chicken forebrain revealed by use of intracranial glutamate. *Brain Research*, 188, 369–382.

Hugdahl, K. (1995). Classical conditioning and implicit learning: The right hemisphere hypothesis. In *Brain Asymmetry*, ed. R.J. Davidson & K. Hugdahl, pp. 235–267. Cambridge, MA: MIT Press.

Hunt, G.R. (2000). Human-like, population-level specialisation in the manufacture of pandanus leaves by New Caledonian crows *Corvus moneduloides*. *Proceedings of the Royal Society London B*, 267, 403–413.

Ingle, D.J. & Hoff, K. vS. (1990). Visually evoked evasive behaviour in frogs. *BioScience* 40, 284–291.

Jarman, P.J. (1972). The development of the dermal shield in impala. *Journal of Zoology*, 166, 349–356.

Kimura, D. (1982). Left-hemisphere control of oral and brachial movements and their relation to communication. *Philosophical Transactions of the Royal Society of London B*, 298, 135–149.

King, V.R. & Corwin, J.V. (1992). Spatial deficits and hemispheric asymmetries in the rat following unilateral and bilateral lesions of parietal or medial agranular cortex. *Behavioural Brain Research*, 50, 53–68.

King, J.E. & Landau, V.I. (1993). Manual preferences in varieties of reaching in squirrel monkeys. In *Primate Laterality: Current Evidence of Primate Asymmetries*, ed. J.P. Ward & W.D. Hopkins, pp. 107–124. New York: Springer-Verlag.

Kosslyn, S.M., Chabris, C.F., Marsolek, C.J. & Koenig, O. (1992). Categorical versus coordinate spatial relations; computational analyses and computer simulations. *Journal of Experimental Psychology: Human Perception and Performance*, 18, 562–577.

Lacreuse, A. & Fragaszy, D.M. (1999). Left hand preferences in capuchins (*Cebus apella*): Role of spatial demands in manual activity. *Laterality*, 4, 65–78.

Lane, R.D. & Jennings, J.R. (1995). Hemispheric asymmetry, autonomic asymmetry and the problem of sudden cardiac death. In *Brain Asymmetry*, ed. R.J. Davidson & K. Hugdahl, pp. 271–304. Cambridge, MA: MIT Press.

Lippolis, G., Bisazza, A., Rogers, I.J. & Vallortigara, G. (2001). Lateralization of predator avoidance responses in three species of toads. *Laterality*, in press.

Lorincz, E. & Fabre-Thorpe, M. (1996) Shift of laterality and compared analysis of paw performances in cats during practice of a visuomotor task. *Journal of Comparative Psychology*, 110, 307–315.

MacNeilage, M.G. (1998). Towards a unified view of cerebral hemispheric specialisations in vertebrates. In *Comparative Neuropsychology*, ed. A. Milner, pp. 167–183. Oxford: Oxford University Press.

MacNeilage, P.F., Studdert-Kennedy, M.G. & Lindblom, B. (1987). Primate handedness reconsidered. *The Behavioural Brain Sciences*, 10, 247–263.

Maren, S. (1999). Long-term potentiation on the amygdala: a mechanism for emotional learning and memory. *Trends in Neuroscience*, 22, 561–567.

McKenzie, R., Andrew, R.J. & Jones, R.B. (1998). Lateralization in chicks and hens: New evidence for control of response by the right eye system. *Neuropsychologia*, 36, 51–58.

Miklósi, A. & Andrew, R.J. (1999) . Right eye use associated with decision to bite in zebrafish. *Behavioural Brain Research*, 105, 199–205.

Miklósi, A., Andrew, R.J. & Dharmaretnam, M. (1996). Auditory lateralization: Shifts in ear use during attachment in the domestic chick. *Laterality*, 1, 215–224.

Mittleman, G., Whishaw, I.Q. & Robbins, T.W. (1988). Cortical lateralization of function in rats in a visual reaction time task. *Behavioural Brain Research*, 31, 29–36.

Nestor, P.G. & Safer, M.A. (1990). A multi-method investigation of individual differences in hemisphericity. *Cortex*, 26, 409–421.

Nottebohm, F. (1970). Ontogeny of bird song. *Science*, 176, 950–956.

Phillips, R.E. & Youngren, O.M. (1986). Unilateral kainic acid lesions reveal dominance of right archistriatum in avian fear behaviour. *Brain Research*, 377, 216–220.

Posner, M.J. & Petersen, S.E. (1990). The attention system of the human brain. *Annual Review of Neuroscience*, 13, 25–42.

Rashid, N. & Andrew, R.J. (1989). Right hemisphere advantage for topographical orientation in the domestic chick. *Neuropsychologia*, 27, 937–948.

Robins, A., Lippolis, G., Bisazza, A., Vallortigara, G. & Rogers, L.J. (1998).
  – Lateralized agonistic responses and hind-limb use in toads. *Animal Behaviour*, 56, 875–881.

Robinson, R.G. (1979). Differential behavioural and biochemical effects of right and left hemispheric cerebral infarction in the rat. *Science*, 205, 707–710.

Robinson, R.G. (1985). Lateralized behavioural and neurochemical consequences of unilateral brain injury in rats. In *Cerebral Lateralization in Nonhuman Species*, ed. S.D. Glick, pp. 135–156. Orlando: Academic Press.

Robinson, R.G. & Downhill, J.E. (1995). Lateralization of psychopathology in response to focal brain injury. In *Brain Asymmetry*, ed. R.J. Davidson & K. Hugdahl, pp. 693–711. Cambridge, MA: MIT Press.

Robinson, R.G. & Price, T.R. (1982). Post-stroke depressive disorders: A follow-up study of 103 patients. *Stroke*, 13, 635–641.

Robinson, R.G., Kubos, K.L., Starr, L.B., Rao, K. & Price, T.R. (1984). Mood disorders in stroke patients: Importance of location of lesion. *Brain*, 107, 81–93.

Rogers, L.J. (1980). Lateralization in the avian brain. *Bird Behaviour*, 2, 1–12.

Rogers, L.J. (1981). Environmental influences on brain lateralization. *The Behavioral Brain Sciences*, 4, 35–36.

Rogers, L.J. (1991). Development of lateralization. In *Neural and Behavioural Plasticity: The Use of the Domestic Chick as a Model*, ed. R.J. Andrew, pp. 507–535. Oxford: Oxford University Press.

Rogers, L.J. (1996). Behavioral, structural and neurochemical asymmetries in the avian brain: A model system for studying visual development and processing. *Neuroscience and Biobehavioral Reviews*, 20, 487–503.

Rogers, L.J. (1997). Early experiential effects on laterality: Research on chicks has relevance to other species. *Laterality*, 2, 199–219.

Rogers, L.J. & Anson, J.M. (1979). Lateralization of function in the chicken forebrain. *Pharmacology, Biochemistry and Behaviour*, 10, 679–686.

Rogers, L.J. & Workman, L. (1993). Footedness in birds. *Animal Behaviour*, 45, 409–411.

Rogers, L.J., Andrew, R.J. & Burne, T.H.J. (1998). Light exposure of the embryo and development of behavioural lateralization in chicks: 1. Olfactory responses. *Behavioural Brain Research*, 97, 195–200.

Rogers, L.J., Zappia, J.V. & Bullock, S.P. (1985). Testosterone and eye–brain asymmetry for copulation in chickens. *Experimentia*, 41, 1447–1449.

Ross, D.A. & Glick, S.D. (1981). Lateralized effects of bilateral frontal cortex lesions in rats. *Brain Research*, 210, 379–382.

Rushworth, M.F.S., Nixon, P.D., Wade, D.T., Renowden, S. & Passingham, R.E. (1998). The left hemisphere and the selection of learned actions. *Neuropsychologia*, 36, 11–24.

Sackeim, H.A., Greenberg, M.S., Weiman, A.L., Gur, R.C., Hungerbuhler, J.P. & Geschwind, N. (1982). Hemispheric asymmetry in the expression of positive and negative emotions: neurological evidence. *Archives of Neurology*, 39, 210–218.

Schiff, B.B. & Gagliese, L. (1994). The consequences of experimentally induced and chronic unilateral pain: Reflection of hemispheric lateralization of emotion. *Cortex*, 30, 255–267.

Schiff, B.B. & Lamon, M. (1989). Inducing emotion by unilateral contraction of facial muscles: A new look at hemispheric specialisation and the experience of emotion. *Neuropsychologia*, 27, 923–935.

Schiff, B.B. & Lamon, M. (1994). Inducing emotion by unilateral contraction of hand muscles. *Cortex*, 3, 247–254.

Tommasi, L. & Vallortigara, G. (1998). Footedness in binocular and monocular chicks. *Laterality*, 4, 89–94.

Tommasi, L. & Vallortigara, G. (2001). Encoding of geometric and landmark information in the left and right hemispheres of the avian brain. *Behavioural Neuroscience*, 115, in press.

Tommasi, L., Vallortigara, G. & Zanforlin, M. (1997). Young chicks learn to localise the centre of a spatial environment. *Journal of Comparative Physiology A*, 180, 567–572.

Tucker, D.M. & Frederick, S.L. (1989). Emotion and brain lateralization. In *Handbook of Social Psychophysiology*, ed. H. Wagner & A. Manstead, pp. 27–70. Chichester: Wiley.

Vallortigara, G. & Andrew, R.J. (1991). Lateralization of response to change in social partner in chick. *Animal Behaviour*, 41, 187–194.

Vallortigara, G. & Andrew, R.J. (1994). Differential involvement of right and left hemisphere in individual recognition in the domestic chick. *Behavioural Processes*, 33, 41–59.

Vallortigara, G., Cozutti, C., Tommasi, L. & Rogers, L.J. (2001). How birds use their eyes: Opposite left–right specialization for the lateral and frontal visual hemifield in the domestic chick. *Current Biology*, 11, 29–33.

Vallortigara, G., Rogers, L.J., Bisazza, A., Lippolis, G. & Robins, A. (1998). Complementary right and left hemifield use for predatory and agonistic behaviour in toads. *NeuroReport*, 9, 3341–3344.

Ward, J.P. (1999). Left hand advantage for prey capture in the Galago (*Galago moholi*). *International Journal of Comparative Psychology*, 12, 173–184.

Watson, N.V. & Kimura, D. (1989). Right-hand superiority for throwing but not for intercepting. *Neuropsychologia*, 27, 1399–1414.

Wittling, W. (1995). Brain asymmetry in the control of autonomic–physiologic activity. In *Brain Asymmetry*, ed. R.J. Davidson & K. Hugdahl, pp. 305–357. Cambridge, MA: MIT Press.

Wittling, W. (1997) The right hemisphere and the human stress response. *Acta Physiologica Scandinavica*, 161 (suppl. 640), 55–59.

Zappia, J.V. & Rogers, L.J. (1987). Sex differences and reversal of brain asymmetry by testosterone in chickens. *Behavioural Brain Research*, 23, 261–267.

4

Advantages and Disadvantages of Lateralization

LESLEY J. ROGERS

The first evidence of lateralization of visual function in a non-human species came from a study of the domestic chick (Rogers and Anson, 1979). At the time it was reported, it was met by disbelief on the grounds that lateralized responses to stimuli on the animal's left and right sides would surely be so disadvantageous that such a characteristic would never have evolved. Although it is now known that visual lateralization is present in a broad range of vertebrate species (as discussed in the preceding chapters), the selective advantages of being so lateralized have yet to be established convincingly. The aim of this chapter is to consider the possible advantages and disadvantages of being lateralized, and to do so in a number of species.

4.1. Disadvantages of Lateralization

It is not difficult to see the disadvantages of being lateralized for visual functions, provided they translate into side biases in latency or effectiveness of response, or even ignoring a particular stimulus when it is on one side and not the other. The first question to address, therefore, is whether differential specialization of the hemispheres does, in fact, lead to side biases.

Chapter 3 by Andrew and Rogers has outlined the differential specialization of the left hemisphere (LHem) to control responses directed toward prey and other food objects, and of the right hemisphere (RHem) to control attack responses. This complementary specialization of the hemispheres has been well established in the chick by administration of pharmacological agents that modify neural development of regions of the left or right hemisphere (summarized in Rogers, 1996). Indeed, localised injection of very low doses of glutamate on day 2 post-hatching has demonstrated that lateralization of responding to food objects, by discriminating them from background, is a function of the left visual Wulst region of the forebrain (Deng and Rogers,

126

1997). Although the details of the mode of action of glutamate are not particularly relevant to this chapter, it is important to point out that the treatment with such a low dose of glutamate does not cause lesioning. Instead, the cellular action of glutamate involved in revealing lateralization is stimulation (not destruction) of neurones (Rogers and Hambley, 1982) at a critical stage of development, which appears to lead to an increase in the arborization of visual neurones projecting to the injected left Wulst (Khyentse and Rogers, 1997). The effect of the glutamate treatment of the left Wulst is long-lasting and would seem to be a functional lesion that is due to the formation of extra new connections that may interfere with visual function in that region of the brain, or cause a deficit in performance due to shifting control from the left Wulst to the less effective right Wulst.

The complementary specializations of the hemispheres translate into differential responsiveness of the left and right eyes in those species with primary visual input to the contralateral side of the brain, since the optic nerve fibres decussate completely at the chiasma. This is the case in lizards, frogs and toads, birds and most mammals with eyes placed laterally on the sides of the head. Monocular testing can, therefore, reveal lateralized responding without difficulty, as has been shown in the chick (Mench and Andrew, 1986; Rogers, 1997), the pigeon (von Fersen and Güntürkün, 1990; see also Chapter 7 by Güntürkün) and also the rat (Cowell, Waters and Denenberg, 1997). It is interesting to note that monocular testing can reveal visual lateralization in mammals that have their eyes positioned laterally despite the existence of the corpus callosum, which connects the hemispheres.

In brief, use of the right eye (RE) has been associated with responses that require the animal to consider the consequences of action, to inhibit some responses and make decisions, whereas the left eye (LE) has been associated with more immediate (or rapid) responses and behaviours that have emotional content. There are many examples of this differential responding in chicks tested monocularly, as discussed in Chapter 3 by Andrew and Rogers. Here it will suffice to remind the reader of just two examples, and ones that will be used as a basis for the arguments to follow. Attack and copulation responses are, for example, elevated in chicks treated with testosterone and then tested using the LE only (i.e. with a patch over the RE) but not when the same chicks are tested using the RE only (Rogers, Zappia & Bullock, 1985). This implicates activation of the RHem in aggressive and sexual responding, whereas the LHem suppresses these responses. In contrast, chicks are able to categorize grain as separate from inedible pebbles when they use the RE (and LHem) and not when they use the LE (and RHem) (Mench and Andrew, 1986; Rogers, 1997). This result is consistent with the demonstrated role of the Wulst of the LHem in controlling the same behaviour. Specialization of the RE and LHem for pecking to feed has also been shown in the pigeon: pigeons tested using the RE peck faster and consume more seeds than those using the LE (Güntürkün and Kesch, 1987) and they also discriminate grain from grit better when using the RE (Güntürkün, 2000).

In species with eyes placed laterally and so with little binocular overlap, lateralized specializations of the hemispheres are manifested as side biases to respond differently to stimuli perceived by on the left and right sides in the lateral, monocular visual fields. For example, chicks direct more pecks at unfamiliar conspecifics on their left side, as shown recently by Vallortigara et al. (2001). Species with eyes placed laterally also turn the head to examine certain visual stimuli, using the laterally directed fovea, as seen readily in birds and lizards. In these species the eye chosen to view a given stimulus is of particular importance because it determines the primary hemisphere to be used in processing the visual information and in controlling the response. It is, perhaps, not surprising therefore that birds exhibit preferences in the eye they use to view different classes of stimuli. Chicks, for example, choose to view an attractive, novel stimulus on which they might imprint with the RE and a neutral stimulus with the LE (McKenzie, Andrew and Jones, 1998). Wild-living kookaburras (*Dacelo gigas*) also show preferential use of their LE to view the ground in search of moving prey (Figure 4.1). Use of the LE would bring the RHem with its capabilities for processing spatial information to bear on the problem of finding moving prey: a snake or lizard. A limited number of observations have noted that, once prey is spotted, the RE is used just before the kookaburra swoops to capture it, in an apparently ballistic flight with the beak pointing at the prey. Use of the RE immediately prior to prey capture is consistent with the hypothesis developed in Chapter 3 (use of the LHem in object-related tasks in which manipulation is required) and in line with the results of chicks (RE for pecking at grain against a background of pebbles).

In the example of the kookaburra, lateralization of the hemispheres is manifested as an eye preference for viewing, and not as a side bias in the animal's responding. In fact, the bird tilts and turns its head into a position allowing viewing in the midline and down so that the bird can view the whole area in which prey might be detected (Figure 4.1). In other words, head turning overcomes any side bias. It is only when stimuli may appear on either side of the head that a problem might be encountered. However, even in such cases, side biases in responding could be avoided by detecting a stimulus initially by one eye and then turning the head to allow the appropriate eye to view the stimulus while a decision is made about responding. Thus, there

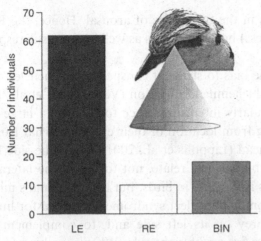

Figure 4.1. _Eye used by kookaburras (*Dacelo gigas*) to view the ground in search of food. Each time a wild kookaburra was spotted sitting on a power wire and looking downward the angle of the head tilt was noted and the eye used to scan the ground was recorded. The number of individuals recorded using the left eye (LE), right-eye (RE) or both eyes (BIN) is plotted. Data collected by L.J. Rogers and G. Kaplan.

would be no problem of the bird missing an important prey stimulus or treating a stimulus inappropriately because it happens to be on the animal's 'incorrect' side. Such absence of side bias would be expected in examples of responding that allow time for head turning and decision making, as opposed to cases in which stimuli must be responded to immediately on detection. Stimuli that elicit rapid responses would, one expects, be more likely to reveal side biases, if they are detected in the lateral field of vision. This appears to be the case in agonistic responses.

Agonistic responses are delivered preferentially to a conspecific in the left lateral field of vision. Not only is the latter seen in chick but also in lizards (*Anolis* spp.; Deckel, 1995) and toads (*Bufo* spp.; Robins et al., 1998; see Figure 3.2 in Chapter 3 by Andrew and Rogers). These leftward aggressive responses are not merely the result of a persistent and general leftward bias for such motor responses, as shown by the fact that non-aggressive pecks of chicks are not biased to the left (Vallortigara et al., 2001) and non-aggressive encounters between lizards are not biased to the left (Deckel, 1995, 1998; also discussed in Chapter 1 by Vallortigara and Bisazza). Agonistic interactions involve intense emotions and are under the control of the RHem. In fact, in *Anolis* lizards such leftward-directed responses are associated with a colour change in the lizard's skin: it lightens. Non-lateralized and non-aggressive responses are associated with darker skin colour. This is probably a reflection of differing hormonal levels in the two states of arousal. Hence, the RHem might have control over (stress) hormone levels as well as agonistic responses directed leftward.

Toads, in fact, show a side bias to strike at conspecifics in their left hemifield and at prey in their right hemifield of vision (Vallortigara et al., 1998). These side biases are particularly interesting since toads have a large binocular field of vision resulting from location of their eyes somewhat frontally and toward the top of the head (Lippolis et al., 2001). Thus, the side biases seen clearly in toads tested binocularly relate, not to use of the lateral and monocular field of vision, as is the case in birds, but to the animal's midline. It is as if the toads suffer from hemineglect syndrome, described for humans (Myslobodsky, 1983), for prey on its left side and, to complement that, neglect of competing conspecifics on its right side. When searching for prey amongst a group of conspecifics, each toad strikes at prey seen to the right of its midline and defends that food source from competing members of its own species by striking at those toads that approach on its left side. Therefore, the toad would tend to miss prey on its left side and fail to drive off competitors on its right side. It is difficult to see how this would be advantageous for survival. Of course, observations of feeding behaviour of different species of toads in their natural habitat are now important to see whether these side biases are manifested or whether they are overcome by other behavioural strategies.

Added to the above, toads would seem to suffer a potential disadvantage by being less reactive to predators advancing from their right side than to those advancing on their left side. This has been shown to be the case by testing toads (three species of *Bufo*) in a circular arena and advancing a mechanical model of a snake towards the toad in the midline (binocular) or in the left or right lateral monocular fields (Lippolis et al., 2001). A toad was more likely to react, chiefly by jumping away from the simulated predator, when the snake stimulus was in its binocular field or in its left monocular field than when the stimulus was in the toad's right monocular field. It can be reasoned, therefore, that toads are more likely to succumb to predators approaching them from their right side. In fact, predators might exploit this side bias, present at a population level (i.e. in most toads), by approaching their prey on its right side and turning to strike leftwards. There has been no research addressing this possibility. Also, it is not yet known whether the lateral bias for reactivity to the simulated predator is present only in the monocular, lateral fields of vision or, as for the lateralization of striking at prey or at a conspecific, the toad's midline is the dividing point so that the left and right hemifields differ in reactivity. Given the large binocular overlap in most species of toads (including the three species tested so far), it would be important to determine whether the lesser reactivity on the right side is confined to the relatively small monocular field or extends over the entire right hemifield of vision. If it is confined to the monocular field of the right eye, it would be tempting to suggest that toads minimize the side bias in reactivity to a predator by maximizing the size of their binocular field in the horizontal plane.

It is relevant to note that the cognitive mechanism of dividing space according to the left and right of the body's midline is obvious in humans, despite the fact that humans have a wide binocular field and, more importantly, each eye sends input to both sides of the brain. It is clearly a higher cognitive phenomenon, independent of direct perceptual inputs. This concept of the midline is evident in hemineglect syndrome following extensive damage, usually caused by stroke, to one of the hemispheres. Neglect of the contralateral side of the body and also the contralateral visual field is more likely to result from damage to the RHem than the LHem (Heilman, Watson and Valenstein, 1985; Mesulam, 1985), a result which in itself demonstrates one form of lateral specialization of the hemispheres. In fact, damage to the RHem often causes neglect, or denial of the existence, of the left half of any object or image conceptualized as a whole. A patient with hemineglect syndrome following a lesion in the RHem, for example, may ignore food on the left side of the plate, dress and groom only the right side of his or her body, read only the right side of a page or draw all of the numbers of a clock face on the right side only of the dial (Bradshaw, 1989). It is often possible to persuade such a patient to attend to events and objects to the left side of his or her midline, but responses are always slower and it is difficult to maintain attention to the left extracorporeal space (Posner et al., 1987). The fact that lesions of the RHem are more likely than lesions of the LHem to cause hemineglect syndrome is consistent with the RHem's role in processing spatial information, here in terms of body maps. The RHem, it seems, subserves a broader area of hemispace than does the LHem. Hemineglect syndrome is interesting to note here in the light of the results with toads. In addition to the points made earlier, the differential responsiveness of toads to prey and conspecifics to the right or left, respectively, of their midline raises the possibility that division of left and right halves is common in animal species, at least with respect to their own midline if not of conceptualized objects (see also Chapter 10 by Andrew and Watkins).

The bias to direct more agonistic responses to conspecifics seen on the animal's left side may be present even in primates, despite the fact that they have frontally placed eyes and that they have both ipsilateral and contralateral projections from each eye to visual centres in the brain. Gelada baboons (*Theropithecus gelada*) direct more agonistic responses to conspecifics on their left side than on their right (Casperd and Dunbar, 1996). Possibly consistent with this finding, Drews (1996) has reported that the pelts of wild baboons (*Papio cynocephalus*) have more injury marks on the right side of the head region. This asymmetry could result from the attacker directing agonistic responses at conspecifics detected in the extreme left peripheral field, provided the animals are moving or facing in the same direction. As well known in humans, the extreme peripheral field of vision is monocular and projects only to the contralateral hemisphere (the RHem in this case). Tachistoscopic presentation of visual stimuli in the extreme lateral fields is, in fact, a standard way of determining hemispheric specialization in humans, and primates, because the visual input is confined to the contralateral hemisphere.

The ubiquity of similar and complementary lateralities for responding to prey and competing with conspecifics across species suggests that lateralization is beneficial for survival but it remains possible that it is a neutral trait that arises as a consequence of some essential process in embryonic development. However, given that lateralization is a characteristic common to a wide range of species, it is unlikely to be neutral to selection. It is worth considering that responding more readily to a predator on the left side, as shown in toads, occurs as a consequence of optimizing neural capacity, and so only in that sense is it advantageous. The brain may have maximized its capacity to detect and avoid predators, or to carry out some other cognitive processes, by becoming lateralized.

4.2. Advantages of Lateralization

Before discussing the possible advantages of having a lateralized brain, it is necessary to distinguish between two kinds of lateralization: (1) lateralization of individuals but without a consistent direction of bias in the majority of the population (referred to as 'individual lateralization'); and (2) lateralization of individuals most, if not all, with the same bias, so that lateralization is present in the population (referred to as 'population lateralization'). Individual lateralization without population lateralization means that the population scores of the particular form of lateralization under examination (i.e. for a particular trait) are distributed bimodally on either side of the no-bias value.

Hand preferences for manipulation fit both of these patterns of lateralization, depending on the species and, in some cases, past experience. Here only a few examples serve to illustrate the point. Hand preferences in mice represent 'individual lateralization' since half of any group tested prefers to retrieve food using the left hand and the other half uses the right hand (Collins, 1985). The same is true of common marmosets (Hook-Costigan and Rogers, 1996; Hook and Rogers, 2000). 'Population lateralization' is, of course, true for (right) handedness in humans, and also (left) footedness in parrots (Rogers, 1980) and for all of the behaviours discussed above.

A bimodal distribution (i.e. one on which individuals are lateralized but not the population) may result from the effects of learned preferences (e.g. learned use of one hand for manipulating food objects) or it may come about by chance simply because the organism has an inability to develop symmetrically. The latter is known as fluctuating asymmetry and, while this is usually discussed in terms of physical features and in terms of chance impacts of disease and parasite attack on physical development (Møller and Swaddle, 1997), it has been considered as a possible cause of bimodal population distributions of hand preferences in mice (Collins, Sargent and Neumann, 1993). If behavioural lateralization at the individual and not population level is, in fact, a fortuitous consequence of fluctuating asymmetry, it may have no particular benefit for the species. Although there is no evidence that this is the case, and no obvious way of testing this hypothesis presents itself, this possibility should be kept in mind as an alternative to the following discussion on the possible advantages of having a lateralized brain. Fluctuating asymmetry, however, applies only to individual lateralization without a population bias, and not to population lateralization. Also, were fluctuating asymmetry an explanation for individual lateralization, we might expect to find large numbers of individuals that are not lateralized or have only a small degree of lateralization. This does not apply to the examples to follow.

If hemispheric specialization is advantageous for the individual, this may occur without selective pressure for all members of the population to be lateralized in the same direction, although the latter may occur because embryological development constrains it to be that way. For example, the genes that determine asymmetry of the internal organs (i.e. of the gastrointestinal tract and cardiac system) may set a baseline of asymmetry, at a population level, on which brain and behavioural asymmetry is based. Of course, selection may act at both of these levels of lateralization (on the viscera and on the brain). There may also be selective pressure to evolve or develop lateralization at a population level because population lateralization benefits some aspects of social behaviour. These possibilities are considered below.

4.2.1. Increased Cognitive Capacity

Any given brain must have an ultimate limit to its neural capacity for processing and storing information, particularly its capacity to carry out a number of these processes simultaneously. One way to increase a brain's capacity to carry out simultaneous processing is to channel or filter the different types of input and/or output so that processing takes place at different regions of the brain. This filtering occurs at different levels in the cognitive hierarchy of the brain (in a vertical sense). It might also occur in a lateral sense, such that each hemisphere carries out a different kind of information processing than the other. We know the latter is the case (Chapter 3 by Andrew and Rogers). The question being asked here is whether we can actually prove that a lateralized brain has a greater *capacity* to process information, rather than simply assume so.

It has been assumed that differential specialization of the hemispheres enhances neural and cognitive capacity (Gazzaniga and Le Doux, 1978; Dunaif-Hattis, 1984) but almost no evidence in support of this hypothesis is available. The problem is finding non-lateralized and lateralized members of the same species to compare. This might be possible if either the presence or absence of lateralization, or at least the degree of lateralization, can be altered by experience. The chick presents itself as a model for such experimentation because it is possible to produce chicks with and without lateralization for attack and feeding responses (see below). The hypothesis could be tested using chicks with lateralized and symmetrical brains with respect to these particular functions.

Chicks hatched from eggs exposed to light just prior to hatching are lateralized for attack and feeding responses (as discussed above) but chicks hatched from eggs incubated in the dark are not lateralized for these particular responses, even though other functions remain lateralized (Rogers, 1982; summarized in Rogers, 1997; discussed in detail in Chapter 6 by Deng and Rogers). Light exposure causes lateralization to develop because the embryo is turned in the egg so that it occludes its left eye, whereas light entering through the shell can stimulate the right eye (Rogers, 1990). The light stimulation enhances the development of visual pathways receiving input from the right eye and projecting to the forebrain (Rogers, 1996) and, as a consequence, the thalamofugal visual projections that cross the midline of the brain become asymmetrical (Rogers and Deng, 1999). There are more projections from the left side of the thalamus to the forebrain than from the right side to the forebrain. Following incubation in the dark during the last few days before hatching, the asymmetry does not develop in these visual projections (Rogers and Bolden, 1991) and there is no lateralization of attack and feeding responses (Rogers, 1990, 1997). The data for feeding responses shows clearly that, after incubation in the dark, there is no lateralization at the group level or at the individual level (Rogers, 1997). Whether or not dark-incubated chicks fail to develop lateralization for attack at the individual level as well as the group level is not entirely certain; an early result (Zappia and Rogers, 1983) gave some suggestion that individual scores for attack might be distributed bimodally. If this is so, the individuals remain lateralized, although the group is not. Recent experiments in my laboratory have, however, found an absence of lateralization for both of these responses at both the individual and group levels in dark-incubated chicks (Deng and Rogers, in preparation).

4.2.2. Reaction Time/Vigilance

It was predicted that the dark-incubated chicks, without lateralization of attack and feeding responses, might perform less successfully than light-incubated, completely lateralized chicks in a task requiring dual attention to entirely different stimuli. The task used tested the chick's vigilance for a simulated aerial predator while the chick was pecking at grain and meal worms on the floor. The task was based on two known and opposite lateralizations of the chick. The first was the use of the RE (and LHem) for pecking to feed, and the second was use of the LE (RHem) to monitor overhead for an aerial predator. In response to being played a conspecific alarm call that signals the presence of an aerial predator, a chicken tilts its head to monitor overhead using the left eye (Evans, Evans and Marler, 1993). This LE preference is consistent with other results showing that the chick uses the LE and RHem to detect and respond to novel stimuli (Rogers and Anson, 1979; Andrew, Mench and Rainey, 1982).

Thus, a completely lateralized chick (exposed to light before hatching) feeding in a situation in which it might expect a predator to appear overhead would be expected to use the LE to monitor overhead and the RE to control feeding pecks. Dark-incubated chicks might find this channelling of separate roles to the hemispheres more difficult, or impossible, and so be less efficient in detecting the overhead predator. In other words, light-exposed, lateralized chicks might detect the overhead predator sooner than dark-incubated chicks, which are not so completely lateralized.

The chicks were tested between days 8 and 10 posthatching and after 4 h deprivation of food (Rogers, 2000). Once they had started to peck at the grains and mealworms, a model predator, shaped to resemble a raptor, was rotated overhead (Figure 4.2). The latency to detect the predator stimulus was scored, detection being indicated by the chick interrupting its pecking and, usually, giving a startle trill and/or a definite twitch of the head. The light-exposed chicks detected the simulated predator after a very short latency when it advanced on their left side (LE) and with a significantly longer latency when it advanced on their right side (RE). This asymmetry was expected. The dark-incubated chicks detected the simulated predator after the same latency for the LE and RE, and the mean time was intermediate between that of the LE and RE scores for the lateralized group of chicks. Therefore, after incubation in the dark, the LE (RHem) fails to show an advantage in detecting the stimulus. The latency scores for the dark-incubated chicks detecting the stimulus with the LE were also more variable than those of the light-incubated chicks using their LE.

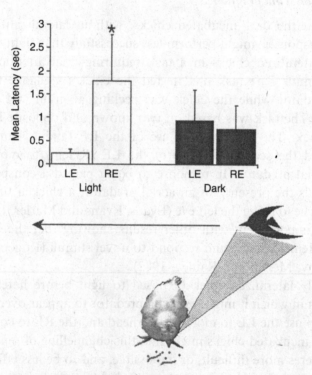

Figure 4.2. The chick is feeding while a model raptor is moved overhead. The latency to detect the model was recorded, as well as the visual field in which it was detected. Detection was indicated by the chick interrupting its pecking and usually giving a startle trill call or a definite twitch of its head. Means and standard errors are plotted. Note that the latency was short when the chicks exposed to light before hatching detected the stimulus with its left eye. LE, left eye; RE, right eye. Adapted from Rogers (2000).

To obtain an indication of the chicks' motivation to feed and respond to the stimulus, the number of pecks made at food and the number of peep vocalizations were also scored. This provided a control to be sure that the above result was not a non-specific effect caused by exposing the chicks to light or incubating them in the dark. Both groups of chicks showed the same decrease in pecking when the stimulus was present. At the same time, peeping increased and it did so to a greater extent in the dark-incubated chicks than in the light-exposed chicks. This result suggests that the dark-incubated chicks were more distressed by the presentation of the simulated predator than the light-exposed chicks even though they did not detect it as soon as the light-exposed chicks. Also, once they had detected the stimulus, both groups of chicks responded to it as fear-inducing, and the response to it was somewhat greater in the dark-incubated chicks than in the light-exposed chicks. Therefore, there was no suggestion that the earlier detection by light-exposed chicks using the LE was due to the chicks being more motivated to respond. Thus it could be concluded that the lateralization that develops as a consequence of exposing the chick embryo to light enhances vigilance in the left lateral field of vision.

It is not difficult to see that the enhanced vigilance measured here for light-incubated chicks using their LE would be an advantage in the chick's natural environment, but the lateral bias in this ability diminishes its potential advantage. Only by tilting the left side of the head upwards to look overhead with the LE could a chicken maximize its ability to detect a predator overhead. In fact, such head tilting to scan overhead using the LE does occur when an adult chicken hears the aerial predator alarm call of another chicken (Evans, Evans and Marler, 1993). Scanning with the left eye would also make use of the RHem's abilities in processing spatial information (Rashid and Andrew, 1989). However, when pecking at food on the ground, as in the task used to test the chicks, the frontal binocular field must be used (summarized in Rogers, 1995) and, since this is confined to approximately 16° on either side of the beak, head tilting would be impossible during bouts of rapid pecking. At such times, the chick with a lateralized brain may have to trade off a lateral bias in vigilance for use of the frontal field in pecking.

A somewhat similar result to that obtained with chicks was reported for cats tested in a task requiring them to track a moving spot of light using one or the other forepaw. Fabre-Thorpe et al. (1993) compared the reaction times of cats with and without a preference to use one paw over the other when performing the task. They found that the cats with lateralization of paw preference had a shorter reaction time to commence tracking than those without lateralization of paw preference, and concluded that this would be a functional advantage of being lateralized. Here, however, the enhanced ability of the lateralized cats must relate to channelling responses to one or the other forepaw: it is lateralization of motor output, rather than perceptual input as is the case in the experiment with chicks.

4.2.3. Efficiency

Not inconsistent with the results obtained with cats (above), McGrew and Marchant (1999) found that wild chimpanzees with completely lateralized hand preferences were more efficient at fishing for termites than those with incomplete lateralization. The exclusively lateralized chimpanzees gathered more prey for a given amount of effort than did incompletely lateralized chimpanzees.

Specialization of the hands, or paws, appears to increase both the speed and efficiency of performing a manipulative task. Both forepaws are required by the cat to track the moving spot, one used for tracking and the other for balance. Likewise both hands are used by the chimpanzees in termite fishing, one to hold the twig used as a probe and the other to act as a stabilizer. The twig covered in termites is rubbed over the stabilizing hand as preparation for the chimpanzee eating them. Consistent specialization of the hands or paws for simultaneous and coordinated use is, the results show, a lateralized advantage. The same may apply in toads that show a right-hand preference to wipe a foreign body from their head or snout (Bisazza et al., 1996), one foot being specialized for wiping and the other for balance. However, there has been no investigation of whether this particular motor lateralization leads to more efficient use of the forepaws in toads.

Returning to the experiment on vigilance in dark-incubated versus light-exposed chicks, it could be asked whether the latter chicks are not only more vigilant but also more efficient in obtaining food while scanning for predators. Although this was not tested directly in the experiment outlined above, there is some indirect evidence that this may be the case. The performance of chicks on a task requiring them to search for grains of chick-mash amongst a background of pebbles has been used to compare the performance of dark-incubated and light-exposed chicks (Rogers, 1997). In the dark-incubated chicks, binocular performance was superior to monocular performance with either the LE or RE, suggesting that performance of this task may depend on integrated use of the two eyes and hemispheres. In fact, the dark-incubated chicks were unable to find the food grains when tested monocularly. The light-exposed chicks were also unable to find the grain when using the LE but could do so with ease when using the RE. This shows the expected LE–RE differences in the two groups, but here we are interested in the efficiency of binocular performance. In the first 20 pecks of the task, some of the dark-incubated chicks made more pecks at pebbles and others made fewer pecks at pebbles than the light-exposed chicks. The scores of the dark-incubated chicks were distributed on either side of those of the light-exposed chicks. In other words, the dark-incubated chicks avoided pecking at pebbles either earlier or later than the light-exposed chicks (earlier or later here referring to numbers of pecks). This bimodal distribution of scores (not just more variable scores) for the dark-exposed chicks in the first 20 pecks may indicate that one or the other eye and its accompanying hemisphere takes control of pecking responses at least temporarily. Some chicks may begin the task by relying on the RE/LHem and so perform well, while the others rely on the LE/RHem and so perform poorly in the first instance. Such decisions to put one or the other hemisphere in charge temporarily have been seen previously in cases in which there is hemispheric conflict as, for example, caused by training pigeons on a brightness discrimination and then testing them in a conflict situation (Palmer, 1972, cited in Zeier, 1975). By making the pigeon wear goggles with a red filter for one eye and a green filter the other eye and presenting them with red and green illuminated keys, the researchers organized the test so that one eye saw, say, the right key as the brighter key and the other eye saw the left key as the brighter key. Thus, the pigeon was asked, simultaneously, to peck at the left key, based on information obtained by one eye, and the right key, with information obtained by the other eye. Pigeons confronted with this dilemma opt to attend to inputs from one or the other eye and so use one hemisphere. Half choose to use the LHem and the other half the RHem. It appears that there is a similar conflict between the two eyes and their respective hemispheres in the dark-incubated chicks, which occurs because they have no clear and consistent lateralized control of pecking responses. This puts half of the dark-incubated chicks at an initial disadvantage because they opt to use the less efficient LE/RHem to control food searching and pecking.

In general, it may be concluded that chicks lacking the particular forms of lateralization caused by exposing the embryo to light on day 19 of incubation have a disadvantage in both vigilance for predators and pecking at food versus inedible objects similar to food. Having a completely lateralized brain may afford more skills, as well as shorter reaction times.

A recent report by Güntürkün et al. (2000) supports these conclusions. In this case pigeons were tested on a task requiring them to discriminate grain from grit, a task which they perform better with the RE than the LE when tested monocularly. The researchers first determined the LE–RE difference in performance for each bird and used that as an index of their degree of lateralization (see Chapter 7 by Güntürkün). They then tested them binocularly to see whether there was an association between their efficiency on the task and the degree of lateralization. There was a mild but significant correlation: the stronger the lateralization, the better the visually guided foraging.

The examples given so far suggest that lateralization may be beneficial for an individual. These advantages would be available to individuals irrespective of the direction in which the lateralization occurs. There would be no particular reason for population lateralization of function to be present. Indeed, if it were possible, it might be advantageous to maintain individual lateralization and avoid population lateralization to prevent predators from exploiting a predictable motor bias (e.g. for escape turning). One should, therefore, ask separately what advantage might be conferred on a species, or a group, by having the majority of individuals lateralized in the same direction.

4.3. Social Predictability

Populations in which the majority, if not all, individuals are lateralized in the same direction may have a certain predictability. Individuals could rely on the predictable lateralization of other members of the group and this could be used to a social advantage. For example, an individual might lower its chances of being attacked by approaching higher-ranking conspecifics on their right side. Another example of lateralization might involve cohesion of a flock, shoal or herd. If all members of the group have the same bias to turn, say, right when they see a predator, the integrity of the group will be maintained during escape avoidance. Thus it is hypothesized that population lateralization is present when there is a need to maintain coordination among individuals in a social context.

4.3.1. Maintaining Integrity of the Group

There is evidence in support of the above hypothesis. Bisazza, Cantalupo, Capocchiano and Vallortigara (2000) have shown that shoaling in fish is associated with a population bias to turn in one direction (either left or right) when faced by a barrier of vertical bars through which the fish could see a model that simulates a predator. Each fish was tested singly. It had to swim down a corridor toward the simulated predator placed behind the barrier. On reaching the barrier the fish turned right or left and this was recorded. Each fish was given ten such trials and from this score an index of lateralization was calculated. Each species was represented by 9–18 individuals and a significant population bias for turning was found in ten of the species. Next the social behaviour of the species was determined in terms of tendency to school. Groups of fish were placed in a tank together and an index of their proximity to each other was determined. Six species were found to be gregarious (i.e. to school) and all six were ones lateralized for turning bias at the population level. Ten species were found to be non-schooling and six of these were not lateralized at the population level (Chapter 1 by Vallortigara and Bisazza). The species not lateralized at a population level were, however, comprised of individuals that were lateralized (i.e. individuals showed a consistent preference to turn in the same direction). As in the case of paw preferences in mice, half of the individuals in these species turned leftward and the other half rightward, and hence there was no group bias. It seems possible, therefore, that a group bias for the majority of individuals to turn in the same direction may be maintained to advantage in those species that school. In schooling species, the group bias must confer an advantage additional to any advantage gained by individuals being lateralized.

In fact, it is not difficult to see that a school in which all individuals have the same bias for turning would maintain its integrity and not dissipate on contact with predators or other features of the environment that evoke a change in the direction of swimming. This would operate effectively provided that the school is moving in one direction or, say, facing upstream so that all of the individuals are facing in the same direction. Once the school remains stationary with the individuals facing in different directions, this mechanism would no longer be able to maintain the integrity of the school and the probability of predation per individual would increase. In fact, cooperative foraging by yellowtail fish (*Seriola lalandei*) on schooling jack mackerel (*Trachurus symmetricus*) uses a method of cornering a school so that it is stationary with individuals facing in different directions. Then the school of mackerel can be dissipated easily by just one of the yellowtails lunging forward into the school, as illustrated by Dugatkin (1997; see Figure 3.6 on p. 57 of the book). In such a case, turning bias is no longer effective to maintain the school.

The example given here is for schooling in fish but a population bias for turning could apply to any species that moves in a herd or flock (e.g. ungulates and birds).

4.3.2. Maintaining the Social Hierarchy

Population lateralization might also have a role in social interactions. As mentioned earlier, if most, or all, individuals in a group are lateralized in the same direction, individuals might use that to advantage in their social interactions. This might be especially the case for attack responses. Since attacks are directed leftwards in a number of species, one individual might avoid provoking attack as it approaches another by moving in on that individual's right side. Such behaviour might be particularly important to avoid aggression and maintain the social hierarchy.

In fact, Rogers and Workman (1989) reported that, in the first 2 weeks of life, dark-incubated chicks form less stable hierarchies than do chicks exposed to light just before hatching. This result suggested that being lateralized at the population level might make social interactions more predictable because, in the light-exposed group, all chicks would have the same side bias for their social responses. An approaching chick might, therefore, avoid attack by not approaching its assailant on its left side, or vice versa. In dark-incubated chicks, there might be a lesser degree of such predictability.

The procedure involved raising young chicks in groups of eight and then scoring the position of each chick in the social hierarchy on a daily basis by allowing them to compete to gain access to food in a small bowl placed in the corner of the cage (Rogers and Workman, 1989). As only two or three chicks could feed from the bowl at a time, the group competed to gain entry to it and a hierarchy could be determined based on the number of entries to the food dish achieved by each individual. An earlier study had shown that the hierarchy so determined is very similar to that determined using other means of group competition (Rogers and Astiningsih, 1991) and, therefore, we may conclude that the measure was a general one reflecting a consistent social order. Six groups of chicks hatched from eggs exposed to light during the final days before hatching (i.e. lateralized at a population level for the visual functions mentioned already) were compared to six groups of chicks incubated in the dark (i.e. not lateralized at a population level for those same visual functions). A rigid (or stable) hierarchy was maintained in the groups that had received exposure to light, as indicated by the fact that the lowest ranking chicks in these groups rarely gained access to the food dish (Figure 4.3). The lowest ranking chicks in the groups hatched from eggs incubated in the dark gained access to the food dish more often than did their light-exposed counterparts, and their entry scores varied more from day to day (not shown in the figure).

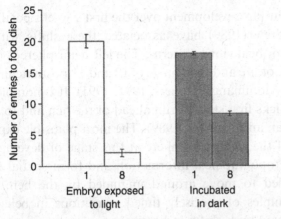

Figure 4.3. Competition for access to a food dish by groups of chicks hatched from eggs exposed to light during the last days of the incubation period (white bars) or groups of chicks incubated in the dark (grey bars). There were eight chicks in each of 12 groups. The chicks were scored daily from days 8 to 16 and after food deprivation for 3 h. The mean number of entries to the food dish for the highest-ranking chick (1) and the lowest-ranking chick (8) is plotted. Note that the lowest-ranking chick of the light-exposed groups gains fewer entries than does the lowest ranking chick of the dark-incubated groups. There is a greater difference between the top- and bottom-ranking chicks in the light-exposed groups than in the dark-incubated groups. Modified from Rogers and Workman (1989).

It is possible that the explanation for maintenance of a more stable hierarchy in the light-exposed groups results from the fact that their population lateralization allows more predictable social interactions, although there is no conclusive evidence in support of this hypothesis. It is impossible to say whether a more or less stable social hierarchy would be advantageous to chicks raised in natural conditions. It should also be mentioned that the effect of light is clearly one of experience, whereas selection may act on the cellular mechanisms underlying sensitivity to light exposure prior to hatching.

4.4. Cognitive Style or Temperament Associated with Lateralization

It is likely that characteristic types of cognitive style or emotional responsiveness are associated with dominance of one or the other hemisphere and that individuals, or even populations, might benefit by expressing one or other of these hemisphere-dependent patterns of information processing and response control.

Some evidence supports an association between specific behaviour patterns and hemispheric dominance. The chick, for example, undergoes shifts in hemispheric dominance during development over the first 2 weeks posthatching. Workman and Andrew (1989) have associated these shifts with the appearance and peaking of behaviour patterns. The left hemisphere is dominant during the first week of life and then on days 10 and 11 posthatching the right hemisphere assumes dominance (Rogers, 1991, 1995). It is notable that on days 10 and 11 the chicks first start to run ahead of the hen and to move out of her sight (Workman and Andrew, 1989). The most plausible explanation is that dominance of the right hemisphere at this stage of development allows the chicks to use topographical information and form spatial memories, as would be needed to move around unguided by the hen. This, together with other examples of precisely timed transitions in behaviour that appear to be associated with shifts in hemispheric dominance over the early stages of development of the chick, suggests that lateralization plays a critical role in the time-course of normal development.

In primates, hand preference might be an indicator of hemispheric dominance and so be associated with cognitive or emotional style. It is possible that consistent use of one hand in a number of tasks either activates the contralateral hemisphere, which controls that hand, or reflects a precondition of higher activity in that hemisphere (Ward et al., 1990; Hellige et al., 1994; King, 1995). To say this in another way, preferential use of a hand is associated with an attentional asymmetry that extends to higher cortical levels in the contralateral hemisphere (Peters, 1995). It might, therefore, be predicted that right-handed individuals would exhibit a cognitive style typical of the LHem and left-handed individuals would exhibit a cognitive style typical of the RHem. Thus left-handed individuals might show diffuse attention and intense emotional responses, including fear responses and withdrawal (Davidson, 1992, 1995). Right-handed individuals might be better able than left-handed ones to inhibit fear and withdrawal to allow exploration of novel objects and situations (see Chapter 3 by Andrew and Rogers for details on such hemispheric differences). Results obtained by testing chimpanzees (Hopkins and Bennett, 1994) and marmosets (Cameron and Rogers, 1999) suggest that this is, indeed, the case.

Hand preferences in primates are usually measured as the preferential use of one hand over the other for simple tasks such as feeding (Ward & Hopkins, 1993). Handedness is used to refer to cases in which the majority of individuals use the same preferred hand over a range of tasks (McGrew & Marchant, 1992). Hopkins and Bennett (1994) tested chimpanzees with novel toys and found that the right-handed chimpanzees approached the toys with a shorter latency than non-right-handed (i.e. left-handed and no preference) chimpanzees. The right-handers also touched more toys than the non-right-handers. Thus approach behaviour and exploration appeared to be associated with greater activation of the left hemisphere and avoidance behaviour with greater activation of the right hemisphere.

A very similar result has been found in common marmosets (*Callithrix jacchus*). Individual marmosets display hand preferences consistent over time and in similar tasks but there is no population bias in this species. Half of the population is left-hand preferent and half right-hand preferent (Box, 1977; Hook and Rogers, 2000; Hook-Costigan & Rogers, 1995, 1996; Rothe, 1973). Cameron and Rogers (1999) found that marmosets with a right-hand preference for holding food displayed a shorter latency to enter, on their own, a novel room containing novel structures and objects, touched more objects and performed more touches and more parallax movements than subjects with a left-hand preference. The results are, again, consistent with specialization of the RHem (left hand) for fear and negative emotional states, and so withdrawal, and specialization of the LHem (right hand) for inhibiting fear in order to approach. Here it is noted that there are at least two independent forms of lateralization in an individual: the basic and consistent specializations of the LHem and RHem, and a bias for one hemisphere to control responding, the latter being revealed by hand preference.

We can conclude that right-handers have higher levels of exploration and lower levels of fear than left-handers. If this association between hand-preference and exploration versus fear and withdrawal extends to primates in their natural environment, and we have no reason to believe that it would not, it is possible to speculate about why some species have a population bias for handedness and others do not. It is possible to hypothesize that a population bias for right-handedness occurs in species for which survival depends on inhibiting fear and exhibiting high levels of exploration, whereas a population bias for left-handedness occurs when survival depends on withdrawal and being more fearful of new environments. Hence right-handedness might evolve or develop in species under pressure to move into new territories or to adapt to changing diets. Left-handedness might come about in species that would be at risk if they colonized new territories or changed from a very specialized diet to a more varied diet.

This hypothesized association between population bias for handedness and exploration must, of course, be qualified by the influences of gender, social position and/or age (Rogers, 1999), as well as the context of testing (Cameron and Rogers, 1999), since all of these factors also influence exploration and fear responses. Nevertheless, it is interesting to see whether the hypothesis is supported, to any extent, by comparing common marmosets and tamarins. King (1995) has provided evidence that tamarins (*Saguinus oedipus*) are right-handed at a population level. Common marmosets have no population bias for handedness. This difference between marmosets and tamarins is of particular interest, as they are closely related primates. Therefore, it is interesting to note that tamarins hold larger territories than common marmosets and that tamarins travel twice as far as marmosets in a day (Tardif, Harrison and Simek, 1993). Although these differences between tamarins and marmosets may be explained by differences in diet, marmosets defending smaller territories around gum sources (Ferrari, 1993), they are also consistent with a population bias for right-handedness in tamarins and not in marmosets.

Whatever the reasons for common marmosets being a mixed population of left- and right-handed individuals, it is possible that groups comprised of both types may benefit from the combined effects of two types of cognitive style and emotional responses. We might also take this hypothesis one step further to suggest that the left-handedness of prosimians is related to their nocturnal existence and the need to be cautious in approaching novel situations. This hypothesis suggests that cognitive and emotional styles are related to species differences in handedness. Moreover, it suggests that the selection of left- or right-handedness in different species may have more to do with cognitive demands than body posture and the mode of eating, as stated by the 'Postural Origins Theory' of McNeilage, Studdert-Kennedy and Lindblom (1987).

4.5. Immune Responses

The hypothesis advanced above is that some expressions of lateralization, such as handedness, may appear at a population level because they are associated with cognitive traits or temperaments that are advantageous to the species in its particular environmental niche. The same argument could be made for particular types of immune response because the profile of immune responses differs according to whether it is under control by the left or right hemisphere. Mice with a lesion located in the right parieto-occipital lobe of the neocortex show depressed mitogen-induced lymphocyte proliferation and enhanced antibody production, whereas mice with an equivalent lesion placed in the left parieto-occipital lobe show no modification of these immune responses (Barnéoud et al., 1987). Another study by Nevue (1988) found that lesioning of the left neocortex depressed T-cell functions, whereas right side lesioning either enhanced T-cell functions or had no effect. Moreover, individual differences in turning bias are associated with different immune responses: rats that circle to the left have higher indices of lymphocyte stimulation than those that circle to the right (Nevue, 1988).

Paw preference in mice is also associated with immune responses, almost certainly because it reflects preferential activation of one or the other hemisphere. Nevue et al. (1991) reported that left-pawed mice (determined for reaching into a tube to obtain food) have higher mitogen-induced T-lymphocyte proliferation than right pawed mice. Also ablation of the left cortex abolished the difference in T-cell function between left- and right-pawed mice while ablation of the right hemisphere had no effect on this difference. In humans also, an association between hand preference and immune reactivity has been suggested, and there is some evidence, albeit not well substantiated, indicating that left-handers have a higher incidence of immune disorders (Geschwind and Behan, 1982). Although the data for humans needs confirmation in well-controlled conditions, which are difficult to achieve in studies of humans (Peters, 1995), considered together with the research on paw preference and immune responses, the evidence makes a reasonably strong case for such an association between hand preference and immune responses.

The lateralization of immune responses is, almost certainly, linked to neurochemical lateralization in the brain and associated lateralized effects on hormonal control (Nevue, 1996). The most likely neurochemical system involved is the dopaminergic one and the most likely hormone system involved is the hypothalamic–pituitary–adrenal axis, the stress hormone system. Nevue (1996) assessed these systems after injecting mice with lipopolysaccharide, a drug known as a stress inducer. An hour or two later, the brain dopamine levels were modulated in a lateralized way and plasma levels of adrenocorticotrophic hormone (ACTH) were elevated in right-pawed but not left-pawed mice. ACTH levels were elevated also in ambidextrous mice. T-lymphocyte proliferation was depressed in right-pawed and ambidextrous mice but not in left-pawed mice. Therefore, it would appear that greater activation of the LHem (right-paw preference) relative to the RHem leads to enhanced physiological reactivity of both the control of stress hormones and the immune system. On face value, this result is not obviously consistent with the association between left-hand preference and behavioural reactivity seen in primates, but it would be premature to draw any conclusions.

The main point of these experiments to be considered here is that paw preference, in mice at least, is associated with immune function. It might be valuable for certain populations to retain approximately equal numbers of left- and right-handed individuals in order to maximize variation in immune responses. Other populations, however, might have been exposed to conditions favouring the immune responses of either the left or right hemisphere, and so would be associated with differences in hemispheric activity at a population level.

To my knowledge, there have been no studies investigating a possible association between paw preference and immune responses in anurans but it presents itself as genuine interest with ecological significance. Right-pawedness has been reported in laboratory studies of three species of toads (Bisazza et al., 1996, 1997) and data showing a population bias in paw, or side, preference collected in a field study of the frog *Litoria latopalmata* is shown in Figure 4.4. As discussed, there may be a number of reasons for the population bias but selection for a particular type of immune response presents itself as a possibility worth investigating in these and other species.

4.6. Conclusion

In conclusion, there appear to be two categories of advantage conferred by lateralization, one at the individual level and the other at the population level. The first refers to enhanced skill performance and faster reaction times in more lateralized compared to less lateralized individuals. The second is a social advantage conferred by the majority of animals in the group being lateralized in the same direction. The social function of lateralization at a population level would, of course, vary with social context and between

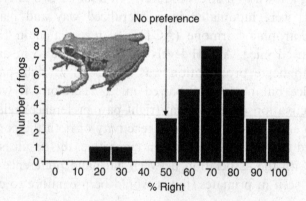

Figure 4.4.   Righting responses in wild frogs (*Litoria latopalmata*). Each frog was given ten trials in which it was laid on its back on a horizontal surface. The forepaw around which its body pivoted in order to right successfully was recorded. The percentage right paw was calculated. The data are presented as a frequency plot. Note the population bias for pivoting around the right paw. In fact, this result indicates active use of the left forelimb by the frog to gain purchase for pivoting and the application of pressure against the substrate by the right forelimb and hand.

species but, hopefully, the controlled experiments with chicks and fish will provide some basis for investigating related functions of lateralization in other species. In addition, there may be other qualities associated with lateralization, including general aspects of reactivity and temperament, and also hormone and immune responses.

It would be pertinent to consider what processes lead to the development of individual lateralization and population lateralization. Interacting phylogenetic and ontogenetic processes are involved.

There are genes known to encode for anatomical asymmetry. These act at an individual level but may be manifested as a consistent population bias. Collins (1985) has shown that it is not possible to select strains of mice on the basis of a group bias for right- or left-pawedness but that the strength of pawedness can be genetically selected (see also Collins, Sargent and Neumann, 1993). However, there is no complete agreement on this point, as some researchers have found that paw preference may be determined genetically in some strains of mice (Biddle, Coffaro, Ziehr and Eales, 1993). Other forms of individual and population-biased lateralization would have to be examined, each in its own right, before making any conclusions about genetic determination or not.

There may be embryological constraints that lead to lateralization of behaviour. The asymmetrical development of the Mauthner cells may be a case in point, as outlined in Chapter 1 by Vallortigara and Bisazza. If the Mauthner cells on one side of the body are larger than the other, as is the case in anurans (Moulton, Jurand and Fox, 1968), population turning biases under stress conditions (C-turns) would result. This appears to be the case in fish and amphibia (Wassersug, Naitoh and Yamashita, 1999).

Ontogenetic factors play a significant role in determining a population bias for lateralization of some perceptual processes and motor responses. The chick model demonstrates the role of light experience in determining population lateralization for certain visual functions (see Chapter 6 by Deng and Rogers for details). Under natural conditions, this outcome could be achieved simply by allowing the eggs to be exposed to light during the final stages of incubation. Genes might then determine only that the embryo orients in the egg so that the right eye is positioned next to the shell and can be stimulated by light, whereas the left eye is occluded. Behavioural lateralization within the population would follow on from this. Nevertheless, light exposure of the late embryo does not influence all forms of lateralization. It is conceivable that some of the other forms of population lateralization, and indeed the basic differences between the hemispheres, are influenced by genes more directly. Future research will need to examine this.

References

Andrew, R.J., Mench, J. & Rainey, C. (1982). Left-right asymmetry of response to visual stimuli in the domestic chick. In *Analysis of Visual Behavior*, ed. D.J. Ingle, M.A. Goodale & R.J.W. Mansfield, pp. 197–209. Cambridge, MA: MIT Press.

Barnéoud, P., Nevue, P.J., Vitiello, S. & LeMoal, M. (1987). Functional heterogeneity of the right and left cerebral neocortex in the modulation of the immune system. *Physiology and Behavior*, 41, 525–530.

Biddle, F.G., Coffaro, C.M., Ziehr, J.E. & Eales, B.A. (1993). Genetic variation in paw preference (handedness) in the mouse. *Genome*, 36, 935–943.

Bisazza, A., Cantalupo, C., Capocchiano, M. & Vallortigara, G. (2000). Population lateralization and social behaviour: A study with sixteen species of fish. *Laterality*, 5, 269–284.

Bizazza, A., Cantalupo, C., Robins, A., Rogers, L.J. & Vallortigara, G. (1996). Right pawedness in toads. *Nature*, 379, 408.

Bizazza, A., Cantalupo, C., Robins, A., Rogers, L.J. & Vallortigara, G. (1997). Pawedness and motor asymmetries in toads. *Laterality*, 2, 49–64.

Box, H. (1977). Observations on spontaneous hand use in the common marmoset (*Callithrix jacchus*). *Primates*, 18, 395–400.

Bradshaw, J.L. (1989). *Hemispheric Specialization and Psychological Function*. Chichester: Wiley.

Cameron, R. & Rogers, L.J. (1999). Hand preference of the common marmoset, problem solving and responses in a novel setting. *Journal of Comparative Psychology*, 113, 149–157.

Casperd, J.M. & Dunbar, R.I.M. (1996). Asymmetries in the visual processing of emotional cues during agonistic interactions in gelada baboons. *Behavioral Processes*, 37, 57–65.

Collins, R.L. (1985). On the inheritance of the direction and degree of asymmetry. In *Cerebral Lateralization in Nonhuman Species*, ed. S.D. Glick, pp. 41–71. New York: Academic Press.

Collins, R.L., Sargent, E.E. & Neumann, P.E. (1993). Genetic and behavioral tests of the McManus hypothesis relating response to selection of lateralization of handedness in mice to degree of heterozygosity. *Behavioral Genetics*, 23, 413–421.

Cowell, P.E., Waters, N.S. & Denenberg, V.H. (1997). Effects of early environment on the development of functional laterality in Morris maze performance. *Laterality*, 2, 221–232.

Davidson, R.J. (1992). Emotion and effective style. *Psychological Science*, 3, 39–43.

Davidson, R.J. (1995). Cerebral asymmetry, emotion, and affective style. In *Brain Asymmetry*, ed. R.J. Davidson & K. Hugdahl, pp. 361–387. Cambridge, MA: MIT Press.

Deckel, A.W. (1995). Laterality of aggressive responses in *Anolis*. *Journal of Experimental Zoology*, 272, 194–200.

Deckel, A.W. (1998). Hemispheric control of territorial aggression in *Anolis carolinensis*: Effects of mild stress. *Brain, Behavior and Evolution*, 51, 33–39.

Deng, C. & Rogers, L.J. (1997). Differential contributions of the two visual pathways to functional lateralization in chicks. *Behavioural Brain Research*, 87, 173–182.

Drews, C. (1996). Context and patterns of injuries in free-ranging male baboons (*Papio cynocephalus*). *Behaviour*, 133, 443–474.

Dugatkin, L.A. (1997). *Cooperation Among Animals: An Evolutionary Perspective.* Oxford: Oxford University Press.

Dunaif-Hattis, J. (1984). *Doubling the Brain.* New York: Peter Lang.

Evans, C.S., Evans, L. & Marler, P. (1993). On the meaning of alarm calls: Functional references in an avian vocal system. *Animal Behaviour*, 46, 23–28.

Fabre-Thorpe, M., Fagot, J., Lorincz, E., Levesque, F. & Vauclair, J. (1993). Laterality in cats: Paw preference and performance in a visuomotor activity. *Cortex*, 29, 15–24.

Ferrari, S.F. (1993). Ecological differentiation in the Callitrichidae. In *Marmosets and Tamarins: Systematics, Behaviour, and Ecology*, ed. A.B. Rylands, pp. 314–328. Oxford: Oxford University Press.

Fersen, von L. & Güntürkün, O. (1990). Visual memory lateralization in pigeons. *Neuropsychologia*, 28, 1–7.

Gazzaniga, M.S. & Le Doux, J.E. (1978). *The Integrated Mind.* New York: Plenum Press.

Geschwind, N. & Behan, P. (1982). Left-handedness: Association with immune disease, migraine, and development of learning disorders. *Proceedings of the National Academy of Science USA*, 79, 5097–5100.

Güntürkün, O., Diekamp, B., Manns, M., Nottlemann, F., Prior, H., Schwarz, A. & Skiba, M. (2000). Asymmetry pays: Visual lateralization improves discrimination success in pigeons. *Current Biology*, 10, 1079–1081.

Güntürkün, O. & Kesch, S. (1987). Visual lateralization during feeding in pigeons. *Behavioural Neuroscience*, 101, 433–435.

Heilman, K.M., Watson, R.T. & Valenstein, E. (1985). Neglect and related disorders. In *Clinical Neuropsychology*, ed. K.M. Heilman & E. Valenstein, pp. 377–401. New York: Oxford University Press.

Hellige, J.B., Bloch, M.I., Cowin, E.L., Eng, T.L., Eviatar, Z. & Sergent, V. (1994). Individual variation in hemisphere asymmetry: Multitask study of effects related to handedness and sex. *Journal of Experimental Psychology: General*, 123, 257–263.

Hook-Costigan, M.A. & Rogers, L.J. (1995). Hand, mouth and eye preferences in the common marmoset (*Callithrix jacchus*). *Folia Primatologica*, 64, 180–191.

Hook-Costigan, M.A. & Rogers, L.J. (1996). Hand preferences in New World primates. *International Journal of Comparative Psychology*, 9, 173–207.

Hook, M.A. & Rogers, L.J. (2000). Development of hand preferences in marmosets (*Callithrix jacchus*) and effects of ageing. *Journal of Comparative Psychology*, 114, 263–271.

Hopkins, W.D. & Bennett, A. (1994). Handedness and approach-avoidance behaviour in chimpanzees. *Journal of Experimental Psychology*, 20, 413–418.

Khyentse, M.D. & Rogers, L.J. (1997). Glutamate affects the development of the thalamofugal visual projections of the chick. *Neuroscience Letters*, 230, 65–68.

King, J.E. (1995). Laterality in hand preferences and reaching accuracy of cotton-top tamarins (*Saguinus oedipus*). *Journal of Comparative Psychology*, 109, 34–41.

Lippolis, G., Bisazza, A., Rogers, L.J. & Vallortigara, G. (2001). Lateralization of predator avoidance responses in three species of toads. *Laterality*, in press.

McGrew, W.C. & Marchant, L.F. (1992). Chimpanzees, tools, and termites: Hand preference or handedness? *Current Anthropology*, 33, 114–119.

McGrew, W.C. & Marchant, L.F. (1999). Laterality of hand use pays off in foraging success for wild chimpanzees. *Primates*, 40, 509–513.

McKenzie, R., Andrew, R.J. & Jones, R.B. (1998). Lateralization in chicks and hens: New evidence for control of response by the right eye system. *Neuropsychologia*, 36, 51–58.

McNeilage, P.F., Studdert-Kennedy, M.G. & Lindblom, B. (1987). Primate handedness reconsidered. *Behavioral and Brain Sciences*, 10, 247–263.

Mench, J. & Andrew, R.J. (1986). Lateralization of a food search task in the domestic chick. *Behavioral and Neural Biology*, 46, 107–114.

Mesulam, M.-M. (1985). Attention, confusional states and neglect. In *Principles of Behavioral Neurology*, ed. M.-M. Mesulam, pp. 125–168. Philadelphia: F.A. Davis.

Møller, A.P. & Swaddle, J.P. (1997). *Asymmetry, Developmental Stability, and Evolution*. Oxford: Oxford University Press.

Moulton, J.M., Jurand, A. & Fox, H. (1968). A cytological study of Mauthner's cells in *Xenopus laevis* and *Rana temporaria* during metamorphosis. *Journal of Embryology and Experimental Morphology*, 19, 415–431.

Myslobodsky, M.S. (1983). *Hemisyndromes: Psychobiology, Neurology, Psychiatry*. New York: Academic Press.

Nevue, P.J. (1988). Cerebral neocortex modulation of immune functions. *Life Sciences*, 42, 1917–1923.

Nevue, P.J. (1996). Lateralization and stress response in mice: Interindividual differences in the association of brain, neuroendocrine, and immune responses. *Behavior Genetics*, 26, 373–377.

Nevue, P.J., Betancur, C., Barnéoud, S., Vitiello, S. & LeMoal, M. (1991). Functional brain asymmetry and lymphocyte proliferation in female mice: Effects of right- and left-cortical ablation. *Brain Research*, 550, 125–128.

Peters, M. (1995). Handedness and its relation to other indices of cerebral lateralization. In *Brain Asymmetry*, ed. R.J. Davidson & K. Hugdahl, pp. 183–214. Cambridge, MA: MIT Press.

Posner, M.I., Walker, J.A., Friedrich, F.A. & Rafal, R.D. (1987). How do the parietal lobes direct covert attention? *Neuropsychologia*, 25, 135–146.

Rashid, N. & Andrew, R.J. (1989). Right hemisphere advantage for topographical orientation in the domestic chick. *Neuropsychologia*, 27, 937–948.

Robins, A., Lippolis, G., Bisazza, A., Vallortigara, G. & Rogers, L.J. (1998). Lateralized agonistic responses and hind-limb use in toads. *Animal Behaviour*, 56, 875–881.

Rogers, L.J. (1980). Lateralization of the avian brain. *Bird Behaviour*, 2, 1–12.

Rogers, L.J. (1982). Light experience and asymmetry of brain function in chickens. *Nature*, 297, 223–225.

Rogers, L.J. (1990). Light input and the reversal of functional lateralization in the chicken brain. *Behavioural Brain Research*, 38, 211–221.

Rogers, L.J. (1991). Development of lateralization. In *Neural and Behavioural Plasticity: The Use of the Domestic Chick as a Model*, ed. R.J. Andrew, pp. 507–535. Oxford: Oxford University Press.

Rogers, L.J. (1995). *The Development of Brain and Behaviour in the Chicken*. Wallingford: CAB International.

Rogers, L.J. (1996). Behavioral, structural and neurochemical asymmetries in the avian brain: A model system for studying visual development and processing. *Neuroscience and Biobehavioral Reviews*, 20, 487–503.

Rogers, L.J. (1997). Early experiential effects on laterality: Research on chicks has relevance to other species. *Laterality*, 2, 199–219.

Rogers, L.J. (1999). Factors influencing exploration in marmosets: Age, sex and hand preference. *International Journal of Comparative Psychology*, 12, 93–109.

Rogers, L.J. (2000). Evolution of hemispheric specialisation: Advantages and disadvantages. *Brain and Language*, 73, 236–253.

Rogers, L.J. & Anson, J.M. (1979). Lateralization of function in the chicken forebrain. *Pharmacology, Biochemistry and Behavior*, 10, 679–686.

Rogers, L.J. & Bolden, S.W. (1991). Light-dependent development and asymmetry of visual projections. *Neuroscience Letters*, 121, 63–67.

Rogers, L.J. & Deng, C. (1999). Light experience and lateralization of the two visual pathways in the chick. *Behavioural Brain Research*, 98, 277–287.

Rogers, L.J. & Hambley, J.W. (1982). Specific and nonspecific effects of neuro-excitatory amino acids on learning and other behaviours in the chicken. *Behavioural Brain Research*, 4, 1–18.

Rogers, L.J. & Workman, L. (1989). Light exposure during incubation affects competitive behaviour in domestic chicks. *Applied Animal Behaviour Science*, 23, 187–198.

Rogers, L.J. & Astiningsih, K. (1991). Social hierarchies in very young chicks. *British Poultry Science*, 32, 47–56.

Rogers, L.J., Zappia, J.V. & Bullock, S.P. (1985). Testosterone and eye–brain asymmetry for copulation in chickens. *Experientia*, 41, 1447–1449.

Rothe, H. (1973). Handedness in the common marmosets (*Callithrix jacchus*). *American Journal of Physical Anthropology*, 64, 417–433.

Tardif, S.D., Harrison, M.L. & Simek, M.A. (1993). Communal infant care in marmosets and tamarins: Relation to energetics, ecology, and social organisation. In *Marmosets and Tamarins: Systematics, Behaviour, and Ecology*, ed. A.B. Rylands, pp. 220–234. Oxford: Oxford University Press.

Vallortigara, G., Cozzutti, C., Tommasi, L. & Rogers, L.J. (2001). How birds use their eyes: Opposite left–right specialization for the lateral and frontal visual hemifield in the domestic chick. *Current Biology*, 11, 29–33.

Ward, J.P. & Hopkins, W.D. (1993). *Primate Laterality: Current Behavioural Evidence of Primate Asymmetries*. New York: Springer-Verlag.

Ward, J.P., Milliken, G.W., Dodson, D.L., Stafford, D.K. & Wallace, M. (1990). Handedness as a function of sex and age in a large population of lemur. *Journal of Comparative Psychology*, 104, 167–173.

Wassersug, R.J., Naitoh, T. & Yamashita, M. (1999). Turning biases in tadpoles. *Journal of Herpetology*, 33, 543–548.

Workman, L. & Andrew, R.J. (1989). Simultaneous changes in behaviour and lateralization during the development of male and female domestic chicks. *Animal Behaviour*, 38, 596–605.

Zappia, J.V. & Rogers, L.J. (1983). Light experience during development affects asymmetry of fore-brain function in chickens. *Developmental Brain Research*, 11, 93–106.

Zeier, H. (1975). Interhemispheric interactions. In *Neural and Endocrine Aspects of Behaviour in Birds*, ed. P. Wright, P.G. Caryl & D.M. Vowles, pp. 163–180. Amsterdam: Elsevier.

Part Two
Development of Lateralization

5

Behavioural Development and Lateralization

RICHARD J. ANDREW

5.1. Evidence from the Domestic Chick

Current evidence suggests that, in the domestic chick, lateralization varies along at least three axes: (1) the specialization of left hemisphere (LHem) and right hemisphere (RHem) for perceptual processing; (2) bias to control by one or other hemisphere, which itself may depend on more than one mechanism; and (3) the dominant pattern of interaction between the hemispheres: collaboration or alternation. It will be argued here that, whilst (1) is predominantly established in early development, as a direct consequence of the appearance of the right–left axis of the body, (2) and (3), whilst perhaps initially so established, are subsequently modulated by factors, which can vary between individuals during development. Variation due to such modulating factors can be marked.

5.1.1. Age-dependent Shifts in Standing Bias to Control by One or Other Hemisphere

In the past there has been considerable interest in the possibility that in humans there may be individual differences in the likelihood that RHem or LHem will take charge, independently of effects of task properties. Attempts to find signs of such 'hemisphericity' have been discouraging (Boles, 1991), partly because it is difficult to decide how to distinguish a standing bias from effects of learned strategies (which themselves might be affected by an earlier period of bias, now over, or be generated in some other way). Dabbs and Choo (1980) tried to avoid this problem by establishing in advance whether human subjects had advantage in verbal or spatial tasks, and then measuring relative blood flow in the two hemispheres whilst the subject was lying relaxed. High spatial scores were accompanied by greater carotid blood flow to the RHem,

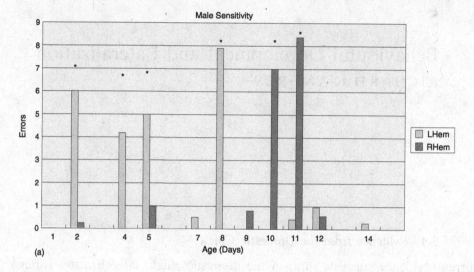

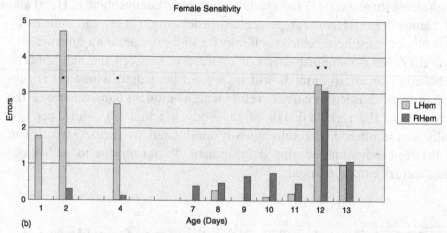

Figure 5.1.   The figure shows, as a measure of long-term disturbance of visual learning, the number of 'errors' (i.e. pecks directed at inedible pebbles rather than food grains) in the last 20 pecks of a 60-peck training session. This 'pebble floor' task is described in the text; note that random direction of pecks would give scores around 10–12.

(a) In males, the left hemisphere (LHem) is sensitive to the injection of the amnestic agent (cycloheximide) on days 2, 4, 5 and 8. The right hemisphere (RHem) is sensitive on days 10 and 11, whilst neither is sensitive on days 1, 7, 9 and 14. Sensitivity requires that the injected hemisphere should be used actively in visual learning, despite disturbances due to injection (see text). This happens in males when there is strong age-dependent bias to the use of the injected hemisphere. A marked transition is revealed between bias to LHem control on day 8 and RHem control on days 10 and 11. On days 7 and 9 there is insufficient standing bias to cause the injected and disturbed hemisphere to continue to be involved in learning. The earlier period of LHem sensitivity (days 2–5) may also depend on task-dependent engagement of the LHem. In these relatively inexperienced chicks the task is likely to be more difficult and so engage LHem abilities more fully (see text).

and high verbal scores by greater flow to the LHem. This finding does not exclude the possibility that mental processes were still determined by learned strategies, rather than by a standing bias; it is difficult to see, however, what further steps could be taken to clinch the matter.

There is much clearer evidence of periods of standing bias to control by one or other hemisphere during development in the domestic chick. I begin with a review of the current state of knowledge concerning these periods.

The first evidence for the age-related changes in bias in the chick came from studies of shifts of sensitivity to intrahemispheric and unilateral injection of cycloheximide (Rogers and Ehrlich, 1983). The fullest version of this time-course is shown in Figure 5.1; see also Rogers (1991). Males show sensitivity to LHem injection on days 2, 4 and 5, with insensitivity (to either RHem or LHem injection) on day 7, a return to LHem sensitivity on day 8, and then a shift to RHem sensitivity on days 10 and 11, preceded and followed by insensitivity on days 9 and 12. The measure used was long-term disturbance of the ability to learn to choose food grains in preference to novel and inedible alternative targets. It was necessary both for the agent to be given to a hemisphere that was currently sensitive, and for the chick to engage in visual processing during the action of the agent (Rogers and Drennen, 1978). The same held for another agent that disturbs memory formation: glutamate (Sdraulig, Rogers and Boura, 1980). It was argued that the disturbance, which resulted from effective action of either agent, was in the ability to carry out visual learning.

Behavioural evidence suggests that the hemisphere, which is sensitive to unilateral application of cycloheximide or glutamate also, tends to control behaviour in the normal chick. Again I will consider male chicks; females are discussed in a later section. Workman and Andrew (1989) found, in broods reared by a broody hen under seminatural conditions, that the shift from LHem sensitivity to RHem sensitivity, between day 8 and day 10, coincided

Caption for Fig. 5.1 (*cont.*)

    (b) Female chicks show far less sensitivity, despite the fact that they undergo the same age-dependent shifts in bias. It is likely that this is because there is a stronger tendency for whichever hemisphere is the more active to take over full control. This would increase the likelihood that the injected and disturbed hemisphere would continue to be involved in learning, and so protect it from long-term disturbance. The occurrence of sensitivity in both RHem and LHem on day 12 may mark a new developmental stage, perhaps a stage when there is unusually strong functional linkage between the hemispheres. Modified from Rogers (1991).

    (Redrawn from L.J. Rogers, 'Development of lateralization', in *Neural and Behavioural Plasticity*, ed. R.J. Andrew, Oxford University Press, 1991. Copyright 1991 Oxford University Press; produced by permission of Oxford University Press.)

with a set of behavioural changes, all of which appeared abruptly on days 10 and 11. These changes were such as would be expected, if the RHem had indeed taken control on the latter 2 days (see below).

Subsequently a much wider range of behaviours have been studied in male chicks raised singly in the laboratory (Andrew et al., in preparation; Rogers, Andrew and Johnston, in preparation). Here complications due to possible changes in the behaviour of the hen were excluded. The results are summarized in Table 5.1. Although, in general, there is agreement between the two bodies of data, there is discrepancy on day 4. The behavioural bias was first found using data for ear use. The side of the head (and so the ear), which was turned towards the sound of hen clucks, varied with age. If it is assumed that the controlling hemisphere was contralateral to the ear that was used, then initially there was LHem control (days 2 and 3), then RHem (day 4), then LHem again (days 5 and 8), with a shift to RHem around day 9. The changes in bias from day 5 to day 11 are those predicted by the sensitivity timecourse (days 9, 10 and 11).

Four other quite different types of test confirmed control by the RHem on day 4. In all cases chicks were tested once only:

1. The side on which a distracting stimulus was most effective in fully interrupting feeding was the right on day 8, and the left on day 10. This suggests that, when the controlling hemisphere sees a distractor, the immediate effect on concurrent feeding is greater than when it is the other hemisphere that sees the distractor. On day 3 a distractor on the right side is more effective, but on day 4 this is true of the left side. This is consistent with bias to RHem control on day 4.

2. In a test of exploration away from the imprinting object, which was placed at the centre of a 'forest' of pillars, chicks went further on day 4 than on day 3 or day 5. This is very like the appearance on days 10 and 11 (see below) of the behaviour of leaving the hen to travel in the general environment.

3. When a potentially frightening stimulus (a pair of large artificial eyes) was shown to chicks with either right or left eye in use, due to an eye patch, fear was greater when it was the eye ipsilateral to the controlling hemisphere which saw the stimulus. This was demonstrated in older chicks, where the direction of hemispheric bias was already known. The scores for younger chicks imply that RHem controlled on day 4 and LHem on day 5.

4. When clucks were played to chicks with either right or left auditory meatus temporarily blocked, two different patterns of behaviour were observed. In one pattern, chicks waited for longer but then went straight to the sound source; in the other, chicks moved readily, but then wandered aimlessly. The latter pattern was commoner with the left ear plugged on day 4 but with the right ear plugged on day 4. The reversal of condition between the two days was marked and suggests that there is shift of bias.

Table 5.1. *A wide range of behavioural measures sensitive to lateralization were taken on different days of life in male chicks that were reared singly in the laboratory. All of the results are summarized by showing the hemisphere that is deduced to be controlling (see text). For example, if injection of the LHem but not the RHem is effective, or if the RE is used in viewing, the LHem (L) is deduced to be controlling. The main discrepancy is for day 4 (D4) when sensitivity implicated the LHem but behavioural tests the RHem. This is discussed in the text*

	D1	D2	D3	D4	D5	D6	D7	D8	D9	D10	D11	D12
SENS		L		L	L		0	L	0	R	R	0
TOPO							L	L			R	R
HEN							L	L	R		R	
WS							L	=			R	
CLUCK	R	L	L	R	L	=	=	L	R	R	R	
DISTR			L	R	=			L		R		
OWL			=	R	L			L		R		
MEAL				L		L	L					
EXPLO			L	R	L							
EARPL		L		R								

The tests are shown as: SENS, sensitivity to unilateral injection; TOPO, ability to use topographical cues (lower with the LHem, higher with the RHem controlling); HEN, eye used by chicks to view a live hen; WS, eye used to view on first exposure to a white sphere (a potential imprinting object); CLUCK, ear turned towards the sound of hen clucks to which the chick is attached); DISTR, degree of distraction by a stimulus on the left or on the right whilst feeding; OWL, level of fear responses evoked by a pair of artificial eyes ('owl eyes'), when using only the RE or the LE; MEAL, intensity of disturbed behaviour at first encounter with a mealworm (larval *Tenebrio* beetle); EXPLO, distance travelled away from the imprinting object in exploration; EARPL, behaviour of chicks with one or other external meatus plugged when attempting to approach a source of clucks. When there is known to be no sensitivity this is shown by 0. When a behavioural measure showed no significant bias to RHem or LHem control, this is shown by =.

It thus seems certain that there is an age-dependent bias to RHem control on day 4. Some explanation for LHem sensitivity to cycloheximide on that day is therefore needed. The most likely is that the behaviour usual on day 4 (feeding and resting in a familiar homecage) put the LHem in charge in these very young chicks, in which interest in learning about food is still high, despite the presence of bias to the RHem.

I now turn to possible functions of age-dependent shifts in bias. Changes in the behaviour patterns shown by broods of chicks raised by a broody hen under seminatural conditions parallel the changes in bias over the period from day 8 to day 11 (Workman and Andrew, 1989; Vallortigara et al., 1997; Workman, Adam and Andrew, 2000).

RE fixation is used when deciding whether to approach a novel imprinting object, suggesting that LHem bias may be appropriate to days when attachment should occur. Day 3 is probably the most usual day for the beginning of attachment to, and learning to distinguish siblings, because it is the first day when group emergences become common, and chicks follow each other in groups around the mother. It is a day of LHem bias, as is day 5 (separated from it by an intercalated RHem day). Day 5 is again a day when attaching to a visual stimulus with novel features may be crucial. Although the mother's head has been seen previously in independent motion, and no doubt the chicks have learned its characteristics and attached to it, day 5 is the usual day on which the mother first moves away from the nest. At this time the chicks see her for the first time in translatory motion, and it is presumably necessary that they should learn her new characteristics and attach to them.

What may be the advantages for learning, if the initial opportunity for learning is done under LHem control, with RHem control following soon after? This is in fact not only favoured by the age-dependent shifts in bias that have just been discussed: it is also the sequence which is followed in the first hour of visual imprinting, judging by patterns of eye use (McKenzie, Andrew and Jones, 1998). Part at least of the advantage may come from the need in very early learning to separate the object of interest from its spatial context, both during perceptual analysis, and in the memory records that are set up. It is also likely to be important to select key properties of the stimulus early in learning. Evidence is reviewed later, which suggests that the LHem is indeed better able to do both these things. On the other hand, the greater ability of the RHem to attend to and record the full range of properties of an object, and to use these properties in recognition (e.g. of siblings) to establish identity or novelty makes valuable its participation in later learning.

Shifts in bias are probably also important ways by which the range of behaviours shown by the chicks is constrained. It is, of course, essential that the chicks follow the mother closely and reliably, as soon as she leaves the nest. Right-eye fixation is used during approach to an object to which a motor response is planned, suggesting special involvement of the LHem in setting up and sustaining a planned motor response under visual control (see Chapter 3 by Andrew and Rogers). Promotion of following by LHem, with attention focused continuously on the hen, may also be served by the continuing LHem bias during the period from day 5 to day 8, in which the dominant behaviour of the chick is following the mother, and feeding where she indicates.

The shift to RHem bias on days 10 and 11 allows movement independent of the mother into the environment, such as is necessary for learning to orient by various features of the environment other than mother and siblings. The skills needed may include knowledge of size constancy for common standard features of the environment, and interpretation of parallax. Local layout may also be learned and perhaps orientation by distant features. Behaviour to siblings also changes: on day 11 frolicking appears, increasing further in frequency on day 12. Frolicking in response to a sibling consists of a run directed at a sibling, which often ends in a collision. It suggests not only a new interest in siblings, but also a withdrawal of the ability to control a vigorous response, especially to check it suddenly at its end. This latter change may reflect progressively less involvement of the LHem in the control of behaviour, during the course of the period of age-dependent bias to RHem control. Similar quantitative shifts in hemispheric involvement during a period of apparent dominance of one or other hemisphere are suggested by human data (see below).

Special behaviour on day 8, which is the day when chicks for the first time begin to stand and stare at objects that move independently in the environment (such as a human observer or a strange conspecific) but without approach, can also be fitted into a scheme of this sort. On day 8, the last day of LHem bias on the above hypothesis, it is likely that the RHem is beginning to be involved. As a result, the chick will tend to pick out as interesting, conspicuous objects that shift position within the environment. However, on day 8, the RHem does this with the LHem in control and inhibiting approach.

Clearly, such examination without approach is advantageous. Objects moving independently in the environment are potentially dangerous. Any opportunity to learn their characteristics safely should be taken, particularly just before days when the chick will commence independent movement through the environment, and so may encounter such objects, without the mother noticing and being able to intervene.

5.1.2. Possible Sex Differences in the Time-course of Development

The sensitivity time-course for females differs markedly from that for males. There is the same early sensitivity to LHem injection of cycloheximide on days 2 and 4, but the day 8 peak of LHem sensitivity and the RHem sensitivity on days 10 and 11 are completely lacking (Rogers, 1991).

It is unlikely that this is because the shifts in bias have quite different timings (or are absent) in females. Firstly, behavioural changes in normal broods show the same changes in both sexes on days 10 and 11. It might be argued that in broods containing males and females – as was true of all broods studied (by Workman and Andrew, 1989) – changes in the behaviour of females were driven by changes in that of males.

However, this cannot apply to evidence, based on the behaviour of singly housed chicks. The best established age-dependent shift, namely from bias to LHem control on day 8 to RHem on days 10 and 11, has been confirmed by two other techniques, where social effects of males on females can be excluded. (1) Regolin and Vallortigara (1995) used tests in which female chicks pecked (due to food reinforcement) at a pair of stimuli, one on the right and one on the left. When one of the stimuli was changed in appearance, pecks tended to be directed at it because of its novelty. On day 8 this effect was more pronounced when the novel stimulus was on the left, but on day 11 it was novel stimuli on the right that tended to be pecked. (2) Dharmaretnam and Andrew (1994) described the use in female chicks of the subordinate eye system to monitor small novel stimuli: on day 5 this was the left eye system, and on day 11 the right eye. Again day 5 and day 11 show reverse eye use.

Secondly, in tests of topographical ability (Rashid, 1988), both male and female left-eyed chicks consistently showed the same poor ability on day 8 to use topographical cues in orientation (as compared to days 10 and 11, when the expected left eye advantage was marked in both). Strong bias to LHem control on day 8 thus seems to interfere with left eye performance in both sexes in this test. The same resemblance of timing between the sexes was also present for a time-course of level of fear (Andrew and Brennan, 1984), with right-eyed, and binocular, males and females all showing the same striking changes between day 8 and day 10.

If the absence of sensitivity to unilateral injection of cycloheximide in females from day 7 onwards is not caused by absence of periods of age-dependent bias, what does cause it? An explanation is suggested by the presence of days of insensitivity in males too (days 7 and 9). Remember that it is necessary for active perceptual processing (and probably learning) to be attempted by the treated hemisphere, if the agent is to induce disturbance of subsequent learning. One likely consequence of injection is general interference with the working of the injected hemisphere. Strong bias to continued use of the affected hemisphere may be necessary, if it is to continue to process perceptual input, rather than allowing control to shift to the other and unaffected hemisphere. In males, at ages when there is little or no such bias (days

7 and 9), the main effect of the agent may be to inactivate the injected hemisphere.

The special property of females, which may explain the sex difference on days 10 and 11, is that females show a greater tendency for one hemisphere at a time to take overall control of attentional strategy (see Section 5.1.5). This may exaggerate the depressive effect of the injection on the injected hemisphere to a point where even strong age-dependent bias to control cannot be sustained.

5.1.3. *Effects of Light Late in Incubation on Chick Lateralization*

In fowl the late embryo consistently holds its head so that the left side faces in to the body and is therefore shielded from entering light (Rogers, 1990). Exposure of the egg to modest levels of light stimulates the right but not the left eye, and generates asymmetry in the projection from eye to the visual Wulst (Rogers and Sink, 1987; discussed in detail in Chapter 6 by Deng and Rogers). This asymmetry is due to a greater development of the projections from the exposed (right) eye to the Wulst that is ipsilateral to it than occurs in the corresponding projection from the unexposed eye.

This, and subsequent studies from Rogers' laboratory, so convincingly established this striking phenomenon as to suggest not only that this effect was the main or only route by which lateralization is established in the domestic fowl, but also (since the anatomical asymmetry is transient) that functional lateralization must also be transient in the fowl. Neither position was put forward by the discoverers (e.g. Rogers, 1991). However, it remains important to stress that the effect of light is only one modulating factor, acting on a pattern of lateralization established much earlier in development.

This was clear from the start, in that the assumption of the standard head position is itself dependent on a pre-existing asymmetry of the central nervous system (CNS): note that the long mobile neck has to be bent in a complex way to bring it about. Liederman and Kinsbourne (1980) make the same point about asymmetries of head position in human infants. Here it seems that the foetus selects a standard position in the uterus with the head down and the left side towards the mother's pelvis and backbone, and so with the left arm relatively more constrained than the right (Michel, 1983). Note too that hens show very similar features of lateralization to chicks (McKenzie, Andrew and Jones, 1998), showing that lateralization is not a transient state in the fowl.

Deng (1998) showed that dark-incubated chicks still showed the left eye (LE) advantage in ability to choose between a familiar and unfamiliar chick, as described by Vallortigara and Andrew (1991). Recently a more extensive series of studies were undertaken by Rogers and collaborators, which were designed to compare chicks that had never been exposed to light during incubation with ones receiving light exposure that was known to be adequate to bring about the anatomical and behavioural effects of light exposure. These studies also compared males and females, so that the two types of modulation of developmental of lateralization could be compared within single experiments.

Rogers, Andrew and Burne (1998) used a test in which response to a novel odorant was measured with either the right or left nostril blocked. The odorant was presented in association with a bead, which was (because of its colour) either modestly effective or strongly effective in evoking visually driven responses like pecking. In the latter case, when the visually evoked response was likely to interfere with the olfactory, there was striking asymmetry: olfactory responses were much higher with right nostril input, than with left. When the visual properties were less effective, there was no significant difference between right and left nostril groups. There was thus no asymmetry in the response to the odorant *per se*.

The olfactory input from the left nostril enters the left hemisphere. The central analysis of left nostril input therefore interacts within a single hemisphere with visual control of pecking. Since the visual input precedes the olfactory, it is processing of the latter that is more likely to be blocked if there is interference. Crucially, it appears that this is more probable when interaction occurs within the same hemisphere. It was already known (Vallortigara and Andrew, 1994) that right nostril input of the odorant, which was used in these experiments, is noticed. Such input affects choice of imprinting object, whereas left nostril input does not appear to be noticed. Asymmetry between the effects of right and left nostril input is thus present for two quite different classes of visual object.

The key finding here is that the presence or absence of light during incubation had no effect on this behavioural asymmetry revealed by occluding the left or right nostril: it was similarly developed in both groups. Differences between RHem and LHem, which are responsible for differences in perceptual processing, are thus established in this instance by some other route than light exposure of the embryo.

A second study (Andrew et al., in preparation) used choice between a familiar imprinting object (a red ball with a white strip) and a transformation of that object (change in orientation of a white strip on the face of the ball). This transformation was known (Vallortigara and Andrew, 1991) to affect choice in chicks that were using their left eye but not in ones that were using the right eye. Differences between right- and left-eyed birds (RE, LE) were expected here for dark-incubated groups (Da), on the basis of earlier work (Deng, 1998). They were indeed obtained, with Da LE choosing familiar, and Da RE failing to choose. More unexpectedly, light-exposed groups (Li) showed no difference between RE and LE; instead, Li LE showed the same behaviour as Li RE (namely, lack of choice). Binocular groups (BIN) behaved like RE in both Da and Li.

The explanation proposed by the authors is that the effect of RE exposure to light is to promote control by the right eye system (LHem). This puts the LHem in control in Li, not only when the chicks use both eyes (BIN) but also when they use only the left or even only the right eye. This hypothesis is supported by the fact that age-dependent bias to LHem control on day 8 interferes with the use of topographical cues in orientation by LE birds (Rashid, 1988). Other instances of effects of a 'non-seeing' dominant hemisphere (i.e. contralateral eye patched) on its partner are considered later, and in Chapter 15 by Andrew.

Both male and female groups behaved similarly in their choice between model (red ball with white strip) and transform (re-oriented white strip). However, the same groups of birds did show sex differences in behaviour, once they had chosen a model and taken up position close by it. The behaviour in which they differed was exploratory pecks directed at features of the model or its surroundings. Only in the chicks tested binocularly was such behaviour common enough to allow analysis. Females pecked at the white strip (the most conspicuous local feature of the imprinting object) more than did males. This suggests that females sustained attention on the imprinting object more than did males, and so were more likely to respond to its features.

A second finding is consistent with this. One group (Da males) responded far more than any other did to a conspicuous feature, which touched the model, but was clearly separate from it. In order to prevent accidental movement of the models, they were stabilized with a blob of blue adhesive, which could be seen only when the chick was standing so close to the model that it could peer behind it. Da males pecked at the adhesive. The chicks had not previously seen anything in association with the ball (with which they had been living). Both on colour and on shape it was more like an environmental feature resting against the ball than a part of it. It thus is likely that Da males tended to shift attention away from the prime focus of attention, the ball, more than other groups. Note that this was the group of chicks least likely to sustain LHem control after reaching the model: *ex hypothesi* they lack the bias to LHem control induced by light exposure, and they lack the female ability by which the controlling hemisphere constrains the attention of its partner (i.e. the other hemisphere).

The next question is clearly what features of lateralization are unaffected by light exposure. The data just reviewed establish that RHem involvement in the detection and assessment of novelty is one such feature. It is likely (but not certain) that the hatchery chicks used in studies at Sussex University are effectively dark incubated (based on reports of hatchery staff). If so, then the use of the RE to fixate a manipulandum (see Chapter 2 by Andrew), as part of the visual control of motor response, is another such feature. The results presented in the previous section for male Da are consistent with this.

The studies of acoustic lateralization that are reported in Chapter 10 by Andrew and Watkins were certainly carried out on Da chicks, since they were incubated in the dark. This means that preferential choice of sounds (hen clucks) first heard in the right hemifield is another feature present in Da; it is comparable with right eye control of response.

In addition, exposure to light late during incubation is almost always excluded or minimized in imprinting studies (McCabe, 1991, p. 274), because it is important to be sure that early hatching does not expose some chicks to the sight of siblings. It is likely therefore that many of the asymmetries in the anatomical and biochemical consequences of imprinting (see Chapter 14 by Johnston and Rose) are not dependent on asymmetric exposure to light. All in all, the evidence suggests that the basic pattern of lateralization is probably present by the end of incubation, and that asymmetric exposure to light has a specific further modulatory effect.

The differences between Li and Da are well explained, if it is assumed that exposure of one eye system to illumination late in incubation sets up a standing bias to control by that eye system. This would be unlike the age-dependent biases, which have already been discussed, in that it is relatively persistent.

Strikingly, it is possible by experimentally providing light input during development to the LE, instead of the RE, to reverse anatomical and behavioural asymmetries (Deng and Rogers, 2000; Rogers, 1990; Chapter 6 by Deng and Rogers). Sensitivity to glutamate injection on day 2 is also reversed (so that it is the RHem that is sensitive). Clearly it is of great interest to establish exactly what properties of lateralization are reversed by this procedure. The approach taken here is to consider whether the full complex of LHem properties must be considered as being reversed or not.

Most data come from measurements of performance following unilateral injection with glutamate. In this case it is possible to invoke a two-step process. If reversed light exposure led to a simple reversal of bias to control by the RHem, instead of the LHem, the second step might be that it would be the RHem that was engaged in visual learning, when the glutamate was administered (on day 2 posthatching). This in turn could cause unilateral disturbance of subsequent visual learning involving the RHem.

A test that has been much used in these studies is the pebble floor test. Chicks are placed on a test floor covered with small pebbles about the size and hue of the familiar food grains that are scattered amongst them. The pebbles have not been seen before and are attached firmly to the floor so that they cannot be swallowed. The measure taken is the number of pecks in which the chick learns to avoid pecking pebbles. In chicks with normal RE exposure to light during incubation, pebbles cease to be pecked sooner when the LHem is functional and controlling (see Chapter 6 by Deng and Rogers). Disturbance of visual learning by the LHem would therefore be expected to affect performance in the pebble-floor test more strongly than corresponding disturbance in the RHem. In chicks with normal exposure it would be necessary to inject the LHem to produce this effect, but in chicks with reversed exposure injection of the RHem would be effective, as is indeed the case (Chapter 6 by Deng and Rogers, Figure 6.5).

Chicks without strong bias to control by either hemisphere would be insensitive to glutamate (as already discussed) because the effects of glutamate would inactivate the injected hemisphere, protecting it from disturbance. Chicks, which receive either no light exposure or bilateral exposure during incubation, apparently show no strong standing bias (see below).

Further, it is also possible to extend such an explanation to effects on copulation and attack. In many vertebrates, copulation and attack are facilitated when the eliciting object is novel (see review by Hinde, 1970, pp. 330–332; fowl: Guhl, 1962). It is likely that this is also true for chicks (where pecks are indeed evoked more strongly by stranger than by partner chicks). Disturbance of ability to learn the visual characters of partners could mean that the partner is continuously novel, and so continuously effective in evoking attack and copulation.

Data are also available for monocular testing on the pebble floor, following different exposure regimes during incubation.

I begin with the findings when dark-incubated chicks and chicks with normal exposure during incubation are compared. It was already known that Dᶏ chicks show the standard better performance with the RE in use, when tested younger than usual, namely on day 3 (Mench and Andrew, 1986). Confirmatory evidence that this can also occur in older chicks, but only under special conditions, has recently been provided (Andrew, Rogers and Johnston, in preparation). Chicks were tested on day 8 or 9 (when there is usually no difference in Da chicks between RE and LE, because both learn well). The chicks were more disturbed by test conditions than usual because of constraints imposed by earlier tests. These included opaque fronts to home cages, so that they were far less exposed to humans than usual, and living with an imprinting object, whose absence at test would also have been disturbing. Da RE chicks differed from their LE counterparts in showing an earlier reduction in pebble pecks. The resulting difference between RE and LE scores was closely similar to the difference between RE and LE Li groups. However, Da RE did differ from Li RE in another way: their overall rates of pecking were much slower. It seemed that Da RE were capable of applying the same strategy as Li RE, but did so under these conditions with considerable difficulty.

The most probable explanation is that the ability of the LHem to sustain control is less in Da chicks, although it is still assigned particularly to the LHem. The appearance of a difference between RE and LE Da groups under these conditions is due to the effects of distraction by the disturbing surroundings, which are more powerful when the RHem is in use. Under the more usual test conditions when the chicks are more fully at ease, this latter effect is absent, and both eye systems are capable of learning not to peck pebbles. Under the same conditions, LE Li chicks show slower learning because of interference from the non-seeing but strongly controlling LHem. Evidence for such effects of non-seeing eye systems, and discussion of the mechanisms is given in Chapter 15 by Andrew.

Taken together, the above findings suggest strongly that the effect of unilateral light exposure is to increase the ability of the exposed eye system to select targets and determine what response should be made to them. This ability is better developed in the LHem even in Da chicks, but it is enhanced by normal RE exposure to light during incubation. It is not yet clear whether the whole complex of abilities, which normally characterize the LHem, shift to the RHem as a result of reversed light exposure. This complex probably includes not only the ability to determine response but to inhibit response whilst a decision is being taken, and also to preferentially learn the consequences of response. This is perhaps to be expected, since these abilities are likely to be intimately linked. This is confirmed by the ability of both RHem and LHem to learn rapidly and at similar rates to inhibit pecks at pebbles in chicks that have received similar exposure on both sides (no light or bilateral light).

On the other hand, there is no clinching evidence that the normal assignment of abilities to the RHem is affected by light exposure. The hypothesis that exposure determines the assignment of control of response and associated abilities, but does not affect the spatial and attentional abilities of the RHem, has the attraction of simplicity. It should be tested.

5.1.4. Sex Differences in Behavioural Lateralization

In chicks, sex differences in lateralization are present in nearly all tests. Explaining them in terms of a single underlying difference has been difficult, but may now be possible. It will be best to begin by setting out the hypothesis: in both sexes, task properties, and standing or age-dependent biases combine to determine whether LHem or RHem tend to take charge. Sex differences in the effects of bias are clear. These probably arise because the controlling hemisphere is better able to impose its strategy of attention on its partner hemisphere in females than in males. As will be seen, this affects both binocular and monocular conditions.

5.1.4.1. Binocular condition

In the binocular condition, the main relevant findings are as follows.

1. Females show much clearer dependence of patterns of spontaneous eye use on age-dependent biases (Dharmaretnam and Andrew, 1994). The changes in direction of bias are the same in both sexes (e.g. the RE tends to be used to examine novel objects on day 8 and the LE on day 11). However, the degree to which the eye appropriate to the day is used, rather than the other, is much greater in females. The strategy of using the eye ipsilateral to the controlling hemisphere to view a small novel stimulus is confined to females. The function of this behaviour is likely to be that of using the subordinate eye system to (literally) keep an eye on a stimulus, about which nothing more for the moment remains to be learned, but which might change again in the future. If so, this is a clear example of the imposition of an attentional strategy by one hemisphere on its partner.

2. When chicks that had been trained to run down a runway to feed in a dish were presented with a conspicuous visual change either in the dish or on the runway walls, males were typically markedly affected by the latter, apparently forgetting the task in hand when distracted in this way. They were much less affected by change in the dish (Klein and Andrew, 1986). Females were far less disturbed than were males by the change in the walls (Andrew, 1991). This suggests greater exclusion in females of sensory input that is not associated with the primary target of attention (see point 3 for discussion).

3. A comparable sex difference in distractibility is present in tests, which used chicks that were accustomed to feed in a hopper. During feeding a

lateral distraction was presented, either to the right or the left of the chick's head (R. McKenzie, pers. comm.). The distraction was the illumination of a small diode. The strongest response was to withdraw the head and to cease feeding for a few seconds; more usually, feeding was only briefly interrupted, without head withdrawal; sometimes the illumination appeared to be ignored. In males, left-side distraction was strikingly more likely to interrupt feeding briefly than was right side; in females there was no such side difference and overall levels of distraction were lower, resulting in significant interactions involving sex and side of presentation in analysis by ANOVA.

This finding (together with point 2) strongly suggests that in males the LHem controls feeding and tends to ignore minor changes in the right visual field, whilst at the same time sustaining feeding. The RHem is relatively free during such periods to assess unexpected change in the left visual field and feeding is interrupted briefly when this happens.

The absence in females of such greater distractibility to left hemifield inputs would be well explained, if the attentional strategy of the LHem (focused on the food source, as in males) were also imposed on the RHem. Note that when more effective distraction is used and full interruptions of feeding become common (see point 2), the direction of difference reverses (at least in males). Naturally, distraction is now likely whether the stimulus is on the left or right; however, with left presentation, feeding is more likely to be resumed after a modest interval. Presumably, when distraction involves the RHem, the LHem is often able to sustain the intention to feed. When distraction is mediated by the LHem, this interferes with the resumption of feeding and a long interruption results.

5.1.4.2. Monocular condition

Evidence from monocular tests shows a similar imposition, in females but not males, of an attentional strategy by the hemisphere put in control by the properties of the task. This occurs, even though the controlling hemisphere is not receiving its normal main input from the contralateral eye. Mechanisms that might allow this are considered later. The first evidence for such effects was from a series of cases (Andrew, Mench and Rainey, 1982) in which female RE and LE groups initially behaved very similarly, whilst male RE and LE behaved very differently. However, in this and later studies, RE–LE differences of the male type appeared eventually in females in the course of testing as follows.

1. In the pebble-floor test (see above), in the second week of life, both RE and LE females tend to show the same fast development of inhibition of

pecks at pebbles, as do RE but not LE males (Zappia and Rogers, 1987). In older males the RE–LE difference disappears, and they show the female type of pattern. One possible reason for this is that, as chicks (whether male or female) grow older, their interest in pecking particles on the floor becomes so confined to familiar food that even LE chicks peck novel objects only briefly.

However this may be, 3-day-old females show the same kind of relative difference between RE and LE, as do males (Andrew, Mench and Rainey, 1982). In one such experiment, female RE and LE initially showed the same steep drop in pebble pecks, as did male RE. However, late in the experiment (in the third and fourth blocks of 20 pecks), female LE suddenly showed a rise in pebble pecks to the levels still being shown by male LE. A significant LE–RE difference resulted in females, like that present in males (Andrew, Mench and Rainey, 1982).

2. Similar delayed appearance in females of LE–RE differences of the male type were usual in tests involving a series of trials, in which chicks pecked spontaneously at coloured beads (Andrew, 1991). The trials were given in pairs, separated by 5 s removal of the bead; the pairs of trials were separated by much longer intervals (120 min). A change in bead appearance occurred between the first and the second pair of trials. In females, response at the first trial of a pair was identical in RE, LE and BIN, and resembled the response shown by male RE at the second trial of the second pair. At the second trial, differences appeared between RE and LE in females as well, which were of the same sort as shown by males.

This pattern was present with stimuli changed in a variety of ways. A good example is provided by the dishabituation of pecking by a change in the direction (from above or below) in which the bead approached (Andrew, 1983, 1991). In males, dishabituation was shown by LE, exactly as would be predicted if RHem processing took spatial context of the point of attention into account. Such dishabituation was completely absent in RE, suggesting that attention was here focused exclusively on the bead and not on the direction in which it was introduced into the cage. In females, both RE and LE showed no dishabituation caused by the change in spatial context at the first of the pair of trials, but dishabituation appeared in LE (only) at the second trial.

It is important to note that LHem control of pecking at objects like beads, in the binocular condition, is predictable from other studies. When a bead is presented sequentially and repeatedly to right or left visual field and then frontally, the evocation of pecks following right presentation correlates markedly and significantly with that at a subsequent frontal presentation. However, there is no correlation between left presentation scores and subsequent frontal scores (Andrew, Mench and Rainey, 1982).

When running to a manipulandum, which it will seize with the bill, a chick fixates the target with the RE. In a similar approach where no such target can be seen, the LE is used instead (Andrew, Tommasi and Ford, 2000). Other things being equal, the LHem would indeed be expected to control in a bead task.

The appearance of RHem-type processing in LE females at the second of a pair of presentations probably occurs because, under these conditions, the short-term memory of the immediately preceding trial is available to the RHem in a way that it is not to the LHem. Note that the LHem did not see with its 'own' eye at the preceding trial. This available short-term memory is likely to allow the RHem to take greater control of processing.

3. One series of bead tests showed that any given degree of novelty had a larger effect in female LE than in male LE (Andrew, 1991). The experimental design was that experience at the first pair of trials varied between groups of chicks: a chick might see (and peck) a violet or red or uncoloured bead, or just the rod which normally carried a bead. Colours were due to internal illumination. At the second pair of trials, all chicks saw the same violet bead.

There were two main patterns of behaviour in the second pair of trials. Small or moderate change resulted in similar dishabituation (elevation of pecking) in both trials. Large enough change resulted in marked dishabituation at the first trial, which was lost at the second. Of necessity, the first trial required the chick to use long-term memory to assess the change. However, with large change in the bead, the second trial of the pair was apparently dominated by short-term memory from the first trial, so that processing established identity with what had just been seen, resulting in low rates of pecking. Modest change, on the other hand, apparently caused continued use of the long-term trace from the first pair of presentations, resulting in continued interest in the character of the change in appearance.

LE males treated the change from red to violet as if no change had occurred (i.e. there was no dishabituation). 'No colour' to 'colour' was treated as a modest change. Only 'no bead' to 'bead' was treated as a large change. On the other hand, for LE females red to violet was a modest change, and no colour to colour was a large one.

It has already been shown that, in LE females, attention appears to be confined to the bead by control from the LHem in a way that does not occur in LE males. As a result, the abilities of the RHem are used to record and to evaluate all the intrinsic properties of the bead accurately. This has been argued to be due to imposition of the strategy of attention appropriate to the LHem. LE males, in the absence of this effect, certainly attend chiefly to the bead, but clearly also take into account its wider spatial context. Note that spatial context (e.g. properties of the rod on which the bead is mounted) did not change between the two pairs of trials, so that inclusion of spatial context would decrease estimates of degree of change in males. The effects of more detailed recording of bead properties by LE female may summate with this latter effect to produce the sex difference.

4. Another study whose results are explicable by greater collaboration between RHem and LHem in females is that of Vallortigara (1996). Chicks were trained to choose one of two boxes for food reward. The reinforced box could be identified both by position and by colour, or by one or the other alone. Retraining followed in which only one type of cue consistently identified the correct box. The sexes did not differ in speed of acquisition of the discrimination when choice was either on position alone (right or left) or colour alone. When both cues were relevant, males and females again performed similarly. However, in retraining, when only position or only colour correctly indicated reinforcement, males retrained more quickly when position was the relevant cue, and females more quickly when the relevant cue was colour.

On the evidence already presented, when a female chick chooses a box, attention of both eye systems will be focused on the chosen box and this will interfere with the RHem recording spatial context. Under such conditions, association of reinforcement with box properties is likely to be more effective for colour than for position. The greater independence of RHem and LHem in males is likely to mean that attention of the RHem will be directed to the relative position of the box not only during approach, but also during response and reinforcement. This should facilitate the use of position as the discriminating cue.

5. Finally, comparable sex differences are present in the control of response to sounds (Chapter 10 by Andrew and Watkins).

5.1.5. *Brain Mechanisms Underlying Sex Differences in Chick Lateralization*

It is likely that more than one mechanism is involved in generating the differences in the functioning of the basic pattern of lateralization that have been discussed here for the chick. This greatly complicates identification of the mechanisms involved, even when there is an anatomical asymmetry suitable to explain the behavioural lateralization.

The first step is to try to decide in what order the differences may develop and how they might relate functionally. It seems (see Section 5.1.4) that at least some sex differences are established independently of the effects of light.

The relative insensitivity of the female chick foetus to unilateral exposure to light is another sex difference, which must be present before, and so be independent of light exposure. One possible explanation of this insensitivity is that it too depends on pre-existing sex differences in the character of interaction between the hemispheres. It has been argued above that female chicks differ from males in that they are more likely to show a single attentional strategy, involving both hemispheres, but imposed by one or other hemisphere. At its simplest, before hatching, this may involve no more than equilibration of activity between left and right visual systems. In an embryo exposed unilaterally to light, this might reduce or abolish the imbalance of activity, which (presumably) generates anatomical and functional asymmetries.

This hypothesis can be better understood if it is put in the context of known functional properties of the vertebrate visual system. Levy (1985) argued, on the basis of evidence for humans with section of the corpus callosum, that a major (perhaps the main) function of this commissure is to equilibrate activation of corresponding structures of right and left hemispheres. Experimental studies show that, without such effects at the forebrain level, capture of a target by one optic tectum (superior colliculus) results in inhibition of the tectum of the other side (due to tecto-tectal inhibition), and consequent unilateral inattention (Sprague and Meikle, 1965). Normally, descending effects from the cerebral hemisphere ipsilateral to the inhibited tectum oppose tecto-tectal inhibition (Sprague, 1966).

Descending effects of this sort from forebrain to tectum are present in the chick. Parsons and Rogers (1993) cut the two commissures linking the tecta (tectal and posterior commissures). Following this, repeated presentation of a small conspicuous object to the RE resulted in progressively increasing levels of pecking, a pattern never seen in normal chicks. Presentation to the LE yielded only low and relatively constant levels. An outflow from the LHem (but not the RHem) that facilitates pecking has been described for the pigeon by Güntürkün and Hoferichter (1985) on the basis of evidence from sectioning the main telencephalic outflow on the left or right. It thus appears that, in the chick, there can be short-term facilitation of pecks at a particular object that are apparently based on descending effects of LHem mechanisms on the associated tectum. In normal chicks this is opposed by intertectal inhibition, which in turn may be modulated by descending effects based on RHem estimates of familiarity.

The sort of real-life situation where facilitation of response to a particular object might be important is the pursuit of a target that appears and disappears, and might fall anywhere in the visual field at its next appearance.

The evidence discussed in the previous section shows that the LHem is likely to be responsible for sustained responsiveness to such a target. The effects of cutting tectal commissures in the Parsons and Rogers study show, firstly, that this is probably executed through effects on the tectum and, secondly, in this particular case, that the RHem exerts an opposing effect, probably by assessment of a stimulus, which was initially interesting because of novelty, as now quite familiar.

Note that we are discussing mechanisms that might allow the attentional strategy of one hemisphere to influence the other in the monocular condition. It has in the past been difficult to understand how, in birds, a 'non-seeing' LHem could affect processing by its partner. Evidence from the control of attention bears directly on this issue.

It is not yet clear how detailed is the control of attention in the partner hemisphere. If it were to prove to include specification of local cues, as well as general position, this would suggest that information from the seeing eye is reaching the 'non-seeing' but controlling hemisphere. One route would be the ascending ipsilateral thalamofugal route from the eye to the forebrain. If the controlling hemisphere had immediate access to information relating to what the uncovered eye was looking at, control of the attentional strategy of its partner (also seeing the same thing) would be simple.

A function of this sort (involvement in interhemispheric control of attention) for the ipsilateral thalamofugal pathways would have interesting implications for the function of the asymmetry in these pathways which is temporarily induced in chicks by unilateral light exposure (Chapter 6 by Deng and Rogers). Note that the thalamofugal route is specially involved in carrying the input from the lateral visual field, including the main lateral fixation point of the retina. It is the two lateral visual fields that are used independently, so that the thalamofugal system must be of great importance in the control of visual attention.

The consequence of the asymmetry in the thalamofugal system is that the LHem receives a less extensive ipsilateral input than does the RHem. This would mean that the LHem would be less able (whilst the asymmetry lasted) to control the attentional strategy of the RHem, since it would receive less information about what the RHem is seeing, than would be the case for the RHem in its relations with its partner. One possible outcome of this asymmetry is that the LHem is more likely to ignore inputs to its partner, and instead to sustain attention on the object to which its attention is directed. In contrast, the RHem would be better able to organize analysis of relations between stimuli in both visual fields, based on shifts of viewing between them.

Both of these patterns of attention are consistent with the special roles of the two hemispheres.

The asymmetry in the thalamofugal pathways is temporary (up to approximately day 21 posthatching) and much more marked in male than female chicks (Chapter 6 by Deng and Rogers). The outcome of the period of asymmetry is likely to be exaggeration of the specializations of the two hemispheres, owing to differences in what is learned by each. This effect is likely to be more marked in males.

It has already been noted that age-dependent bias to control by one or other hemisphere shifts so quickly during development that it can hardly depend on asymmetries of connection. Shifts in the balance of activity of brainstem nuclei providing monoaminergic inputs to higher centres is a promising candidate mechanism: asymmetries of activity at forebrain level are present in adult humans for noradrenalin, dopamine and 5HT (or serotonin) (Wittling, 1995). If unilateral light exposure in chicks has as its major or only consequence a relatively long-term bias to control by the illuminated hemisphere (see above), then this bias also may be mediated by a similar mechanism. The involvement of the nigrostriatal system in the initiation of motor response is consistent with the idea that dopaminergic systems act as modulators of bias to control by one or other hemisphere.

5.1.6. *Chick Lateralization: A Summary*

Before turning to other vertebrates, I review previous sections relating to the chick as a guide for effects to be looked for elsewhere. Firstly, the chick shows a series of precisely timed shifts in bias to control by RHem or LHem, which are superimposed on the effects of other factors. These shifts are closely correlated with changes in behaviour of broods of chicks reared under seminatural conditions. They probably function to constrain the behaviour of the chicks (e.g. to make it unlikely that they will wander off into the environment when very young). They may also allow each hemisphere in turn to learn free of constraints by its partner.

Secondly, environmental factors, which vary in their degree between individuals during development, strongly affect some aspects of lateralization. Exposure of the eggs to light in the last days of incubation must be affected by the nest position, weather and position of egg relative to others (which in turn might be affected by egg turning by the mother). The gonadal steroid content of the egg may depend on the order of lay (at least in the clutch laying ancestral jungle fowl), as it does in some passerines (Schwabl, 1993). Deng and Rogers (Chapter 6) have shown that corticosterone reduces the number of neurones in the ipsilateral route to the RHem (in chicks hatched after normal RE exposure to light), and thereby abolishes the expected asymmetry. As they note, this means that stress experienced by the embryo (e.g. due to unusual cooling) could affect the pattern of lateralization shown after hatching.

Two general issues are raised by these findings for the chick. Firstly, it is possible that some at least of these environmental effects have evolved because there is correlation between the environmental factor and the conditions that the animal will face in later life. Perhaps the most common case (see below) is that bodily condition (especially body size) may determine which of two or more pre-programmed strategies it would be best to follow. There is extensive evidence suggesting that stress during development affects some aspects of lateralization in some vertebrates (see below; Chapter 8 by Cowell and Denenberg). Stress during development seems likely to be a predictor of subsequent poor condition of the developing animal and, in some cases, of an environment in which it may be difficult to make a living. Appropriate modulation of lateralization may be a very widespread way of adjusting to such conditions.

The modulation of the development of lateralization by gonadal steroids is clearly related to the generation of lateralization patterns appropriate to male and to female behaviour. However, it is likely that variations in their levels also generate interindividual differences within each sex. One case where this may hold is that of satellite and territorial males in fish and birds (see below). Note that in fowl subordinate males usually live within the territory of a dominant cock (McBride, Parer and Foenander, 1969).

Secondly, most such environmental effects are likely to have a strong randomness in their incidence (i.e. they are very imperfect predictors of future conditions). Some may even be purely random; this could well be true of the degree of illumination of eggs late in incubation. The result of random effects is the generation of phenotypic variability independently of genetic variation.

My interpretation of the chick data to date is that the modulation of lateralization by factors like light or hormones affects bias to use of one or other hemisphere, and the degree of collaboration or independence of their functioning, rather than the assignment of abilities to RHem or LHem. The effects of reversed exposure to light that have already been discussed are unlikely to occur under natural conditions. They do suggest that LHem abilities can be moved experimentally by an environmental effect.

5.2. Age-dependent Shifts in Bias During Development in Vertebrates

The only species for which there is direct evidence of such shifts other than the chick are rats and humans. In the case of the rat there are no more than preliminary hints. Bianki (1988, pp. 57–63) used an 'extrapolation' task, in which the rat had to follow a food trough that could disappear behind a screen either on the left or on the right. The experimental manipulation was to inactivate either RHem or LHem temporarily by spreading depression. Unfortunately for present purposes, the outcome was affected by a variety of factors, in particular the extent of prior experience with the test. There were also conditions, particularly with only the LHem functional, when there was choice of the wrong side. The clearest feature of the data was that a functional RHem was more important when prior experience had been brief, and a functional LHem when prior experience had been extensive. This makes it difficult to interpret the finding that the LHem was more important in young rats (35–40 days), but the RHem was more important in the same rats tested after puberty. It may indeed represent an age-dependent shift, as Bianki suggests; on the other hand, prior experience may have affected adult performance.

There is far more evidence available for human development. There appears to be an improvement in the recognition of 'unfamiliar' (i.e. perceived only once before) human faces and voices, which peaks at 10 years and then falls to a trough around 12–13 or 14 years, before rising again to adult levels. This trough is important in that it argues against an explanation based on progressive improvement in performance due to maturation. Recognition of unfamiliar faces is known to be a task for which there is RHem advantage when the face is upright (rather than inverted), and this holds also for the improvement at 10 years (Levine, 1985). Recognition of an unfamiliar voice shows almost exactly the same time-course as that for faces (except for what may be a slightly shorter trough), and again there is special RHem involvement in the ability (Mann, Diamond and Carey, 1979; Carey and Diamond, 1980).

A different change also occurs soon after 10 years: Ladavas (1982) analysed the asymmetry of posed expressions, blocking data into 2-year blocks (10 and 11, 12 and 13 years being the relevant ones). More intense expression in the left hemiface (the usual adult condition) appeared suddenly between the two age blocks. Rothbart, Taylor and Tucker (1989) argued that this asymmetry is predominantly due to LHem inhibition of the right side of the face. Ladavas' data would suggest, on this hypothesis, an assertion of LHem control somewhat after the 10-year point (at which time the face and voice recognition evidence suggest bias to control by the RHem). This body of results is thus consistent with a peak in RHem abilities at about 10 years, which is followed by a trough, because of increasing bias to LHem control.

Thatcher, Walker and Giudice (1987) used variations in coupling between electroencephalograms (EEGs) recorded at different sites to show that between 8 and 10 years the most conspicuous changes were in the RHem, where temporal–frontal coupling increased progressively, catching up with similar coupling within the LHem. This is consistent with the evidence (see above) for development of improved RHem ability in this period, culminating at 10 years. Subsequent development was predominantly in left–right frontal coupling, the first phase of which occupied the period of the trough (11–14 years), and might well explain interference of the LHem in RHem tasks.

The LHem also shows progressive increase in frontal–occipital coupling from 3 to 6 years, a period in which comparable change in the RHem is much less clear. Since the comparable later changes in coupling within the RHem, at 8–10 years, appear to be associated with a period of RHem control, it is possible that the coupling changes in the LHem are also accompanied by LHem control. If present, this period of LHem control would centre around 6 years.

Still earlier shifts in hemispheric bias may be present: Young (1977) interprets earlier work on the incidence of temporal lobe epilepsy as indicating a shift from relative inactivity of the LHem in the first year to relative inactivity of the RHem in the second and third. Rothbart, Taylor and Tucker (1989) found very clear changes in the first year in asymmetry of expression: at 10 months, but not 6.5 or 13.5 months, facial expressions were more marked on the right side of the face. This direction of asymmetry is the reverse of that found in adults (see above), and Rothbart, Taylor and Tucker suggest that early in development RHem control inhibits rather than facilitates emotional behaviour. This could come about if, at this age, the most intense expressions occur with no cerebral control, so that even RHem involvement reduces the intensity. It has to be assumed, on this hypothesis, that the LHem is involved little if at all throughout the period.

The possibility of a brief period of enhanced bias to RHem control at about 10 months is further supported by the fact that the normal bias to right-hand reaching, which is present at 6.5 and 13.5 months, disappears at 10 months (Rothbart, Taylor and Tucker, 1989).

In summary, it is likely that changes somewhat like those considered earlier in the chapter for the chick occur also in humans. It is even possible that there are more detailed resemblances: both species may show a period of bias to LHem control when the young are part of a small social unit to which it is best to confine social interactions (3–6 years in humans; 5–8 days in chicks). This is followed by a period of RHem bias, when the young would normally begin to interact with a wider environment and with a wider range of fellows (10 years and days 10–11, respectively).

A somewhat similar suggestion was made by Bever (1980), who argued that the development of face recognition in humans is constrained by the number of individuals that have to be recognized; however, he saw the causation as proceeding in the opposite direction, with the problems of learning driving shifts in hemispheric dominance. He saw these latter as therefore constrained by the ages at which children transferred between schools. In the case of the chick such a direction of causation can certainly be excluded, since the shifts in bias can be demonstrated in their full form in chicks reared under unchanging conditions and with no interaction with conspecifics. No such demonstration is possible for humans, but it seems unlikely that such consistent timings could be generated solely by changes in social experience, which must vary substantially between individuals and countries.

If bias shifts are present in two species of vertebrates belonging to widely separated systematic groups, it is worth considering whether they could occur more generally. Sharp changes in behaviour during development might in some cases be so generated. Although shifts in bias remain to be clearly demonstrated in rodents (see above), there are sharp shifts of some interest here: in mice, a series of spurts in brain growth have been reported (0–6, 8–12 and 17–23 days). At least the second two of these coincide with enhanced ability to learn passive avoidance tasks, in comparison with learning ability during the plateaux in growth that separate, and follow the spurts (Lavooy et al., 1981). Rose (1980) argued for a correspondence between changes at 4.5–5 years in humans, and changes in behaviour occurring at about 30 days in the rat, which are probably associated with development of prefrontal control of the hippocampus. The changes include the appearance of spontaneous alternation and improved use of contextual cues in conditional discriminations.

There is even some hint of links with changes in anatomical asymmetry. In rats, there is a very sudden increase in the ability to use a landmark cue between 24 and 38 days (Schenk et al., 1994), which is superimposed on a relatively unchanging ability to use general spatial cues in orienting. Strikingly, significant asymmetry in the prefrontal cortex appears at about the same time: there is no difference at 24 days but the left prefrontal becomes larger by 39 days (van Eden, Uylings and van Pelt, 1984). It is thus possible that the behavioural change depends on greater involvement of LHem mechanisms in orientation at a point between 24 and 38 days.

In altricial birds, leaving the nest is a sudden and often largely irreversible act. Even if fledglings of species like *Emberiza citrinella* are carefully returned at once to the nest, after leaving, they will not remain there (pers. obs.). The transition may be comparable with the development of locomotion independent of the mother in precocial species like the domestic fowl, and would be well explained by an age-dependent shift of bias away from the LHem and an associated ability to inhibit response at about the normal time of fledging.

5.3. Sex Differences in Lateralization

5.3.1. Sex Differences in Hemispheric Interaction

The corpus callosum is larger in male rats, and this sex difference is known to be affected prenatally and perinatally by androgens and oestrogens (Fitch et al., 1990, 1991a,b; Chapter 8 by Cowell and Denenberg). It seems likely that this particular sex difference is in some way linked to differences in hemispheric interaction. It is therefore interesting to ask whether sex differences in the rat resemble those already considered for the chick, where I have already argued that females show greater collaborative involvement of both hemispheres in tasks, thanks to the readier imposition of the attentional strategy of the controlling hemisphere on its partner.

Bianki (1988, pp. 46–49) showed that inactivation of one hemisphere after partial or full acquisition of a conditioned active avoidance had different effects in males and females. In males, early in acquisition (100 pairings) the RHem was more important to performance; later, both hemispheres were clearly necessary, but disruption was now maximal without the LHem. Females showed somewhat similar shifts in overall performance (early good performance with one hemisphere inactivated, later very poor); however, there were either no differences between the effects of inactivating the RHem and LHem or, in one instance (conditioning to a sound, late in acquisition), a slight difference in the same direction as in males (that is, better performance with the LHem active).

In a range of studies, Bianki (1988, p. 69) found that this was the usual pattern, with females showing less or no difference between LHem and RHem inactivation. Collaboration between hemispheres in tasks is more likely to result in both hemispheres being involved in learning about the task; even if different aspects of the task are involved, the result might be expected to be greater residual competence when the partner is experimentally incapacitated. It is interesting that a similar lesser disturbance by unilateral lesions has been reported (McGlone, 1980) for women, as compared to men, in verbal performance (after LHem injury) and probably for constructional tasks (after RHem injury).

The only marked exception to this sex difference in rats that was reported by Bianki was in tests in which rates of reinforcement varied between stimuli, and the measure of interest was whether choice frequency was adjusted to match the probability of reinforcement. In both sexes, LHem involvement (i.e. RHem inactivation) tended to give choice matching the probability of reinforcement, whereas RHem involvement gave random choice.

However, this effect was significantly larger in females. This may perhaps be compared with the clearer lateralization of motor asymmetries (rotation, neonate tail posture) which is shown by females in at least two strains of rats (Bradshaw and Rogers, 1993, pp. 112–113). Control of motor response and recording the consequences of response (e.g. reinforcement) may well go together.

One striking sex difference in the effects of right frontal lesions on locomotor activity (Lipsey and Robinson, 1986) is probably only an apparent exception to the clearer lateralization of motor control in female rats. In males, the resulting rise in activity, which is absent for corresponding left frontal lesions, is one of the clearest indices of consistent lateralization at the population level in rats. In females there is no difference between the effects of right and left lesions on running wheel activity. However, this is not because consistent asymmetry is absent in females; another effect typical of males, namely that only right frontal lesions cause depletion of noradrenalin in both right and left frontal cortex, whereas left frontal lesions have no effects of this sort, was present in females.

It is thus likely that the basic sex difference is in the control of wheel running. This is consistent with the fact that male and female sham-lesion groups also differed in the Lipsey and Robinson (1986) study: males showed a standing depression of postoperative activity levels below preoperative, whilst females progressively recovered from an initial depression (as did female lesion groups). One obvious difference between the factors affecting wheel running in males and females is that the females, both before and after the operation, showed the usual 4-day oestral cycle in running wheel activity.

The clearer lateralization of motor control in female rats and humans (Annett, 1972) is probably another aspect of the greater ability of a controlling hemisphere to influence its partner in females: when a motor response is likely, the controlling hemisphere may be better able in females to prevent any initiation of motor activity by its partner.

I turn now to cognitive processes. Rats show sex differences in maze learning: males tend to use room cues (general layout of environment), whilst females use selected landmarks. The difference depends on effects of testosterone that normally occur perinatally in male 'rats (Williams, Barnett and Meck, 1990).

There appear to be comparable human sex differences: Kimura (1999) has recently reviewed and assessed evidence for sex differences in cognitive tests in humans. In general, the clearer sex differences are found in tests that involve relatively direct interaction with the real world. My interpretation of the findings is as follows: a group of such tests show males to be more likely to separate local cues, which identify an object of interest, from spatial context. As a result, males are more likely to record points by specifying absolute position in, and direction relative to the environment in activities like map reading or maze solving. Local spatial context has also less effect on judgements such as verticality (Rod-in-Frame: Pitblado, 1979).

Females, on the other hand, make more use of immediate spatial context, storing movement through the environment as a series of descriptions of spatial layout at choice points. Greater female ability to detect which objects have been exchanged in a test (Kim's game) in which a variety of very different objects are laid out in an array, also suggests that object characteristics are intimately associated with spatial context.

All of this is a probable outcome of a strategy of close collaboration between the hemispheres in females and is (surprisingly) reminiscent of the condition described above for the chick. There I argued that, in female chicks, imposition of a LHem strategy on RHem analysis of stimulus properties and position resulted in analysis and recording of the detailed properties of the stimulus, as well as its position in the overall environment.

A sex difference in hemispheric interaction is strongly suggested by the consistently better ability of women to fuse Julesz patterns presented separately to the two eyes (Kimura, 1999, pp. 86–87), a difference that may already be present in infants, since the ability can be detected earlier in girls than in boys. Greater interhemispheric cooperation in women than in men is also suggested by studies showing greater asymmetry in brain activity in men. This holds for both spatial and verbal tasks, with men showing higher alpha in the hemisphere less suited to the task, suggesting that it is relative inactive (Ray et al., 1981). In rhyming tasks, men show left frontal activation, whilst activation is bilateral in women (Shaywitz et al., 1995). McGlone (1980) summarizes evidence for greater asymmetry between right and left visual field presentation both of spatial and verbal material, which would also be consistent with involvement of one hemisphere at a time being more usual in men.

Finally, direct comparison between humans and rats is possible for one measure: the direction of turning in free locomotion. Bracha et al. (1987) showed that, when consistently right-handed individuals, with corresponding right foot and eye dominance, were compared, men turned to the right and women to the left in free locomotion. This striking result would be well explained if right-handed women were to make more use of RHem spatial abilities during free locomotion, attending to, and so turning towards objects detected in the left hemifield, despite the involvement of the LHem in control of approach to targets of interest and response. If men were to make less use of hemispheric cooperation, this would tend to increase the effect on turning of approaching targets of response under LHem control. Sex differences in the overall direction of bias in rotation have been reported for two strains of rats (Robinson, 1985; Robinson et al., 1985). However, the differences were slight and in the opposite direction to those shown by humans.

Another very characteristic human sex difference, female advantage in tasks involving a rapid series of different patterns of movement of distal musculature (Kimura, 1999), seems likely to be related to the above sex difference in motor behaviour. Broadly, the ability to use abilities of both hemispheres, but with one (presumably here usually the LHem) in continuous overall charge, seems likely to underlie the female advantage.

It is thus possible that in birds and mammals (at least) the extent of interaction between cerebral hemispheres, which allows one hemisphere to affect the pattern of attention shown by its partner, is an important way of modulating lateralization. The presence of such modulation commonly depends on the action of gonadal steroids early in development, and so may also appear as a sex difference.

Any such sex difference in collaboration is likely to involve differences in effects that act through forebrain commissures. It is therefore intriguing that there is evidence of sex differences in corpus callosum (and in humans in anterior commissure) dimensions (see Chapter 8 by Cowell and Denenberg). Unfortunately, the difference is probably in opposite directions in rat and human. There has been controversy over the possibility that the forebrain commissures are larger, relatively or absolutely, in women than in men. On balance, the female corpus callosum may be relatively larger (Driesen and Raz, 1995), whilst the female anterior commissure is more certainly larger than the male (Allen and Gorski, 1991). In rats, the experimental evidence is unambiguous (see above): testosterone acting perinatally results in a corpus callosum, which is both relatively and absolutely larger in cross-section in normal males than in females or in males in which the action of testosterone is prevented. Oestrogens have the opposite effect. It is at present impossible to say for certain what is the significance of such size differences, particularly in the rat, in which the corpus callosum grows more after infancy in female rats, abolishing the sex difference (Berrebi et al., 1988).

The site of the action of testosterone, which affects the corpus callosum, remains to be established. Asymmetric action on the cortex seems certain in mammals in view of the higher level of androgen receptors in the right frontal cortex than in the left during foetal development (rhesus: Sholl and Kim, 1990). Marked differences between corresponding right and left populations of cortical neurones, such as certainly are present in male rats (van Eden, Uylings and van Pelt, 1984) and mice (Ward and Collins, 1985) would be expected to affect the development of fibre tracts that connect them. These cortical differences are also known to depend on perinatal effects of androgens in rats (Lewis and Diamond, 1995). Despite all these unknowns, an important implication for current purposes remains: sex differences exist in commissural properties, and these are likely to be related in some way to sex differences in hemispheric interaction.

Finally, in mammals at least, another effect of gonadal steroids needs brief mention. At present they are best considered as a quite separate type of effect, but it is, of course, possible that the same effects, acting earlier in development, are also involved in the generation of the sex differences that have already been considered.

They are extensively documented for humans, based both on natural fluctuations in hormone levels, such as occur through the menstrual cycle, and on clinical manipulations of such levels. Bias change is suggested by a weakening of the normal rightward turning bias at the midluteal point of high oestrogen levels (Mead and Hampson, 1997), which the authors interpreted as perhaps acting on dopaminergic function to reduce the normally more active supply to the LHem.

Kimura (1999, pp. 116–123) summarizes more recent work by herself and colleagues on both men and women. At the high oestrogen point, women show enhanced performance on tasks where female advantage is usual, such as verbal fluency and fine manual dexterity, whilst at the low oestrogen point advantage shifts to spatial tasks. In men the normal good performance on spatial tasks is reduced when testosterone levels are high, either because of daily or seasonal fluctuations, or because of clinical intervention. Tasks, in which females normally perform better, tend to show decrements of performance when testosterone levels are high, so that (as in the case of women) shifts in overall level of performance are not responsible. It is possible that comparable effects are present in female rats: spatial performance (in a water maze) is at its worst at the high oestrogen point (Warren and Juraska, 1997).

These effects of changes in hormone levels cannot be explained easily by shifts in bias to the use of one or other hemisphere that are directly induced by hormonal effects. The variation in performance in both sexes is consistent with the hypothesis that the pattern of abilities characteristic of each sex is shown most clearly in a particular hormonal state: high oestrogen for women and low androgen for men (but in both cases within the normal range of variation). For the present they are perhaps best treated as reflecting changes in mechanisms which do not directly act on lateralization, but affect abilities that are themselves lateralized.

5.3.2. *Commissures and Hemispheric Interaction*

Perhaps the greatest obstacle to advances in our understanding of the role of hemispheric interaction in lateralization is the great divergence of opinion about the functions served by commissures like the corpus callosum. The two extreme positions are, firstly, that the main function is the adjustment of patterns of activation in the partner hemisphere (as has already been discussed for the chick). The other extreme position is that the main function is to transfer information from perception or recall between the hemispheres.

The issues are well exemplified by human studies examining the use of visual information supplied to left or right visual fields (LVF, RVF) in tasks (e.g. recognition) carried out very soon after perception. Different time-courses of availability follow inputs (complex abstract patterns) to LVF and RVF (Bevilacqua et al., 1979). When both the first and the recognition trial involved presentation to the same eye, performance was initially better for RVF inputs. By 60 s performance for RVF had fallen so much that it was considerably worse than LVF performance (which changed little over this period). Bevilacqua et al. (1979) suggest that progressive 'coding' by the LHem may explain its more complex time-course.

Availability to the hemisphere ipsilateral to the eye that receives the initial input also changes with time (Coney and MacDonald, 1988). Verbal material was used. At short delays (12 s) it was found that access was equally good for LE to RE change as for LE to LE. The same was true for RE to LE and RE to RE. In all cases the use of the RE in the second recognition test produced faster response than use of the LE, presumably because of the verbal abilities of the LHem. However, access to the original material was equally good, whether or not eyes were changed between trials, despite this left–right difference in current processing during the test trial. Access may depend (as Coney and MacDonald suggest) on the establishment of a second trace in the hemisphere ipsilateral to the eye in use at the first presentation. If so, this ipsilateral trace becomes rapidly unavailable: by 32 s use of the same eye at recall as at recognition produced much better performance than when the eye was changed.

However, there is a second way of modelling such 'interocular transfer' tests: namely that unilateral inputs produce unilateral traces, and interhemispheric access depends on the extent to which material held in the other hemisphere, which is the one responsible for further processing or recall, is linked to such a unilateral trace. On this sort of model, access by the hemisphere responsible for recall to material in its partner immediately after perception by that partner is explained by the saliency given by the activated state of the trace, and perhaps also by its specific labelling by attentional mechanisms, acting between hemispheres. If the ipsilateral material remains available to later recall, it is because links between trace fragments, such as are required by theories of distributed memory, associate different material held in each hemisphere. The whole topic is discussed further in Chapter 15 by Andrew.

There is no consensus as to whether this model is correct or whether models invoking the bulk transfer of information by commissures are more likely to be correct. Kucharski and Hall (1987) postulate unilateral trace production as the norm following unilateral input, and review a literature which has assumed that commissures can function to confine trace formation to one side, even with bilateral inputs. Equally, failure of access following section of the corpus callosum is commonly attributed to blocking the transfer of information, which would otherwise occur (e.g. Eaton and Gaffan, 1990).

It is clearly necessary that both models be borne in mind when considering hemispheric interaction. However, the linkage model has the virtue that it postulates processes that are also likely to be needed as part of the moment by moment control of vision, in vertebrates with independent eye movements.

The first such strategy that I will consider is the alternate use of the two lateral points of fixation, which in most birds requires the head to be turned through a relatively large angle. It seems likely that capture of the target after the turn depends on both precise pre-programmed control of the head turn to the correct position, and the ability of the hemisphere initially viewing the target to confirm its identity, after its capture by the other hemisphere. Passerine birds commonly use such alternation of fixation when identifying a target (as is obvious when watching blackbirds hunting on a lawn). Its interest from the point of view of lateralization is that it may allow each hemisphere to apply its strategy of analysis independently to an object, which is unambiguously identical to the object that is currently being analysed by its partner.

If, instead of alternating use of the eyes, one eye is used sustainedly, there are two intergrading possibilities. (1) Both hemispheres analyse what they see through the contralateral eye. One probable example in the chick is the use of the RE to fixate the target, which is to be grasped by the bill, whilst the LE is used to monitor position in the environment. The strategy would allow useful recall, only if the two records, which were established, were to be linked, so that both were activated together at recall. Note that there would be no requirement for information transfer at recall: each hemisphere could use its own record to evaluate its own perceptual input.

(2) Only the 'seeing' hemisphere is active in a way relevant to the task. Here sustaining fixation is equivalent to sustaining a single strategy of analysis and response. There is evidence that in the rat the corpus callosum may be important in such an effect: Denenberg et al. (1986) showed that callosal section released mouse killing (under appropriate conditions), and that this probably depended on the removal of a net LHem inhibitory effect on RHem control of response to a mouse.

5.3.3. Degree of Hemispheric Specialization

Perhaps the best known hypothesis concerning the mode of action of steroids on lateralization during development is that of Geschwind and Behan (1982), who postulated that testosterone slowed the development of the human LHem, thus increasing RHem control early in development. The evidence for thicker cortex in the RHem of male but not female rats (e.g. Diamond, 1985) was specifically cited in support of this hypothesis. It is likely that both hemispheric interaction and the functioning of each hemisphere are very closely linked and should be regarded as complementary, rather than alternative. Both the functional organization of, and what is learned by, a hemisphere is likely to be strongly constrained by its relations with its partner.

A hypothesis which concerns itself only with differences in hemisphere properties does not readily explain the overall pattern of human sex differences in lateralization. This is particularly clear when the greater bilateral involvement of the hemispheres in cognitive tasks in women is considered together with their more consistent motor lateralization (see above).

However, the idea that the male RHem is in some way more specialized for spatial analysis has been much used to explain both human and animal findings. Casey, Pezaris and Nuttall (1992) showed that the strategy used to solve mathematical problems (rather than level of ability to find some solution) differed between right-handed and non-right-handed girls, with the former showing little use of spatial abilities and the latter showing the same dependency on spatial ability as boys. The same laboratory (Casey and Nuttall, 1990) linked non-right-handedness in girls to masculinizing action of steroids during development of the sort postulated by Geschwind and Behan, on the basis of effects on masculinity measured on the Tomboy scale.

Equally, Bradshaw and Rogers (1993, p. 132) point out that a larger right hippocampus in male rats is likely to be linked with greater demands from territoriality in males. This seems entirely reasonable: the conclusion to be drawn is that both the separate properties of the two hemispheres and their interaction need to be studied to understand sex (and other) differences in lateralization.

5.4. Phenotypic Variation

I discuss here the causes and possible functions of the widespread occurrence of marked phenotypic variation in some aspects of lateralization.

5.4.1. Dopaminergic Systems

A very striking aspect of lateralization in the rat is that at least one aspect of it, the bias on motor response imposed by the nigrostriatal system, varies independently of other aspects. These other aspects, which are allocated much more consistently at the population level than the nigrostriatal bias, include differing strategies of perceptual analysis (Chapter 3 by Andrew and Rogers), and frontal asymmetry, measured by metabolic activity (Glick and Ross, 1981) or by the effect of frontal lesions (Robinson, 1985). Glick and Ross (1981) concluded that the relationship between the direction of frontal and nigrostriatal functional asymmetries had important effects on behaviour. They argued that the reason why rats that tended to rotate to the right during free locomotion showed more asymmetry in rotation was because, in such rats, frontal and nigrostriatal asymmetries coincided in their effects. However, in the same study, it became clear that these effects of concordance were largely due to the behaviour of rats in which the nigrostriatal input to the LHem was the more active (and so coincided with the side of greater right frontal activity, which is consistently the left).

It is not yet clear whether this type of asymmetry is confined to the nigro-striatal system or whether it also holds for any (or all) of the other structures receiving substantial dopaminergic supplies (e.g. in the rat, prefrontal, entorhinal and primary visual cortex: Berger, Gaspar and Verney, 1991).

The striatal asymmetry is likely (Dunnett and Robbins, 1992) to affect which hemisphere is involved in preparing for motor response. The same hemisphere is likely to predominate in processing the consequences of response. Overall, this is consistent with either LHem or RHem tending to control motor plans and record their outcomes in different rats.

Shifts in the balance of activity between right and left parts of ascending systems like the dopaminergic (or of course other monoamine systems) would allow the rapid age-dependent shifts in hemispheric bias, which have been discussed earlier.

5.4.2. *Functions of Phenotypic Variation in Lateralization*

Variation in levels of gonadal steroids during foetal and perinatal development of rodents strikingly modulates sexual, aggressive and parental behaviour in adults, producing differences between individuals of the same sex, as well as between sexes. In addition to steroids from the foetus' own gonads or entering from the mother, an important source of variation in steroid levels in rodents is steroid production by uterine neighbours. The convention is followed here of classing animals as 0M if they have no male neighbours, and 2M if they have two.

vom Saal and Bronson (1980) argue that in female mice 2M individuals would perform better under highly crowded conditions: they are highly aggressive to females, and urine mark in a way that may help to exclude female competitors. They reach puberty sooner under crowded conditions and defend their pups fiercely. 0M females are more suited to low densities: they reach puberty sooner when at low densities than 2M, and are more attractive to males.

The same argument can be applied to males but here it is probably 0M individuals that would do better under crowded conditions. They show (vom Saal, 1983) more sexual behaviour and more infanticide, but less parental behaviour and less aggression than 2M males: it is probably pointless for males to try to maintain territories under very crowded conditions, but they need to copulate readily with as many females as possible. Any offspring encountered are likely to be ones sired by another male, and so infanticide is adaptive. Conversely, 2M males will be effective territory holders at low density; they do not need to copulate with females other than the one or two that may be ready to breed at any one time and offspring encountered are likely to be the males' own, and so should be cared for rather than killed.

It is likely, at least in rats, that these behavioural changes are accompanied by similarly caused modulation of lateralization. Ross, Glick and Meibach

(1981) showed that direction of neonate tail deflection, which is an index of motor bias, was dependent in female rats on the number of males in the litter. Clearly, the more male litter mates, the more likely it is that a female would have had male neighbours in the uterus. It is, of course, unlikely that the behavioural changes are all caused by changes in lateralization, but such changes would be expected to be congruent. It is possible, for example, that some degree of territoriality in females requires RHem function for topographical analysis to approach the male pattern.

Species differences in rodents as to whether males and females differ in spatial tasks have been held to be adaptive. Thus it has been argued that enhanced maze learning in males may be related to a need for good topographical ability. In voles that do not show the rat sex difference in this ability, this has been explained as being due to similar requirements for such ability in both sexes of a monogamous species, in which males do not travel widely in search of females (Gaulin, Fitzgerald and Wartell, 1990).

Perhaps the most extraordinary effect of variations in testosterone derived from intrauterine partners is the generation in Mongolian gerbils of sterile males from foetuses situated between two female sibs (0M); such males also show enhanced parental behaviour (Clark and Galef, 2000). The authors suggest that these nurses may increase survival of young in the field sufficiently to increase their own fitness by rearing kin or that they may be generated by the mother to increase her fitness (or both).

One key property of the effects of the sex of intrauterine partners is that they are likely to be random in their distribution between individuals. In rats and mice this would be compatible with the unpredictability of the conditions which the litter members may face. Densities reached depend very much on the character of future food supply; only if it is freely available for some time in a particular locality will high density be attained. If lack of food causes dispersal, then future conditions are even more unpredictable. As already noted, patterns of lateralization are amongst the aspects of behaviour that will be affected.

5.4.3. Genetically Generated Interindividual Variation

Roth (1979) demonstrated intense selection during early life on interindividual variation in CNS asymmetries in the goldfish. The anatomical measures that were used cannot at present be translated into functions. They were whether the left optic nerve or the right lies dorsally at the chiasma, and whether the fibre from the left Mauthner cell is dorsal or not at its decussation with the fibre from its partner. In both cases, the dorsal component tended to originate from the smaller eye or the smaller side of the medulla (where the Mauthner cells are situated). It is thus possible that the earlier maturing fibres take the ventral route, in both cases.

Immediately after hatching, the two configurations are almost equally frequent in the optic chiasma, whereas left above right is somewhat commoner (and significantly so) for the Mauthner cell decussation. After development under normal pond conditions, populations of juveniles and young adults show a shift to marked and significant preponderance of left over right for the optic chiasma as well. Cannibalism amongst fry is the main reason for the differential mortality that brings this about. Although Roth discusses the findings as perhaps related to concordance or otherwise of the two asymmetries that were measured, he notes that the only clear change was in the optic chiasma. In aquaria, fry with the optic chiasma pattern that is selected against in ponds, tended to swim nearer the bottom, suggesting that they might choose different parts of a natural environment.

There is one hint as to the reason for the existence of the phenotypic variation: although no data are given, the right over left configuration of the optic chiasma is said to be more common in fry from smaller eggs, and such larvae grow more slowly. This suggests that, whatever its nature, the pattern of lateralization in such fry may serve to make the best of a bad job, and be appropriate to small body size. Such strategies are widespread amongst male teleosts as adults, and result, for example, in the adoption of satellite status and/or the strategy of sneaker fertilization.

However, a study of two other teleosts (trout and codfish: Larrabee, 1906) showed that the configuration of the optic chiasma was distributed almost exactly in a ratio of 1:1 and was insensitive to one generation of selection. It seems that here, as in the goldfish, it is advantageous that both phenotypes should be common at hatching.

The matter is complicated by the fact that selection for reversal of some measures of behavioural lateralization is surprisingly effective within a single generation in at least two species of teleost (*Girardinus*: Bisazza, Facchin and Vallortigara, 2000; Chapter 1 by Vallortigara and Bisazza; *Brachydanio*: Miklósi and Andrew, in preparation). This contrasts with the much quoted failure of selection for paw use in mice to affect bias to one side or the other (Collins, 1985). The rapid effectiveness of selection suggests that there is substantial genetic variation for control of left–right allocation of function in normal outbred populations of some teleosts. This is also strongly suggested by the goldfish study.

A final fish example involves the repeated reversal of the direction of selection. Takahashi and Hori (1994) showed that mouth asymmetry in a cichlid that eats the scales of other fish is under genetic control, and that the commoner side for the mouth to open varies with about a 5-year periodicity. They explain this by postulating that the less frequent phenotype is at an advantage because it attacks on the unexpected side.

The balance between genetic and environmental modulation of patterns of lateralization is considered further below.

It should finally be noted here that it is possible that sometimes there are advantages in diversity of patterns of lateralization that arise solely from properties of social interactions: specifically, it may be that there are 'niches' for a certain proportion of individuals that cannot be adaptively occupied by all the members of the group. This possibility is supported by evidence (Marchetti and Drent, 2000) that there is individual variation in great tits (*Parus major*) in patterns of use of the environment during foraging in large flocks. Birds at the ends of the continuum of variation have been characterized as 'fast' and 'slow'. Fast birds rapidly cover all parts of a new environment, but once a good feeding site has been discovered, they are inflexible in abandoning it when food ceases to be available. They are also much more influenced by the behaviour of a feeding tutor in their selection of site. Importantly, such birds can be either dominant or completely subordinate to social fellows, so that the differences in foraging are not a simple consequence of social position. Slow birds are thus likely to find new sites, uninfluenced by the behaviour of others, whilst fast birds are likely to concentrate on obviously profitable sites.

There is no evidence as to whether differences in lateralization are involved, although the variations in behaviour are such as might well be generated in this way (see also Chapter 4 by Rogers). My point is simply that phenotypic variation in behaviour is here clearly adaptive, or it would not be sustained in natural populations.

Other examples of modulation may well remain to be found, in addition to those already discussed. A common pattern of evolution seems to have been to use mechanisms evolved to generate sex differences in lateralization to generate in addition interindividual variation within each sex. In many reptiles, external temperature during egg development affects sex determination. Incubation temperature also affects patterns of adult behaviour within individuals of one sex (Gutzke and Crews, 1988; Crews and Gans, 1992). It would not be surprising in view of the evidence cited above (and the evidence for the existence of lateralization in reptiles: Chapter 3 by Andrew and Rogers) also to find effects here of incubation temperature on lateralization.

Asymmetries of eye use have been suggested as affecting development of lateralization in mammals: Rogers (Bradshaw and Rogers, 1993, pp. 145–146) cites studies that show the left eye to open first in (handled) rat pups, whereas the right eye opens first in rabbits, and argues that this might 'contribute to the establishment of asymmetry in the brain'.

In general, this sort of control puts the advantage in any conflict between the interests of parent and offspring on the side of the parent. Thus it would be possible for parental behaviour, physiology or anatomy to evolve to change the steroid content of eggs or the distribution of blood flow in the uterus, or the illumination of eggs or of infant rodents in the nest. It is more difficult to see how offspring could evolve so as to manipulate such variables.

If we put aside cases where the environmental variable (whether manipulated by parent or not) may be involved because it to some extent predicts future conditions, the remaining cases may be best explained by supposing that the involvement of an environmental variable, whose effectiveness varies between individuals, protects the genetic control of lateralization from change by strong but varying selective pressures. In the clearest example, the production of behavioural complexes adapted to low or high population densities in rodents like rats and mice, selection pressures certainly reverse frequently. However, the genetic basis of the complexes is insulated from disruption, since random environmental effects determine which complex occurs in any individual.

Recent evidence (reviewed by Goodnight and Stevens, 1997) for the great (and unexpected) effectiveness of group selection suggests that insulation of the genetic basis of lateralization from change due to selection may be widely important. The group selection, whose consequences were studied experimentally, was based on the performance of separate small populations. These populations were, or were not allowed to proceed into the next generation, based on measures like rate of population growth in the previous generation. In the flour beetle, *Tribolium*, the variable that was strongly affected by selection was the extent of cannibalism. Work with coexisting species showed that the effectiveness of group selection depended, in part, on changing interactions between individuals belonging to different species.

Lateralization would be expected to be particularly vulnerable to alteration by selection whose direction depended on the phenotypes of the conspecifics with which the animal had to interact. Selection, which changes the genetic basis of the standard pattern of allocation of abilities, is likely to have deleterious results. This is for two types of reason. Firstly, some genetic changes in the establishment of the left–right axis also change the consistency between visceral asymmetries and so result in enhanced mortality (Goldstein, Ticho and Fishman, 1998). Note that genes involved in generation of visceral asymmetries also express asymmetrically in the developing brain (zebrafish: Thisse and Thisse, 1999), and presumably are involved in the generation of cerebral lateralization. However, in humans (as no doubt in other vertebrates) the genetic control of visceral asymmetries is not identical with that of CNS asymmetry. Tanaka et al. (1999) showed for nine cases of *situs inversus* that there was no accompanying reversal of right ear advantage in dichotic tests. Finally, the most likely reason for the preservation of a standard pattern of lateralization is that appropriate allocation of abilities between LHem and RHem is necessary for the most effective functioning of behaviour.

Fish may be different. It has already been noted that selection rapidly reverses the direction of behavioural asymmetries in at least two species of teleost. This implies that there at least two genes with different effects on lateralization that are simultaneously common in at least some outbred populations of teleosts. At the same time, there is direct evidence in the goldfish of powerful selection against one common direction of CNS asymmetry. It is therefore likely that opposing selection pressures maintain the presence of two (or more) genes that affect lateralization in different ways.

Some teleosts thus may use genetic variation to generate phenotypic variation in lateralization. At least two factors may have made it easier to evolve such a control system. Firstly, it is common for teleosts to produce very large numbers of fry, amongst which there is great selective mortality. It may have been impossible to avoid the evolution of effective genetic control of different patterns of lateralization. Secondly, the polyploidy usual in teleosts, and the opportunities that this offers for the evolution of families of alleles, may have allowed deleterious pleiotropic effects to be minimized.

5.4.4. Interindividual Variation in Humans

One index of reversal of some aspect of human lateralization, left-handedness, has an incidence of around 10% in the absence of social constraint (Bryden, 1982, pp. 157–159). Genetic effects are strongly suggested by the existence of correlations between anatomical asymmetries such as dermatoglyphic patterns and direction of handedness (Levy, 1977). It is thus likely that selection pressures have shaped the relatively high frequency of left-handedness.

One powerful reason for this is likely to be the complex and differing involvement of both hemispheres in the measure of handedness. Different factors favour the predominant use of right or of left hand, such as the need for the emission of series of skilled movements of the fingers or the need for visual control of movements. Striking evidence for the latter sort of effect is provided by a study, which showed that right-handers tended to use stored 'acting-on-object' representations of hands (presumably chiefly records of their own hand positions) when analysing pictures of grasping hands (Gentilucci, Daprati and Gangitano, 1998). Left-handers, in contrast, tended to use 'pictorial' representations, suggesting preponderant LHem visual control of hand posture in right-handers, and RHem in left-handers.

A recent study appears to demonstrate a new route for maternal effects on human behaviour, somewhat comparable to the intrauterine effects described above for rodents (Williams et al., 2000). The measure of individual variation, which was used, was the ratio of the length of the index finger to that of the fourth digit. This is lower in men than in women and the difference is argued to arise from effects of androgens in male foetuses. The more older brothers a man has, the more male this ratio becomes, suggesting that the effects of androgens during uterine life increase the more sons a pregnant woman has had. This may allow adaptive adjustment of the physical and behavioural properties of later sons. Further evidence for the validity of the finger ratio measure itself is given by Manning et al. (1999). Evidence that suggests that some aspects of human lateralization are affected by gonadal steroids during development has already been reviewed; admittedly it is not yet clear just what the induced changes are.

If there prove to be effects on lateralization by the route described by Williams et al. (2000), then this is of especial interest in the present context. Most possible selection pressures that might lead to the evolution of developmental modulation of adult patterns of lateralization do not seem to be applicable to humans. There seems little advantage in making random changes in lateralization, with the payoff that at least one child will fit an unpredictable environment (cf. effects of intrauterine neighbours in rodents), when very few children are available for randomization. Devoting some offspring to a role that reduces or prevents breeding, with fitness being provided by rearing siblings (cf. helpers in gerbils) seems very unlikely. Helpers may well have been important in our past evolution, but the long period of association of children with the family group means that helpers can be obtained without any need to reduce their subsequent potential for direct breeding. Only a strategy where the existing composition of the family can be taken into account seems appropriate for humans.

References

Allen, L.S. & Gorski, R.A. (1991). Sexual dimorphism of the anterior commissure and massa intermedia of the human brain. *Journal of Comparative Neurology*, 312, 143–153.

Andrew, R.J. (1983). Lateralization of emotional and cognitive function in higher vertebrates, with special reference to the domestic chick. In *Advances in Vertebrate Neuroethology*, ed. J.P. Ewert, R.R Capranica & D.J Ingle, pp. 477–510. New York: Plenum.

Andrew, R.J. (1991). The nature of behavioural lateralization in the chick. In *Neural and Behavioural Plasticity: The Use of the Domestic Chick as a Model*, ed. R.J. Andrew, pp. 536–554. Oxford: Oxford University Press.

Andrew, R.J. & Brennan, A. (1984). Sex differences in lateralization in the domestic chick. *Neuropsychologia*, 22, 503–509.

Andrew, R.J., Mench, J. & Rainey, C.J. (1982). Right–left asymmetry of response to visual stimuli in the domestic chick. In *Analysis of Visual Behaviour*, ed. D.J. Ingle, M.A Goodale & R.J.W. Mansfield, pp. 197–209. Cambridge, MA: MIT Press.

Andrew, R.J., Rogers, L.J., Johnston, A.N.B. & Robins, A. Light experience and the development of behavioural lateralization in chicks II: Choice of familiar versus unfamiliar. (in preparation).

Andrew, R.J., Tommasi, L. & Ford, N. (2000). Motor control by vision and the evolution of cerebral lateralization. *Brain and Language*, 73, 220–235.

Annett, M. (1972). The distribution of manual asymmetry. *British Journal of Psychology*, 63, 343–358.

Berger, B., Gaspar, P. & Verney, C. (1991). Dopaminergic innervation of the cerebral cortex: unexpected differences between rodents and primates. *Trends in Neurosciences*, 14, 21–27.

Berrebi, A.S., Fitch, R.H., Ralphe, D.L., Denenberg, J.O., Friedrich, V.L. & Denenberg, V.H. (1988). Corpus callosum: Region-specific effects of sex, early experience and age. *Brain Research*, 438, 216–224.

Bever, T.G. (1980). Broca and Lashley were right: Cerebral dominance is an accident of growth. In *Biological Studies of Mental Processes*, ed. D. Caplan, pp. 186–230. Cambridge, MA: MIT Press.

Bevilacqua, L., Capitani, E., Luzzatti, C. & Sinnler, H.R. (1979). Does the hemisphere stimulated play a specific role in delayed recognition of complex abstract patterns? A tachistoscopic study. *Neuropsychologia*, 17, 93–97.

Bianki, V.L. (1988). *The Right and Left Hemispheres of the Animal Brain: Cerebral Lateralization of Function*. New York: Gordon and Breach.

Bisazza, A., Facchin, L. & Vallortigara, G. (2000). Heritability of lateralization in fish: Concordance of right–left asymmetry between parents and offspring. *Neuropsychologia*, 3, 907–912.

Boles, D.B. (1991). Factor analysis and the cerebral hemispheres: pilot study and parietal functions. *Neuropsychologia*, 29, 59–91.

Bracha, H.S., Seitz, D.J., Otemaa, J. & Glick, S.D. (1987). Rotational movement (circling) in normal humans: Sex difference and relationship to hand, foot and eye preference. *Brain Research*, 411, 231–235.

Bradshaw, J.L. & Rogers, L.J. (1993). *The Evolution of Lateral Asymmetries, Language, Tool Use and Intellect*. Englewood Cliffs, NJ: Prentice Hall.

Bryden, M.P. (1982). *Laterality: Functional Asymmetry in the Intact Brain*. New York: Academic Press.

Carey, S. & Diamond, R. (1980). Maturational determination of the developmental course of face encoding. In *Biological Studies of Mental Processes*, ed. D. Caplan, pp. 60–93. Cambridge, MA: MIT Press.

Casey, M.B. & Nuttall, R.L. (1990). Differences in feminine and masculine characteristics as a function of handedness: Support for the Geschwind–Galaburda theory of brain organisation. *Neuropsychologia*, 28, 749–754.

Casey, M.B., Pezaris, E. & Nuttall, R.L. (1992). Spatial ability as a predictor of math achievement: The importance of sex and handedness patterns. *Neuropsychologia*, 30, 35–46.

Clark, M.M. & Galef, B.G. (2000). Why some male Mongolian gerbils may help at the nest: Testosterone, asexuality and alloparenting. *Animal Behaviour*, 59, 801–806.

Collins, R.L. (1985). On the inheritance of direction and degree of asymmetry. In *Cerebral Lateralization in Nonhuman Species,* ed. S.D. Glick, pp. 41–71. Orlando: Academic Press.

Coney, J. & MacDonald, S. (1988). The effect of retention interval upon hemispheric processes in recognition memory. *Neuropsychologia*, 26, 287–295.

Crews, D. & Gans, C. (1992). The interaction of hormones, brain and behaviour: An emerging discipline in herpetology. In *Biology of the Reptilia*, Vol. 18. *Hormones, brain and behaviour*, ed. D. Crews & C. Gans, pp. 1–23. Chicago: University of Chicago Press.

Dabbs, J.M. & Choo, G. (1980). Left–right carotid blood flow predicts specialised mental ability. *Neuropsychologia*, 18, 711–713.

Deng, C. (1998). Organisation of the visual pathways and visual lateralization in the chick. Ph.D. thesis, University of New England, Armidale, Australia.

Deng, C. & Rogers, L.J. (2000). Organisation of intratelencephalic projections to the visual Wulst of the chick. *Brain Research*, 856, 152–162.

Denenberg, V.H., Gall, J.S., Berrebi, A. & Yutzey, D.A. (1986). Callosal mediation of cortical inhibition in the lateralized rat brain. *Brain Research*, 397, 327–332.

Dharmaretnam, M. & Andrew, R.J. (1994). Age- and stimulus-specific effects on the use of right and left eyes by the domestic chick. *Animal Behaviour*, 48, 1395–1406.

Diamond, M.C. (1985). Rat forebrain morphology: right–left; male–female; young–old; enriched–impoverished. In *Cerebral Lateralization in Nonhuman Species,* ed. S.D. Glick, pp. 73–87. Orlando: Academic Press.

Driesen, N.R. & Raz, N. (1995). The influence of sex, age, and handedness on corpus callosum morphology. A meta-analysis. *Psychobiology*, 23, 240–247.

Dunnett, S.B. & Robbins, T.W. (1992). The functional role of mesotelencephalic dopamine systems. *Biological Reviews*, 67, 491–518.

Eaton, M.J. & Gaffan, D. (1990). Interhemispheric transfer of visuomotor conditional learning via the anterior corpus callosum. *Behavioural Brain Research*, 38, 109–116.

Fitch, R.H., Berrebi, A.S., Cowell, P.E., Schrott, L.M. & Denenberg, V.H. (1990). Corpus callosum: Effects of neonatal hormones on sexual dimorphism in the rat. *Brain Research*, 515, 111–116.

Fitch, R.H., Cowell, P.E., Schrott, L.M. & Denenberg, V.H. (1991a). Corpus callosum: Ovarian hormones and feminisation. *Brain Research*, 542, 313–317.

Fitch, R.H., Cowell, P.E., Schrott, L.M. & Denenberg, V.H. (1991b). Corpus callosum: Demasculinisation via perinatal antiandrogen. *International Journal of Developmental Neuroscience*, 1, 35–38.

Gaulin, S.J.C., Fitzgerald, R.W. & Wartell, M.S. (1990). Sex differences in spatial ability and activity in two vole species. *Journal of Comparative Psychology*, 104, 88–93.

Gentilucci, M., Daprati, E. & Gangitano, M. (1998). Right-handers and left-handers have different representations of their own hands. *Cognitive Brain Research*, 6, 185–192.

Geschwind, N. & Behan, P. (1982). Lefthandedness: Association with immune disease, migraine and developmental learning disorder. *Proceedings of the National Academy of Sciences USA*, 79, 5097–5100.

Glick, S.D. & Ross, D.A. (1981). Right-sided population bias and lateralization of activity in normal rats. *Brain Research*, 205, 222–225.

Goldstein, A.M., Ticho, B.S. & Fishman, M.C. (1998). Patterning the heart's left–right axis: from zebrafish to man. *Developmental Genetics*, 22, 278–287.

Goodnight, C.J. & Stevens, L. (1997). Experimental studies of group selection; what do they tell us about group selection in nature? *American Naturalist*, 150, S59–79.

Guhl, A.M. (1962). The behaviour of chickens. In *The Behaviour of Domestic Animals*, ed. E.S.E. Hafez. London: Baillière, Tindall and Cox.

Güntürkün, O. & Hoferichter, H. (1985). Neglect after section of a left telencephalic tract in pigeons. *Behavioural Brain Research*, 18, 1–9.

Gutzke, W.H.N. & Crews, D. (1988). Embryonic temperature determines adult sexuality in a reptile. *Nature*, 332, 832–834.

Hinde, R.A. (1970). *Animal Behaviour*, 2nd edn. New York: McGraw-Hill.

Kimura, D. (1999). *Sex and Cognition*, Cambridge, MA: MIT Press.

Klein, R.M. & Andrew, R.J. (1986). Distraction, decisions and persistence in runway tests using the domestic chick. *Behaviour*, 99, 139–156.

Kucharski, D. & Hall, W.G. (1987). New routes to old memories. *Science*, 238, 786–788.

Ladavas, E. (1982). The development of facedness. *Cortex*, 18, 535–545.

Larrabee, A.P. (1906). The optic chiasma of teleosts: a study of inheritance. *Proceedings of the American Academy of Arts and Sciences*, 42, 217–230.

Lavooy, M.J., Lavooy, J., Hahn, M.E. & Simnel, E.C. (1981). Passive avoidance during brain-growth spurts and plateaus. *Bulletin of the Psychonomic Society*, 17, 153–155.

Levine, S.C. (1985). Developmental changes in right-hemisphere involvement in face recognition. In *Hemisphere Function and Collaboration in the Child*, ed. C.T. Best, pp. 157–191. Orlando: Academic Press.

Levy, J. (1977). The origins of lateral asymmetry. In *Lateralization in the Nervous System*, ed. S. Harnad, R.W. Doty, L. Goldstein, J. Jaynes & G. Krauthamer, pp. 195–209. London: Academic Press.

Levy, J. (1985). Interhemispheric collaboration; single-mindedness in the asymmetric brain. In *Hemisphere Function and Collaboration in the Child*, ed. C.T. Best, pp. 11–31. Orlando: Academic Press.

Lewis, D.W. & Diamond, M.C. (1995). The influence of gonadal steroids on the asymmetry of the cerebral cortex. In *Brain Asymmetry*, ed. R.J. Davidson & K. Hugdahl, pp. 31–50. Cambridge, MA: MIT Press.

Liederman, J. & Kinsbourne, M. (1980). Rightward turning biases in neonates reflect a single neural asymmetry in motor programming: A reply to Turkewitz. *Infant Behaviour and Development*, 3, 239–244.

Lipsey, J.R. & Robinson, R.G. (1986). Sex dependent behavioural responses to frontal cortical suction lesions in the rat. *Life Sciences*, 38, 2185–2192.

McBride, G., Parer, I.P. & Foenander, F. (1969). The social organisation and behaviour of the domestic fowl. *Animal Behaviour*, 2, 125–181.

McCabe, B.J. (1991). Hemispheric asymmetry of learning-induced changes. In *Advances in Vertebrate Neuroethology*, ed. J.P Ewert, R.R. Capranica & D.J Ingle, pp. 262–276. New York: Plenum.

McGlone, J. (1980). Sex differences in human brain asymmetry. *Behavioural and Brain Sciences*, 3, 215–263.

McKenzie, R., Andrew, R.J. & Jones, R.B. (1998). Lateralization in chicks and hens: New evidence for control of response by the right eye system. *Neuropsychologia*, 36, 51–58.

Mann, V.A., Diamond, R. & Carey, S. (1979). Development of voice recognition: Parallels with face recognition. *Journal of Experimental Child Psychology*, 27, 153–165.

Manning, J.T., Trivers, B.L., Singh, D. & Thornhill, R. (1999). The mystery of female beauty. *Nature*, 399, 214–215.

Marchetti, C. & Drent, P.J. (2000). Individual differences in the use of social information in foraging by captive great tits. *Animal Behaviour*, 60, 131–140.

Mead, L.A. & Hampson, E. (1997). Turning bias in humans is influenced by phase of the menstrual cycle. *Hormones and Behavior*, 31, 65–74.

Mench, J.A. & Andrew, R.J. (1986). Lateralization of a food search task in the domestic chick. *Behavioural and Neural Biology*, 46, 107–114.

Michel, G.F. (1983). Development of hand-use preference during infancy. In *Manual Specialisation and the Developing Brain*, ed. G. Young, S.J. Segalowitz, C.M. Corter & S.E. Trehub, pp. 33–70. New York: Academic Press.

Miklósi, A. & Andrew, R.J. The effects of selection of lateralization in the zebrafish (*Brachydanio rerio*). (in preparation).

Parsons, C.H. & Rogers, L.J. (1993). Role of the tectal and posterior commissures in lateralization of the avian brain. *Behavioural Brain Research*, 54, 153–164.

Pitblado, C. (1979). Visual field differences in perception of vertical with and without a visible form of reference. *Neuropsychologia*, 17, 381–392.

Rashid, N.Y. (1988). Lateralization of topographical learning and other abilities in the chick. D.Phil. thesis, University of Sussex, Brighton.

Ray, W.J., Newcombe, N., Semm, J. & Cole, P.M. (1981). Spatial abilities, sex differences and EEG functioning. *Neuropsychologia*, 19, 719–722.

Regolin, L. & Vallortigara, G. (1995). Lateral asymmetries during responses to novel-coloured objects in the domestic chick: A developmental study. *Behavioural Processes*, 37, 67–74.

Robinson, R.G. (1985). Lateralized behavioural and neurochemical consequences of unilateral brain injury in rats. In *Cerebral Lateralization in Nonhuman Species*, ed. S.D Glick, pp. 135–156. Orlando: Academic Press.

Robinson, T.E., Becker, J.B., Camp, D.M. & Mansour, A. (1985). Variation in the pattern of behavioural and brain asymmetries due to sex differences. In *Cerebral Lateralization in Nonhuman Species*, ed. S.D. Glick, pp. 185–231. Orlando: Academic Press.

Rogers, L.J. (1990). Light input and the reversal of functional lateralization in the chicken brain. *Behavioural Brain Research*, 38, 211–221.

Rogers, L.J. (1991). Development of lateralization. In *Neural and Behavioural Plasticity: The Use of the Domestic Chick as a Model*, ed. R.J. Andrew, pp. 507–535. Oxford: Oxford University Press.

Rogers, L.J. & Drennen, H.D. (1978). Cycloheximide interacts with visual input to produce permanent slowing of visual learning in chickens. *Brain Research*, 158, 479–482.

Rogers, L.J. & Ehrlich, D. (1983). Asymmetry in the chicken forebrain during development and a possible involvement of the supraoptic decussation. *Neuroscience Letters*, 37, 123–127.

Rogers, L.J. & Krebs, G.A. (1996). Exposure to different wavelengths of light and the development of structural asymmetries in the chicken. *Behavioural Brain Research*, 80, 65–73.

Rogers, L.J. & Sink, H.S. (1987). Transient asymmetry in the projections of the rostral thalamus to the visual hyperstriatum of the chicken, and reversal of its direction by light exposure. *Experimental Brain Research*, 70, 378–398.

Rogers, L.J., Andrew, R.J. & Burne, T.H.J. (1998). Light exposure of the embryo and the development of behavioural lateralization in chicks I: Olfactory responses. *Behavioural Brain Research*, 97, 195–200.

Rogers, L.J., Andrew, R.J. and Johnston, A.N.B. Light experience and the development of behavioural lateralization. III: Pebble versus grain pecking. (in preparation).

Rose, D. (1980). Some functional correlates of the maturation of neural systems. In *Biological Studies of Mental Processes*, ed. D. Caplan, pp. 27–43. Cambridge, MA: MIT Press.

Ross, D.A., Glick, S.D. & Meibach, R.C. (1981). Sexually dimorphic brain and behavioural asymmetries in the neonatal rat. *Proceedings of the National Academy of Sciences USA*, 78, 1958–1961.

Roth, R.L. (1979). Decussation geometries in the goldfish nervous system; correlation with probability of survival. *Proceedings of the National Academy of Sciences USA*, 76, 4131–4135.

Rothbart, M.K., Taylor, S.B. & Tucker, D.M. (1989). Right-sided facial asymmetry in infant emotional expression. *Neuropsychologia*, 27, 675–687.

Schenk, F., Grobéty, M.C., Lavenex, P. & Lipp, H.-P. (1994). Dissociation between basic components of spatial memory in rats. In *Behavioural Brain Research in Naturalistic and Seminaturalistic Settings*, ed. E. Alleva, A. Fasola & H.-P. Lipp, pp. 177–300. Dordrecht: Kluwer Academic Publisher.

Schwabl, H. (1993). Yolk is a source of maternal testosterone for developing birds. *Proceedings of the National Academy of Sciences USA*, 90, 11446–11450.

Sdraulig, R., Rogers, L.J. & Boura, A.L.A. (1980). Role of the supraoptic decussation in the development of asymmetry of brain function in the chicken. *Developmental Brain Research*, 28, 33–39.

Shaywitz, B.A., Shaywitz, S.E., Pugh, K.R., Constable, R.T., Skudlarski, P., Fulbright, R.K., Bronen, R.A., Fletcher, J.M., Shankweiler, D.P., Katz, L. & Gore, J.C. (1995). Sex differences in the functional organisation of the brain for language. *Nature*, 373, 607–609.

Sholl, S.A. & Kim, K.L. (1990). Androgen receptors are differentially distributed between right and left cerebral hemispheres of the foetal male rhesus monkey. *Brain Research*, 516, 122–126.

Sprague, J.M. (1966). Interaction of cortex and superior colliculus in mediation of visually guided behaviour. *Science*, 153, 1544–1547.

Sprague, J.M. & Meikle, T.H. (1965). The role of the superior colliculus in visually guided behaviour. *Experimental Neurology*, 11, 115–146.

Takahashi, S. & Hori, M. (1994). Unstable evolutionarily stable strategy and oscillation: A model of lateral asymmetry in scale-eating cichlids. *American Naturalist*, 144, 1001–1020.

Tanaka, S. et al. (1999). Dichotic listening in patients with situs inversus: brain asymmetry and situs inversus. *Neuropsychologia*, 37, 869–874.

Thatcher, R.W., Walker, R.A. & Giudice, S. (1987). Human cerebral hemispheres develop at different rates and ages. *Science*, 236, 1110–1113.

Thisse, C. & Thisse, B. (1999). Antivin, a novel and divergent member of the TGFβ superfamily, negatively regulates mesoderm induction. *Development*, 126, 229–240.

Vallortigara, G. (1996). Learning of colour and position cues in domestic chicks: Males are better at position, females at colour. *Behavioural Processes*, 36, 289–296.

Vallortigara, G. & Andrew, R.J. (1991). Lateralization of response to change in social partner in chick. *Animal Behaviour*, 41, 187–194.

Vallortigara, G. & Andrew, R.J. (1994). Olfactory lateralization in the chick. *Neuropsychologia*, 32, 417–423.

Vallortigara, G., Andrew, R.J., Sertori, L. & Regolin, L. (1997). Sharply timed behavioural changes during the first 5 weeks of life in the domestic chick (*Gallus gallus*). *Bird Behaviour*, 12, 29–40.

van Eden, C.G., Uylings, H.B.M. & van Pelt, J. (1984). Sex-difference and left–right asymmetries in the prefrontal cortex during postnatal development in the rat. *Developmental Brain Research*, 12, 146–153.

vom Saal, F.S. (1983). Models of early hormonal effects on intrasex aggression in mice. In *Hormones and Aggressive Behaviour*, ed. B.B. Svare, pp. 197–222. New York: Plenum Press.

vom Saal, F.S. & Bronson, F.H. (1980). Sexual characteristics of adult female mice are correlated with their blood testosterone levels during prenatal development. *Science*, 208, 597–599.

Ward, R. & Collins, R.L. (1985). Brain size and shape in strongly and weakly lateralized mice. *Brain Research*, 328, 243–249.

Warren, S.G. & Juraska, J.M. (1997). Spatial and nonspatial learning across the rat oestrous cycle. *Behavioural Neuroscience*, 111, 259–266.

Williams, C.L., Barnett, A.M. & Meck, W.H. (1990). Organisational effects of early gonadal secretions on sexual differentiation in spatial memory. *Behavioural Neuroscience*, 104, 84–97.

Williams, T.J., Pepitone, M.E., Christensen, S.E., Cooke, B.M., Huberman, A.D., Breedlove, N.J., Breedlove, T.J., Jordan, C.L. & Breedlove, S.M. (2000). Finger-length ratios and sexual orientation. *Nature*, 404, 455–456.

Wittling, W. (1995). Brain asymmetry in the control of autonomic–physiologic activity. In *Brain Asymmetry*, ed. R.J. Davidson & K. Hugdahl, pp. 305–357. Cambridge, MA: MIT Press.

Workman, L. & Andrew, R.J. (1989). Simultaneous changes in behaviour and in lateralization during the development of male and female domestic chicks. *Animal Behaviour*, 38, 596–605.

Workman, L., Adam, J. & Andrew, R.J. (2000). Opportunities for visual experience which might allow imprinting in chicks raised by broody hens. *Behaviour*, 137, 221–231.

Young, G. (1977). Manual specialisation in infancy: Implications for lateralization of brain function. In *Language Development and Neurological Theory*, ed. S.J. Segalowitz & F.A. Griber, pp. 289–311. New York: Academic Press.

Zappia, J.V. & Rogers, L.J. (1987). Sex differences and reversal of brain asymmetry by testosterone in chickens. *Behavioural Brain Research*, 23, 261–267.

6

Factors Affecting the Development of Lateralization in Chicks

CHAO DENG AND LESLEY J. ROGERS

6.1. Introduction

Over the last two decades, it has been well established that there are a variety of forms of structural and functional lateralization in a number of avian species, including chicks (reviewed by Andrew, 1988, 1991; Rogers, 1995, 1996), pigeons (Güntürkün, 1997b), canaries (Nottebohm, 1977), zebra finches (Alonso, 1998) and marsh tits (Clayton and Krebs, 1994). Among these avian species, the chick, *Gallus gallus domesticus*, has been used as a model to study how hormones and early experience interact to influence the development of brain lateralization. In this chapter we review a number of studies that have examined the influence of light exposure prior to hatching and of steroid hormones on the development of asymmetry of the visual pathways and lateralization of visual behaviour.

Although the chick has been used extensively in studies of learning, memory formation, visual lateralization and visual neurone development, knowledge of the organization of the visual pathways in the chick is limited. Until recently, most knowledge of the organization of avian visual pathways came only from studies of the pigeon (reviewed by Güntürkün et al., 1993; Bischof and Watanabe, 1997), and it was generally assumed that the visual system of chicks had the same organization as that of the pigeon or one very similar to it. However, our recent studies have shown there are some clear differences between the chick and the pigeon in the organization of the central visual pathways (Deng and Rogers, 1998a, 1998b). Therefore, we begin this chapter by discussing the organization of the visual pathways in the chick and later discuss lateralization in the visual system and factors influencing the development of visual lateralization. Comparison is made with the organization of these pathways in the pigeon. We recognize that we are comparing data from young chicks with data from adult pigeons and, therefore, the differences could depend on age differences rather than species differences alone, but this cannot be resolved until further research is conducted on adult chicks and young pigeons. In the meantime, it is important to point out the species/age differences because they are so frequently ignored.

6.2. Bilateral Projections in the Two Visual Pathways

Birds have two main visual pathways to the telencephalon: the thalamofugal and tectofugal pathways (Figure 6.1). In the thalamofugal pathway, the nucleus geniculatus lateralis pars dorsalis (GLd) receives visual afferents from the contralateral retina and then projects to the visual Wulst in each hemisphere of the telencephalon (Deng and Rogers, 1998b). The visual Wulst is a multilayered structure with at least four layers, including the hyperstriatum accessorium (HA), the nucleus intercalatus of hyperstriatum accessorium (IHA), the hyperstriatum intercalatum superior (HIS) and the hyperstriatum dorsale (HD), arranged in this order from dorsal to ventral regions (Deng and Rogers, 2000a). In the tectofugal pathway, each optic tectum (TeO) receives contralateral retinal afferents and then projects to the nucleus rotundus (Rt) on each side of the thalamus and, in turn, each Rt projects to the ectostriatum (E) in the telencephalon (Deng and Rogers, 1998a).

An important feature of the avian visual system is the complete decussation of the optic nerves in the optic chiasma; therefore, each GLd and TeO receives only contralateral retinal afferents (Ehrlich and Mark, 1984). The GLd sends the majority of its efferents to its ipsilateral visual Wulst but also sends efferents to the contralateral visual Wulst by fibres that recross the midline in the supraoptic decussation, pars dorsalis (Repérant et al., 1974; Deng and Rogers, 1998b). Each TeO projects to both the ipsilateral and contralateral Rt (Bischof and Niemann, 1990; Ngo et al., 1994; Deng and Rogers, 1998b). Thus, further forward than the first relay station in each visual pathway, visual projections are sent to both sides of the brain. There are, however, more projections from each eye to its contralateral hemisphere (Rt, E and visual Wulst regions) than to its ipsilateral hemisphere (Figure 6.1).

6.2.1. Thalamofugal System of the Chick

Using a double-labelling technique, we have revealed a distinct difference between the bilaterally projecting neurones in the thalamofugal and tectofugal pathways of the chick (Deng and Rogers, 1998b). To examine the thalamofu-

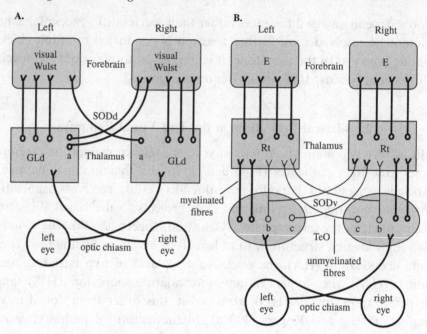

Figure 6.1.   (A) The thalamofugal visual pathway of the chick. Note that each eye projects to the contralateral nucleus geniculatus lateralis pars dorsalis (GLd) in the thalamus and that each GLd has both contralateral and ipsilateral projections, via myelinated fibres, to the visual Wulst. The contralateral projections (a) cross the midline in the dorsal supraoptic decussation (SODd) and more cross from the left GLd to right Wulst than vice versa. (B) The tectofugal visual pathway. Note that there are neurones in the optic tecta (TeO) that project, via myelinated fibres, to the ipsilateral nucleus rotundus (Rt) on each side of the brain and others that cross the midline in the ventral supraoptic decussation (SODv). Those that cross the midline are of two types: b, collateral branches of neurones that also project ipsilaterally; c, neurones without collaterals that project only to the contralateral Rt; both the bilaterally projecting neurones and the contralaterally projecting neurones (c) of TeO send out unmyelinated fibres to Rt. Projections from the Rt go to the ipsilateral ectostriatal region (E) of the forebrain. The bilaterally projecting neurones, which have axon collaterals projecting to both sides of the brain, have been shown only in the TeO–Rt projections. The thin lines indicate unmyelinated axons, and the thick lines indicate myelinated axons. Modified from Deng and Rogers (1998b).

(Reprinted from *Brain Research*, 794, Deng, C. & Rogers, L.J., Bilaterally projecting neurones in the two visual pathways in the chick, pp. 281–290, © 1998, with permission from Elsevier Science.)

gal projections, injections of the fluorescent retrograde tracers fluorogold (FG) and rhodamine B isothiocyanate (RITC) were made into the Wulst of each hemisphere separately. Labelled neurones in GLd were counted. Although the distribution areas of ipsilaterally and contralaterally labelled neurones overlap partly, very few double-labelled neurones were found in GLd (only 0.01% of all labelled neurones). This means that very few GLd neurones have collaterals projecting bilaterally to both the ipsilateral and contralateral visual Wulst. As shown in Figure 6.1A, apart from an extremely small subpopulation, each GLd neurone, after receiving inputs from the contralateral eye, projects solely to either the ipsilateral or the contralateral Wulst. Furthermore, since the axons from the GLd to the contralateral Wulst are large myelinated fibres (Saleh and Ehrlich, 1984), the visual information from GLd is transmitted through these fibres to the contralateral Wulst without a delay any greater than occurs in the projections from the GLd to the ipsilateral Wulst, also via myelinated axons (Denton, 1981; Deng and Rogers, 1998b). Thus each eye sends simultaneous visual inputs to the Wulst regions of both hemispheres, albeit via different neuronal populations (ipsilaterally or contralaterally projecting neurones). Considering this, together with the fact that there are fewer contralateral than ipsilateral projections from each GLd to the visual Wulst, it seems that the Wulst regions in each hemisphere may simultaneously receive different visual inputs and may be involved in processing different visual information. It is likely that the Wulst ipsilateral to an eye receives more sketchy information than does the Wulst contralateral to that eye.

In addition to this differential ipsilateral and contralateral input to the visual Wulst, in many birds, including chicks, each eye can focus independently of the other and move independently to scan the environment (Wallman and Pettigrew, 1985; Wallman and Letelier, 1993). Thus, each eye sends different information to the forebrain even in the frontal visual field (Wallman and Pettigrew, 1985; Vallortigara and Andrew, 1994). Furthermore, Hart et al. (2000) have found, in the European starling, *Sturnus vulgaris*, that there is a significant difference between the left and right retinae in terms of the proportion of four types of photoreceptors. The left retina contains a significantly higher proportion of long-wavelengthsensitive (LWS, 18%) and medium-wavelength-sensitive (MWS, 18%) single cones than the right retina (LWS, 17%; MWS, 17%). The right retina also contains a higher proportion of the double cones (56%) than the left retina (52%). A similar pattern of asymmetry has been found in the blue tit, *Parus caeruleus* (Hart et al., 2000). Since the single cones are used for colour vision and the double cones for movement detection (Bowmaker et al., 1997; Campenhausen and Kirschfeld, 1998), it is possible that visual inputs from the left eye are more involved in processing of colour cues than those from the right eye, whereas visual inputs from the right eye are more involved with processing movement cues of the stimulus. Some of this different information transmitted from the eyes may, therefore, be processed differentially in the visual regions of left and right forebrain hemispheres (in the visual Wulst regions or E).

If there are significant differences in the nature of visual input from each eye to the visual regions of the hemispheres, it would be important for a bird to chose to view a stimulus using the eye most appropriate for the task at hand. As McKenzie, Andrew and Jones (1998) have shown, chicks do choose to view certain stimuli with the right eye and others with the left eye (discussed in Chapter 3 by Andrew and Rogers). The choice of eye may depend on lateralized memory processes in the hemispheres (Chapter 14 by Johnston and Rose, and Chapter 15 by Andrew), as well as on differences in perceptual information received by each hemisphere.

Differential use of the bird's laterally placed eyes raises the question of how information from the two eyes is integrated, apart from in binocular vision. When, for example, a stimulus is fixated by the right eye, in the monocular field, the left Wulst would have the primary role in processing the visual input. The left Wulst would also receive sketchy input from the left eye, which views an entirely different visual scene. The left eye might, therefore, be used in panoramic scanning that provides information monitored as background. At the same time inputs from the left eye would be received in more detail by the right hemisphere and these would be monitored against more sketchy ipsilateral input from the right eye. Integration of the differential inputs from each eye would, therefore, be an essential aspect of wide-angle monocular vision, and not solely subserve binocular vision.

6.2.2. Comparison with the Thalamofugal System of the Pigeon

It is worth noting that, compared to the chick, the pigeon has a somewhat different organization of the projections from GLd to the Wulst. The GLd of the pigeon contains more bilaterally projecting neurones (9–38% in nucleus dorsolateralis anterior thalami pars lateralis pars dorsalis, DLLd, and 18–46% in nucleus superficialis parvocellularis, SPC, both of these nuclei being subnuclei of GLd) (Miceli and Repérant, 1982; Güntürkün et al., 1993). These neurones have collaterals and so send the same visual information simultaneously to both the ipsilateral and contralateral Wulst (Miceli and Repérant, 1982). Therefore, in the pigeon, the visual Wulst regions in each hemisphere may process information from the same GLd neurones at the same time. Compared to the chick, the pigeon has an additional bilaterally projecting population of neurones in GLd and, therefore, the thalamofugal inputs to the Wulst in each hemisphere of the pigeon may be more similar than in chicks. This might indicate that the Wulst regions of the pigeon are more integrated than in the chick and that they receive concordant information, whereas this does not occur in the Wulst of the chick. We are tempted to speculate that this species difference may be associated with the fact that the pigeon is better adapted for flight than is the chick, which is primarily ground living. The species differences might also relate to differences in cognitive abilities.

The visual Wulst of the chick receives not only the afferents from GLd but also inputs from many telencephalic regions, including the neostriatum frontale pars lateralis, the neostriatum intermedium, the dorsolateral neostriatum and the archistriatum (Deng and Rogers, 2000a; Figure 6.2). Through these intratelencephalic afferents to the visual Wulst, visual information transmitted in the thalamofugal pathway may be modulated by processing in other telencephalic areas. The visual Wulst also sends efferents to these telencephalic regions (Figure 6.2). Similar reciprocal connections between the visual Wulst and other telencephalic areas have been found in the pigeon (Bagnoli and Burkhalter, 1983; Shimizu, Cox and Karten, 1995). Therefore, it is likely that the visual Wulst is not merely a primary visual area, as

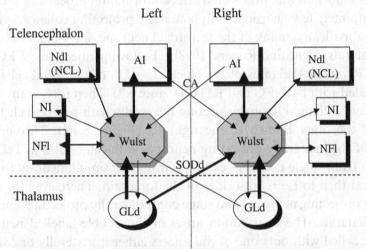

Figure 6.2. A schematic diagram demonstrates the reciprocal connections between the visual Wulst and other telencephalic areas and afferents from the thalamus. Note the asymmetrical projections from the GLd to the visual Wulst. No investigation of possible asymmetry in the other projections has been made. AI, archistriatum intermedium; CA, commissura anterior; GLd, nucleus geniculatus lateralis pars dorsalis; NCL, neostriatum caudolaterale; Ndl, dorsal–lateral neostriatum; NFl, neostriatum frontale pars lateralis; NI, neostriatum intermedium; SODd, dorsal supraoptic decussation. The size of arrows indicates the relative number of projections: the larger arrows indicate more projections and vice versa. Modified from Deng and Rogers (2000a).

(Reprinted from *Brain Research*, 856, Deng, C. & Rogers, L.J. Organisation of intratelecephalic projections to the visual Wulrt of the chick, pp. 152–162, © 2000, with permission from Elsevier Science.)

suggested previously (Karten, 1979) but also an integration area of the fore-brain. This is consistent with the known fact that lesioning the visual Wulst impairs cognitive function. For example, marked deficits in reversal learning are caused by lesioning the visual Wulst, as shown in bobwhite quails (Stettener and Schultz, 1967), chicks (Benowitz and Lee-Teng, 1973) and pigeons (Macphail, 1976; Shimizu and Hodos, 1989). Marked deficits have also been found in delayed matching-to-sample performance after visual Wulst lesions in pigeons (Pasternak, 1977). In addition, one of our studies (Deng and Rogers, 1997) found that the visual Wulst of the chick is involved in categorizing food from non-food and allows the chick to avoid pecking at pebbles (more details are given below).

6.2.3. Tectofugal System of the Chick

The tectofugal projections of the chick differ from the thalamofugal projections in a fundamental way that would affect information processing: whereas there are extremely few neurones with bilaterally projecting collaterals in the thalamofugal pathway, many of the tectofugal neurones have bilaterally projecting collaterals (Deng and Rogers, 1998b). Following injections of FG and RITC into Rt in the thalamus of the chick, 25–45% of the tectal cells were double-labelled with both FG and RITC (Figure 6.3). Therefore, many tectal neurones have collateral axons projecting bilaterally with one branch to the left Rt and the other branch to the right Rt (Figure 6.1B). Through the collaterals of the bilaterally projecting neurones with cell bodies in TeO, the information from single neurones can be transmitted simultaneously to both Rt nuclei and then to E on both sides of the forebrain. There are also many ipsilaterally projecting neurones and some contralaterally projecting neurones without collaterals. The distribution areas of the double-labelled neurones and those labelled with only one of the tracers either ipsilaterally or contralaterally were found to overlap completely (Figure 6.3). To date, no data on the number and distribution of bilaterally projecting tectofugal neurones are available in other avian species, including the pigeon.

It is most interesting that, in the chick, nearly all of the axons projecting to the contralateral Rt (99.5%) are small, unmyelinated fibres, no matter whether they are collaterals of bilaterally projecting neurones or fibres of neurones projecting solely contralaterally from TeO (Saleh and Ehrlich, 1984; Ngo et al., 1994). By contrast, the TeO neurones projecting to the ipsilateral Rt have large myelinated axons (Ngo, Egedy and Tömböl, 1992; Ngo et al., 1994). This organization suggests that there are two stages of visual transmission in the tectofugal pathway of the chick: the first involves

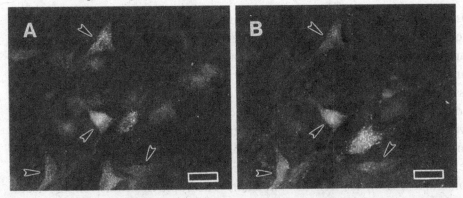

Figure 6.3. Fluorescence photomicrographs of the same section through the optic tectum showing rhodamine B isothiocyanate (RITC) labelled neuronal cell bodies (A) and fluorogold (FG) labelled neuronal cell bodies (B) in the layer SGC (stratum griseum centrale) after injecting these tracers into the left or right Rt separately. The same field is viewed with the appropriate filters for RITC (A) and FG (B). Neurones double labelled with RITC and FG are indicated by arrowheads. Scale bar: 20 μm. From Deng and Rogers (1998b).

(Reprinted from *Brain Research*, 794, Deng, C. & Rogers, L.J., Bilaterally projecting neurones in the two visual pathways in the chick, pp. 281–290. © 1998, with permission from Elsevier Science.)

rapid transmission via large myelinated axons and the second involves delayed transmission via small unmyelinated fibres (Figure 6.1B). In the rapid transmission component, the visual information would be transmitted through the pathway TeO to its ipsilateral Rt and then to its ipsilateral E. From each eye this well-developed pathway of rapid transmission goes to the contralateral hemisphere. In the delayed transmission component, visual information would be transmitted through the pathway from TeO to its contralateral Rt and then to E ipsilateral to that Rt and also via the pathway TeO to its ipsilateral Rt and then to the ipsilateral E. From each eye, therefore, these less well-developed and slower transmitting pathways go to both hemispheres.

In summary, considering that the optic nerves project only to the contralateral TeO, visual information is transmitted rapidly solely to the Rt and E contralateral to a given eye. Next, visual information is transmitted slowly from a given eye to both the ipsilateral and contralateral Rt in the unmyelinated TeO to Rt projections and then to both E. Therefore, the proposed first stage, rapid transmission, would involve unilateral processing of information from each eye in its contralateral hemisphere, whereas the second stage involves bilateral processing of information from both eyes, and coordinated activity of the left and right Rt and left and right E. There are two possible implications for this organization. First, it is possible that the rapid neural transmission is used for detecting the stimulus and priming the neuronal activities of the particular forebrain area(s). Then the delayed neural transmission following may be used for further binocular information processing, including the information used to perceive depth and make decisions about responding. However, so far, no studies have been carried out to investigate the physiological characteristics of the proposed two-stage neural transmission or the properties of possible binocular neurones in Rt or E in the chick, or in any other avian species. The second possibility involves monocular vision, as well as binocular. The rapid transmission of visual input to the hemisphere contralateral to the eye may be used for initial detection of the stimulus and to direct head movements to allow appropriate fixation. Following a shift of fixation from one eye to the other, the slow transmission, which goes to both hemispheres, may ensure continued attention to that stimulus. The delayed arrival of the slower transmission for both hemispheres would be optimal for such a function.

6.3. Asymmetry of the Visual Projections to the Forebrain

Structural asymmetry of the visual system has been reported in the thalamofugal projections of the chick and no obvious asymmetry has been found in the tectofugal pathways. In two strains of domestic chicks, *Gallus gallus domesticus* (Boxer and Stanford, 1985; Rogers and Sink, 1988) and a strain of feral chicks *Gallus gallus* (Adret and Rogers, 1989), asymmetry of the thalamofugal projections has been revealed using fluorescent tracers or horseradish peroxidase. For example, after injecting FG into the visual Wulst on one side of the forebrain and true blue (TB) into the visual Wulst on the other side, the GLd neurones labelled on both sides of the thalamus were counted. The ratio of the number of GLd cells labelled contralaterally to the injection site to the number labelled in GLd ipsilaterally to the injection site (c/i ratio) was determined for each tracer. The ratio was used to control for variation in the amount of these tracer dyes injected, which is an unavoidable aspect of using these tracers, since they are injected as suspensions which tend to precipitate in the injecting syringe. The c/i ratio obtained after injecting tracer into the right Wulst was found to be 70% higher than that obtained by injecting into the left Wulst (Rogers and Sink, 1988; Adret and Rogers, 1989). Recently, we have used RITC, which is completely soluble in dimethyl sulfoxide/water mixture, and so it is possible to control the amount injected more precisely. Therefore, we were able to use the data for absolute numbers of contralaterally and ipsilaterally labelled cells.

Using this technique, we were able to confirm the presence of asymmetry in the thalamofugal projections and to locate it in the contralateral GLd–Wulst projections, not in the ipsilateral projections (Rogers and Deng, 1999). There are approximately 60% more contralateral projections from the left GLd to the right visual Wulst than contralateral projections from the right GLd to the left visual Wulst (Figure 6.1A), whereas the left and right visual Wulst regions receive equal numbers of ipsilateral GLd inputs. Location of the asymmetry in the contralateral GLd to Wulst projections only is consistent with the earlier HRP study of Boxer and Stanford (1985). Also, consistent with the right visual Wulst receiving more contralateral projections than the left Wulst, the synaptic density per unit volume is significantly (22%) higher in the right HA than in the left HA, as shown in 2-day-old male chicks (Stewart et al., 1992). We are, of course, aware that the higher synaptic density in the right Wulst is unlikely to be merely a reflection of increased inputs from the left GLd. It should be noted that the asymmetry in the GLd to Wulst projections of the chick is present only for the first 3 weeks post-hatching (Rogers and Sink, 1988). This transient asymmetry of the visual projections may, however, leave residual and persistent asymmetries in higher order integration regions of the forebrain. In other words, it may have a snowball effect in determining behavioural and neural asymmetries that persist into adulthood (Rogers, 1991).

No clear asymmetry has been found in the GLd–Wulst projections of the pigeon (Güntürkün, pers. comm.). This suggests that there is a clear difference between the species. Asymmetry in the TeO–Rt projections has, however, been reported in the pigeon. There are about twice as many contralateral projections from right TeO to the left Rt than from the left TeO to the right Rt (Güntürkün et al., 1998). Significant left–right differences have been found also in the soma size of the TeO neurones: cell bodies in layers 2–12 are larger in the left TeO, whereas cell bodies in layers 13–15 are larger in the right TeO of the adult pigeon (Güntürkün, 1997a; Chapter 7 by Güntürkün).

In contrast to the pigeon, chicks have no obvious asymmetry in the tectofugal projections to Rt (Rogers and Deng, 1999). Since FG and TB are unable to label the contralaterally projecting neurones in the tectofugal pathway (Deng and Rogers, 1999), only RITC was used as the tracer dye to label cell bodies. After injecting RITC into the left or right Rt of the chick, the labelled neurones were counted in both the ipsilateral and contralateral TeO. There was no indication of asymmetry in the contralateral TeO–Rt projections determined in terms of absolute counts of labelled neurones. There was also no significant asymmetry in the c/i ratio calculated after counting the labelled cells in TeO as a whole. However, when the c/i ratios were calculated separately for the ventral and dorsal regions of TeO, a slight but significant asymmetry emerged in the c/i ratio calculated for the ventral regions of the TeO. The c/i ratio obtained after injecting tracer into the right Rt was 20% higher than that obtained by injecting tracer into the left (Rogers and Deng, 1999). Therefore, if there is any asymmetry in the tectofugal projections in the chick, it is marginal and this contrasts to the consistent and clear asymmetry in the thalamofugal projections of the chick, as well as to the clear asymmetry in the tectofugal projections of the pigeon.

The apparent anomaly of asymmetry in only the thalamofugal visual projections of the chick and of asymmetry in only the tectofugal visual projections of the pigeon could depend on an interactive effect of light exposure of the late embryo and orientation of the embryo in the egg. First we will explain the role of light exposure in establishing the asymmetry and then we will relate it to the precocial (chick) versus altricial (pigeon) stages of development of the two species at hatching.

6.4. Effect of Light Stimulation Prior to Hatching on Asymmetry of the Visual Pathways

Asymmetry of the thalamofugal projections in the chick is induced by lateralized light stimulation of the right eye during the later stages of incubation (Rogers and Sink, 1988; Rogers and Bolden, 1991). This results because, during this period, the chick embryo tilts its head to the left side of the body and so the left eye is occluded by the body while the right eye is positioned next to the air sac. As a result, the right eye, but not the left eye, is able to receive light input reaching it through the egg shell and membranes. The eyelids alternate frequently between being open and closed from day E17 to hatching on day E21 but, since the eyelid is transparent, the right eye receives light even when the eyelid is closed (Freeman and Vince, 1974; Rogers, 1995).

The asymmetrical light input during the period from day E18 to E20 of incubation is particularly important for development of visual lateralization in the chick because, throughout this period, the visual projections to the forebrain are becoming functional (Rogers and Bell, 1989; Rogers, 1995). Visually evoked potentials can be detected in the TeO on day E17 and they mature on day E18 (Peters et al., 1958; reviewed by Rogers, 1995). Visually evoked potentials can be recorded from the forebrain for the first time on day E19 (Sedláček, 1967). Also on day E19/20 both the ectostriatum and the visual Wulst (HD and HA) of the chick forebrain have high levels of metabolic activity, as indicated by the amount of uptake of 2-deoxyglucose (Rogers and Bell, 1994).

Rogers and Sink (1988) were able to reverse the direction of the structural asymmetry in the thalamofugal projections by withdrawing the embryo's head from the egg on day E19/20 of incubation, applying a patch to the right eye and exposing the left eye to light. The eyepatch was removed at hatching and thereafter both eyes received exposure to light. On day 2 post-hatching, FG and TB were injected into the left and right visual Wulst separately and then the GLd–Wulst projections were analysed as discussed above. Control chicks with the left eye occluded, which mimics the normal condition, retained the normal pattern of asymmetry in the GLd–Wulst projections. In the chicks with the right eye occluded and with the left eye receiving exposure to light during the last days of incubation (opposite to the normal condition), the asymmetry was reversed so that there were more contralateral projections from the right GLd to the left visual Wulst than from the left GLd to the right visual Wulst. After incubation in dark conditions, irrespective of whether the hatched chicks were reared in the dark or exposed to light, there was symmetry of the thalamofugal projections (Rogers and Bolden, 1991; summarized in Rogers and Adret, 1993).

The light-dependent structural asymmetry in the visual projections to the Wulst is matched by asymmetries in the density of various neurotransmitter receptors. By manipulating light exposure on day E19 of incubation, it has been possible to show that some neurochemical asymmetries are induced by light stimulation prior to hatching (Johnston et al., 1997). Chicks with light exposure of the right eye (mimicking the normal condition) have [^{3}H]MK-801 binding asymmetry in HA, HIS and HD subdivisions of the visual Wulst: these regions in the left hemisphere were found to have significantly higher [^{3}H]MK-801 binding density than their equivalent regions in the right hemisphere. [Note that [^{3}H]MK-801 binds to N-methyl-D-aspartate (NMDA) receptors.] The reverse direction of receptor binding asymmetry was found in the chicks with the left eye exposed to light. Asymmetrical muscimol binding density also occurs in the visual Wulst of chicks following exposure of the right eye of the embryo to light, but muscimol binding asymmetry is removed, rather than reversed, in HA, HIS and HD of the chicks with left eye exposure to light before hatching (Johnston et al., 1997). These results suggest that glutamate (NMDA) and gamma-aminobutyric acid (GABA) receptors may play an important role in light-induced visual lateralization in the chick.

Light intensities of around only 100 lux during the sensitive period from day 19/20 to hatching are sufficient to cause the development of the asymmetry in the thalamofugal projections (Rogers, 1996). In addition, both red and green wavelengths of light are as effective as broad-spectrum (white) light in establishing asymmetry in these projections (Rogers and Krebs, 1996). This lack of wavelength specificity for establishing the development of asymmetry is consistent with the finding that the thalamofugal pathway is not involved in colour vision (Pritz, Mead and Northcutt, 1970).

The effect of light exposure of the embryo triggers asymmetrical growth of the thalamofugal projections, but this asymmetry is not apparent until some days after the time of exposure to light. Whereas as little as 24 h of light exposure on day E19/20 of incubation is sufficient to establish asymmetry detected between 2 and 6 days posthatching, its effects are not apparent prior to and around the time of hatching. It has been possible to inject fluorescent tracers into the visual Wulst of two groups of chick embryos on day E19 of incubation (E refers to embryonic age) (Rogers, Adret and Bolden, 1993): one group was incubated in the dark and the other exposed to light for 24 h on day E18. The injected chicks were allowed to survive for 4 days (i.e. until day 2 posthatching). Unlike chicks injected with the tracer on day 2 post-hatching and sacrificed on day 6 posthatching, the injected embryos had no asymmetry of the thalamofugal projections, irrespective of whether they had received light exposure or were kept in darkness. Therefore, the development of asymmetry induced by light exposure on day E19/20 before hatching is not manifested until day 2 posthatching and it persists until the third week post-hatching.

To summarize, asymmetrical light stimulation during the last stages of incubation (a critical period) induces the asymmetry of the contralateral thalamofugal projections in the chick. In the normal condition, since the left GLd receives more visual inputs than the right GLd, owing to the left eye being occluded by the embryo's body, the left GLd has more contra-lateral projections to the Wulst than the right GLd.

Asymmetrical stimulation of the chick embryo's eyes by light does not induce any obvious asymmetry in tectofugal visual projections. Light stimu-lation before hatching does, by contrast, play an important role in establish-ing asymmetry of the tectofugal pathway of the pigeon (Güntürkün, 1993; Chapter 7 by Güntürkün). Pigeons incubated in the dark have no morpho-logical asymmetries in the tectal layers 2–7, whereas pigeons incubated with light exposure (the right eye receiving the light stimulation) have the usual asymmetry in that cell body sizes in layers 2–7 of the left tectum are larger than those in the same layers of the right tectum (Güntürkün, 1993). The direction of asymmetry can also be reversed after hatching by occluding the right eye and allowing the left eye to receive light stimulation from day 1 to day 10 posthatching (Manns and Güntürkün, 1999a). Therefore, in the pigeon, the sensitive period for the development of the morphological asymmetries in the tectofugal pathway extends to at least the first 10 days posthatching. To date, no study has been performed to investigate whether asymmetry of the TeO–Rt projections in pigeons is also induced by asymmetrical light exposure of the embryo, although this is most likely since morphological asymmetries of cell body sizes in both TeO and Rt are modified by asymmetrical light exposure of the embryo (Güntürkün, 1993; Manns and Güntürkün, 1999b).

6.5. Developmental Differences Between Precocial and Altricial Species

It is interesting to speculate why asymmetrical light stimulation of the embryo does not induce asymmetry of the thalamofugal visual system of the pigeon, whereas it does in the chick, and why it induces asymmetry in tectofugal visual system of the pigeon and not the chick. The pigeon, an altricial species, hatches at a far more immature stage of development than the chick. The chick is a precocial species (Fontanesi et al., 1993; Starck and Ricklefs, 1998a). Therefore, at the stage of development when the embryo is oriented in the egg so that only its right eye receives stimulation by light, the visual pathways are less mature in the pigeon than in the chick (Figure 6.4). As discussed above, the tectofugal and thalamofugal pathways of the chick are functional on day E18/19 of incubation and the tectofugal pathway is probably more advanced than the thalamofugal pathway (Rogers and Bell, 1989; Bell and Rogers, 1992). The adult pattern of the retinotectal projections is reached between day E11 and E18 of incubation (Fontanesi et al., 1993) and the cytoarchitectonic differentiation of the tectum is completed by day E18 in the chick (Mey and Thanos, 1992). Therefore, asymmetrical light stimulation during the last stages of incubation (day E18 to day E21) may have little effect on the development of the already established TeO–Rt projections in the chick. By comparison, the thalamofugal projections of the chick are still immature on day E19 (Figure 6.4A) and this may be the reason why asymmetrical light stimulation induces the asymmetry in the GLd–Wulst projections in the chick.

Johnston et al. (1997) found that light exposure of the left or right eye of the chick embryo caused asymmetry of [^{3}H]MK-801, [^{3}H]AMPA and [^{3}H]muscimol binding densities in the forebrain region of the tectofugal visual system (i.e. in E), thus showing asymmetries in glutamate and GABA receptors. It is likely that the E matures later than TeO and Rt, and so it is influenced by the asymmetrical light stimulation. This means

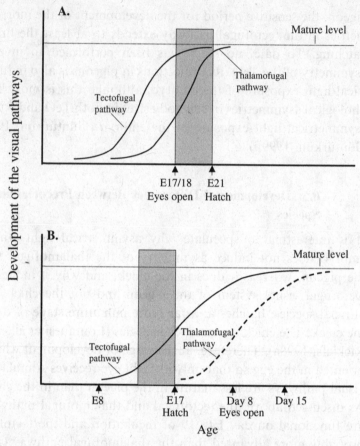

Figure 6.4.   Comparison of the developmental time-courses of the visual pathways in the chick (A, precocial) and the pigeon (B, altricial). At present, there are no complete data for development of the visual pathways in the chick or the pigeon, or any other avian species, but this figure represents the most likely course of events. For the chick, the evidence is based on Peters, Vonderahe and Powers (1958), Freeman and Vince (1974), Mey and Thanos (1992), Rogers and Bell (1994), and Rogers (1995). For the pigeon, the time-course for development of the tectofugal pathway is based on Fontanesi et al. (1993) and Manns and Güntürkün (1997). There are no reported data on the developmental course of the thalamofugal pathway of the pigeon. The dashed line shows the proposed developmental time-course of the thalamofugal pathway in pigeon based on data of three altricial species, the great tit, redstart, and common blackbird (Khayutin, 1985). Thalamofugal pathway refers to the GLd–Wulst projections and the tectofugal pathway to retino–TeO–Rt projections. The shaded rectangle indicates the stage of development when the embryo is oriented so that the right eye receives light stimulation and the left eye is occluded by the embryo's body. Note that, at this stage of development, only the thalamofugal visual projections of the chick are undergoing rapid development, and so are only the tectofugal projections of the pigeon.

that the Rt to E projections of the tectofugal pathway might develop asymmetry in response to light exposure of the chick embryo, even though the TeO–Rt projections of the chick do not. This possibility has not yet been investigated.

Hatching in the pigeon occurs at an earlier stage of development than in the chick. The tectofugal pathway of the pigeon does not become fully functional until after hatching and it continues to develop for at least the first 2 weeks posthatching (Fontanesi et al., 1993; Manns and Güntürkün, 1997). Although Manns and Güntürkün (1997) report that the optic fibres penetrate the layers of TeO quite early in embryonic development, and they have suggested that the tectofugal visual system of the pigeon may be able to process some aspects of visual information prior to hatching, the final organization of retinotectal projections is not reached until day 5 or 6 after hatching (Fontanesi et al., 1993). Differentiation of the tectum is not completed before day 15 posthatching in the pigeon (Manns and Güntürkün, 1997). It is clear, therefore, that prior to hatching the tectofugal pathway of the pigeon is much more immature than that of the chick (Figure 6.4). This may explain its vulnerability to asymmetrical light exposure both prior to hatching and up to day 10 posthatching (Güntürkün, 1993; Manns and Güntürkün, 1999a, 1999b).

The thalamofugal pathway develops later than the tectofugal pathway and, in the pigeon, this may shield it from the effects of asymmetrical (right eye only) light exposure of the embryo. In other words, prior to hatching, the pigeon's thalamofugal projections may be too immature to be affected by light experience (Figure 6.4). Khayutin (1985) has found that visually evoked responses in the Wulst are not established until shortly before the eyes open in three altricial species: great tits, redstarts and common blackbirds. If this is also the case in the pigeon, the thalamofugal pathway of the pigeon should start to respond to light stimulation between days 6 and 9 posthatching, at the time when the eyes open. The eyes of the chick open much earlier; in fact, they open before hatching, just before light exposure causes the development of asymmetry (Figure 6.4; Rogers, 1995). Light exposure of the chick embryo prior to day E17 does not generate asymmetry in the thalamofugal projections (Rogers, 1982). Hence, there is a precise timing of the onset of the sensitive period as well as of its offset. We assume that, in the pigeon, the sensitive period for effects on the thalamofugal projections does not commence until days 6–9 posthatching; monocular eye occlusion in the second week posthatching, we predict, would cause asymmetry to develop in the thalamofugal pathway in the pigeon.

Although the course of development of the thalamofugal pathway in the pigeon has not been delineated clearly, it is clear that the thalamofugal pathway remains plastic and immature even after hatching. Fontanesi et al. (1993) have found that unilateral removal of the retina immediately after hatching alters the relative contribution of the GLd afferents to the visual Wulst of the pigeon. Early monocular deprivation posthatching reduces ipsilateral GLd–Wulst projections from the deprived visual thalamus and increases contralateral GLd–Wulst projections from the thalamus on the side innervated by the intact eye (Fontanesi et al., 1993). Therefore, it is possible that visual stimulation after hatching, but not prior to hatching, plays a key role in the development of the ipsilateral and contralateral GLd–Wulst projections of the pigeon, and not the chick.

An alternative explanation for the existence of asymmetry in the tectofugal visual system of the adult pigeon, and not the young chick, may be that it results from differential use of the eyes after hatching and so develops with age. Although it is dependent on light exposure of the embryo, it may emerge as a secondary consequence of prolonged use of one eye (and hemisphere) to view certain stimuli. There is a strong possibility that different patterns of use of the left and right eyes will affect lateralization, as well as the eyes themselves.

A third possible explanation for the absence of asymmetry in the thalamofugal projections of the adult pigeon is that, as in the chick, asymmetry in these projections is transient, occurring only in early life. Therefore, it may simply have been overlooked.

6.6. The Altricial–Precocial Continuum and Evolution of Lateralization

We believe that the role of light exposure of the eggs in determining some forms of structural and functional lateralization in the avian brain needs further consideration in the context of altricial versus precocial species. As we have discussed, the effect of light relies on the asymmetrical orientation of the avian embryo just before hatching (referred to as the tucking position). This hatching posture is adopted by altricial and precocial species alike but altricial birds adopt this posture at an earlier stage of development than do precocial species (Rogers, 1990, 1995; Fontanesi et al., 1993; Manns and Güntürkün, 1997, 1999b). This earlier stage of development concerns the visual pathways as well as other brain systems and general physiology. It may, as we have suggested, explain the presence of asymmetry in the tectofugal but not the thalamofugal system of the pigeon and vice versa in the chicken (Figure 6.4). Since the relative timing of hatching, and so of adopting the tucking position, varies between species, whether or not asymmetry is found in the tectofugal system, the thalamofugal system or both systems may depend on where on the altricial–precocial continuum a species lies. As gulls and terns are semi-precocial and herons are semi-altricial (Ricklefs, 1983; Starck and Ricklefs, 1998b), it would be worth assessing asymmetry in their visual pathways for comparison to the chick and the pigeon. Hawks and owls are also semi-altricial but, at least in the case of owls, the frontally directed eyes may minimize or eliminate any lateralized stimulation of the eyes just prior to hatching. Of course, the incubation and nesting condition typical of a species would modify the effect of light. We will discuss this later but here it is worth mentioning that the eggs of the 'superprecocial' megapods, the Australian bush turkey (*Alectura lathami*) and the mallee fowl (*Leipoa ocellata*) are incubated in mounds of earth and leaf litter, and so are not exposed to light (Dekker and Brom, 1992). Megapode embryos, we note also, do not tilt their heads during the final stages of incubation (Baltin, 1969; Dekker and Brom, 1992), perhaps reflecting an absence of evolutionary selection for developmental processes that lead to visual lateralization.

The evolution of lateralization in birds might be considered from a precocial versus altricial perspective. Precocial development is considered to be more primitive than altricial, the latter allowing more brain growth after hatching (Starck, 1998). This evolutionary transition to altricial development might, therefore, have carried with it a shift from asymmetry in the thalamofugal visual system to asymmetry in the tectofugal system. Hence, the shift from precocial to altricial development might have been associated with a shift from asymmetry of visual functions involving the thalamofugal system to asymmetry of functions involving the tectofugal system. Precocial and altricial species follow the same pattern of brain development but the timing at which hatching is superimposed on this programme differs. The lasting effects that this relative shift in the time of hatching has on visual lateralization may have been an aspect of the evolution of the altricial form of development, along with other potential advantages stemming from an increase in adult brain size and a longer period of early learning. Alternatively, the transition in visual lateralization that may have occurred with the evolution of altricial development may be of no particular consequence in itself and so merely a byproduct of the changed stage of development at hatching. However, given that the forms of lateralization affected by light in the chick include attack and copulation responses, the possibility that it is merely a byproduct of other developmental demands seems unlikely. We recognize that lateralization as a general characteristic of the brain is likely to rely on selection processes that are independent of those influencing precocial–altricial characteristics, but we are suggesting that those aspects of visual lateralization that develop as a result of asymmetrical light stimulation may be closely associated with the precocial–altricial characteristic: light exposure of the embryo influences some, but far from all, aspects of lateralization (discussed below).

6.7. Behavioural Lateralization Dependent on Light Stimulation Prior to Hatching

Consistent with asymmetry of the visual pathways, functional lateralization has been found in visually guided behaviour in the chick (Andrew, 1988, 1991; Rogers, 1996, 1997; Chapter 3 by Andrew and Rogers) and the pigeon (Güntürkün, 1997b; Manns and Güntürkün, 1999a; Chapter 7 by Güntürkün). By localized injection of 0.5 µl 100 mM glutamate into various forebrain areas of the chick, it has been possible to show that the visual Wulst of the thalamofugal pathway and E of the tectofugal pathway contribute differentially to lateralized performance of a food-searching task, and to attack and copulation responses (Deng and Rogers, 1997). Chicks treated with glutamate in the left visual Wulst were found to make more errors in a task requiring them to search for grains of food scattered on a background of pebbles (pebble-floor task) than did chicks treated in the right visual Wulst. Glutamate injection into the left visual Wulst also elevated attack and copulation scores, but this did not occur following injection of the right visual Wulst. By contrast, glutamate injections of the left E affected only attack behaviour and not performance in the pebble-floor test or copulation responses. Glutamate treatment of the right E had no effect on performance in the pebble-floor test or on attack and copulation responses. This experiment provides direct evidence that asymmetry of the thalamofugal projections is associated with lateralization of function of the Wulst in the chick and that the two pathways have differential contributions to behavioural asymmetries.

The left Wulst, which receives fewer contralateral projections from the right GLd of the thalamus than that does the right Wulst from the left GLd, has a dominant role in pebble-floor performance. The result is consistent with an early demonstration of right eye dominance for performing the pebble-floor task (Mench and Andrew, 1986; Zappia and Rogers, 1987) and right eye dominance in memory recall of visual discrimination (Gaston and Gaston, 1984).

Clearly, one of the functions of an intact left Wulst is suppression of attack and copulation responses because these responses can be evoked prematurely after glutamate treatment of the left Wulst (also see Bullock and Rogers, 1986). This result is consistent with the fact that testosterone-treated chicks tested monocularly show elevated attack and copulation responses when they use the left eye and not the right eye (Rogers, Zappia and Bullock, 1985). Together these results indicate that the right Wulst is involved in activation of attack and copulation responses, whereas the left Wulst is involved in inhibiting these responses. Of course, other regions of the forebrain, especially the E, may also be involved.

As discussed previously in the chick, each Wulst receives different visual inputs from the left and right GLd (coming from the right and left eyes, respectively) because GLd contains almost no bilaterally projecting neurones, and ipsilateral and contralateral projecting neurones are located separately. Hence, different information from the two eyes may be integrated in each Wulst region. Although there is no difference between the left and right visual Wulst in the number of ipsilateral GLd–Wulst projections, the existence of more contralateral projections from left GLd to the right Wulst than vice versa may allow better integration of information from the two eyes in the right visual Wulst. This may be an underlying reason why attack, which depends on the use of both eyes, is controlled by the right hemisphere.

Lateralization of visual behaviour controlled by the thalamofugal pathway (the Wulst) depends on light exposure of the chick embryo, as is the case for asymmetry in the visual projections. It was shown some years ago that chicks hatched from eggs incubated in darkness during the last stages of incubation have no visual lateralization (at the group level) for performance on the pebble-floor task, attack or copulation: glutamate treatment of the left hemisphere or the right hemisphere did not reveal asymmetry in dark-incubated chicks (Rogers, 1982; Zappia and Rogers, 1983; Rogers and Krebs, 1996). Exposure of the embryo's left eye alone to light reverses the lateralization, so that chicks treated with glutamate in the right hemisphere now perform better on the pebble-floor task and have elevated levels of attack and copulation (Rogers, 1990).

Recently, we have manipulated light exposure by applying an eyepatch for 24 h on day E19/20 (of incubation) and studied the effects of the asymmetrical light stimulation on the lateralization revealed by placing a localized injection of glutamate into the left or right visual Wulst on day 2 posthatching (Deng and Rogers, 2000b). When the left eye was occluded by the eyepatch (light exposure of the right eye), mimicking the natural condition, the same direction of lateralization for pebble-floor performance, attack and copulation was shown as in chicks hatched from unoperated eggs exposed to light before hatching (Deng and Rogers, 1997): glutamate treatment of the left visual Wulst impaired the chick's ability to categorize pebbles from grains (Figure 6.5A) and elevated attack and copulation, whereas treatment of the right visual Wulst had no effect on those behavioural responses. In contrast, when the right eye of the embryo was occluded by the eyepatch, leaving the left eye exposed to light (the reverse of the natural condition), the direction of

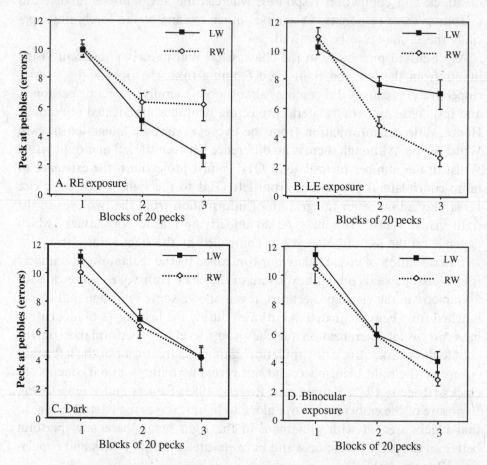

Figure 6.5. Performance in the pebble-floor test is presented for chicks hatched from eggs incubated in four different conditions of light exposure and receiving glutamate treatment of the left Wulst (LW) or right Wulst (RW). (A) Light exposure of the right eye of embryos, mimicking the normal condition. (B) Light exposure of the left eye of embryos. (C) Dark incubation. (D) Binocular exposure: exposure of both the right and left eyes of the embryo to light. The mean number of errors (pecks at pebbles) in each block of 20 pecks is plotted with standard error scores. Note the errors made by chicks in the third block of 20 pecks and the reversal of lateralization in (A) and (B).

functional lateralization was reversed: glutamate treatment of the right visual Wulst impaired pebble-floor performance (Figure 6.5B) and elevated attack and copulation, whereas treatment of the left visual Wulst had no effect. Moreover, chicks hatched after dark incubation have no lateralization of these behaviours (Figure 6.5C). In the dark-incubation group some individuals had high scores and the others had low scores, which suggests a bimodal distribution. In addition, some chicks had both the right and left eyes exposed to light stimulation on day E19/20. It is interesting to note that these chicks receiving light exposure of both eyes also lacked lateralization of all these behaviours, and they performed similarly to the chicks of the dark group (Figure 6.5D).

This experiment provides clear evidence for the role of asymmetrical light stimulation in determining the direction of particular forms of functional lateralization in the chick brain and, in particular, it ties the effect of light to the thalamofugal visual pathway of the chick.

It has been reported recently that turning bias, in the chick and bobwhite quail, *Colinus virginianus,* is also influenced by light stimulation prior to hatching (Casey and Lickliter, 1998; Casey and Karpinski, 1999). When tested in a T-maze, young birds of both species show a bias to turn to the left side (85% in quail and 90% in chick). Manipulating light stimulation of embryo's eyes prior to hatching by applying a patch to one eye demonstrated that the left side turning bias is determined by unilateral light stimulation of the right eye. Chicks and quails with both eyes exposed to light just before hatching were found to have no turning bias at group level, although individuals had turning biases (Casey and Lickliter, 1998; Casey and Karpinski, 1999). Reversed (right side) turning bias resulted from exposing the left eye only to light in the quail (Casey and Lickliter, 1998), but the same exposure of the left eye did not cause reversal in the chick (Casey and Karpinski, 1999). This failure to reverse the direction of turning bias in the chick may have occurred because the left eye exposure was on day E20 of incubation, after the eggs had been exposed to some light and so the lateralization may have been already established (Zappia and Rogers, 1983). Thus, exposure of the left eye to visual stimulation on day E20 would have had to reverse the normal lateralization.

Although the role of light stimulation prior to hatching has a clear effect on turning bias after hatching, in the same way that it affects pebble-floor performance and attack and copulation responses, it is not certain whether the T-maze test reveals lateralization of visual, auditory or motor performance. The quail chicks were tested in a T-maze with a stuffed bobwhite quail hen and a speaker playing maternal calls at the end of each arm (Casey and Lickliter, 1998). Although these conditions were balanced at the choice point, the quail chick could choose to approach the stimuli on its left or right side based on hemispheric preference for information processing. Leftward turning might, therefore, reflect a left ear preference (Miklósi, Andrew and Dharmaretnam, 1996), a left eye preference (see Chapter 3 by Andrew and Rogers) or a motor turning bias. The domestic chicks were tested in a T-maze with maternal calls but it seems they were not presented with stuffed models of the hen at each end of the runway. In their case, therefore, choice might be based on auditory preference or motor turning bias, but visual cues for spatial location might also be used. Since light exposure prior to hatching does not appear to affect sensory modalities other than vision (auditory, Figure 6.6A; olfactory, Figure 6.6B; and see Rogers, Andrew and Burne, 1998), it is likely that visuomotor biases were measured in the T-maze. It would be interesting to see whether they too rely on asymmetry in the thalamofugal system.

6.8. Sensitive Period

The sensitive period for the effect of light exposure on functional lateralization has been investigated (Rogers, 1990, 1997). The onset of the sensitive period begins after day E17 of incubation in the chick. Light exposure of the embryo prior to day E17 has no effect on functional lateralization (Rogers, 1982). The end of the sensitive period was determined by monocular occlusion of the chick posthatching (Rogers, 1990). After incubation in the dark, the chicks were kept undisturbed in darkness until early on the first day post-hatching. At this time the right eye was occluded and the left eye was exposed to light for 24 h. Unilateral treatment of the hemispheres with glutamate on day 2 demonstrated that the direction of lateralization for attack and copulation was reversed but not that for lateralization of performance of the pebble-floor task. When the monocular exposure to light was performed on day 3 posthatching, after the chicks had been incubated and raised in darkness, there was no longer any asymmetry in attack and copulation scores following injection with glutamate to the left or right hemisphere (Rogers, 1990). Therefore, the sensitive period for the effect of light exposure on attack and copulation ends by day 3 posthatching, whereas the sensitive period for light exposure on pebble-floor performance is over by day 1 posthatching (Rogers, 1990, 1991). To date, we do not know why there is 1 or 2 days difference between the sensitive period for the effect of light on lateralization of attack–copulation and on pebble-floor performance, although different neural mechanisms are involved in controlling attack and copulation (the right

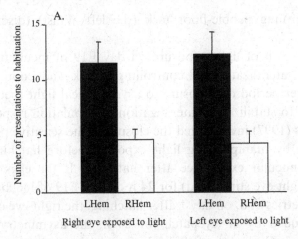

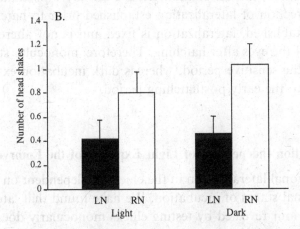

Figure 6.6. Light exposure prior to hatching has no effect on auditory and olfactory lateralization. (A) Auditory lateralization: the mean number of presentations of an auditory stimulus (a broad frequency sound) necessary for the chick to habituate is plotted with standard error values. The treatment (5 μl 100 mM glutamate; for methods, see Howard, Rogers and Boura, 1980) given to each hemisphere to reveal lateralization is indicated. LHem refers to glutamate treatment of the left hemisphere and RHem refers to glutamate treatment of the right hemisphere. No matter which eye was exposed to the light during incubation, glutamate treatment of the left hemisphere only caused slower auditory habituation. N = 4 per group in a pilot experiment. (B) Olfactory lateralization: the mean numbers of headshakes that are given on presentation of clove oil odour together with a blue bead are presented with standard errors. LN refers to use of the left nostril (right nostril occluded) and RN refers to use the right nostril (left nostril occluded). Higher levels of head shaking were expressed in chicks using the RN compared to those using the LN, irrespective of whether the chicks were hatched from eggs incubated in the dark or exposed to light during late incubation. N = 35–36 per group. Modified from Rogers, Andrew and Burne (1998). All left–right differences are significant.

Wulst) and the performing pebble-floor task (the left Wulst) (discussed earlier).

Although as little as 2 h of light exposure on day E19 of incubation is sufficient to establish lateralization of controlling attack and copulation (Rogers, 1982), a longer period of exposure to asymmetrical light (between 2.5 and 6 h) is needed to stabilize the lateralization of copulation responses (Rogers, 1990). Rogers (1997) investigated the closure of the sensitive period for the effect of light by manipulating light exposure before hatching in combination with monocular experience after hatching. If the eggs were exposed to light (the right eye stimulated) for 24 h on day E19 of incubation to generate the asymmetries and, on day 1 after hatching, the right eye of the same chicks was occluded by an eyepatch for 24 h, the asymmetry was unchanged from normal. Occlusion of the right eye on day 1 posthatching did not reverse the direction of lateralization established prior to hatching, showing that, once established, lateralization is fixed and is not altered by monocular exposure of the eyes after hatching. Therefore, monocular stimulation by light closes the sensitive period, whereas dark incubation extends the sensitive period into the early posthatching period.

6.9. Lateralization Independent of Light Exposure of the Embryo

Not all forms of functional lateralization in the chick are dependent on light exposure during the final stages of incubation. We have found that lateralization of choice behaviour revealed by testing chicks monocularly does not depend on light stimulation during incubation (Deng, 1998). Eggs were incubated in a dark incubator (dark group) or incubated with light exposure for 24 h on day E18/19 of incubation (light group). After hatching, each chick was paired with another chick from the same batch and they lived together for 2 days. On day 3 posthatching, each chick was given a choice test between the familiar chick and an unfamiliar chick placed at either end of a runway. Lateralization was found in the choice scores of chicks in both the light group and the dark group. In both groups, the chicks tested using the left eye had higher choice scores (meaning they chose to approach one or other of the two chicks) than the chicks tested using the right eye (Figure 6.7). The chicks hatched from eggs incubated in the dark did differ from the light group; they had a longer latency for starting to approach the stimuli and spent longer in the centre of the runway than the chicks exposed to light. Therefore, light experience prior to hatching does not influence the lateralization of choice behaviour, even though it does affect latency to approach the stimuli.

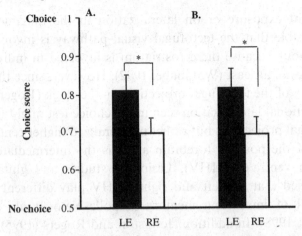

Figure 6.7. The choice scores (mean ± SE) measured in a choice test in which the 3-day-old chick can approach either a familiar chick (cagemate) or an unfamiliar chick. (A) Results obtained from the dark-incubated chicks. (B) Results obtained from the chicks exposed to light just prior to hatching. Note: if a chick approached and stayed with either only the familiar or only the unfamiliar chick, it was given a score of 1 (choice); if a chick alternated between approaching the familiar and the unfamiliar chicks, or stayed in the centre of the runway, it was given a score of 0.5 (no choice). LE, tested using the left eye; RE, tested using the right eye (*, $p < 0.05$, two-tailed t-test).

Similarly, Andrew et al. (paper in preparation; also discussed in Chapter 5 by Andrew) have found that the left eye and right eye differences in choice between a familiar red ball (with a horizontal white bar) and an unfamiliar red ball (with a vertical white bar) are independent of light exposure before hatching. Eggs were incubated either in the dark or with exposure to light from day 17 to hatching. After hatching, the chicks were reared with a red ball with a horizontal white bar and they were given a choice test between the familiar ball and an unfamiliar ball (with a vertical white bar) on days 3 and 4. The lateralization of the choice was clear in chicks hatched from eggs incubated in the dark. Choice was made by dark-incubated chicks using the left eye but not by their counterparts using the right eye. Light exposure of the embryos somewhat reduced the left eye and right eye difference. Similarly, Vallortigara and Andrew (1991) have also found that, even following incubation in the dark, chicks using the left eye, but not ones using the right eye, respond to small changes of the stimulus in a choice test between a familiar and an unfamiliar ball.

Since lateralization of choice behaviour is present in dark-incubated as well as light-exposed chicks, we can say that this particular lateralized behaviour does not rely on light exposure or on lateralization in the thalamofugal projections. It is possible that the tectofugal visual pathway is involved in this particular behaviour because the ectostriatum is involved in individual recognition, in the pigeon at least (Watanabe, 1992). However, since there is only minor asymmetry of the tectofugal projections in the chick (Rogers and Deng, 1999), the functional lateralization seen in the choice test might not be located in the tectofugal projections but rather in lateralized higher centres in the forebrain. One of the possible forebrain areas is the intermediate and medial hyperstriatum ventrale (IMHV). Lesioning studies and glutamate treatment have revealed that the left and right IMHV play different roles in storage and recall of imprinting memory (Cipolla-Neto et al., 1982; Johnston and Rogers, 1998). In addition, Johnston and Rogers (1995) have reported that there is asymmetry of [^{3}H]MK-801 binding in IMHV of the dark-incubated chicks and this asymmetry is reversed by imprinting learning. However, direct evidence is needed to confirm this speculation.

Light exposure of the chick embryo, therefore, affects a number of visually guided behaviours but not all of them. Moreover, light exposure has no effect on lateralization in other sensory modalities. Chicks respond to a noxious odour stimulation associated with presentation of a blue bead by shaking their head, when they are using the right nostril (left nostril blocked with wax) but not when they are using the left nostril (Figure 6.6; Rogers et al., 1998). Lateralization of auditory habituation is also unaffected by light exposure of the chick embryo. Auditory lateralization can be revealed posthatching by injecting the left or right hemisphere with cycloheximide (Rogers and Anson, 1979) or with glutamate (Howard, Rogers and Boura, 1980). Injection of the left hemisphere shows habituation to a novel sound, whereas injection of the right hemisphere has no effect. The same lateralization is present in chicks in which the left or right eye is exposed to light before hatching (Figure 6.6 reports data obtained in a pilot experiment).

6.10. Hormones and the Development of the Visual Lateralization

Although asymmetry of the GLd–Wulst projections has been found in both male and female chicks, the asymmetry is of a lesser degree in the female than in the male (Rajendra and Rogers, 1993). This sex difference may reflect a role of sex hormones during the sensitive period for light experience (Rogers, 1996). In both male and female chick embryos, plasma testosterone level increases until day E13 (in males) and day E15 (in females). Then there is a decline in the testosterone level from E13 to E17 in males, but only a slight decline from E15 to E17 in females (Woods, Simpson and Moore, 1975;

summarized in Rogers, 1995). Tanabe, Saito and Nakamura (1986) did not find this decrease in plasma testosterone level in males and females. They also found that there was no sex difference in plasma testosterone level throughout embryonic development. However, irrespective of whether there is a sex difference in testosterone level during the sensitive period for development of asymmetry of the GLd–Wulst projections, the oestrogen levels of female embryos are several-fold higher than those of male embryos (Woods and Brazzill, 1981; Tanabe et al., 1986). After injecting oestrogen into eggs on day E16, no asymmetry of the GLd–Wulst projections was found in males or females (Rogers and Rajendra, 1993). Therefore, it is possible that a higher level of oestrogen in female embryos causes a lesser degree of asymmetry to develop in the GLd–Wulst projections, compared to males. In other words, higher levels of oestrogen suppress the effects of light stimulation during the sensitive period.

Low levels of testosterone in the final stages of incubation may be essential for the development of asymmetry of the GLd–Wulst projections in males. Schwarz and Rogers (1992) have injected testosterone into eggs on day E16 in a slow-release form (thereby elevating the hormone level until after hatching) and found that there is no asymmetry of the GLd–Wulst projections in either the males or females, even though the eggs were exposed to light.

These studies have shown that the development of the asymmetry in the thalamofugal projections of the chick is modulated by an interaction of the sex hormone levels and lateralized light input. Other steroid hormones, such as the glucocorticoids, may also influence the development of asymmetry of the GLd–Wulst projections. Glucocorticoids, including corticosterone and cortisol, are detectable in plasma of chick embryos from day E9 of incubation (Kalliecharan and Hall, 1974; Figure 6.8). Levels of both corticosterone and cortisol start to increase on day E14 of incubation (Wise and Frye, 1973; Kalliecharan and Hall, 1974). On day E15, the level of cortisol reaches a peak and then decreases over the later stages of incubation and the first week posthatching (Kalliecharan and Hall, 1974; Figure 6.8). However, Tanabe, Saito and Nakamura (1986) did not find a peak in plasma cortisol level on day E15. They found that plasma cortisol was at low levels (less than 1 ng/ml) throughout the incubation period. The level of corticosterone continues to increase and reaches a peak on day E19/E20 of incubation and declines to remain at lower levels in the first week posthatching (Kalliecharan and Hall, 1974; Tanabe, Saito and Nakamura, 1986; Figure 6.8). Therefore, corticosterone is the dominant glucocorticoid in the chick embryo during the last stages of incubation. Considering the fact that corticosterone levels are higher during the last stages of incubation than in the first two-thirds of

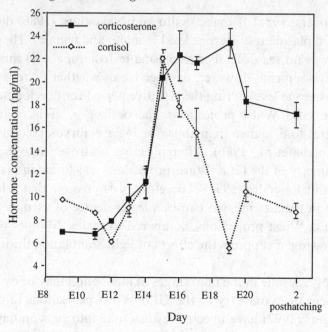

Figure 6.8.   Mean plasma corticosterone and cortisol levels are plotted for chick embryos at various ages from day E9 of incubation to day 2 post-hatching. Based on data from Kalliecharan and Hall (1974).

incubation and the first week posthatching, this hormone may play a role in development of chick embryos. In fact, corticosterone is involved in regulation of anterior pituitary cell differentiation and growth hormone secretion on day E16 of incubation (Dean and Porter, 1999; Dean, Morpurgo and Porter, 1999).

Recently, we have investigated the effect of injecting corticosterone into the egg on the development of asymmetry of the GLd–Wulst projections. On day E18/19 of incubation, under dim light illumination, a small hole was drilled through the eggshell above the position of the embryo. A dose of 0.5 ml (60 μg) corticosterone solution or saline (control) was injected into eggs. Then the eggs were kept in an incubator with light illumination (200–300 lux) until hatching. On day 2 after hatching, the chicks were injected with RITC and FG into the left or right visual Wulst to label the projecting neurones in the GLd. In the control group, the asymmetry of the GLd–Wulst projections was confirmed: the c/i ratio obtained after injecting tracer into the right Wulst was significantly higher than that obtained by injecting into the left Wulst. However, in the corticosterone-treated group, there was no significant difference in the c/i ratios obtained after injecting the left or right Wulst (Figure 6.9). By comparing the absolute number of labelled neurones in GLd between the corticosterone-treated chicks and control group injected with saline, we found that corticosterone-treated chicks had fewer contralaterally projecting neurones in the left GLd compared to the control group. Therefore, it appears that corticosterone (at the dose reaching the embryo in this study) suppresses growth of contralateral projections from the left GLd to the right visual Wulst induced by light stimulation of the right eye only (Rogers and Bolden, 1991; Rogers and Deng, 1999). Corticosterone has the same effect in removing asymmetry as the sex steroid hormones, but it is not known whether these hormones interact with light stimulation in exactly the same way or in the same population of neurones.

These steroid hormones are likely to have long-lasting effects on lateralized processing of visual information, but their effects on behavioural lateralization have yet to be tested. Sui, Sandi and Rose (1997) have reported that corticosterone treatment of chick embryos on day E19/20 during dark incubation improves retention of a weak passive avoidance task after hatching

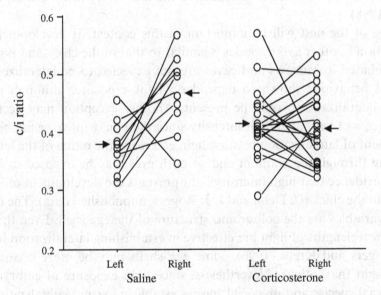

Figure 6.9. The c/i ratio (proportion of labelled cells in the side of the thalamus contralateral to the injection site relative to labelled cells in the side of the thalamus ipsilateral to the injection site) is presented for individual chicks treated with saline or corticosterone on day E19 of incubation. Lines connect the c/i ratio calculated from dye injections in the left Wulst and the right Wulst for individual chicks. Arrows indicate the mean left and right c/i ratio. Only in the saline treatment group, the mean right c/i ratio is significantly different from the mean left c/i ratio ($p < 0.01$, two-tailed t-test).

but lateralization was not investigated in this study. Schwabl (1993) has reported that, in the canary, there is a positive correlation between testosterone levels in eggs and social aggression posthatching but, again, lateralization was not considered, even though the left and right hemispheres are differentially involved in aggressive responses in the chick at least (Rogers, Zappia and Bullock, 1985).

6.11. Lateralization and Development in the Natural Environment

The intensity and duration of light exposure necessary to establish lateralization is likely be available in natural conditions of incubation. As mentioned above, in the laboratory, light intensities as low as 100 lux, and exposure periods of as brief as 2 h are sufficient to establish lateralization (Rogers, 1982, 1996). Such a short period of light exposure during incubation should be easily met in the natural environment when the maternal bird leaves the nest to forage for food or take part in other activities. In fact, pigeons and zebra finches incubated naturally (without controlled light conditions) have visual lateralization similar to that seen in the laboratory (Güntürkün, 1985; Alonso, 1998).

The site of the nest will be important in this context. If development of lateralization in other avian species is similar to that of the chick and pigeon, birds incubated in burrows and caves are not expected to be lateralized for the visual behaviour shown to depend on light exposure, although other forms of lateralization would be present. Another exception may occur in environments of very high light intensity since these may mask or inhibit the development of lateralization because light can reach the retina of the left eye by passing through the cranium and so both eyes may be exposed to light. We have evidence that high intensity light prevents the development of lateralization in the chick (C. Deng and L.J. Rogers, unpublished data). The other possible variables are the colour and structure of the egg shell. Even though different wavelengths of light are effective in establishing lateralization in the chick (Rogers and Krebs, 1996), some egg shells may be more opaque to incident light than others. Nevertheless, since light exposure of embryos in both altricial species and precocial species establishes visual lateralization, it would appear that the role of light in establishing visual lateralization for some functions may be widespread in avian species.

The source of steroid hormones acting even in the later stages of incubation may be of maternal origin as well as from the embryo itself. Schwabl (1993) has reported that the maternal bird deposits testosterone in the egg, in the canary and zebra finch. Maternal testosterone has also been found in red-winged blackbird eggs (Lipar, Ketterson and Nolan, 1999). This maternal testosterone enhances postnatal growth and the development of aggressive behaviour of the offspring (Schwabl, 1993, 1996). Maternal estradiol has been found in the egg yolk of Japanese quail (Adkins-Regan et al., 1995). Treatment of laying hens with oestradiol benzoate results in a high deposition of this hormone in the yolk and morphological alteration of the reproductive system in some female offspring (Adkins-Regan, Ottinger and Park, 1995). These hormones deposited in the egg by the maternal bird may also influence the development of lateralization, provided they persist until hatching. In fact, in canaries and red-winged blackbirds, yolk testosterone levels vary with laying order: eggs laid at the beginning of the clutch contain lesser amounts of maternal testosterone than those laid later (Schwabl, 1993; Lipar, Ketterson and Nolan, 1999). These variations in sex hormone levels between individuals may lead to individual differences in degree of lateralization and, if we can extrapolate, lateralization may be greater in first-hatched canaries and red-winged blackbirds compared to later hatched ones. However, in another species, the cattle egret (*Bubulcus ibis*), testosterone concentration decreases with laying order (Schwabl, Mock and Gieg, 1997), so that in this species lateralization may be lesser in first-hatched birds compared to later hatched ones. First-hatched egrets practise siblicide (Mock and Parker, 1997), and this is likely to be related to their levels of testosterone and aggression. The aggression involved in siblicide might also be related to a lesser degree of lateralization, since dominance of the left hemisphere suppresses aggression, as known for the chick at least (Bullock and Rogers, 1986).

The hypothalamo–hypophyseal–adrenal cortex axis in chick embryos is functional from day E14 to day 16 of incubation (Woods, De Vries and Thommes, 1971; Wise and Frye, 1973). At this stage of development, the hypothalamus does not appear to control normal resting levels of corticosterone, although it is essential for the stress response (Wise and Frye, 1973). Stress experienced by the embryo (e.g. heat or cold stress, or even some forms of toxic chemical stress) may increase the corticosterone level during the last stages of incubation. As this may occur during the sensitive period for the effect of light on lateralization, stress experienced by the embryo could bring about individual differences in the degree of lateralization. Furthermore, stress experienced by the hen before laying (at the time the egg is forming) may also change the corticosterone level of embryo. Recently, Schwabl (1999) has reported that birth order within broods is related to corticosterone levels in 15-day-old (nestling stage) and 23-day-old (fledgling stage) canaries. The first-hatched canary (hatched from the first-laid egg) had higher corticosterone levels than the last-hatched one (hatched from the last-laid egg), while second-hatched canaries had intermediate levels of corticosterone. Since hatching asynchrony results from the laying order of eggs, it is possible that the laying order determines variation of the corticosterone level in nestling and fledgling canaries (Schwabl, 1999). Therefore, it is possible that corticosterone levels in the last stages of incubation in chicks may be influenced by laying order of the eggs or other maternal effects causing stress. The stress level of both the hens and the embryos could, therefore, affect the development of lateralization.

6.12. Summary: A Gene–Environmental Model of Avian Visual Lateralization

We have reviewed the roles of various environmental factors known to influence the development of asymmetries of the avian visual system and certain types of functional lateralization in visual behaviour. It is clear that genetic influences alone do not determine asymmetry of the visual pathways or visual behaviour. Genes must have an interactive role, since the left-turned orientation of the embryo is essential for these asymmetries to develop.

Recent studies have shown that morphological left–right asymmetries such as direction of the embryo's rotation and the side on which the heart, lungs and other visceral organs are positioned are determined by asymmetrical expression of *Shh, nodal, lefty* and *FGF8* genes during and after the gastrulation stage of embryonic development (Levin et al., 1995; Collignon, Varlet and Robertson, 1996; Boettger, Wittler and Kessel, 1999; Meyers and Martin, 1999). For example, in the chick embryo, *Shh* functions as a 'left determinant', which is expressed on only the left side of Hensen's node during gastrulation (stage 4–6). Then, Shh protein stimulates expression of the downstream left-specific genes *nodal, lefty* and *Pitx2*, and determines the direction of heart looping and embryo turning to the left side (Levin et al., 1995; Collignon, Varlet and Robertson, 1996; Meyers and Martin, 1999). In contrast, *FGF8* functions as a 'right determinant', which is expressed to the right of the node and determines right side specification of embryo (Boettger et al., 1999; Meyers and Martin, 1999). If FGF8 protein is applied to the left side of Hensen's node, the direction of heart looping is randomized (Boettger, Wittler and Kessel, 1999). Therefore, it is clear that the asymmetrical orientation of the avian embryo in the egg is caused by the expression of left–right genes.

Here we have proposed a model for ontogenetic development of light-dependent visual lateralization in birds. As shown in Figure 6.10, differential

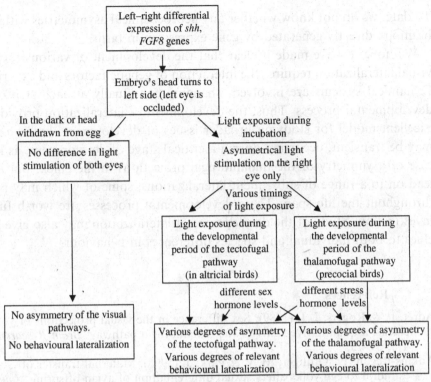

Figure 6.10. A proposed model to explain the influence of genetic and environmental factors on development of the light-dependent visual lateralization in birds.

expression of left- and right-specific genes (such *shh* and *FGF8*) determines the left-side turning of embryo's head during the last stages of incubation (e.g. on day E18 in chicks). This prenatal orientation asymmetry of embryo is a precursor for some visual asymmetries appearing after hatching, since it brings about asymmetrical light stimulation of the embryo's right eye only (and also of the neural structures connected to the right eye). However, whether visual lateralization of pebble-grain performance, and of attack and copulation behaviour, develops depends on experience (incubation in the dark or exposure to light, and the timing of the light exposure) prior to hatching. In addition to light stimulation, other environmental factors generate variations in visual lateralization between individuals. Testosterone levels, oestrogen levels and stress-related corticosterone levels in the embryos may also influence the degree of lateralization and so affect behaviour after hatching.

Gene expression may cause some neurochemical asymmetries in some brain regions of embryos, and then these vulnerable asymmetries may be modified (enlarged or diminished) by experience or environmental factors.

To date, we do not know whether any neurochemical asymmetries within the brain are directly generated by gene expression in birds.

We hope to have made it clear that the development of various forms of visual lateralization requires the interaction of genetic factors and experience. Dynamical systems are involved, changing continually at each step of the developmental process. Thus, the development of lateralization provides an excellent model for studying broader issues of development. Lateralization may be transient, occurring during a critical stage of development, as in the case of asymmetry of the thalamofugal projections in the chick, but it can lead-on to a range of subsequent lateralizations, some of which may persist throughout the life span. These developmental processes are worth further investigation. Study of the development of lateralization may also give some clues to both individual and species differences in behaviour.

References

Adret, P. & Rogers, L.J. (1989). Sex difference in the visual projections of young chicks: a quantitative study of the thalamofugal pathway. *Brain Research*, 478, 59–73.

Adkins-Regan, E., Ottinger, M.A. & Park, J. (1995). Maternal transfer of estradiol to egg yolks alters sexual differentiation of avian offspring. *Journal of Experimental Zoology*, 271, 466–470.

Alonso, Y. (1998). Lateralization of visual guided behaviour during feeding in zebra finches (*Taeniopygia guttata*). *Behavioural Processes*, 43, 257–263.

Andrew, R.J. (1988). The development of visual lateralization in the domestic chick. *Behavioural Brain Research*, 29, 201–209.

Andrew, R.J. (1991). The nature of behavioural lateralization in the chick. In *Neural and Behavioural Plasticity: The Use of the Domestic Chick as a Model*, ed. R.J. Andrew, pp. 536–554. Oxford: Oxford University Press.

Bagnoli, P. & Burkhalter, A. (1983). Organization of the afferent projections to the Wulst in the pigeon. *Journal of Comparative Neurology*, 214, 103–113.

Baltin, S. (1969). Zur Biologie und Ethologie des Talegalla-Huhns (*Alectura lathami* Gray) unter besonderer Berücksichtigung des Verhaltens während der Brutperiode. *Zeitschrift für Tierpsychologie*, 6, 524–572.

Bell, G.A. & Rogers, L.J. (1992). Metabolic activity in the hyperstriatum of 2-day-old chicks during optomotor and contrasting visual stimulation. *Behavioural Brain Research*, 50, 177–183.

Benowitz, L. & Lee-Teng, E. (1973). Contrasting effects of three forebrain ablations on discrimination learning and reversal in chicks. *Journal of Comparative and Physiological Psychology*, 81, 391–397.

Bischof, H.-J. & Niemann, J. (1990). Contralateral projections of the optic tectum in the zebra finch (*Taenopygia guttata castanotis*). *Cell and Tissue Research*, 262, 307–313.

Bischof, H.-J. & Watanabe, S. (1997). On the structure and function of the tectofugal visual pathway in laterally eyed birds. *European Journal of Morphology*, 35, 246–254.

Boettger, T., Wittler, L. & Kessel, M. (1999). FGF8 functions in the specification of the right body side of the chick. *Current Biology*, 9, 277–280.

Bowmaker, J.K., Heath, L.A., Wilkie, S.E. & Hunt, D.M. (1997). Visual pigments and oil droplets from six classes of photoreceptor in the retinas of birds. *Vision Research*, 37, 2183–2194.

Boxer, M.I. & Stanford, D. (1985). Projections to the posterior visual hyperstriatal region of the chick: an HRP study. *Experimental Brain Research*, 57, 494–498.

Bullock, S.P. & Rogers, L.J. (1986). Glutamate-induced asymmetry in the sexual and aggressive behaviour of young chickens. *Pharmacology, Biochemistry and Behavior*, 24, 549–554.

Campenhausen, M.V. & Kirschfeld, K. (1998). Spectral sensitivity of the accessory optic system of the pigeon. *Journal of Comparative Physiology A: Sensory, Neural and Behavioral Physiology*, 183, 1–6.

Casey, M.B. & Karpinski, S. (1999). The development of postnatal turning bias is influenced by prenatal visual experience in domestic chicks (*Gallus gallus*). *Physiological Record*, 49, 67–74.

Casey, M.B. & Lickliter, R. (1998). Prenatal visual experience influences the development of turning bias in bobwhite quail chicks (*Colinus virginianus*). *Developmental Psychobiology*, 32, 327–338.

Cipolla-Neto, J., Horn, G. & McCabe, B.J. (1982). Hemispheric asymmetry and imprinting: The effect of sequential lesions to the hyperstriatum ventrale. *Experimental Brain Research*, 48, 22–27.

Clayton, N. S. & Krebs, J. R. (1994). Lateralization and unilateral transfer of spatial memory in marsh tits: Are two eyes better than one? *Journal of Comparative Physiology A: Sensory, Neural and Behavioral Physiology*, 174, 769–773.

Collignon, J., Varlet, I. & Robertson, E.J. (1996). Relationship between asymmetric nodal expression and the direction of embryonic turning. *Nature*, 382, 155–158.

Dean, C.E. & Porter, T.E. (1999). Regulation of somatotroph differentiation and growth hormone (GH) secretion by corticosterone and GH-releasing hormone during embryonic development. *Endocrinology*, 140, 1104–1110.

Dean, C.E., Morpurgo, B. & Porter, T.E. (1999). Induction of somatotroph differentiation *in vivo* by corticosterone administration during chicken embryonic development. *Endocrine*, 11, 151–156.

Dekker, R.W.R.J. & Brom, T.G. (1992). Megapode phylogeny and the interpretation of incubation strategies. *Zoologische Verhandelingen*, 278, 19–31.

Deng, C. (1998). *Organization of the visual pathways and visual lateralization in the chick*. Ph.D. thesis. University of New England, Armidale, Australia.

Deng, C. & Rogers, L.J. (1997). Differential contributions of the two visual pathways to functional lateralization in chicks. *Behavioural Brain Research*, 87, 173–182.

Deng C. & Rogers, L.J. (1998a). Organisation of the tecto-rotundal and SP/IPS-rotundal projections in the chick. *Journal of Comparative Neurology*, 394, 171–185.

Deng, C. & Rogers, L.J. (1998b). Bilaterally projecting neurones in the two visual pathways in the chick. *Brain Research*, 794, 281–290.

Deng, C. & Rogers, L.J. (1999). Differential sensitivities of the two visual pathways of the chick to labelling by fluorescent retrograde tracers. *Journal of Neuroscience Methods*, 89, 75–86.

Deng, C. & Rogers, L.J. (2000a). Organization of intratelencephalic projections to the visual Wulst of the chick. *Brain Research*, 856, 152–162.

Deng, C. & Rogers, L.J. (2000b). Early light exposure affects functional lateralization of the thalamofugal visual pathway. In *Proceedings of the Australian Neuroscience Society*, Vol. 11, pp. 143. Melbourne: The Australian Neuroscience Society.

Denton, C.J. (1981). Topograph of the hyperstriatal visual projection area in the young domestic chicken. *Experimental Neurology*, 74, 482–498.

Ehrlich, D. & Mark, R.F. (1984). An atlas of the primary visual projections in the brain of the chick *Gallus gallus*. *Journal of Comparative Neurology*, 223, 592–610.

Fontanesi, G., Casini, G., Ciocchetti, A. & Bagnoli, P. (1993). Development, plasticity, and differential organisation of parallel processing of visual information in birds. In *Vision, Brain, and Behavior in Birds*, ed. H.P. Zeigler & H.-J. Bischof, pp. 195–205. Cambridge, MA: MIT Press.

Freeman, B.M. & Vince, M.A. (1974). *Development of the Avian Embryo*. London: Chapman and Hall.

Gaston, K. & Gaston, M.G. (1984). Unilateral memory after binocular discrimination training: left hemisphere dominance in the chick? *Brain Research*, 303, 190–193.

Güntürkün, O. (1985). Lateralization of visually controlled behaviour in pigeons. *Physiology and Behavior*, 34, 575–577.

Güntürkün, O. (1993). The ontogeny of visual lateralization in pigeons. *German Journal of Psychology*, 17, 276–287.

Güntürkün, O. (1997a). Morphological asymmetries of the tectum opticum in the pigeon. *Experimental Brain Research*, 116, 561–566.

Güntürkün, O. (1997b). Visual lateralization in birds: From neurotrophins to cognition? *European Journal of Morphology*, 35, 280–302.

Güntürkün, O., Hellmann, B., Melsbach, G. & Prior, H. (1998). Asymmetry of representation in the visual system of pigeons. *NeuroReport*, 9, 4127–4230.

Güntürkün, O., Miceli, D. & Watanabe, M. (1993). Anatomy of the avian thalamofugal pathway. In *Vision, Brain, and Behaviour in Birds*, ed. H.P. Zeigler & H.-J. Bischof, pp. 115–135. Cambridge, MA: The MIT Press.

Hart, N.S., Partridge, J.C. & Cuthill, I.C. (2000). Retinal asymmetry in birds. *Current Biology*, 10, 115–117.

Howard, K.J., Rogers, L.J. & Boura, A.L.A (1980). Functional lateralization of the chicken forebrain revealed by use of intracranial glutamate. *Brain Research*, 188, 369–382.

Johnston, A.N.B. & Rogers, L.J. (1995). [^{3}H]MK-801 binding asymmetry in the IMHV region of dark-reared chicks is reversed by imprinting. *Brain Research Bulletin*, 37, 5–8.

Johnston, A.N.B. & Rogers, L.J. (1998). Right hemisphere involvement in imprinting memory revealed by glutamate treatment. *Pharmacology, Biochemistry and Behaviour*, 60, 863–871.

Johnston, A.N.B., Bourne, R.C., Stewart, M.G., Rogers, L.J. & Rose, S.P.R. (1997). Exposure to light prior to hatching induces asymmetry of receptor binding in specific regions of the chick forebrain. *Developmental Brain Research*, 103, 83–90.

Kalliecharan, R. & Hall, B.K. (1974). A developmental study of the levels of progesterone, corticosterone, cortisol, and cortisone circulating in plasma of chick embryos. *General and Comparative Endocrinology*, 24, 364–372.

Karten, H.J. (1979). Visual lemniscal pathways in birds. In *Neural Mechanism of Behavior in the Pigeon*, ed. A.M. Granda & J.H. Maxwell, pp. 409–430. New York, London: Plenum Press.

Khayutin, S.N. (1985). Sensory factors in the behavioral ontogeny of altricial birds. *Advances in the Study of Behavior*, 15, 105–152.

Levin, M., Johnson, R.L., Sten, C.D., Kuehn, M. & Tabin, C. (1995). A molecular pathway determining left-right asymmetry in chick embryogenesis. *Cell*, 82, 803–814.

Lipar, J.L., Ketterson, E.D. & Nolan, V., Jr. (1999). Intraclutch variation in testosterone content of red-winged eggs. *The Auk*, 116, 231–235.

Macphail, E.M. (1976). Effects of hyperstriatal lesions on within-day serial reversal performance in pigeons. *Physiology and Behavior*, 16, 529–536.

Manns, M. & Güntürkün, O. (1997). Development of the retinotectal system in the pigeon: A cytoarchitectonic and tracing study with cholera toxin. *Anatomy and Embryology*, 195, 539–555.

Manns, M. & Güntürkün, O. (1999a). Monocular deprivation alters the direction of functional and morphological asymmetries in the pigeon's visual system. *Behavioral Neuroscience*, 113, 1257–1266.

Manns, M. & Güntürkün, O. (1999b). 'Nature' and artificial monocular deprivation effects on thalamic soma sizes in pigeons. *NeuroReport*, 10, 3223–3228.

McKenzie, R., Andrew, R.J. & Jones, R.B. (1998). Lateralization in chicks and hens: new evidence for control of response by the right eye system. *Neuropsychologia*, 36, 51–58.

Mench, J.A. & Andrew, R. J. (1986). Lateralization of a food search task in the domestic chick. *Behavioral and Neural Biology*, 46, 107–114.

Mey, J. & Thanos, S. (1992). Developmental of the visual system of the chick – a review. *Journal für Hirnforschung*, 33, 673–702.

Meyers, E.N. & Martin, G.R. (1999). Differences in left–right axis pathways in mouse and chick: Functions of FGF 8 and SHH. *Nature*, 285, 403–406.

Miceli, D. & Repérant, J. (1982). Thalamo-hyperstriatal projections in the pigeon (*Columba livia*) as demonstrated by retrograde double-labelling with fluorescent tracers. *Brain Research*, 245, 365–371.

Miklósi, A., Andrew, R.J. & Dharmaretnam, M. (1996). Auditory lateralization: Shifts in ear use during attachment in the domestic chick. *Laterality*, 1, 215–224.

Mock, D.W. & Parker, G.A. (1997). *The Evolution of Sibling Rivalry*. Oxford: Oxford University Press.

Ngo, T.D., Egedy, G. & Tömböl, T. (1992). Golgi study on neurones and fibers in Nucl. rotundus of the thalamus in chicks. *Journal für Hirnforschung*, 33, 203–214.

Ngo, T.D., Davies, D.C., Egedi, G.Y. & Tömböl, T. (1994). A phaseolus lectin anterograde tracing study of the tectorotundal projections in the domestic chick. *Journal of Anatomy*, 184, 129–136.

Nottebohm, F. (1977). Asymmetries in neural control of vocalisation in the canary. In *Lateralization in the Nervous System*, ed. S. Harnard, R.W. Doty, L. Goldstein, J. Jaynes, & G. Krauthamer, pp. 23–44. New York: Academic Press.

Pasternak, T. (1977). Delayed matching performance after visual Wulst lesions in pigeons. *Journal of Comparative and Physiological Psychology*, 91, 472–484.

Peters, J.J., Vonderahe, A.R. & Powers, T.H. (1958). Electrical studies of functional developmental of the eye and optic lobes in the chick embryo. *Journal of Experimental Zoology*, 160, 255–262.

Pritz, M.B., Mead, W.R. & Northcutt, R.G. (1970). The effects of Wulst ablations on color, brightness and pattern discrimination in pigeons (*Columbia livia*). *Journal of Comparative Neurology*, 140, 81–100.

Rajendra, S. & Rogers, L.J. (1993). Asymmetry is present in the thalamofugal visual projections of female chicks. *Experimental Brain Research*, 92, 542–544.

Repérant, J., Raffin, J.-P. & Miceli, D. (1974). La voie rétino-thalamo-hyperstriatale chez le Poussin (*Gallus domesticus* L.). *Comptes Rendus-Academie des Sciences Paris, serie III*, 279, 279–282.

Ricklefs, R.E. (1983). Avian postnatal development. In *Avian Biology*, Vol. 7, ed. D.S. Farner, J.R. King & K.C. Parkes, pp. 1–83. New York: Academic Press.

Rogers, L.J. (1982). Light experience and asymmetry of brain function in chickens. *Nature*, 297, 223–225.

Rogers, L.J. (1990). Light input and the reversal of functional lateralization in the chicken brain. *Behavioural Brain Research*, 38, 211–221.

Rogers, L.J. (1991). Development of lateralization. In *Neural and Behavioural Plasticity: The Use of the Domestic Chick as a Model*, ed. R.J. Andrew, pp. 507–535. Oxford: Oxford University Press.

Rogers, L.J. (1995). *The Development of Brain and Behaviour in the Chicken*. Wallingford: CAB International.

Rogers, L.J. (1996). Behavioral, structural and neurochemical asymmetries in the avian brain: A model system for studying visual development and processing. *Neuroscience and Biobehavioral Reviews*, 20, 487–503.

Rogers, L.J. (1997). Early experiential effects on laterality: Research on chicks has relevance to other species. *Laterality*, 2, 199–219.

Rogers, L.J. & Adret, P. (1993). Developmental Mechanisms of lateralization. In *Vision, Brain, and Behavior in Birds*, ed. H.P. Zeigler & H.-P. Bischof, pp. 227–242. Cambridge, MA: MIT Press.

Rogers, L.J., Adret, P. & Bolden, S.W. (1993). Organization of the thalamofugal visual projections in chick embryos, and a sex difference in light-stimulated development. *Experimental Brain Research*, 97, 110–114.

Rogers, L.J., Andrew, R.J. & Burne, T.H.J. (1998). Light exposure of the embryo and development of behavioural lateralization in chicks, I: olfactory responses. *Behavioural Brain Research*, 97, 195–200.

Rogers, L.J. & Anson, J.M. (1979). Lateralization of function in the chicken forebrain. *Pharmacology, Biochemistry and Behavior*, 10, 679–686.

Rogers, L.J. & Bell, G.A. (1989). Different rates of functional development in the two visual systems of the chicks revealed by [^{14}C]2-deoxyglucose. *Developmental Brain Research*, 49, 161–172.

Rogers, L.J. & Bell, G.A. (1994). Changes in metabolic activity in the hyperstriatum of the chick before and after hatching. *International Journal of Developmental Neuroscience*, 12, 557–566.

Rogers, L.J. & Bolden, S.W. (1991). Light-dependent development and asymmetry of visual projections. *Neuroscience Letters*, 121, 63–67.

Rogers, L.J. & Deng, C. (1999). Light experience and lateralization of the two visual pathways in the chick. *Behavioural Brain Research*, 98, 277–287.

Rogers, L.J. & Krebs, G.A. (1996). Exposure to different wavelengths of light and the development of structural and functional asymmetries in the chicken. *Behavioural Brain Research*, 80, 65–73.

Rogers, L.J. & Rajendra, S. (1993). Modulation of the development of light-initiated asymmetry in chick thalamofugal projections by oestradiol. *Experimental Brain Research*, 93, 89–94.

Rogers, L.J. & Sink, H.S. (1988). Transient asymmetry in the projections of the rostral thalamus to the visual hyperstriatum of the chicken, and reversal of its direction by light exposure. *Experimental Brain Research*, 70, 378–384.

Rogers, L.J., Zappia, J.V. & Bullock, S.P. (1985). Testosterone and eye brain asymmetry for copulation. *Experientia*, 41, 1447–1449.

Saleh, C.N. & Ehrlich, D. (1984). Composition of the supraoptic decussation of the chick (*Gallus gallus*). *Cell and Tissue Research*, 236, 601–609.

Schwabl, H. (1993). Yolk is a source of maternal testosterone for developing birds. *Proceedings of National Academy of Sciences USA*, 90, 11446–11450.

Schwabl, H. (1996). Maternal testosterone in the avian egg enhances postnatal growth. *Comparative and Biochemical Physiology*, 114A, 271–276.

Schwabl, H. (1999). Developmental changes and among-sibling variation of corticosterone levels in an altricial avian species. *General and Comparative Endocrinology*, 116, 403–408.

Schwabl, H., Mock, H.D. & Gieg, J.A. (1997). A hormonal mechanism for parental favoritism. *Nature*, 386, 231.

Schwarz, I.M. & Rogers, L.J. (1992). Testosterone: A role in development of brain asymmetry in the chick. *Neuroscience Letters*, 146, 167–170.

Sedlácek, J. (1967). Development of optic evoked potentials in chick embryos. *Physiologia Bohenmoslovenica*, 13, 268–273.

Shimizu, T., Cox, K. & Karten, H.J. (1995). Intratelencephalic projections of the visual Wulst in pigeons (*Columba livia*). *Journal of Comparative Neurology*, 359, 551–572.

Shimizu, T. & Hodos, W. (1989). Reversal learning in pigeons: Effects of selective lesions of the Wulst. *Behavioral Neuroscience*, 103, 262–272.

Starck, J.M. (1998). Structural variants and invariants in avian embryonic and postnatal development. In *Avian Growth and Development: Evolution within the Altricial–Precocial Spectrum*, ed. J.M. Starck & R.E. Ricklefs, pp. 59–88. New York: Oxford University Press.

Starck, J.M. & Ricklefs, R.E. (1998a) *Avian Growth and Development: Evolution within the Altricial–Precocial Spectrum*. New York: Oxford University Press.

Starck, J.M. & Ricklefs, R.E. (1998b). Patterns of development: The altricial–precocial spectrum. In *Avian Growth and Development: Evolution within the Altricial–Precocial Spectrum*, ed. J.M. Starck & R.E. Ricklefs, pp. 3–30. New York: Oxford University Press.

Stettener, L.J. & Schultz, W.J. (1967). Brain lesions in birds: Effects on discrimination acquisition and reversal. *Science*, 15, 1689–1692.

Stewart, M.G., Rogers, L.J., Davies, H.A. & Bolden, S.W. (1992). Structural asymmetry in the thalamofugal visual projections in 2-day old chicks is correlated with a hemispheric difference in synaptic number in the hyperstriatum accessorium. *Brain Research*, 585, 381–385.

Sui, N., Sandi, C. & Rose, S.P.R. (1997). Interactions of corticosterone and embryonic light deprivation on memory retention in day-old chicks. *Developmental Brain Research*, 101, 269–272.

Tanabe, Y., Saito, N. & Nakamura, T. (1986). Ontogenetic steroidogenesis by testes, ovary, and adrenal of embryonic and postembryonic chickens (*Gallus domesticus*). *General and Comparative Endocrinology*, 63, 456–463.

Vallortigara, G. & Andrew, R.J. (1991). Lateralization of response by chicks to change in a mode partner. *Animal Behaviour*, 41, 187–194.

Vallortigara, G. & Andrew, R.J. (1994). Differential involvement of right and left hemisphere in individual recognition in the domestic chick. *Behavioural Processes*, 33, 41–58.

Wallman, J. & Letelier, J.-C. (1993). Eye movements, head movements, and gaze stabilization in birds. In *Vision, Brain, and Behaviour in Birds*, ed. H.P. Zeigler & H.-J. Bischof, pp. 245–263. Cambridge, MA: MIT Press.

Wallman, J. & Pettigrew, J.D. (1985). Conjugate and disjunctive saccades in two avian species with contrasting oculomotor strategies. *Journal of Neuroscience*, 5, 1418–1428.

Watanabe, S. (1992). Effects of lesions in the ectostriatum and Wulst on species and individual discrimination in pigeons. *Behavioural Brain Research*, 49, 197–203.

Woods, J.E. & Brazzill, D.M. (1981). Plasma 17β-estradiol levels in chick embryo. *General and Comparative Endocrinology*, 44, 37–43.

Woods, J.E., De Vries, G.W. & Thommes, R.C. (1971). Ontogenesis of pituitary–adrenal axis in the chick embryo. *General and Comparative Endocrinology*, 17, 407–415.

Woods, J.E., Simpson, R.M. & Moore, P.L. (1975). Plasma testosterone levels in the chick embryo. *General and Comparative Endocrinology*, 27, 543–547.

Wise, P.M. & Frye, B.E. (1973). Functional development of the hypothalamo–hypophyseal–adrenal cortex axis in the chick embryo, *Gallus domesticus*. *Journal of Experimental Zoology*, 185, 277–292.

Zappia, J.V. & Rogers, L.J. (1983). Light experience during development affects asymmetry of forebrain function in chickens. *Developmental Brain Research*, 11, 93–106.

Zappia, J.V. & Rogers, L.J. (1987). Sex differences and reversal of brain asymmetry by testosterone in chickens. *Behavioural Brain Research*, 23, 261–267.

7

Ontogeny of Visual Asymmetry in Pigeons

ONUR GÜNTÜRKÜN

7.1. Introduction

John Daniel led a life of the upper classes at the end of the British Empire. During parties at his home in 15 Sloane Street, London, he was known for his perfect manners; at 5 o'clock he never missed drinking a cup of tea and after dinner he always asked for a coffee. Apart from that he was known to be right-handed. This aspect of him was, probably, noted by his contemporaries only because John Daniel was not a human being but a gorilla. His life was described by Cunningham (1921), and his brain by LeGros Clark (1927), who discovered a conspicuous asymmetry in the anteroposterior extent of the hemispheres with a larger size on the left side. Of course, LeGros Clark was not able to draw a causal relationship between this morphological asymmetry and John Daniel's handedness.

Since the life and death of John Daniel, we have come a long way in understanding how brains are asymmetrical. We now know that a large number of vertebrate species are lateralized and we are slowly starting to understand that these asymmetries seem to form a coherent pattern, which may indicate that several of the left–right differences observed in the brains of humans and other animals can be traced back to common ancestors (Vallortigara, Rogers and Bisazza, 1999; see also Chapter 1 by Vallortigara and Bisazza). It was Cunningham's dream to uncover the details of the ontogeny and neurobiology of cerebral asymmetries and he was sure that animal studies could help to trace the details of lateralized systems (Cunningham, 1892, 1902). Now we seem to be closer than ever to achieving Cunningham's dream. Several promising animal models are under study world-wide and the advance of scientific techniques provides unprecedented tools to analyse a large number of neuronal details that are asymmetrical. Avian visual lateralization is one of these models. It not only allows us to investigate experimentally the interplay of the neurobiological substrate and behavioural functions but also provides us with the opportunity to study the ontogenetic events leading to asymmetries (see Chapter 6 by Deng and Rogers).

The following account describes the ontogeny of the functional architecture of the pigeon's visual system, one of the avian models presently under study. The main emphasis of this chapter is a recapitulation of the ontogenetic events leading to a lateralized visual system. Therefore, I will start with an overview of the genetic mechanisms that determine a slight torsion of the embryo to the right side within its egg. Then it will be shown that this tilt brings the right eye close to the translucent egg shell while the left eye is covered by the body – a condition that results in asymmetrical stimulation of the two eyes with light. This lateralized light input comes just in time to mould the developing tectofugal visual system into an anatomically asymmetrical organization. Thus, the right eye dominance for visual pattern analysis in pigeons results from a tight interplay between genetic and epigenetic factors. The anatomical and physiological asymmetries result in a lateralized functional architecture of the tectofugal visual system. This asymmetry develops in a couple of weeks and maintains its function for more than a decade – and thus the whole life span of a pigeon. During this period, the visual lateralization not only alters learning, memorizing and recognition processes of objects in an asymmetrical fashion, but also determines complex cognitive processes of the visual system.

7.2. The Way it Starts

All vertebrates exhibit a left–right (LR) asymmetry in the position of their visceral organs, as in the case of the heart, which invariantly loops to the right side. Experiments with chicks have revealed the mechanisms that determine the embryological events leading to this asymmetry (Ramsdell and Yost, 1998). Three distinct processes are a requisite of normal LR patterning. First, an initial asymmetry must become established in the embryo, creating a global basis for LR axis formation. A failure to break symmetry results in randomized LR development, whereby the heart and viscera become oriented independently of each other in a stochastic fashion. Without a consistent alignment of these asymmetries, a wide range of defects occurs and can be lethal (Levin et al., 1997). Second, LR asymmetries must be consistently oriented with respect to the other two body axes. Reversed orientation of the LR axis result in *situs inversus*, a harmless condition in which heart and viscera are mirror-reversed (Bowers et al., 1996). Third, global LR patterning information must be transmitted to organ primordia, which in turn must correctly interpret these positional cues to execute an appropriate morphogenetic response. Aberrations in this process can result in conditions, such as reversal of individual organs, with fatal consequences (Ramsdell and Yost, 1998).

In chicks, the LR axis is initiated together with the dorsoventral axis prior to gastrulation (Yost, 1995). Thus, tissues lateral to the embryo's node are already biased in their LR identities and confer this identity to the node – a mechanism that is still not completely understood. Once this event has taken place, the node then directs LR development by signalling back to adjacent tissues (Pagan-Westphal and Tabin, 1998). This exchange of positional information is mediated by a sequential molecular pathway of activin and sonic hedgehog (shh) signalling, whereby right-sided expression of activin in the node induces asymmetric expression of the activin receptor ActRII and shh (Levin et al., 1995) (Figure 7.1). Subsequently, nodal, lefty2 and the bicoid-related homeobox gene, Pitx2, is expressed within the left lateral plate mesoderm (Hyatt, Lohr and Yost, 1996; Meno et al., 1997). On the right side, a gene encoding a zinc finger protein of the Snail family, cSnR, is expressed in

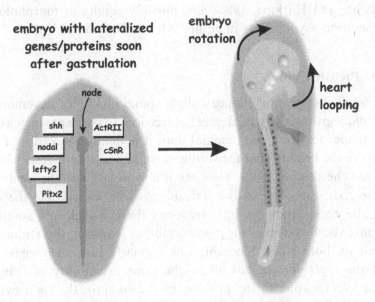

Figure 7.1.   Schematic overview of asymmetrically expressed proteins during early embryogenesis that contribute to heart looping and right-sided torsion of the body axis. Both embryos, which are not shown in scale, are seen from the dorsal orientation. ActRII, activin receptor; cSnR, snail-related zinc-finger transcription factor; shh, sonic hedgehog. Adapted from Robertson (1997).

the lateral mesoderm (Isaac, Sargent and Cooke, 1997). Perhaps nodal represses cSnR on the left side of the axis, resulting in a right-side dominance of expression (Robertson, 1997). During all these events, midline structures of the embryo seem to function as a barrier, allowing inductive interactions to occur on one side of the embryo while preventing such activity from occurring on the contralateral side. Possibly, however, the story is slightly more complicated. Lateral plate mesoderm explants derived from the left or the right embryo both express nodal when cultured in isolation from the midline (Lohr, Danos and Yost, 1997). Therefore, it is conceivable that the midline serves both to induce and repress different signal cascades.

These processes result in a rightward looping of the heart, a counterclockwise looping of the gut and finally in a slight torsion of the embryo with the forehead pointing to the right (Ramsdell and Yost, 1998). This last point is just what is needed to induce visual lateralization. The rightward spinal torsion is, by the way, also true for human embryos that have a preference for sucking on their right thumb, partly due to an embryonic right turn of their head (Hepper, Shahidullah and White, 1991). Most newborns still have a preference for a right turn of their head when being in a supine position (Michel, 1981), and this preference seems to correlate with subsequent handedness (Michel and Harkins, 1986) and possibly results in morphological hand–motoneuron asymmetries in adults (Melsbach et al., 1996).

7.3. Prehatch Vision

All events detailed up to now induce a slight spinal torsion of the embryo in the egg. Rapidly growing in size, degrees of freedom to move are increasingly becoming smaller for the young animal until it is finally trapped in a fixed position where the head is bent forward and rotated so that it rests with the left eye on the chest (Kuo, 1932). The beak is now pointing to the right, so as to bring the right eye close to the translucent shell. Every time the adult incubating the eggs stands up, light traverses the egg shell and stimulates the developing visual system of the pigeon embryo. However, this stimulation is not equal on both sides. Since pigeons regularly turn their eggs, each portion of the egg's surface has about the same probability of being on top. On the average, due to the position of the embryo, the right eye will be stimulated by light to a larger extent than the left. As outlined below, this stimulation asymmetry induces left–right differences in the ascending tectofugal visual pathway, which then result in lateralized visual behaviour.

However, for light stimulation to exert any effects, it first has to be perceived by the developing visual system of the young pigeon. Unlike precocial chicks, which have a rather developed system before hatching (Mey and Thanos, 1992), the embryonic visual pathways of altricial pigeons are far less functional (see Chapter 6 by Deng and Rogers). Bagnoli et al. (1985) showed that the first photosensitive discs in the outer segments of the retinal photoreceptors appear at about the fourth posthatch day (Ph4), the time when the first electroretinograms and the first visually evoked tectal potentials could be recorded (Bagnoli et al., 1985, 1987). However, only the foveal region of the retina was examined in this ultrastructural study, leaving the possibility that mature receptor cells may exist in the periphery. This assumption is strengthened by the observation that the pupillary light reflex can be elicited from embryonic day 15 (E15) onwards, which is about 2 days before hatching (Heaton and Harth, 1974; but see Bagnoli et al., 1985). The neural circuit of this reflex receives its input from a small population of retinal ganglion cells, which mainly cluster in the more peripheral retinal areas (Gamlin et al., 1984). Thus, these early-maturing cells are located outside the two fovea-like areas of enhanced vision in pigeons, namely the red field in the superiotemporal retinal quadrant and the area centralis. Indeed, the central retina seems not to be the site where retinal ganglion cells are generated first (Rager, Rager and Frei, 1993). Since the neural elements of the retina develop from columnar-organized clones, processing information from the same region of visual space (Cepko, 1993), it is conceivable that receptor cells outside the central fovea also develop before hatching. This is supported by a recent study, demonstrating spatial and temporal differences in the expression of photopigment molecules in the chick retina, with short-wavelength opsin first detected in the area centralis and the temporal retina, but rhodopsin first detected in the inferior retina (Bruhn and Cepko, 1996). Since morphological differentiation precedes the expression of photopigment molecules, this expression pattern indicates that photoreceptor differentiation does not start only in the central retina. Additionally, this process begins in only a few isolated cells of the presumptive photoreceptor layer (Saha and Grainger, 1993; Stiemke and Hollyfield, 1995), opening the possibility that some elements of each retinal area may start functioning well before their direct neighbours. Thus, it is likely that light is perceived by some elements of the pigeon's visual system, although full visual functions probably do not develop until long after hatching.

About 90% of all retinal ganglion cells project to the tectum in pigeons (Remy and Güntürkün, 1991). The optic tectum is a highly complex neural entity in which even simple histological techniques, like Nissl-staining, reveal 15 laminae (Cajal, 1911). More modern immunocytochemical or receptor autoradiographic techniques allow one to identify even more than 20 layers.

As outlined several paragraphs below, the tectum is the starting point of the tectofugal pathway, which ascends to the forebrain and which to a large extent determines visual lateralization in pigeons.

The maturation of the tectal lamination is not complete before Ph15 (Manns and Güntürkün, 1997). This is about twice the time needed by chicks (E16; LaVail and Cowan, 1971) and quails (E14; Senut and Alvarado-Mallart, 1986). However, the pigeon's tectum displays a strong heterochrony in its different components and not all tectal areas are as slow to develop. The rostrolateral tectum develops especially fast and at E16, 24 h before hatching, the lamination of the retinal fibres in this area already exhibit the adult pattern (Manns and Güntürkün, 1997). In the chick, the penetration of the deeper retinorecipient layers by the first retinal axons coincides with the development of the first synaptic contacts with tectal cells (Rager and von Oeynhausen, 1979; McLoon, 1985). The growth of these first synapses marks the point in time, when the first postsynaptic responses from the tectal surface can be recorded following electrical stimulation of the optic nerve in chicks (Rager, 1976). Since the development of the retinotectal system in the pigeon occurs in the same temporal order, the formation of functional synapses in the pigeon's tectum may take place after E15/16, when the first retinal fibres enter the deep retinorecipient layers.

To summarize this part, it can be concluded that the development of the visual system of the pigeon, an altricial species, is dramatically slower than that of the precocial chick. However, distinct retinal and retinotectal mechanisms develop rapidly enough to transmit visual signals 24–48 h before hatching to the embryonic rostrolateral tectum. This is exactly what was needed to initiate asymmetry in the tectofugal visual system.

7.4. From Light Asymmetry to Synaptic Plasticity

A considerable number of studies demonstrate that, in chicks and pigeons, visual lateralization is triggered by light exposure of the embryo in the egg shortly before hatching. The asymmetrical light stimulation is effective in modulating synaptic patterns of the ascending pathways during a short time window prior to and shortly after hatching. As little as 2 h of prehatch light stimulation (100 lux) has proven to be sufficient for asymmetrical development in domestic chicks and it determines their right eye superiority in grain-pebble distinction (Rogers, 1982; Zappia and Rogers, 1983; see Chapter 6 by Deng and Rogers). The direction of lateralization can even be reversed by experimentally exposing the left eye to light for 24 h shortly before hatching (Rogers, 1990). If chicks are incubated and hatched under dark conditions, the usual lateralization pattern for certain visual behaviours disappears (Rogers, 1982). Together with the behavioural asymmetries, also the anatomical left–right differences of the thalamofugal pathway (see Chapter 6 by Deng and Rogers) can be reversed or altered in chicks depending on the experimental manipulations of the light stimulation prehatching (Rogers and Bolden, 1991; Rogers, 1996).

The situation in pigeons seems to be remarkably similar. Dark incubation prevents the formation of behavioural and anatomical tectal asymmetries involving a superiority of the right eye and larger soma sizes in the retino-recipient tectal layers on the left (Güntürkün, 1993). Ten days of posthatch monocular deprivation of the right eye reverses visual lateralization (Manns and Güntürkün, 1999a). Therefore, light stimulation asymmetry before hatching seems to be the trigger for visual asymmetry in pigeons, as it is in chicks.

In principle, these results are in accordance with findings from monocular deprivation studies in mammals (Sherman and Spear, 1982) and zebra finches (Herrmann and Bischof, 1986a,b), which all reported smaller soma sizes of neurones receiving afferents from the deprived eye. Thus, at first glance, these results might point to close similarities of the mechanisms that govern natural monocular deprivation due to the asymmetrical embryonic posture in birds and postbirth monocular deprivation in mammals. However, although seemingly similar, different mechanisms must be involved. A first hint as to these separate mechanisms is provided by comparing the anatomical locations where the effects of deprivation occur. Morphological soma size effects of monocular deprivation in mammals are restricted to the binocular portion of the lateral geniculate nucleus and are absent in the retina and visual cortex (Sherman and Spear, 1982). These effects are regarded as secondary consequences from synaptic competition at cortical level between geniculate fibres representing the deprived and the non-deprived eye (Rauschecker, 1991; Movshon and Van Sluyters, 1981). While effects of embryonic lateralized light input or posthatch monocular deprivation are also absent in the retinae of birds (Güntürkün, 1997c; Herrmann and Bischof, 1993), they can be found in the optic tectum (Güntürkün, 1997c; Manns and Güntürkün, 1999a), the nucleus rotundus (Manns and Güntürkün, 1999b; Herrmann and Bischof, 1986a), and in the ectostriatum (Herrmann and Bischof, 1986b) (Figure 7.2). Thus, prehatch or posthatch monocular deprivation affects perikaryal sizes along the whole tectofugal system. While inputs of both eyes could compete at rotundal level (Güntürkün et al., 1998), a comparable competition is absent in the tectum and unlikely in the ectostriatum (Güntürkün, 2000). This suggests that visual deprivation effects in birds are mediated through

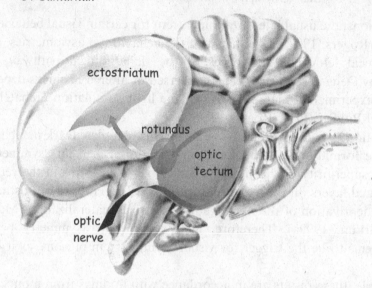

Figure 7.2.   Three-dimensional overview of the pigeon's tectofugal system. The optic nerve crosses completely at the optic chiasm to innervate the contralateral optic tectum. From there, fibres lead to the n. rotundus in the diencephalon which finally projects to the ectostriatum in the forebrain. Adapted from Dubbeldam (1998).

activity-correlated and eventually trophic deprivation effects within one hemisphere, and they possibly operate without direct synaptic competition between neurones representing deprived and non-deprived eyes.

A further clue for differences between mammalian and avian visual deprivation mechanisms results from the fact that in the mammalian geniculocortical system, only the unilateral absence of contoured visual patterns induces significant deprivation effects. Asymmetries of luminance alone do not lead to alterations (Movshon and Van Sluyters, 1981). This supports the assumption that fibre competition is mediated by a Hebbian mechanism, which requires correlated activity of presynaptic and postsynaptic cells for stabilization or retraction of synapses (Rauschecker, 1991). In chicks and pigeons, the situation must be different since light has to shine through the egg shell and the closed lid of the embryo to induce cerebral asymmetries (Güntürkün, 1993; Rogers, 1982). Therefore, natural monocular deprivation in pigeons has to be induced by brightness and not by contoured visual pattern differences. Brightness differences are probably coded by mere activity differences between the eyes and could induce asymmetries by the release of neurotrophins between the stimulated and the deprived hemisphere (Güntürkün, 1997b). Such activity-dependent trophic effects could generate the morphological left–right differences between rotundal cells observed in the present study.

7.5. Visual Lateralization in Pigeons: A Behavioural Analysis

The optic nerves of birds cross completely at the optic chiasm and only a small number of fibres recross via mesencephalic and thalamic commissures (Weidner et al., 1985). Since the primary visual projections remains remarkably crossed in birds, the performance of one half brain can easily be studied using eyecaps with which sight can temporarily be restricted to one eye, and so mainly directing the information to the contralateral hemisphere. With this procedure, visual lateralization can be demonstrated using a wide range of techniques. Using the right eye, pigeons are superior in discriminating two-dimensional artificial patterns (Güntürkün, 1985) (Figure 7.3) and three-dimensional natural objects (Güntürkün and Kesch, 1987). This greater visual processing capacity of the right eye system in pattern discrimination also leads to a higher degree of illusion of this side when being confronted with geometrical optic illusions (Güntürkün, 1997b). In visual memory tasks in which 725 abstract patterns have to be memorized, the animals are able to remember most of them with their right eye, but are barely above chance level with their left (Fersen and Güntürkün, 1990).

Figure 7.3. Schematic view of a pigeon in a skinner box during a simultaneous pattern discrimination task. Two patterns are back projected by microprojectors on to pecking keys. Correct choices are reinforced, the food being delivered in a food hopper. During the reinforcement period grains are accessible through the hole in front of the animal.

It is probably this asymmetry in memorizing visual stimuli that results in a significant right eye advantage when homing from a distant release site over known territory to the loft (Ulrich et al., 1999). Homing makes great demands on spatial orientation. To find a left hemisphere advantage during homing is therefore astonishing considering the large body of data showing a right hemisphere dominance for visuospatial tasks (chicks: Rashid and Andrew, 1989; Vallortigara et al., 1996; marsh tits: Clayton and Krebs, 1994; cats: Lorincz and Fabre-Thorpe, 1996; baboons: Vauclair and Fagot, 1993; humans: Maguire et al., 1997). However, spatial orientation is a multi-component feature in which several cognitive processes with diverse cerebral asymmetries interact (Hellige, 1995). It is therefore conceivable that the pigeons used a cognitive strategy that is more left-hemisphere based. As discussed by Ulrich et al. (1999), it is likely that the birds utilized visual memory-based snapshot tracking to pursue visual features along their pre-learned route. Since, as outlined above, pigeons are known to be left hemisphere dominant for memorizing and discriminating visual features, it is indeed likely that the homing task was, in part, performed by a succession of visual feature discriminations. If pigeons are tested in a maze where they cannot utilize this strategy, the left hemisphere advantage vanishes (Prior et al., 1998).

The visual lateralization also affects cognitive processes of the animals. Diekamp et al. (1999) tested pigeons under monocular conditions in successive colour reversals. Thus, the animals learned to favour green (S+) over red (S−) using their right eye until reaching learning criterion. Then the conditions were changed and red was now rewarded (S+) while green was not (S−). As soon as the pigeons successfully learned the reversal, conditons were altered again, and so on. One group of animals performed 30 reversal under right-eye seeing conditions and another group under left-eye seeing conditions. After a couple of reversals both groups showed a 'learning to learn' effect such that each reversal was achieved with fewer trials (Figure 7.4). Reversal learning can be described best on a mathematical basis by an exponential function of the type $y = a + \exp^{(b-cx)}$, with a representing the asymptote (i.e. the error rate around which the performance oscillates after several reversals), b determining the starting value of the function for the first reversal and c representing the steepness of the curve (i.e. the rate of error reduction). Both for a and for c, Diekamp, Prior and Güntürkün (1999) could reveal a right eye superiority. Thus, using the right eye–left hemisphere, the animals were faster in understanding the basic principle of this experiment (c) and exhibited a higher level of performance after reaching asymptote (a). Visual lateralization in

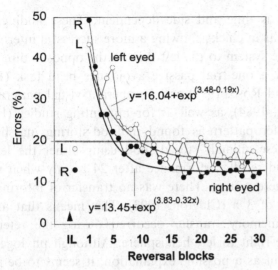

Figure 7.4. Visual asymmetry in a 'learning to learn' effect during a successive colour-reversal task. The pigeons completed 30 reversals in a red–green discrimination experiment. With both monocular seeing conditions the birds acquired a concept of the successive reversals resulting in an exponential reduction of errors over sessions ($y = a + \exp^{(b-cx)}$). In this equation, a represents the asymptote, i.e. the error rate around which the performance oscillates after several reversals, b determines the starting value of the function for the first reversal and c represents the steepness of the curve, i.e. the rate of error reduction. Both for a and for c, Diekamp, Prior and Güntürkün (1999) could reveal a right eye superiority, indicating both faster achievement of the concept as well as higher final performance with the right eye–left hemisphere. Adapted from Diekamp, Prior and Güntürkün (1999).

pigeons, therefore, not only consists of asymmetries in simple pattern recognition and memorization processes but also affects 'cognitive' systems, which extract general properties of the visual world.

Visual asymmetry also seems to affect interhemispheric interactions. Skiba et al. (2000) conditioned pigeons under monocular conditions to learn a simple simultaneous colour discrimination. As soon as learning criterion was acquired, the animals were taken back to their cages to be tested after a certain time in the same colour discrimination with the other eye. The results showed that interhemispheric transfer of a colour discrimination occurs in both directions but asymmetrically within the first hour after acquisition. Transfer was slower from the left to the right eye up to 50 min after acquisition than vice versa. For intervals longer than 3 h, no interocular transfer differences were found. These data show that each hemisphere shifts the stored information to the contralateral side, but the efficiency of this process is time and side dependent. These findings are fairly consistent with results in chicks showing a more successful interocular transfer from the right eye system to the left than in the opposite direction. This has been shown for a one-trial passive avoidance bead task (Rose, 1991; Sandi, Patterson and Rose, 1993) and an operant visual pattern discrimination task (Gaston, 1984), as well as for imprinting studies (Horn, 1991). A deviating transfer pattern is found in food-storing marsh tits. Experiments in which these animals stored food using either the left or the right eye showed good recall performance after 24 h only when using the right eye–left hemisphere system. There was no transfer of information after a retention interval of 3 h (Clayton, 1993). This means that a unilateral interhemispheric memory transfer occurred during the retention interval of 24 h from the right to left hemisphere. Although phylogenetic factors cannot be ruled out as a possible explanation, it seems to be more likely that differences in interocular transfer patterns can be explained by different types of cognitive processes required in these tasks. The food-storing tasks demand the animals to utilize spatial and thus right-hemisphere-based cues to find the correct site. The preferential transfer of storing information from the right to the left hemisphere in marsh tits, therefore, represents a shift of engrams from the (spatially) dominant to the subdominant half brain. In principle, this resembles the results of the colour discrimination task in pigeons in which the initial transfer occurs from the left hemisphere, which is dominant for visual object features, to the right.

As outlined in the beginning, the complete decussation of the optic nerve makes it likely that this right eye superiority is related to a left hemisphere dominance for visual object analysis. This assumption is supported by the fact that lateralized performance is not caused by peripheral factors such as differences in visual acuity, wavelength discrimination or depth resolution (Martinoya, Le Houezec and Bloch, 1988; Remy and Emmerton, 1991; Güntürkün and Hahmann, 1994). Further support of a left hemispheric superiority is shown by the behavioural results demonstrating that unilateral left hemisphere lesions cause severe deficits, while right-sided lesions have minor impact (Güntürkün and Hoferichter, 1985; Güntürkün and Hahmann, 1999). As testified by other chapters in this book, this left hemisphere superiority for visual analysis (categorization *sensu* Andrew, 1991) is not restricted to pigeons but has been also described with various methods in domestic chicks (Mench and Andrew, 1986; Vallortigara et al., 1996; Rogers, 1995), zebra finches (Alonso, 1998) and food-storing and non-storing parids and corvids (Clayton and Krebs, 1994).

7.6. Differential Behavioural Effects of Light Stimulation During Ontogeny

Light stimulation asymmetry during early ontogeny could induce a left hemisphere superiority for object discrimination by increasing performance of left hemisphere processes, by decreasing those of the right, or by differently adjusting distinct neural circuits on both sides (see also Chapter 5 by Andrew). In a recent study, these alternatives were tested by comparing dark- and light-incubated pigeons in two visual experiments that both yield a right eye advantage but probably tap different kinds of visual processes (Skiba et al., submitted). One of them was grit–grain discrimination in which the animals have to peck within 30 s 30 grains from a trough filled with about 1000 pieces of grit resembling the grains in colour and shape (Güntürkün and Kesch, 1987). Since not the number of pecks, but the number of grains eaten, differs between left and right monocular conditions, visual asymmetry as tested in grain–grit discrimination seems to depend largely on the discrimination accuracy and not on visuomotor speed (Güntürkün and Kesch, 1987; Manns and Güntürkün, 1999a). This is different from successive pattern discriminations using variable ratio schedules with low rates of reinforcement like VR32 (Güntürkün, 1985). Here, the animals have to distinguish the correct pattern right at the beginning of a trial but their reinforcement success largely depends on their speed of pecking. Consequently, their right eye superiority is due to the number of pecks emitted, while the two monocular conditions generally do not differ with respect to the discrimination scores (Güntürkün, 1985; Güntürkün and Hoferichter, 1985; Güntürkün and Böhringer, 1987; Güntürkün and Kischkel, 1992). Thus, pattern discrimination with these variable ratio schedules seems mainly to reveal an asymmetry in visuomotor speed.

Using these two behavioural paradigms, the mechanism by which embryonic light stimulation induces visual lateralization was tested by analysing adult pigeons that were either dark- or light-incubated. As shown in Figure 7.5, in neither of the two test paradigms did dark-incubated animals display any differences between the two seeing conditions. In contrast, light-incubated pigeons had higher discrimination scores in the grain–grit and higher pecking scores in the successive pattern discrimination, using their right eye compared with the left. Thus, as shown previously by Güntürkün (1993), visual lateralization was dependent on light incubation. However, a closer inspection of the data reveals that the right eye dominance arises due to two different mechanisms in these two paradigms. The right eye data of light-incubated animals are significantly higher than those of the dark-incu-

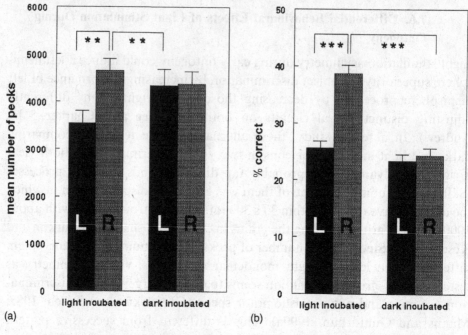

Figure 7.5.   Results of a successive pattern and a grain–grit discrimination task under monocular conditions. Both experiments were performed with the same group of light- or dark-incubated pigeons. In the successive pattern discrimination (a), lateralization emerges due to a higher visuomotor speed using the right eye. In the grain–grit task (b), visual asymmetry is due to a higher percentage discrimination score with the right eye seeing. The data pattern shows that both tasks demonstrate a significant asymmetry (asterisks) in light-incubated but not in dark-incubated animals. However, left–right differences arise due to a decrease of left-eye processes in visuomotor speed and an increase of right-eye mechanisms in visual discrimination. ** $p < 0.01$, *** $p < 0.001$. Adapted from Skiba, Diekamp and Güntürkün (submitted).

bated pigeons, indicating enhanced left hemisphere performance in lateralized birds. By contrast, in the pattern discrimination, the left eye results of light-incubated animals are significantly lower than those of the dark-incubated pigeons. Thus, light incubation seems to increase left hemispheric performance in visual discrimination, but decreases right hemispheric accomplishment in visuomotor speed (Figure 7.5). This data set therefore reveals that neural mechanisms in both hemispheres are differently adjusted when visual asymmetry is established in ontogeny.

7.7. The Anatomical Substrate for Visual Asymmetry in Pigeons

As outlined above, visual information ascending to the forebrain is processed by two parallel pathways, the thalamofugal and the tectofugal system, suggested to be equivalent to the geniculo-cortical and the extrageniculo-cortical visual pathways of mammals, respectively (Shimizu and Karten, 1993). The pigeon's thalamofugal pathway mainly processes visual input from the lateral monocular fields of the laterally placed eyes (Remy and Güntürkün, 1991; Güntürkün and Hahmann, 1999). In the asymmetry experiments with pigeons mentioned earlier, however, the stimuli were viewed by the frontal binocular visual field, which is mainly analysed by the tectofugal pathway (Güntürkün and Hahmann, 1999; Hellmann and Güntürkün, 1999). The tectofugal system is composed of optic nerve fibres projecting to the contra-lateral optic tectum, from which fibres project bilaterally to the rotundal nuclei on each side of the thalamus. Each nucleus rotundus by itself projects to the ipsilateral ectostriatum of the forebrain (Engelage and Bischof, 1993) (Figure 7.6). The thalamofugal and the tectofugal pathways have been shown to constitute structural asymmetries related to lateralized visual behaviour in chicks (Rogers, 1996) and pigeons, respectively.

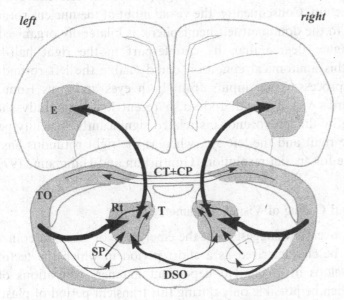

Figure 7.6. Schematic overview of the pigeon's tectofugal system in frontal view. CP, commissura posterior; CT, commissura tectalis; DSO, dorsal supraoptic decussation; E, ectostriatum; Rt, n. rotundus; SP, n. subpretec-talis; T, n. triangularis; TO, tectum opticum. Adapted from Güntürkün (2000).

In pigeons, a morphometric study of tectal perikarya size revealed morphological asymmetries with the superficially located retinorecipient cells being larger on the left side, contralateral to the dominant eye (Güntürkün, 1997a). This is also the case for the nucleus rotundus, the next tectofugal entity (Manns and Güntürkün, 1999b). Thus, the pigeon's tectofugal system displays significant morphological asymmetries, which might be related to the behavioural lateralization of the animals.

Relay neurones of the tectal lamina 13 project bilaterally on to the n. rotundus (Hellmann and Güntürkün, 1999). The bilaterality of this projection should lead to representations of both the ipsilateral and the contralateral eye in the tectofugal system of each hemisphere. Indeed, Engelage and Bischof (1988, 1993) were able to show that binocular input is represented in the ectostriatum in zebra finches. In pigeons, Güntürkün et al. (1998) demonstrated with anterograde and retrograde tracers that the ratio of ipsilateral to contralateral tectorotundal projections is asymmetrically composed. While the quantity of ipsilateral tectorotundal projections is about equal, the number of neurones projecting contralaterally from the right tectum to the left rotundus are about twice in number than vice versa (Güntürkün et al., 1988). As a result, the nucleus rotundus on the dominant left side receives, beside a massive ipsilateral tectal input, also a large number of afferents from the contralateral tectum. Consequently, the visual input of the nucleus rotundus, which projects to the dominant left hemisphere, is bilaterally organized to a significantly higher degree than its counterpart in the right half-brain. Functionally, this anatomical condition could enable the left rotundus to integrate and process visual inputs from both eyes and thus from both sides of the bird's visual world. Indeed, a recent lesioning study showed that processing by the left rotundus is related significantly to acuity performance with the right and the left eye, whereas the right rotundus has only minor relevance for spatial resolution (Güntürkün and Hahmann, 1999).

7.8. Dual Coding of Visual Asymmetry

Up to now this overview suggests that the emergence of visual lateralization in pigeons can be characterized as a short period in which the tectofugal asymmetry develops in an activity-dependent manner. Alterations of the system would then be possible only during this transient period of plasticity, while asymmetry would be 'static' and unmodifiable for the remaining lifetime of the bird. Several lines of evidence suggest this assumption of a static asymmetry to be incomplete. If the tectal and the posterior commissures, which connect the tecta of both hemispheres, are transected, visual lateralization reverses to a left eye dominance. This reversal of laterality is proportional to the number of transected fibres (Güntürkün and Böhringer, 1987). If hemispheric asymmetry is reversed by tectal commissurotomy, it is likely that this asymmetry was maintained previously, at least in part, by asymmetrical interactions between the tecta (see also Parsons and Rogers, 1993), which are known to inhibit each other primarily (Robert and Cuénod, 1969; Hardy et al., 1984). Keysers, Diekamp and Güntürkün (2000) tested this hypothesis by recording field potentials from intratectal electrodes in response to a stroboscope flash to the contralateral eye and an electrical stimulation of the contralateral tectum. They found that the left tectum was able to modulate the flash-evoked potential of the right tectum to a larger extent than vice versa. This lateralized interhemispheric cross-talk could thus constitute an important 'dynamic' component of asymmetrical visual processing (Figure 7.7).

This result makes it likely that the emergence of visual asymmetry in pigeons is related to a dual coding of left–right differences. Thus, visual

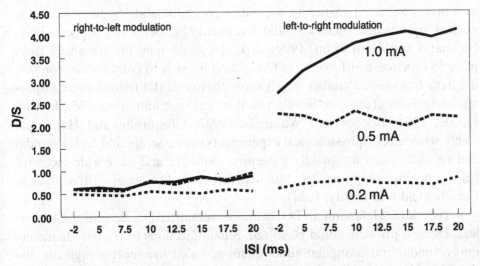

Figure 7.7.   Average tectotectal modulation from right to left and from left to right side. D/S represents the difference (D) between the voltage of the visually evoked tectal potential with contralateral stroboscope flash only (S) and with stroboscope flash plus stimulation of the contralateral tectum. ISIs represent the interstimulus intervals between the contralateral tectum stimulation and the contralateral stroboscope flash. The tectum contralateral to the recorded side was stimulated with three different amplitudes (0.2 mA, 0.5 mA, 1.0 mA). The data reveal a clear asymmetry of tectotectal modulation. At higher stimulation amplitudes, the dominant left tectum modulates visually evoked potentials of the right tectum to a much higher extent than vice versa. Adapted from Keysers, Diekamp and Güntürkün (2000).

lateralization cannot be explained entirely by the anatomical differences between left and right components of the tectofugal pathway. Obviously, a second, more dynamic component exists, which is able to modulate neural processes of the optic tecta in an asymmetrical manner. Altering this second dynamic component, as in the experiment of Güntürkün and Böhringer (1987), results in an important alteration of visual asymmetry.

7.9. The Tectofugal Pathway: From Structure to Function

The most important challenge to understand the processing principles of the tectofugal pathway is to unravel the tectorotundal junction. While the outer tectal layers constitute a two-dimensional map of the visual surroundings (Remy and Güntürkün, 1991), the nucleus rotundus is composed of functional subregions that process different visual features, including movement, colour or luminance in parallel (Wang and Frost, 1992; Wang, Jiang and Frost, 1993; Sun and Frost, 1998). Additionally, retinotopic place coding seems to be absent within the nucleus rotundus, since each point of the tectum is connected to nearly the entire rotundus and its dorsal cap, the nucleus triangularis (Benowitz and Karten, 1976; Deng and Rogers, 1998; Hellmann and Güntürkün, 1999). It is still open how the transition from place to function coding occurs. The second quest is to combine the conflicting data from lesion studies which show that rotundal lesions cause important deficits in acuity and subtle visual object discriminations (Macko and Hodos, 1984; Bischof and Watanabe, 1997; Güntürkün and Hahmann, 1999), while electrophysiological experiments demonstrate that tectorotundal and rotundal units are mainly movement sensitive and have wide receptive fields encompassing up to 180° (Frost and DiFranco, 1976; Jassik-Gerschenfeld and Hardy, 1984).

Recent data (Hellmann and Güntürkün, submitted) might indicate how at least the first problem could be solved. Small injections of retrograde tracers into rotundal and triangular subdomains revealed five morphologically distinct tectal layer 13 cell populations, which together establish the tecto-rotundal/triangular system (Karten, Cox and Mpodozis, 1997; Luksch et al., 1998; Hellmann and Güntürkün, submitted). Each type was characterized by its location on the tectal map, position of somata within different layer 13 depths, soma size, specific input from tectal laminae 3-11 and projections onto separate subregions of the rotundo-triangular system (Hellmann and Güntürkün, submitted). The cell type-specific differences of dendritic arborizations in the retinorecipient tectal laminae could enable these five cell classes to collect different aspects of the visual scenery, since different classes of avian retinal ganglion cells terminate in a lamina-specific manner within the tectal layers 2–7 (Repérant and Angaut, 1977; Yamagata and Sanes, 1995; Karten, Cox and Mpodozis, 1997). Thus, the different types of tecto-rotundal relay cells probably collect and process different aspects of vision.

Since these five different cell classes innervate distinct rotundal subregions, they seem to establish the spatially segregated information processing of different visual features, which is typical for the nucleus rotundus (Jassik-Gerschenfeld and Guichard, 1972; Frost and DiFranco, 1976; Revzin, 1979; Wang and Frost, 1992; Wang, Jiang and Frost, 1993; Deng and Rogers, 1998; Laverghetta and Shimizu, 1999). Therefore, these tectorotundal cell types seem to constitute the transition from retinotopic to 'functionotopic' coding principles taking place at the tectorotundal junction.

Still, a second problem hinders an understanding of the processing principles of the tectofugal system. This problem is the discrepancy between two observations. On one side are the lamina 13 neurones with extensive dendritic ramifications and wide receptive fields, which might be suited ideally for movement analysis but poorly for fine-grained vision (Jassik-Gerschenfeld, Minios and Cond-Courtine, 1970; Frost and DiFranco, 1976; Frost and Nakayama, 1983; Wang, Jiang and Frost, 1993; Karten, Cox and Mpodozis, 1997; Luksch, Cox and Karten, 1998). On the other side are the behavioural studies in pigeons, which demonstrate that the tectofugal pathway contributes importantly to visual acuity and pattern discrimination (Hodos and Karten, 1966; Hodos, 1969; Hodos and Bonbright, 1974; Macko and Hodos, 1984; Watanabe, 1991; Güntürkün and Hahmann, 1999). A possible solution for this discrepany are coarse coding principles as suggested by Hinton (1981) and Hinton, McClelland and Rumelhart (1986). In this scheme, the resolution is determined by the number of different firing patterns in the neural population as a stimulus crosses the sensory space. This leads to the result that large receptive fields yield a high resolution as long as they overlap extensively (but not completely) and therefore show a high number of encodings (Eurich and Schwegler, 1997). If coarse coding principles are applied using the tecto-reticular system of salamanders as an example, about 800 neurones with an average receptive field of about 40° suffice to localize targets of 0.5° (Eurich, Schwegler and Woesler, 1997). If coarse coding principles also apply to the tectorotundal system, it could explain how lamina 13 neurones with their wide dendritic arbors could contribute to the high visual resolution performance of the tectofugal system. This, then, would enable neural processes by which fine visual patterns are differentiated and stored – a core aspect of visual asymmetry in pigeons.

7.10. Conclusion

Visual asymmetry in birds results from a tight interplay of genetic and epigenetic factors. Genetic factors orchestrate the LR patterning of the embryo and initiate the torsion of the embryo's head to the right. Genetic factors induce various asymmetries but visual lateralization seems only to be initiated if the epigenetic factor 'light' interacts with the embryo's asymmetrical position. If this happens, a complex chain of molecular factors, which we have just started to unravel, transforms the asymmetrical light stimulation into a morphological and physiological lateralization of the ascending visual pathways. As a result, the animal displays left–right differences in various visual learning, discrimination, memorization and cognition tasks.

Does this asymmetry provide the animal any kind of advantage or is it just *l'art pour l'art*, a worthless, but also harmless side-effect of ontogeny? To analyse the possible benefit of visual asymmetry, Güntürkün et al. (2000) tested 108 pigeons for their discrimination performance under binocular as well as under left and right monocular conditions in a grain–grit discrimination. For each bird the absolute individual degree of lateralization was expressed as the asymmetry index (AI) which was calculated as:

$$|AI| = |\text{right} - \text{left eye performance (\%)}|$$

As usual, most pigeons showed a right eye advantage. The most important result was, that the correlation between AI and discrimination level under *binocular* conditions was highly significant ($r = 0.50$, $p < 0.001$), showing that the level of lateralization was positively related to visual object discrimination performance. Thus, an individual increase in visual cerebral asymmetry was accompanied by an increase in discrimination success. Therefore, a strengthening of left hemisphere functions seems to accelerate visual discrimination and thereby increase lateralization. For our pigeons, a 10-point rise in asymmetry resulted in a 10% rise of discrimination success. Thus, asymmetry pays.

References

Alonso, Y. (1998). Lateralization of visual guided behaviour during feeding in zebra finches (*Taeniopygia guttata*). *Behavioural Processes*, 43, 257–263.

Andrew, R.J. (1991). The nature of behavioural lateralization. In *Neural and Behavioural Plasticity: The Use of the Domestic Chick as a Model*, ed. R.J. Andrew, pp. 536–554. Oxford: Oxford University Press.

Bagnoli, P., Porciatti, V., Lanfranchi, A. & Bedini, C. (1985). Developing pigeon retina: Light-evoked responses and ultrastructure of outer segments and synapses. *Journal of Comparative Neurology*, 235, 284–394.

Bagnoli, P., Porciatti, V., Fontanesi, G. & Sebastiani, L. (1987). Morphological and functional changes in the retinotectal system of the pigeon during the early posthatching period. *Journal of Comparative Neurology*, 256, 400–411.

Benowitz, L.I. & Karten, H.J. (1976). Organization of tectofugal visual pathway in pigeon: Retrograde transport study. *Journal of Comparative Neurology*, 167, 503–520.

Bischof, H.-J. & Watanabe, S. (1997). On the structure and function of the tectofugal visual pathway in laterally eyed birds. *European Journal of Morphology*, 35, 246–254.

Bowers, P.N., Brückner, M. & Yost, H.J. (1996). The genetics of left–right development and heterotaxia. *Seminars in Perinatology*, 20, 577–588.

Bruhn, S. & Cepko, C. L. (1996). Development of the pattern of photoreceptors in the chick retina. *Journal of Neuroscience*, 16, 1430–1439.

Cajal, S.R.Y. (1911). *Histologie du Systeme Nerveux de l'Homme et des Vertebres*. Paris: Maloine.

Cepko, C.L. (1993). Retinal cell fate determination. *Progress in Retinal Research*, 12, 1–12.

Clayton, N.S. (1993). Lateralization and unilateral transfer of spatial memory in marsh tits. *Journal of Comparative Physiology A*, 171, 799–806.

Clayton, N.S. & Krebs, J.R. (1994). Memory for spatial and object-specific cues in food-storing and non-storing birds. *Journal of Comparative Physiology A*, 174, 371–379.

Cunningham, D.F. (1892). *Contribution to the Surface Anatomy of the Cerebral Hemispheres*. Dublin: Royal Irish Academy.

Cunningham, D.J. (1902). Right-handedness and left-brainedness. *Journal of the Anthropological Institute of Great Britain and Ireland*, 32, 273–296.

Cunningham, D.J. (1921). A gorilla's life in civilization. *Zoological Society Bulletin*, 24, 118–124.

Deng, C. & Rogers, L.J. (1998). Organisation of the tectorotundal and SP/IPS-rotundal projections in the chick. *Journal of Comparative Neurology*, 394, 171–185.

Diekamp, B., Prior, H. & Güntürkün, O. (1999). Lateralization of serial color reversal learning in pigeons (*Columba livia*). *Animal Cognition*, 2, 187–196.

Dubbeldam, J.L. (1998). Birds. In *The Central Nervous System of Vertebrates*, ed. R. Nieuwenhuys, H. J. Ten Donkelaar, & C. Nicholson, pp. 1526–1636. Berlin: Springer.

Engelage, J. & Bischof, H.-J. (1988). Enucleation enhances ipsilateral flash evoked responses in the ectostriatum of the zebra finch (*Taeniopygia guttata castanotis* Gould). *Experimental Brain Research*, 70, 79–89.

Engelage, J. & Bischof, H.-J. (1993). The organization of the tectofugal pathway in birds: A comparative review. In *Vision, Brain, and Behavior in Birds*, ed. H.P. Zeigler & H.-J. Bischof, pp. 137–158. Cambridge, MA: MIT Press.

Eurich, C.W. & Schwegler, H. (1997). Coarse coding: Calculation of the resolution achieved by a population of large receptive field neurones. *Biological Cybernetics*, 76, 357–363.

Eurich, C.W., Schwegler, H. & Woesler, R. (1997). Coarse coding: Applications to the visual system of salamanders. *Biological Cybernetics*, 77, 41–47.

Fersen, L. & Güntürkün, O. (1990). Visual memory lateralization in pigeons. *Neuropsychologia*, 28, 1–7.

Frost, B.J. & DiFranco, D.E. (1976). Motion characteristics of single units in the pigeon optic tectum. *Vision Research*, 16, 1229–1234.

Frost, B.J. & Nakayama, K. (1983). Single visual neurones code opposing motion independent of direction. *Science*, 220, 744–745.

Gamlin, P.D.R., Reiner, A., Erichsen, J.T., Karten, H.J. & Cohen, D.H. (1984). The neural substrate for the pupillary light reflex in the pigeon (*Columba livia*). *Journal of Comparative Neurology*, 226, 523–543.

Gaston, K.E. (1984). Interocular transfer of pattern discrimination learning in chicks. *Brain Research*, 310, 213–221.

Güntürkün, O. (1985). Lateralization of visually controlled behavior in pigeons. *Physiology and Behavior*, 34, 575–577.

Güntürkün, O. (1993). The ontogeny of visual lateralization in pigeons. *German Journal of Psychology*, 17, 276–287.

Güntürkün, O. (1997a). Avian visual lateralization – a review. *NeuroReport*, 6, iii–xi.

Güntürkün, O. (1997b). Visual lateralization in birds: From neurotrophins to cognition? *European Journal of Morphology*, 35, 290–302.

Güntürkün, O. (1997c). Morphological asymmetries of the tectum opticum in the pigeon. *Experimental Brain Research*, 116, 561–566.

Güntürkün, O. (2000). Sensory physiology: vision. In *Sturkie's Avian Physiology*, ed. G.C. Whittow, pp. 1–19. Orlando: Academic Press.

Güntürkün, O. & Böhringer, P.G. (1987). Reversal of visual lateralization after midbrain commissurotomy in pigeons. *Brain Research*, 408, 1–5.

Güntürkün, O., Diekamp, B., Manns, M., Nottelmann, F., Prior, H., Schwarz, A. & Skiba, M. (2000). Asymmetry pays: Visual lateralization improves discrimination success in pigeons. *Current Biology*, 10, 1079–1081.

Güntürkün, O. & Hahmann, U. (1994). Visual acuity and hemispheric asymmetries in pigeons. *Behavioural Brain Research*, 60, 171–175.

Güntürkün, O. & Hahmann, U. (1999). Functional subdivisions of the ascending visual pathways in the pigeon. *Behavioural Brain Research*, 98, 193–201.

Güntürkün, O., Hellmann, B., Melsbach, G. & Prior, H. (1998). Asymmetries of representation in the visual system of pigeons. *NeuroReport*, 9, 4127–4130.

Güntürkün, O. & Hoferichter, H.H. (1985). Neglect after section of a left telencephalotectal tract in the pigeon. *Behavioural Brain Research*, 18, 1–9.

Güntürkün, O. & Kesch, S. (1987). Visual lateralization during feeding in pigeons. *Behavioral Neuroscience*, 101, 433–435.

Güntürkün, O. & Kischkel, K.-F. (1992). Is visual lateralization sex-dependent in pigeons? *Behavioural Brain Research*, 47, 83–87.

Hardy, O., Leresche, N. & Jassik-Gerschenfeld, D. (1984). Postsynaptic potentials in neurones of the pigeon's optic tectum in response to afferent stimulation from the retina and other visual structures. *Brain Research*, 311, 65–74.

Heaton, M.B. & Harth, M. (1974). Developing visual function in the pigeon embryo with comparative reference to other avian species. *Journal of Comparative and Physiological Psychology*, 86, 151–156.

Hellige, J.B. (1995). Hemispheric asymmetry for components of visual information processing. In *Brain Asymmetry*, ed. R. J. Davidson & K. Hugdahl. Cambridge, MA: MIT Press.

Hellmann, B. & Güntürkün, O. (1999). Visual field specific heterogeneity within the tectofugal projection of the pigeon. *European Journal of Neuroscience*, 11, 1–18.

Hellmann, B. & Güntürkün, O. The structural organization of parallel information processing within the tectofugal visual system of the pigeon. (submitted).

Hepper, P.G., Shahidullah, S. & White, R. (1991). Handedness in the human fetus. *Neuropsychologia*, 29, 1107–1111.

Herrmann, K. & Bischof, H.J. (1986a). Effects of monocular deprivation in the nucleus rotundus of zebra finches: A Nissl and deoxyglucose study. *Experimental Brain Research*, 64, 119–126.

Herrmann, K. & Bischof, H.J. (1986b). Monocular deprivation affects neurone size in the ectostriatum of the zebra finch brain. *Brain Research*, 37, 143–146.

Herrmann, K. & Bischof, H.J. (1993). Development of the tectofugal visual system of normal and deprived zebra finches. In *Vision, Brain and Behavior in Birds*, ed. H.P. Zeigler & H.J. Bischof, pp. 207–226. Cambridge, MA: MIT Press.

Hinton, G.E. (1981). Shape representation in parallel systems. In *Proceedings of the 7th International Joint Conference on Artificial Intelligence*, Vancouver, pp. 211–245.

Hinton, G.E., McClelland, J.L. & Rumelhart, D.E. (1986). Distributed representations. In *Parallel Distributed Processing*, ed. D.E. Rumelhart & J.L. McClelland, Vol. 1, pp. 77–109. Cambridge, MA: MIT Press.

Hodos, W. (1969). Color-discrimination deficits after lesions of the nucleus rotundus in pigeons. *Brain Behavior Evolution*, 2, 185–200.

Hodos, W. & Bonbright, J.C. (1974). Intensity difference thresholds in pigeons after lesions of the tectofugal and thalamofugal visual pathways. *Journal of Comparative and Physiological Psychology*, 87, 1013–1031.

Hodos, W. & Karten, H.J. (1966). Brightness and pattern discrimination deficits after lesions of nucleus rotundus in the pigeon. *Experimental Brain Research*, 2, 151–167.

Horn, G. (1991). Imprinting and recognition memory; a review of neural mechanisms. In *Neural and Behavioral Plasticity: The Domestic Chick as a Model*, ed. R.J. Andrew, pp. 219–261. Oxford: Oxford University Press.

Hyatt, B.A., Lohr, J. L. & Yost, H. J. (1996). Initiation of vertebrate left–right axis formation by maternal Vg1. *Nature*, 384, 62–65.

Isaac, A., Sargent, M.G. & Cooke, J. (1997). Control of vertebrate left–right asymmetry by a snail-related zinc finger gene. *Science*, 275, 1301–1304.

Jassik-Gerschenfeld, D. & Guichard, J. (1972). Visual receptive fields of single cells in the pigeon's optic tectum. *Brain Research*, 40, 303–317.

Jassik-Gerschenfeld, D. & Hardy, O. (1984). The avian optic tectum: neurophysiology and behavioural correlations. In *Comparative Neurology of the Optic Tectum*, ed. H. Vanegas, pp. 649–686. New York: Plenum Press.

Jassik-Gerschenfeld, D., Minios, F. & Cond-Courtine, F. (1970). Receptive field properties of directionally selective units in the pigeon's optic tectum. *Brain Research*, 24, 407–421.

Karten, H.J., Cox, K. & Mpodozis, J. (1997). Two distinct populations of tectal neurones have unique connections within the retinotectorotundal pathway of the pigeon (*Columba livia*). *Journal of Comparative Neurology*, 387, 449–465.

Keysers, C., Diekamp, B. & Güntürkün, O. (2000). Evidence for asymmetries in the phasic intertectal interactions in the pigeon (*Columba livia*) and their potential role in brain lateralization. *Brain Research*, 852, 406–413.

Kuo, Z.Y. (1932). Ontogeny of embryonic behavior in aves. III. The structural and environmental factors in embryonic behavior. *Journal of Comparative Psychology*, 13, 245–271.

LaVail, J.H. & Cowan, W.M. (1971). The development of the chick optic tectum. I. Normal morphology and cytoarchitectonic development. *Brain Research*, 28, 391–419.

Laverghetta, A.V. & Shimizu, T. (1999). Visual discrimination in the pigeon (*Columba livia*): Effects of selective lesions of the nucleus rotundus. *NeuroReport*, 10, 981–985.

LeGros Clark, W.I. (1927). Description of cerebral hemispheres of the brain of the gorilla. *Journal of Anatomy*, 61, 467–475.

Levin, M., Johnson, R.L., Stern, C.D., Kuehn, M. & Tabin, C. (1995). A molecular pathway determining left–right asymmetry in chick embryogenesis. *Cell*, 82, 803–814.

Levin, M., Pagan, S., Roberts, D.J., Cooke, J., Kuehn, M.R. & Tabin, C.J. (1997). Left/right patterning signals and the independent regulation of different aspects of situs in the chick embryo. *Developmental Biology*, 189, 57–67.

Lohr, J.L., Danos, M.C. & Yost, H.J. (1997). Left–right asymmetry of a nodal-related gene is regulated by dorsoanterior midline structures during Xenopus development. *Development*, 124, 1465–1472.

Lorincz, E. & Fabre-Thorpe, M. (1996). Shift of laterality and compared analysis of paw performances in cats during practice of a visuomotor task. *Journal of Comparative Psychology*, 110, 307–315.

Luksch, H., Cox, K. & Karten, H.J. (1998). Bottlebrush dendritic endings and large dendritic fields: Motion-detecting neurones in the tectofugal pathway. *Journal of Comparative Neurology*, 396, 399–414.

Macko, K.A. & Hodos, W. (1984). Near-field acuity after visual system lesions in pigeons I. Thalamus. *Behavioural Brain Research*, 13, 1–14.

Maguire, E.A., Frackowiak, R.S.J. & Frith, C.D. (1997). Recalling routes around London: Activation of the right hippocampus in taxi drivers. *Journal of Neuroscience*, 17, 103–110.

Manns, M. & Güntürkün, O. (1997). Development of the retinotectal system in the pigeon: A choleratoxin study. *Anatomy and Embryology*, 195, 539–555.

Manns, M. & Güntürkün, O. (1999a). Monocular deprivation alters the direction of functional and morphological asymmetries in the pigeon's visual system. *Behavioral Neuroscience*, 113, 1–10.

Manns, M. & Güntürkün, O. (1999b). 'Natural' and artificial monocular deprivation effects on thalamic soma sizes in pigeons. *NeuroReport*, 10, 3223–3228.

Martinoya, C., Le Houezec, J. & Bloch, S. (1988). Depth resolution in the pigeon. *Journal of Comparative Physiology*, 163, 33–42.

McLoon, S.C. (1985). Evidence for shifting connections during development of the chick retinotectal projection. *Journal of Neuroscience*, 5, 2570–2580.

Melsbach, G., Spieß, M., Wohlschläger, A. & Güntürkün, O. (1996). Morphological asymmetries of motoneurons innervating upper extremities – clues to the anatomical foundations of handedness? *International Journal of Neuroscience*, 86, 217–224.

Mench, J.A. & Andrew, R.J. (1986). Lateralization of a food search task in the domestic chick. *Behavioral and Neural Biology*, 46, 107–114.

Meno, C., Ito, Y., Saijoh, Y., Matsuda, Y., Tashiro, K., Kuhara, S. & Hamada, H. (1997). Two closely-related left–right asymmetrically expressed genes, lefty-1 and lefty-2: Their distinct expression domains, chromosomal linkage and direct neuralizing activity in Xenopus embryos. *Genes Cells*, 2, 513–524.

Mey, J. & Thanos, S. (1992). Development of the visual system of the chick. A review. *Journal für Hirnforschung*, 33, 673–702.

Michel, G.F. (1981). Right-handedness. A consequence of infant supine head-orientation preference? *Science*, 212, 685–687.

Michel, G.F. & Harkins, D.A. (1986). Postural and lateral asymmetries in the ontogeny of handedness during infancy. *Developmental Psychobiology*, 19, 247–258.

Movshon, J.A. & Van Sluyters, R.C. (1981). Visual neural development. *Annual Review of Psychology*, 32, 477–522.

Pagan-Westphal, S. M. & Tabin, C. J. (1998). The transfer of left–right positional information during chick embryogenesis. *Cell*, 93, 25–35.

Parsons, C.H. & Rogers, L.J. (1993). Role of the tectal and posterior commissures in lateralization of the avian brain. *Behavioural Brain Research*, 54, 153–164.

Prior, H., Ulrich, C., Lipp, H. P. & Güntürkün, O. (1998). Zur Lateralization der visuell-räumlichen Orientierung bei Tauben. In *Experimentelle Psychologie*, eds. H. Lachnit, A. Jacobs, & F. Rösler, pp. 265–266. Lengerich: Pabst Science Publishers.

Rager, G. (1976). Morphogenesis and physiogenesis of the retino-tectal connection in the chicken. II. The retinotectal synapses. *Proceedings of the National Academy of Sciences*, 192, 353–370.

Rager, U., Rager, G. & Frei, B. (1993). Central retinal area is not the site where ganglion cells are generated first. *Journal of Comparative Neurology*, 334, 529–544.

Rager, G. & von Oeynhausen, B. (1979). Ingrowth and ramification of retinal fibers in the developing optic tectum of the chick embryo. *Experimental Brain Research*, 35, 213–227.

Ramsdell, A.F. & Yost, H.J. (1998). Molecular mechanisms of vertebrate left–right development. *Trends in Genetics*, 14, 459–465.

Rashid, N. & Andrew, R.J. (1989). Right hemisphere advantage for topographic orientation in the domestic chick. *Neuropsychology*, 27, 937–948.

Rauschecker, J.P. (1991). Mechanisms of visual plasticity: Hebb synapses, NMDA receptors, and beyond. *Physiological Reviews*, 71, 587–613.

Remy, M. & Emmerton, J. (1991). Directional dependence of intraocular transfer of stimulus detection in pigeons (*Columba livia*). *Behavioral Neuroscience*, 105, 647–652.

Remy, M. & Güntürkün, O. (1991). Retinal afferents of the tectum opticum and the nucleus opticus principalis thalami in the pigeon. *Journal of Comparative Neurology*, 1991, 305, 57–70.

Repérant, J. & Angaut, P. (1977). The retinotectal projections in the pigeon. An experimental optical and electron microscope study. *Neuroscience*, 2, 119–140.

Revzin, A.M. (1979). Functional localization in the nucleus rotundus. In *Neural Mechanisms of Behavior in the Pigeon*, ed. A.M. Granda & H.J. Maxwell, pp. 165–175. New York: Plenum Press.

Robert, F. & Cuénod, M. (1969). Electrophysiology of the intertectal commissures in the pigeon II. Inhibitory interaction. *Experimental Brain Research*, 9, 123–136.

Robertson, E. J. (1997). Left–right asymmetry. *Science*, 275, 1280.

Rogers, L.J. (1982). Light experience and asymmetry of brain function in chickens. *Nature*, 297, 223–225.

Rogers, L.J. (1990). Light input and the reversal of functional lateralization in the chicken brain. *Behavioural Brain Research*, 38, 211–221.

Rogers, L.J. (1995). *The Development of Brain and Behavior in the Chicken*. Wallingford: CAB International.

Rogers, L. (1996). Behavioral, structural and neurochemical asymmetries in the avian brain: A model system for studying visual development and processing. *Neuroscience and Biobehavioral Reviews*, 20, 487–503.

Rogers, L.J. & Bolden, S.W. (1991). Light-dependent development and asymmetry of visual projections. *Neuroscience Letters*, 121, 63–67.

Rose, S.P.R. (1991). How chicks make memories: The cellular cascade from c-fos to dendritic remodelling. *Trends in Neurosciences*, 14, 390–397.

Saha, M.S. & Grainger, R.M. (1993). Early opsin expression in *Xenopus* embryos precedes photoreceptor differentiation. *Molecular Brain Research*, 17, 307–318.

Sandi, C., Patterson, T.A. & Rose, S.P. (1993). Visual input and lateralization of brain function in learning in the chick. *Neuroscience*, 52, 393–401.

Senut, M.C. & Alvarado-Mallart, R.M. (1986). Development of the retinotectal system in normal quail embryos: Cytoarchitectonic development and optic fiber innervation. *Developmental Brain Research*, 29, 123–140.

Sherman, S.M. & Spear, P.D. (1982). Organization of visual pathways in normal and visually deprived cats. *Physiological Reviews*, 62, 738–855.

Shimizu, T. & Karten, H. J. (1993). The avian visual system and the evolution of the neocortex. In *Vision, Brain and Behavior in Birds*, ed. H. P. Zeigler & H.-J. Bischof, pp. 103–114. Cambridge, MA: MIT Press.

Skiba, M., Diekamp, B. & Güntürkün, O. Embryonic light stimulation asymmetry induces divergent lateralized visual effects in pigeons. (submitted).

Skiba, M., Diekamp, B., Prior, H. & Güntürkün, O. (2000). Lateralized interhemispheric transfer of color cues: Evidence for dynamic coding principles of visual lateralization in pigeons. *Brain and Language*, 73, 254–273.

Stiemke, M.M. & Hollyfield, J.G. (1995). Cell birthdays in *Xenopus laevis* retina. *Differentiation*, 58, 189–193.

Sun, H. & Frost, B.J. (1998). Computation of different optical variables of looming objects in pigeon nucleus rotundus. *Nature Neuroscience*, 1, 296–303.

Ulrich, C., Prior, H., Duka, T., Leshchins'ka, I., Valenti, P., Güntürkün, O. & Lipp, H.-P. (1999). Left-hemispheric superiority for visuospatial orientation in homing pigeons. *Behavioural Brain Research*, 104, 169–178.

Vallortigara, G., Regolin, L., Bortolomiol, G. & Tommasi, L. (1996). Lateral asymmetries due to preferences in eye use during visual discrimination learning in chicks. *Behavioural Brain Research*, 74, 135–143.

Vallortigara, G., Rogers, L.J. & Bisazza, A. (1999). Possible evolutionary origins of cognitive brain lateralization, *Brain Research Reviews*, 30, 164–175.

Vauclair, J. & Fagot, J. (1993). Manual and hemispheric specialization in the manipulation of a joystick by baboons (*Papio papio*). *Behavioral Neuroscience*, 107, 210–214.

Wang, Y. & Frost, B.J. (1992). Time to collision is signalled by neurones in the nucleus rotundus of pigeons. *Nature*, 356, 236–238.

Wang, Y., Jiang, S. & Frost, B.J. (1993). Visual processing in pigeon nucleus rotundus: Luminance, color, motion, and looming subdivisions. *Visual Neuroscience*, 10, 21–30.

Watanabe, S. (1991). Effects of ectostriatal lesions on natural concept, pseudoconcept, and artificial pattern discrimination in pigeons. *Visual Neuroscience*, 6, 497–506.

Weidner, C., Repérant, J., Miceli, D., Haby, M. & Rio, J. P. (1985). An anatomical study of ipsilateral retinal projections in the quail using radioautographic, horseradish peroxidase, fluorescence and degeneration techniques. *Brain Research*, 340, 99–108.

Yamagata, M. & Sanes, J.R. (1995). Target-independent diversification and target-specific projection of chemically defined retinal ganglion cell subsets. *Development*, 121, 3763–3776.

Yost, H. J. (1995). Vertebrate left–right development. *Cell*, 82, 689–692.

Zappia, J.V. & Rogers, L.J. (1983). Light experience during developmental affects asymmetry of forebrain function in chickens. *Developmental Brain Research*, 11, 93–106.

8

Development of Laterality and the Role of the Corpus Callosum in Rodents and Humans

PATRICIA E. COWELL AND
VICTOR H. DENENBERG

8.1. The Origins of Brain Lateralization in Animals

From the perspective of a new century, it seems difficult to comprehend that only 30 years ago no one believed that animals had lateralized brains and asymmetrical behaviours. Indeed, the conventional dogma was that humans were uniquely different from all other animals in having handedness, brain laterality and speech. The first two functions have now been well established in animals and the third is under serious challenge (Savage-Rumbaugh et al., 1993).

Before addressing these issues, it is necessary to make a distinction between brain lateralization for *individual* animals and brain lateralization at the *population* level (Denenberg, 1981). If the population is lateralized (e.g. handedness in humans), then it is obvious that the individuals within that population are lateralized. However, individual animals can be lateralized even when there is no population asymmetry. Thus, Collins (1977) found that 50% of mice used their right paws in reaching for food and 50% used their left. Each animal was given 50 trials and most were found to be strongly biased one way or the other, thereby showing strong asymmetry at the individual level, but no asymmetry at the population level.

The first definitive evidence for population behavioural asymmetry in nonhumans was Nottebohm's paper in 1970 reporting that song production in the chaffinch was under the control of the left half of the syrinx (the vocal organ of birds). Later work found this to be true for the canary, the white-crowned sparrow and the white-throated sparrow as well (Lemon, 1973; Nottebohm and Nottebohm, 1976). Further work, lesioning the right or left brain centres that control song production, established that the left hemisphere was dominant (Nottebohm, 1977, 1979; Nottebohm, Stokes and Leonard, 1976).

The first report of behavioural asymmetry in a non-human mammal was by Denenberg et al. (1978), who showed that differing early experiences (handling between birth and weaning, and enriched environment after weaning) induced brain and behavioural asymmetry in rats. This paper was followed a few months later by a report showing that Japanese macaque monkeys were more effective in discriminating 'coo' sounds (obtained from field recordings of Japanese macaque vocalizations) when heard by the right ear than when heard by the left ear (Beecher et al., 1979; Petersen et al., 1978). Since the right ear sends its information primarily to the left hemisphere, this work provided evidence that the left hemisphere was preferentially involved in processing conspecific vocalizations that carried communicative information.

The following year, Rogers and Anson (1979) showed that 2-day-old chicks could learn a visual discrimination (peck for starter crumbs scattered randomly on a background of small pebbles stuck to a plexiglass floor) only if they had a fully functioning left forebrain. They injected cycloheximide, an antibiotic that inhibits ribosomal protein synthesis, into one or the other hemispheres. Injection of the left hemisphere, but not the right, impaired learning.

Shortly thereafter, Rogers (1980) used the cycloheximide procedure with emotional stimuli and found that chicks with a functional right hemisphere and an immobilized left brain showed much greater attack and copulatory behaviour, compared to chicks with cyclohexidime injected into the right or both hemispheres. The right hemisphere is also preferentially sensitive to fear-producing stimuli on the first day of life, though the development of fear by each hemisphere over the course of the first 2 weeks of life follows a complex pattern (Andrew and Brennan, 1983).

During this same time period, studies with rats receiving handling stimulation in infancy found that animals with an intact right hemisphere and a lesioned left hemisphere, when compared to those with an intact left hemisphere and a lesioned right: (1) had a higher muricide incidence (Garbanati et al., 1983); (2) had better long-term memory of a severe taste aversion experience (association of sweetened milk solution with later visceral upset) (Denenberg et al., 1980); and (3) had a stronger spatial bias (Sherman et al., 1980).

Finally, in a follow-up of work on 'handedness' in parrots, originally reported in 1938 by Friedman and Davis, Rogers (1980) observed parrots from nine different species with respect to which foot they used to manipulate food and objects. She found that eight of the species had a strong left-foot preference (ranging from 87% to 100%), and one species had a 77% incidence of using its right foot.

In summary, in roughly a 10-year span, research went from virtually no knowledge of population laterality in animals to multiple findings showing that the left hemisphere in a variety of species was involved in learning and social communication, and that the right hemisphere was involved in spatial processing and a wide range of affective behaviours. The parallels to human brain laterality are obvious. These findings establish that brain laterality has an evolutionary history and is not a phenomenon unique to humans. It is for this reason that Denenberg, in his review, concluded that

> functional lateralization, when present, will be similar across species: the left hemisphere will tend to be involved in communicative functions while the right hemisphere will respond to spatial and affective information; both hemispheres will often interact via activation–inhibition mechanisms when affective or emotional responses are involved (Denenberg, 1981, p. 1).

8.1.1. Laterality Mechanisms in Development

The genetic mechanisms involved in producing a lateralized brain remain poorly understood. However, it is known that in at least two species, the rat and the chicken, the genetic substrates must be acted upon by an appropriate set of environmental stimuli, occurring at a restricted time in development, for the organism to have a lateralized brain.

This phenomenon is most clearly seen in Rogers' studies of the chicken brain. The chick embryo is situated in the egg so that the right eye receives light input through the shell, while the left eye does not (Rogers and Anson, 1979). Rogers (1990) reversed the light distribution during the latter phase of development in the egg by occluding the right eye and arranging for the left to receive light stimulation. This caused a reversal of brain laterality, with the left hemisphere controlling copulation and the right controlling visual discrimination learning (see Chapter 6 by Deng and Rogers). A final experiment found that behavioural asymmetry was not exhibited on either measure when chick embryos had received no light stimulation.

Studies from Denenberg's laboratory compared adult rats that, as infants, had been reared as undisturbed controls or had received daily handling experience between birth and weaning. In some of these studies the handled and non-handled groups were split at weaning, with half being pair-housed in standard laboratory cages and the remainder being housed in enriched environments (Denenberg, 1977). Handling consisted of removing pups from the home cage and placing them individually into gallon-sized tin cans for 3 min daily. Environmental enrichment consisted of socially housing animals in groups of up to 12 in large cages containing 'playthings' (Krech, Rosenzweig and Bennett, 1962; Rosenzweig, Bennett and Diamond, 1972). Complex interactive effects upon laterality were observed when the two manipulations were combined, and these are described in detail below. However, if one looks for a unifying result across these studies, which examined a variety of behavioural measures, what is clear is that handling in infancy induced brain laterality in all of them (Denenberg et al., 1978, 1980; Sherman et al., 1980; Garbanati et al., 1983).

Early stimulation can only induce asymmetries in brain and behaviour if the potential for lateralization is already present. Thus, the studies described above are a prime example of the interactional effects of genetics and early experiences in a number of non-human species. The findings from this body of work may also be viewed as a model for the study of individual differences, a topic to which we now turn.

8.1.2. Early Experience, Brain Laterality and Individual Differences

A series of experiments were designed that tested and confirmed the hypothesis that early experiences can generate stable individual differences (Denenberg, 1970; Whimbey and Denenberg, 1967). The purpose of these studies was to document that stable individual differences, of the kind commonly thought of as being generated by genetic variability, can be as effectively generated by differential early experiences. The application to human work suggests that, in a population of genetically heterogeneous individuals growing up and developing in complex multidimensional environments, the emerging neurocognitive individual differences would be the interactive result of both long-term (genetic) and immediate (environmental) sources of variance.

The ideas behind this research can be simply illustrated by considering the data presented in Table 8.1. In two independent experiments, rat pups were handled or left undisturbed between birth and weaning (21 days), and were then reared in laboratory cages or enriched environments from 21 to 50 days of age, in a 2 × 2 factorial design. In both experiments, four males were used per litter: one was given sham surgery, another was given no surgery, the third was given a right hemisphere lesion, and the last was given a left hemisphere lesion. Hemispheric lesions were performed using pipette suction to remove all of the neocortex extending from the frontal pole to the sagittal sinus medially, the rhinal fissure laterally and the caudal border of the hemisphere posteriorly. After a period of recovery, all animals were tested behaviourally. In one experiment (Denenberg et al., 1978), rats were tested in the open field and their activity was recorded. In the second experiment

Table 8.1. *Effects of four early experience conditions on data obtained from the animals that received a right hemisphere lesion. The results in the second column represent mean activity scores as a function of number of squares entered in the open field (Denenberg et al., 1978). The results in the third column represent the percentage of mouse kills over a period of 5 days (Garbanati et al., 1983)*

Early experience	Open field	Mouse killing
Non-handled Laboratory cage	22.33	68.8
Non-handled Enriched environment	32.89	94.7
Handled Laboratory cage	36.27	67.6
Handled Enriched environment	3.00	57.1

(Garbanati et al., 1983), a mouse was placed into each cage with a rat and the incidence of muricide was measured. Table 8.1 lists the four early experience conditions and the data obtained for the animals that received a right hemisphere lesion.

Consider the open-field activity data column. The rats given both handling and enrichment have a score of 3 squares entered, while the group which only received handling had a score of 36 squares entered and the group that only had enrichment had a score of 33. Thus, the combined experience of handling plus enrichment interacted to markedly and significantly depress activity.

Turning now to the muricide data, the non-handled control group reared in enrichment after weaning had a remarkably high incidence of 95% muricide, but the group that had handling and enrichment only had a 57% occurrence, again a significant difference. Here, too, one finds that the two early experience variables interact with each other, but in a manner different from that found for the activity measure.

The kinds of interaction effects just described are often found in animal studies, and their interpretations remain a challenge. Our knowledge of the antecedent conditions for each group is critical for any interpretation offered. But suppose the researcher had no knowledge of the early experiences of the subjects? Suppose she was given a group of adult rats and told to study the effects of brain lesions upon behaviour, but was told nothing about their prior experiential history? In such a situation the researcher would only be able to look at the 'main effects' of the surgery variable. The 'within group' variance would be calculated from the data of the handled, non-handled, enriched and cage-reared animals pooled. Thus, the systematic variance attributed to the early experience variables in Table 8.1 would be incorporated with the error variance, yielding an inflated error term and reducing the likelihood of even finding a significant main effect of lesion.

Although this appears to be a bizarre and contrived example, we submit that it is prototypical of the kinds of research problems that confront researchers who choose to study human subjects, especially adults. Because of these difficulties, the researcher working with humans has developed procedures to compensate for lack of direct control on subjects' development, including gathering detailed demographic information and case histories, giving test batteries to measure key behaviours, and interviewing relatives, teachers, doctors, etc. The researcher then selects a variable to investigate what is psychometrically meaningful and is sufficiently robust that significant effects are likely to be found.

The analytical procedures available to the researcher include at least correlation–regression analyses, and analysis of variance procedures, and may include more advanced techniques as well. The experienced researcher knows that most analytical techniques are excellent at extracting linear relationships. They may also be effective in the study of certain forms of non-linearity (e.g. circadian rhythms, polynomial regression). However, they are very weak at solving problems of the kind presented in Table 8.1, which involves interactions of two (or more) prior experiential variables. Yet these kinds of interactions are likely to be quite common in human development.

8.1.3. Multifactorial Interactions and Sex Differences

This section reviews additional laterality research in non-human species and the interactive effects of early environment with other factors, notably sex. The aim is to build upon the evidence reviewed above, to further develop the view that behavioural and related neurobiological aspects of laterality are emergent properties of a plastic and environmentally sensitive system. As the review of the animal literature is continued, particular emphasis is placed on insights gained through the study of male–female differences, as a means to better understand the nature of laterality in humans. Together with the ideas presented in the sections above, this approach is offered to account for what has been viewed as a 'failure to replicate' certain effects, notably sex differences, in neurocognitive studies of laterality in humans. Supporting evidence from behaviour, neurobiology and brain–behaviour relationships is presented below.

An extensive literature on the factors that affect the development, organization and expression of lateralized neurobehavioural features exists for numerous non-human species, notably rodents, but also avian and primate species. Behavioural asymmetry in adult rats is affected by prenatal and early postnatal manipulations of both internal biochemistry (Zimmerberg and Riley, 1988; Rodriguez, Santana and Alonso, 1994) and external environment (Alonso, Castellano and Rodriguez, 1991; Cowell, Waters and Denenberg, 1997). For example, early postnatal handling was found to induce visuospatial lateralization as measured by monocular testing during water escape (Cowell, Waters and Denenberg, 1997). Male rats left undisturbed during the first 21 days of life performed equally well with either the right or left eye covered while learning to swim to a submerged escape platform. However, male rats handled in infancy showed differential performance depending on the eye used during testing. Handled animals who swam the maze with the left eye open (i.e. the right eye patched) took less time and distance to reach the escape platform than animals who swam the maze with the right eye open (i.e. the left eye patched). In the rodent visual system, the majority of retinal cells project to the contralateral hemisphere (Mittelman, Wishaw and Robbins, 1988; Nakahara and Ikeda, 1984; Okada et al., 1991) such that visual projections are predominantly crossed (Hayhow, Sefton and Webb, 1962; Lund, 1965). Therefore, the results of the eye-patch study suggest that handling induced a right-hemisphere lateralization for spatial navigation.

Some behavioural asymmetries have been shown to have a genetic basis (Bulman-Fleming, Bryden and Rogers, 1997; Collins, 1977), including both direction and degree of paw preference in mice (Collins, 1985; Waters and Denenberg, 1994). Yet, as described above in Section 8.1.1, these effects may be modulated by variation in early environmental conditions such as the immune status of the mother's uterine environment (Denenberg et al., 1991b). This work is consistent with the view that neurobehavioural laterality is relatively plastic in nature and has the potential to emerge along a variety of paths depending on the developmental history of the organism.

A similar case for the plastic and multipotential aspects of laterality can be made by looking at neurobiological measures. Neuroanatomical effects have been demonstrated to be quite sensitive to individual differences. Riddle and Purves (1995) showed that, although there was no evidence for a population level asymmetry, a high degree of individual variation was present in the lateralization of primary sensory cortex of the male rat (Riddle and Purves, 1995). Approximately half of the subjects studied had sensory areas that were larger in the left hemisphere and half had larger right sensory areas. Additionally, lateralization of somatic subfields differed both across and within individual subjects. For example, a given animal might show a rightward asymmetry in the whisker pad region and a leftward asymmetry in the forepaw region, or vice versa. Other work has linked the sensitivity of anatomical asymmetries to genetic, behavioural and environmental factors. Female mice bred to demonstrate strongly lateralized paw preference were shown to have greater asymmetry in cortical thickness and hippocampal volume measures (Lipp, Collins and Nauta, 1984). In male rats, preweaning handling combined with postweaning enrichment was observed to increase a 'left greater than right' pattern of cortical thickness asymmetry in occipital zones and decrease this asymmetry pattern in frontal regions (Isasa Chueca, Lahoz Gimeno and Redondo Marco, 1992). Although it is not clear which differences in the study by Isasa Chueca et al. (1992) were due to the handling and enrichment components, clearly, regional cortical laterality is affected by the interaction of the two factors, which is consistent with similar findings at the behavioural level (Denenberg et al., 1978; Garbanati et al., 1983).

The development and plasticity of asymmetries in cortical thickness in the rat has been shown to be highly sex dependent. Male–female differences (Diamond, Dowling and Johnson, 1981; Diamond et al., 1983; Diamond, 1991), as well as interactions between prenatal stress and sex hormones (Diamond, Dowling and Johnson, 1981; Fleming et al., 1986; Stewart and Kolb, 1988), have been demonstrated. Notably, the 'right greater than left' pattern observed in males can be disturbed by stress in early prenatal development (Fleming et al., 1986; Stewart and Kolb, 1988) and appears to diminish with advanced age in adulthood (Diamond et al., 1983; Diamond, 1991). Furthermore, the 'left greater than right' pattern observed in adult females is present only after 7 days of age, can be reversed with neonatal ovariectomy and is susceptible to reversal patterns in old age (Diamond, Dowling and Johnson, 1981; Diamond, 1991).

Additional compelling evidence for the plasticity of structural and functional asymmetries comes from the study of brain–behaviour relationships. As summarized above, early postnatal handling has been shown to affect basic motor asymmetries and spatial preference, as well as more complex learned and emotionally motivated behaviours, measured after unilateral cortical lesions (Denenberg et al., 1978, 1980; Sherman et al., 1980; Garbanati et al., 1983; Maier and Crowne, 1993). Not only manipulations during early life, but also relatively short-term behavioural experiences in adulthood, can affect the expression of lateralized brain–behaviour systems. Reaching experience in adult rats has been shown to alter asymmetries in motor cortex cells (Diaz, Pinto-Hamuy and Fernandez, 1994), and lateralized hypothalamic reward affected paw preference in adults (Hernandez-Mesa and Bures, 1985). Moreover, exposure to substances commonly prescribed to humans, such as antidepressant drugs, has been shown to influence lateralized motor behaviour and related prefrontal dopamine function in adult rats (Carlson et al., 1996).

The lateralized expression of some neurobehavioural systems also varies as a function of sex and the interaction of sex hormones with environmental and genetic factors. In a variety of rodent species, dopaminergically modulated motor behaviours (Camp et al., 1984) and hypothalamically driven mating vocalizations (Holman and Hutchison, 1991) differed as a function of sex and neonatal sex hormone manipulations. In the cat, Tan and Kutlu (1993) showed that the relationship between Sylvian fissure asymmetry and paw preference was sex dependent. In male cats, positive asymmetry coefficients (indicative of higher Sylvian fissure morphology in the right hemisphere) were more common in right- than left-pawed animals. The opposite pattern was seen in female cats, where positive asymmetry coefficients were more common in left- than right-pawed animals. Such patterns of individual differences provide support for the idea that lateralized brain–behaviour relationships seen in adults are the result of complex developmental interactions between hormones, and other, as yet unknown, factors responsible for determining feline paw preference.

Evidence in animals suggests overwhelmingly that neurobehavioural lateralization remains responsive to change across the life span. Variation has been demonstrated in response to early environment, learning, pharmacological intervention and interactions of these factors with genetics. Furthermore, many of these factors have been shown to differ in males and females, providing evidence of hormonal effects on the development and organization of laterality. Despite a high degree of sensitivity to a wide range of factors, a comprehensive literature with general agreement on the developmentally plastic nature of laterality has emerged. This consensus is due in part to the ability of researchers to place tight experimental controls on their research subjects' development and manipulations, thereof. In contrast, the study of developmental effects on laterality in humans is usually conducted indirectly through the comparison of subject groups with different developmental histories (e.g. developmental language disorders), behavioural profiles (e.g. handedness) or clinical symptoms (e.g. psychiatric disorders). In humans, it is rarely possible to control or manipulate environmental and genetic backgrounds as one would do routinely in animal experiments. Therefore, variation in highly sensitive measurements of laterality is, in many cases, probably obscured by the influence of numerous interacting variables.

It is not surprising that scientific opinion about laterality and its development in humans differs in many respects from views held about similar characteristics in non-human species. For example, the notion that lateralized neurobehavioural systems are relatively invariant has been put forward by some (for a review, see Hiscock, 1998). While this perspective may not be commonly held as an explicit opinion *per se*, it does manifest in a subtle and pervasive form throughout the literature. The notion that lateralized systems are relatively developmentally invariant (after a prescribed prenatal growth period) and non-plastic (with respect to the numerous factors that differentiate one person's life history from another's) is a tacitly held assumption that underlies the procedures used by many researchers when investigating unifactorial group differences in laterality such as male–female differences.

If the findings of the animal studies on laterality are generalized across species, they suggest that researchers studying parallel phenomena in human subjects must control for numerous demographic and environmental factors if they hope to isolate sex differences reliably, or hormonally mediated effects. To look for a main effect of a variable such as sex in a lateralized neurobehavioural measure without attempting to control for the interactive effects of environment (through recruitment of homogeneous samples of subjects, and/or use of multifactorial statistical analysis) represents, in effect, a methodology consistent with the view that these variables have little or no effect on the dependent measures. As argued above, this would be akin to an experiment where one group of male and female rats are handled, another group of male and female rats are left undisturbed in infancy, and both groups are pooled for analysis of sex differences in laterality effects in adulthood.

Numerous issues must be considered if one is systematically to address the discrepancies between the fields of laterality research in humans and non-humans. They include re-examination of: (1) the research design and methods used in human research; (2) the viewpoints for interpretation of results; and (3) the nature of the research questions asked. These issues will be considered through a cross-species discussion of the corpus callosum, the largest interhemispheric commissure in the primate brain, which is estimated by cytoarchitectural studies in monkeys to contain axons from approximately 2% of cortical neurones (LaMantia and Rakic, 1990). This structure is relevant to discussion of laterality since it has been shown, both anatomically and functionally, to modulate interhemispheric relations within asymmetrical neurobehavioural systems (Rosen, Sherman and Galaburda, 1989; Witelson,

1989; Aboitiz, Scheibel and Zaidel, 1992; Clarke and Zaidel, 1994; Geffen, Jones and Geffen, 1994; Aboitiz and Ide, 1998; Clarke et al., 1998).

8.2. Review of the Animal Literature on Corpus Callosum Anatomy with Focus on Rodents

Inspired by early indications that measurements of the callosum taken from midline anatomical sections could reveal sex differences in humans (DeLacoste-Utamsing and Holloway, 1982), and supported by a well-established research program studying neurobehavioural asymmetries in rodents, Denenberg and colleagues set out to use the rat as an animal model to study sex differences in the corpus callosum. In their first study, Berrebi et al. (1988) found that the size of the corpus callosum was larger in adult males compared to adult females, and that this difference was affected by early environmental manipulations. Rats from litters that had been handled during the first 21 days of life showed a larger sex difference than rats from non-handled litters (Berrebi et al., 1988). A key message from this research was that sex differences in the corpus callosum must not be looked at in isolation from other environmental and developmental factors. As discussed above, direction and degree of asymmetry in rodent behaviour is developmentally plastic and may be affected by even the most subtle, and as shown by Stewart and Kolb (1988), unintentional, environmental influences. Indeed, the findings of Berrebi et al. (1988) provided further confirmation at a structural level for Denenberg's longstanding thesis that lateralized behaviours and interhemispheric systems were sensitive to early developmental factors (Denenberg, 1970, 1977, 1980; Denenberg and Yutzey, 1985).

In the 8 years following the report by Berrebi et al., Denenberg and colleagues continued to demonstrate that manipulations of gonadal hormones affected development of the rat corpus callosum, and that the size of hormonal effects seen in adulthood was moderated by early environmental manipulation (i.e. handling). Both testosterone and oestrogen manipulations were found to influence callosal development in experiments using animals that were handled for the first 21 days of life (Fitch et al., 1990a, 1991; Mack et al., 1993). Furthermore, Denenberg et al. (1991a) showed that testosterone injections on day 4 led to an increase in female callosum size only in adult animals that had been handled in infancy. While sex differences and hormone manipulations affected overall callosal area, region-specific effects were also shown (Denenberg et al., 1991a; Fitch et al., 1990a, 1991). This indicated that some areas of cortex were more sensitive to the interactive effects of hormones and environment than others.

At the time that Denenberg's group was reporting reliable sex differences and hormonally mediated effects in callosal size, controversy was emerging in a related area of research. The original sex difference in humans, as published by De Lacoste-Utamsing and Holloway (1982), was not being replicated by numerous independent studies (for a review, see Bishop and Wahlsten, 1997). However, independent laboratories studying the callosum in rats have shown evidence of sex-mediated effects at the gross anatomical (Zimmerberg and Mickus, 1990; Nunez and Juraska, 1998) and cellular levels (Juraska and Kopcik, 1988). With respect to the interactive effects of sex and handling, other groups were also reporting that sex differences in the rat corpus callosum were modulated by environmental variation. Juraska and colleagues found that rearing rats in complex or isolated environments affected both between-sex and within-sex variation in callosal cytoarchitecture (Juraska and Kopcik, 1988), and later showed that sex differences were influenced by early developmental factors such as litter sex ratios and cryoanaesthesia (Nunez and Juraska, 1998). Zimmerberg and Mickus (1990) showed that sexual dimorphisms in the callosum were susceptible to prenatal alcohol. Decreased callosal size (in overall size and in one anterior region) was observed in male, but not female, rats exposed to alcohol during prenatal development by way of the maternal diet.

Dimorphisms in the rat callosum reported across laboratories were, therefore, consistent at a conceptual level that was significant above and beyond the simple demonstration of similar patterns of male–female differences. It appeared, as was demonstrated in the study of behavioural and anatomical asymmetries, that the structure of the corpus callosum was affected by hormone and environment interactions. This is noteworthy for two reasons, particularly in light of the lack of interlaboratory reliability of sex differences in the human corpus callosum. First, the animal work suggests that stable sex differences can be demonstrated if the animals examined are reared in comparable, standard laboratory environments. More importantly, however, the work suggests that variations in factors analogous to those known to vary across human development (early stress levels, prenatal diet and alcohol exposure, postnatal environmental enrichment) can enhance, reverse or remove sex differences in callosal size, its cytoarchitectural components (Juraska and Kopcik, 1988) and behavioural correlates (Zimmerberg and Mickus, 1990). Thus, if such factors were not accounted for through screening of subjects or analysis of the data, sex differences would almost certainly be obscured.

Having set the stage, we now turn to a discussion of human corpus callosum research where sex effects are examined in light of the animal literature.

A point we will emphasize is that some of the seemingly non-reproducible findings reported in humans may well be the product of interaction effects that are yet to be isolated.

8.3. Consideration of the Human Literature on Corpus Callosum Anatomy in Light of Findings in Rodents

Shortly after the publication of the study by Berrebi et al. (1988) showing sex by handling interactions in the rodent corpus callosum, Witelson (1989) published a paper that stood out among the growing list of controversial reports on sex differences in the corpus callosum. She reported that sex differences in the isthmus of the human corpus callosum varied as a function of right hand consistency. The area of the callosal isthmus was larger in non-consistent right-handed men than in consistent right-handed men, but this difference was not seen in women. This study indicated support for two findings that researchers had previously shown in animals. First, it indicated that sex effects in callosal anatomy were unlikely to represent straightforward, developmentally invariant dimorphisms, but were probably the result of a complex constellation of interactions across development. Second, it supported the view that sex differences were region specific, possibly representative of localized sex differences in outlying cortical regions.

Witelson's (1989) study showed that sex differences could be affected by the proportion of consistent and non-consistent right-handed men included in a given study. Several other reports in support of this finding included: (1) independent replications of the sex by handedness interaction (Denenberg, Kertesz and Cowell, 1991; Habib et al., 1991; Clarke and Zaidel, 1994); (2) findings that degree of handedness consistency moderated sex effects differently in right- and left-handers (Cowell, Kertesz and Denenberg, 1993); and (3) reports that size of regions in frontal and temporal cortices were affected by degree of right-handedness in men but not women in a manner paralleling the callosum effect (Witelson and Kigar, 1992). Furthermore, studies that used a disparate classification system for handedness, did not replicate Witelson's finding (Jancke et al., 1996; also see Clarke and Zaidel, 1994). Despite the relative stability of this body of findings (especially in light of the growing controversy surrounding sex differences in the corpus callosum) and its implication that sex differences varied as a function of a lateralized behavioural characteristic, a primary focus in the study of the human corpus callosum continued to be the search for a straightforward, between-group, male–female difference in callosal area or regional callosum size (see Cowell, 2002).

Several studies were published in the early and mid-1990s that looked at sex differences in the corpus callosum with respect to age. The rodent research had clearly suggested that hormonally induced effects observed in the adult callosum were the net result of divergent developmental pathways in males and females (Berrebi et al., 1988; Fitch et al., 1990b). In other words, the between-sex difference *per se* was not as neurodevelopmentally meaningful as the unique within-sex changes in callosal anatomy that characterized females and males. In this light, the between-sex differences in adults may be conceptualized as merely the result of observing the development of callosal anatomy in male and female rats at a time point where it happened to diverge. Furthermore, the view of sex differences in callosal size as emergent and flexible is consistent with the hypothesis that the corpus callosum and related interhemispheric systems, particularly lateralized ones, are developmentally sensitive to environmental effects.

Given that the ontogenic paths of males and females appear to be fundamentally different (a concept introduced by Goldman et al. in 1974 as sex differences in 'developmental tempo'), it is possible that a researcher might sample sex differences in callosal size at a point when gross size measures were the same in males and females. However, lack of sex differences at a single developmental point would not necessarily indicate that sex differences in developmental pathways (and at time points other than the one sampled) did not exist. It could simply be indicative of a study that observed the crossing point of disparate developmental trajectories. Without further developmental examination, a lack of sex effects at one age (or in a group of subjects pooled across age) could be mistaken for no sex differences whatsoever. This was the logic that drove reanalysis by Cowell et al. (1992) of the life span database of human callosa by Allen et al. (1991).

In 1991, Allen et al. studied a large sample of male and female callosa taken from magnetic resonance images (MRIs). Their study focused on an attempt to replicate De Lacoste-Utamsing and Holloway's (1982) report of a larger callosal splenium in women compared to men, and to see whether this effect was also present in children. The Cowell et al. (1992) reanalysis of the data of Allen et al. (1991) drew from the animal research for its approach. As such, it did not seek to find *a* sex difference, but aimed to investigate the presence of regional sex differenc*es* in the developmental patterns of the male and female callosum across the life span. Results from studies by Fitch et al. (1990b) had shown that female rats treated with testosterone not only had similar adult profiles with respect to callosal size but that they had similar patterns of development to male rats. Furthermore, the effects of removing the ovaries led to size differences in adulthood that mimicked the effects of testosterone but that had a unique developmental time-course (Fitch et al., 1990b). Likewise, in humans, the anteriormost and posteriormost regions of the callosum in men increased in size from the first through the third decades of life, and decreased in size from that point onwards. In women, however, the same callosal regions did not reach maximum size until the fifth decade of life, and decreased in size thereafter. This finding was supported by similar trends seen in the callosal cytoarchitecture (Aboitiz et al., 1996) and in dichotic listening performance (Hugdahl, 2002). Witelson (1991) also demonstrated a dimorphic pattern of age effects in gross corpus callosum structure in men and women.

Cowell et al. (1992) analysis in humans indicated that any search for the main effects of sex in callosal size, particularly in human studies that did not control for age, would be unlikely to yield reproducible results. Moreover, if one considers the range of environmental factors shown to interact with sex hormones with respect to corpus callosum development in rodents (Berrebi et al., 1988; Juraska and Kopcik, 1988; Zimmerberg and Mickus, 1990; Denenberg et al., 1991a; Nunez and Juraska, 1998), it is not surprising that researchers who studied the human callosum found it difficult to replicate sex effects. Many explanations have been put forward to account for this lack of consistency, including small sample sizes, failure to match subjects for factors such as age or handedness, lack of regionally sensitive measurement techniques, and the highly variable and developmentally sensitive nature of the callosum itself (for a review, see Cowell, 2002).

Although the factors named above probably played some part in the controversy surrounding sex differences, they constitute methodological impediments that are relatively straightforward to overcome. Yet, one key theoretical point has been overlooked, specifically, that the nature of the search itself may have been flawed. To address this point, human research would need to embrace the idea, derived mainly from work in animals, that sex differences in adulthood are not necessarily the most critical biological endpoint of interest. Instead, sex differences are markers or indicators that hormones played a role in the development and organization of the neurobehavioural system in question.

Vast literatures have been compiled that document the effect of gonadal steroids on the growth, function, organization and dysfunction of neurobehavioural systems in a number of animal species (for reviews, see Becker, Breedlove and Crews, 1992). The issue of sex differences has ceased to be a controversial one, and it is becoming widely accepted that male and female developmental trajectories are unique in many important ways. Despite this realization, and the growing level of sophistication about the nature and relative importance of hormonal effects in mammalian cortical development and organization, the search for the elusive male–female difference in the human callosum continued to gather momentum through the 1980s and well into the 1990s (for a review, see Bishop and Wahlsten, 1997). Indeed, an opinion often voiced at scientific conferences was that the controversy over male–female differences in the human callosum simply reflected that sex was not an important variable in the study of this structure (for a review, see Cowell, 2002).

Some researchers have moved away from the search for a reliable sex difference and toward the search for a link between sex differences in callosal size and sex differences in head size (Jancke et al., 1996; Bishop and Wahlsten, 1997, 1999). This line of inquiry does not seek to reject the fact that sex differences exist, but rather aims to ascribe them to the general realm of sex differences associated with cognitively neutral factors such as body size. However, this approach circumvents a key message of the animal work, which is that hormonal effects on callosal development may have *multiple* manifestations. Some may be relatively straightforward effects of hormones on body tissue that lead to sex differences in body size, head size and overall brain size. Others may be the result of complex interactions between environmental and hormonal factors on neural development. These two types of effects may coexist, such that removing effects of head size does not affect the male–female differences revealed in developmental trends (for a review, see Cowell, 2002). However, the search for sex differences in the human callosum and neurobehavioural measures of laterality remains focused on the former approach and relatively indifferent to the latter.

8.4. Implications for Neurobehavioural Laterality and Interhemispheric Relationships and Future Research

To summarize, the above review of the developmental effects on laterality and the corpus callosum supports several related ideas:

- Lateralized neurobehavioural systems and related interhemispheric structures are sensitive to multiple genetic and environmental factors that influence their development and organization.
- These influences vary in their effects depending on the sex of the organism or on experimental manipulations of gonadal steroids.
- In order to examine systematic variation in laterality or the callosum, one must consider these neurobehavioural endpoints within the context of diverse and multifactorially determined developmental pathways.

- Examination of how laterality or the corpus callosum are affected by a single factor of interest (e.g. the effects of environment or testosterone, etc.) must be done within the framework of a tightly controlled developmental study or analysed in a multifactorial manner.

The above points provide a challenge for animal research that has been met, to a large extent, by scientists working within the field of developmental psychobiology and cognate disciplines. Rigorous controls on early environment, nutrition, genetics and age are relatively standard considerations for developmental experimentation on animals. Imposing such controls *directly* poses nearly impossible hurdles for scientists trying to investigate laterality and interhemispheric relationships in humans. However, there are several tactics supported by the animal literature that could be applied to enable researchers indirectly to accomplish similar ends. These will be examined in a brief review of key studies in the field of neurobehavioural lateralization and interhemispheric relationships in humans.

8.4.1. Embracing Interactions

Ethical limitations on human research restrict the control that scientists can impose on the backgrounds of their subjects. This is a generic issue in human research, but it has specific implications for dependent measures such as neurobehavioural asymmetries and the corpus callosum that are particularly sensitive to environmental influences. It is not possible to influence or restrict the developmental course of human subjects, as is done in animal research, but it is possible to place *post hoc* 'controls' via multifactorial analysis of the data. For example, if one refers to the study of factors that influence the gross regional morphology of the corpus callosum, one can survey the literature and see evidence of both handedness effects and age effects, each of which interacts with sex in its own way. However, there has not been sufficient definitive work to examine how the sex by handedness interactions unfold with age across the life span. Thus, we know a lot about which single factors and one-way (two-factor) interactions are potentially involved in creating variation across individuals. However, we know much less about how lateralized manual behaviour develops or ages in relationship to the corpus callosum, and outlying cortical structures, and in the context of changes in sex hormones across the life span. Considerable resources would be needed to collect data from sufficient numbers of human subjects to perform such a study. Yet, given the controversies over the stability of laterality effects in neurodevelopmental disorders such as dyslexia and schizophrenia (for a review, see Cowell, Fitch and Denenberg, 1999), such efforts would undoubtedly have wide-ranging clinical applications in addition to adding to basic research on normative processes.

The controversy over sex differences in the corpus callosum literature (reviewed in Section 8.3) is paralleled in the study of lateralized auditory perceptual processing. A number of demographic and methodological factors have been reported to affect functional asymmetries as measured by dichotic listening performance, a measure believed to be an indicator of laterality in speech processing (for a review, see Hugdahl, 1995). This includes factors observed to affect neurobehavioural asymmetries in animals such as sex, age, behavioural profile, test conditions and developmental abnormalities. The tendency for human research studies to explore one or two key factors at a time (against the backdrop of inadequate/insufficient controls for other factors) has affected the literature in two important ways. First, it has led to inconsistencies and inability to reproduce certain effects. In addition, we hypothesize that, as with the corpus callosum literature, perception of inconsistencies by the scientific community may lead to the premature rejection of some variables (e.g. sex or age) as factors relevant to the study of lateralized neurobehavioural development and function. These issues are discussed below in relationship to male–female differences in dichotic listening performance in humans and related work in rodents.

In light of the growing controversy over sex differences in dichotic listening performance, Hiscock et al. (1994) conducted a review of relevant studies published in six major journals and considered sex effects reported in main effects and higher order interactions. They concluded that there was evidence for a weak population-level sex difference in hemispheric specialization. Sex differences (main effects or interactions) were present in approximately 35% of studies that provided information about the sex of their subjects; a similar percentage was found in studies of adults and children. At first glance, this seems to indicate that sex may be a key factor in human auditory lateralization. However, the authors adjusted the approach of their analysis and perspective in two important ways that moderated the impact of these sex effects. First, they narrowed their search to focus specifically on the hypothesis that males show a greater right ear advantage (REA; indicative of left hemisphere specialization) than females. When they examined two-way, three-way, four-way and within-group comparison studies, a significant proportion of the outcomes did show sex effects in this direction. They also discounted the highest order multifactorial and within-group comparison effects, stating that these are 'unlikely to provide convincing evidence of a sex difference' (Hiscock et al., 1994, p. 426). These were valid steps in terms of evaluating the specific hypothesis in contention, that men are more later-alized in terms of a REA in dichotic listening, but it ignored the larger and neurodevelopmentally relevant issue regarding the role of sex hormones in the organization and function of lateralized systems.

A more developmentally and multifactorially driven investigation of sex differences in dichotic listening would encompass both the complex, higher order interaction effects as well as the studies showing effects in the direction opposite to that put forward in the review by Hiscock et al. (1994). One would also want to consider conditions that might lead to increased REA in females, or no sex differences in REA. There is evidence that the REA in dichotic listening is affected by fluctuations in hormone levels measured within women across the menstrual cycle (Altemus, Wexler and Boulis, 1989; Sanders and Wenmoth, 1998). Thus, it is possible that the effects observed in some studies were influenced by this factor.

A recent analysis examined dichotic listening performance in adults as a function of sex, handedness, age, family history of developmental language or learning disability and test conditions. The experiment studied subjects' abilities to report consonant-vowel syllables (e.g. pa, ba, ta, ka, ga) presented in 36 pairs (e.g. pa to the left ear and ka to the right ear) under three sets of instructions (i.e. test conditions). In the first condition, subjects were instructed to report the sound they heard most clearly (baseline asymmetry). In the two focused attention conditions, subjects were asked to report the sound they heard from the right ear (right focused condition) and the one they heard most clearly from the left ear (left ear focused condition). Results showed that these measures of lateralized function were sensitive to numer-ous cognitive and demographic effects that interacted with sex (Cowell and Hugdahl, 2000). Indeed, statistical significance of the greater REA in men across all conditions was found to be sensitive to both direction of handed-ness and family history of developmental language or learning disability. These findings suggested that sex effects manifest themselves in the context of higher order interactions representative of different developmental his-tories. Further analyses showed that sex differences, with men having a greater REA than women in the baseline condition, were more evident with increasing age. Age and sex by age effects took on different manifesta-tions in the other two test conditions. The interactions observed between sex, age and test condition indicated that questions such as 'are men more later-alized than women' should be modified to 'at which stage(s) across the life span, and under what test condition(s), do human beings from specific neu-rodevelopmental backgrounds show evidence that hormones affect latera-lized auditory processing?'. This view is supported by research in rats, showing that laterality in processing rapidly changing auditory stimuli is different in males and females (Fitch, Brown and Tallal, 1993), and that it is differentially affected across the sexes by both neonatal brain lesions and exposure to the neuroprotectant agent MK-801 (a non-competitive *N*-methyl-D-aspartate receptor antagonist injected 30 min prior to brain lesion surgery) prior to brain lesioning (Fitch et al., 1997a).

As an additional note, it should be mentioned that, even when one does not find evidence of a sex difference in the *magnitude* of a laterality effect between two groups, e.g. men and women, this does not necessarily mean that important organizational differences do not exist. For example, it is possible that even if men and women show similar neurocognitive asymmetries or appear to be equivalent in corpus callosal size, the relationships among these measures may be distinct within each sex. Significant differences in the patterns of neurobehavioural correlations in men and women have been shown for a number of lateralized neurobehavioral measures including: (1) handedness, the corpus callosum and outlying cortex (Witelson, 1989; Witelson & Kigar, 1992); (2) cortical asymmetry, callosal size and callosal cytoarchitecture (Aboitiz et al., 1992; Aboitiz and Ide, 1998); and (3) the corpus callosum and various measures of lateralized behaviour and cognition (Clarke and Zaidel, 1994; Clarke et al., 1998).

8.4.2. Examining Assumptions about the Stability/Plasticity of Laterality Across the Life Span

As discussed above, the laterality and callosum literatures in humans have stirred much controversy on the issue of sex differences. In humans, failure to replicate sex differences is often found and is attributed to the highly sensitive, environmentally plastic and individually different nature of these measures. In contrast, there is strong evidence from the animal literature that sex differences can be replicated when patterns of variation linked to differences in early environment are controlled. Thus, the researcher studying humans concludes that sex-difference findings cannot be replicated because of uncontrolled environmental and developmental events, whereas those studying animals conclude that sex differences can be replicated when environmental and developmental events are controlled. In a fundamental sense, the animal literature highlights and confirms the conclusion reached by researchers studying humans, and also points the way towards the resolution of the commonly observed 'failure to replicate' phenomenon.

How could these similar research paradigms have led to such disparate views about the nature of lateralized systems across human and non-human species? One reason may be the differences in the perspectives that are held about the developmental plasticity of laterality across these traditionally separate research areas. Such a fundamental gap has multiple historic roots, not the least of which is the fact that a key impetus for laterality research in humans has been to *understand lateralization of human language and handedness.* Until very recently, these were behaviour-cognitive domains that were believed by many to be unique in all respects to humans (Denenberg, 1988). Lateralization of other behaviours within these domains were, therefore, also believed to have no true analogue or homologue in animals.

For example, in the context of human language development, early investigations of key cortical structures examined in infant brains reported signs of lateralization in the expected adult pattern of left greater than right size measures (Witelson and Pallie, 1973; Wada, Clarke and Hamm, 1975; Chi, Dooling and Gilles, 1977). This, and similar patterns of dichotic listening asymmetry patterns observed in adults, children (for a review, see Hiscock et al., 1994) and infants (Bertoncini et al., 1989) seemed to indicate that in normative development, many key laterality factors are fixed early on (for a review, see Hiscock, 1998). The view that '... anatomical asymmetry precedes any learning effects, since the postnatal age of the infants precluded little, if any, environmental experiences, such as language acquisition or preferred hand usage...' (Witelson and Pallie, 1973, p. 644) had a pervasive and lasting effect on this research area. Given this influence, it was unlikely that scientists in the areas of human laterality research, particularly in areas related to normative language asymmetries, would have turned to their colleagues doing animal work for confirmatory or alternative interpretations of their results.

This mind set probably explains why much of the research investigating development and ageing of human laterality has remained relatively unaffected by the growing literature in animals documenting key factors that influenced its emergence and organization. Even within the human literature, there is growing evidence that postnatal development of cortical asymmetries may be more plastic than was originally believed. For example, examination of the raw data in reports on foetal and infant planum temporale in humans shows trends in the direction of a greater leftward asymmetry in the brains of female foetuses and infants (compared to male foetuses and infants) (Witelson & Pallie, 1973; Wada, Clarke and Hamm, 1975), whereas there is a greater leftward asymmetry in brains of adult males (compared to adult females) (Wada, Clark and Hamm, 1975). This developmental discrepancy has recently been confirmed in two independent MRI studies. 'Left greater than right' asymmetry in the planum temporale was found to be greater in female than male children (aged 3–14 years) (Preis et al., 1999) and the opposite effect has been reported in the adult brain (aged 20–35 years) (Kulynych et al., 1994).

Given that Preis et al. (1999) did not show any developmental changes in planum asymmetry in their sample, and barring any age-related morphology effects that might bias measurement, several theoretical interpretations are possible: (1) a significant change in lateralization of the planum temporale takes place during late adolescence; (2) the adult and child samples from the various studies were different in terms of some other factor that affected the direction of planum asymmetry; or (3) both. All three explanations are consistent with the animal literature reviewed above that lateralized systems are highly plastic in response to environmental and genetic history. For example, Diamond et al. demonstrated in 1983 that cortical thickness asymmetry patterns observed in adult female rats were not present at 7 days of age.

It is of interest to note that several areas of human research have embraced the notion of plasticity in the development of lateralized systems. Some work comes from the study of the ageing process, which found neuroanatomical changes after the age of 40 to be sex-, region- and hemisphere-specific (Cowell et al., 1994). Other work comes from the study of lateralized neurophysiological and neuroanatomical changes seen in the progression of schizophrenia (Gur et al., 1998; for a review, see Cowell et al., 1999). With respect to the study of handedness, recent work has provided evidence that emergence of this lateralized behavioural function is linked to cytoarchitectural changes in motor cortex asymmetry (Amunts et al., 1997). Several additional areas of research, from the speech and language domain, have also supported the notion of plasticity in cerebral and related behavioural asymmetries. These are discussed in detail below.

One body of evidence comes from research on dichotic listening testing that enables the experimenter to assess the effects of focusing attention to the right or the left ear on auditory processing asymmetries (this procedure is explained in Section 8.4.1). Several reports have been published showing that the ability to override the baseline asymmetry in the left-focused attentional condition does not fully mature until late childhood. One set of results was found with consonant-vowel syllables: 8-year-old boys were unable to attenuate their REA in the left focused condition, although girls at ages 8 and 9, and boys at age 9, were able to do this successfully, although the asymmetry patterns were governed by different mechanisms in each group (i.e. ability to increase left ear response vs. decrease right ear responses in the left focused condition) (Andersson and Hugdahl, 1987). Similar developmental sex differences in dichotic listening of words have been found using variations of this technique (Moulden and Persinger, 2000). Thus, although the baseline asymmetry (REA) appears to be present in children, integration of this lateralized system with higher order, and perhaps interhemispherically based, attentional mechanisms, seems to exhibit plasticity between the period of 6–9 years of age (Orbzut, Horgesheimer and Boliek, 1999).

Other evidence of changes in laterality as a function of sex across development comes from anatomical analyses of Brodmann's areas 44 and 45 (Broca's area) in a sample of postmortem cases ranging from the newborn to 33 years of age (Uylings et al., 1999). Area 45, in particular, showed a left greater than right asymmetry in volume that was more lateralized in men than women. The developmental pattern showed that females up to and including age 7 had symmetrical, and in one case, rightwardly asymmetrical regional brain volumes in area 45. From age 12.75 to 33, a left greater than right pattern was apparent in females, but it was less pronounced than in the males aged 12–32 years. Males showed a rightward asymmetry in early life and a leftward asymmetry from age 7 onward. This work, together with the findings from the dichotic listening literature, suggests that, although some aspects of the behavioural and anatomical bases for language asymmetry may be in place at birth, others remain plastic throughout development and sometimes into adulthood. Furthermore, differences between development of laterality often differ between the sexes.

Another prominent example of human research that has contributed to the study of plasticity in laterality and interhemispheric relations is from the area of neurodevelopmental language disorders. This body of work has incorporated and relied heavily on rodent and primate models throughout its history – an approach that has paid rich dividends in terms of deriving an in-depth understanding of the mechanisms involved in human neurobehavioral asymmetries in both normative and disordered development. The tradition began with Geschwind and colleagues' study of neurobehavioural measures related to childhood language disorders, which provided some of the strongest early indications that variation in human neurocognitive lateralization was sensitive to environmental and hormonal effects in perinatal development (Geschwind and Behan, 1982; Geschwind and Galaburda, 1986). Geschwind's early ideas were developed further and carried forward by Galaburda and colleagues in work that involved using rodent and human models together to advance the understanding of dyslexia (Galaburda et al., 1985, 1994; Rosen et al., 1989, 1995; Denenberg et al., 1991b, 1991c, 1996; Livingstone et al., 1991; Rosen, Sherman and Galaburda, 1992; Schrott et al., 1992; Galaburda, 1993; Waters et al., 1997). In that work, the neurobehavioural mechanisms contributing to laterality and disordered laterality featured prominently, and findings on laterality in animals led directly to advancements in dyslexia research that have, in turn, informed the study of normative lateralization in humans (Galaburda et al., 1985, 1994; Rosen et al., 1989; Rosen, Sherman and Galaburda, 1992; Denenberg et al., 1991b, 1991c, 1996; Galaburda, 1993).

A related set of behaviourally based animal models for developmental language disability has been developed by Fitch and colleagues (for a review, see Fitch, Miller and Tallal, 1997). Neurobehavioural asymmetry, its development and plasticity has also featured prominently in this research program, which investigated numerous neurodevelopmental correlates to lateralized auditory temporal processing in rats (Fitch et al., 1994, 1997; Herman et al., 1997). In addition to shedding much light on the origins of sensory-perceptual deficits in developmental language and learning impairment, this animal model has provided some key insights into the developmental mechanisms that may contribute to variability in dichotic listening performance in humans (reviewed in Section 8.4.1).

8.5. Conclusions: Minimum Requirements for Studying Laterality and Interhemispheric Relations in Humans

Over the past 30 years, great progress has been made in understanding the neurodevelopmental processes that affect laterality in animals. Although there are a few notable exceptions, most of the research on laterality and interhemispheric relations in humans has been done in relative isolation from work in other species. This has resulted in a human literature that does not consistently control or analyse for factors known to lead to individual variation in both functional and structural measures of laterality.

Research on laterality in humans would benefit from a more multidimensional and neurodevelopmentally sensitive approach such as that used in developmental psychobiological work in animals. Such an approach demands that studies of neurocognitive asymmetry in humans take the following issues into account:

- The neurobiological and neurobehavioural bases of laterality and interhemispheric function begin to develop prenatally and continue to develop and change across the life span, from birth through old age.
- There are genetic determinants of lateralized measures, but they can be modulated by environmental and hormonal factors.
- Multiple environmental factors may interact to influence the development and organization of laterality.

Certain areas of the human literature on laterality and the corpus callosum are fraught with inconsistent results. We propose that this is due, in large part, to a failure to account for the above three issues. In order to take these issues into account in the context of human research, where tight experimental controls on development are not possible, scientists must: (1) carefully screen subjects for multiple measures known to affect laterality and the callosum; (2) conduct multifactorial analyses in search of interaction effects; and (3) replace research questions such as 'is group A or group B more lateralized' with 'in what important ways does the developmental history of lateralized systems in group A differ from that in group B'. Finally, those who conduct laterality research with human subjects are encouraged to look to the animal literature, and to develop animal models whenever possible, to understand, gain insight and seek confirmation of their findings.

References

Aboitiz, F. & Ide, A. (1998). Anatomical asymmetries in language-related cortex and their relation to callosal function. In *Handbook of Neurolinguistics*, ed. B. Stemmer & H.A. Whitaker, pp. 393–404. San Diego, CA: Academic Press.

Aboitiz, F., Rodriguez, E., Olivares, R. & Zaidel, E. (1996). Age-related changes in fibre composition of the human corpus callosum. *NeuroReport*, 7, 1761–1764.

Aboitiz, F., Scheibel, A. & Zaidel, E. (1992). Morphometry of the Sylvian fissure and the corpus callosum, with emphasis on sex differences. *Brain*, 115, 1521–1541.

Allen, L.S., Richey, M.F., Chai, Y.M. & Gorski, R. (1991). Sex differences in the corpus callosum of the living human being. *Journal of Neuroscience*, 11, 933–942.

Alonso, J., Castellano, M.A. & Rodriguez, M. (1991). Behavioral lateralization in rats: Prenatal stress effects on sex differences. *Brain Research*, 539, 45–50.

Altemus, M., Wexler, B.E. & Boulis, N. (1989). Changes in perceptual asymmetry with the menstrual cycle. *Neuropsychologia*, 27, 233–240.

Amunts, K., Schmidt Passos, F., Scheiler, A. & Zilles, K. (1997). Postnatal development of interhemispheric asymmetry in the cytoarchitecture of human area 4. *Anatomy and Embryology*, 196, 393–402.

Andersson, B. & Hugdahl, K. (1987). Effects of sex, age and forced attention on dichotic listening in children: A longitudinal study. *Developmental Neuropsychology*, 3, 191–206.

Andrew, R.J. & Brennan, A. (1983). The lateralization of fear behaviour in the male domestic chick: A developmental study. *Animal Behaviour*, 31, 1166–1176.

Becker, J.B., Breedlove, S.M. & Crews, D. (eds.) (1992). *Behavioral Endocrinology*. Cambridge MA: MIT Press.

Beecher, M.D., Petersen, M.R., Zoluth, S.R., Moody, D.B. & Stebbins, W.C. (1979). Perception of conspecific vocalization by Japanese macaques. *Brain, Behavior and Evolution*, 16, 443–460.

Berrebi, A.S., Fitch, R.H., Ralphe, D.L., Denenberg, J.O., Friedrich, V.L. & Denenberg, V.H. (1988). Corpus callosum: Region-specific effects of sex, early experience and age. *Brain Research*, 438, 216–224.

Bertoncini, J., Morais, J., Bijeljac-Babic, R., McAdams, S., Peretz, I. & Mehler, J. (1989). Dichotic perception and laterality in neonates. *Brain and Language*, 37, 591–605.

Bishop, K.M. & Wahlsten, D. (1997). Sex differences in the human corpus callosum: myth or reality? *Neuroscience and Biobehavioral Reviews*, 21, 581–601.

Bishop, K.M. & Wahlsten, D. (1999). Sex and species differences in mouse and rat forebrain commissures depends on the method of adjusting for brain size. *Brain Research*, 815, 358–366.

Bulman-Fleming, M.B., Bryden, M.P. & Rogers, T.T. (1997). Mouse paw preference: Effects of variations in testing protocol. *Behavioural Brain Research*, 86, 79–87.

Camp, D.M., Robinson, T.E. & Becker, J.B. (1984). Sex differences in the effects of early experience on the development of behavioral and brain asymmetries in rats. *Physiology and Behavior*, 33, 433–439.

Carlson, J.N., Visker, K.E., Nielsen, D.M., Weller, R.W. & Glick, S.D. (1996). Chronic antidepressant drug treatment reduces turning behavior and increases dopamine levels in the medial prefrontal cortex. *Brain Research*, 707, 122–126.

Chi, J.G., Dooling, E.C. & Gilles, F.H. (1977). Left–right asymmetries of the temporal speech areas of the human fetus. *Archives of Neurology*, 34, 346–348.

Clarke, J.M. & Zaidel, E. (1994). Anatomical–behavioral relationships: Corpus callosum morphometry and hemispheric specialization. *Behavioural Brain Research*, 64, 185–202.

Clarke, J.M., McCann, C.M. & Zaidel, E. (1998). The corpus callosum and language: Anatomical–behavioral relationships. In *Right Hemisphere Language Comprehension*. ed, M. Beeman & C. Chiarello, pp. 27–50. Mahwah, NJ: Lawrence-Erlbaum.

Collins, R.L. (1977). Origins of the sense of asymmetry: Mendelian and non-Mendelian models of inheritance. *Annals of the New York Academy of Sciences*, 299, 283–305.

Collins, R.L. (1985). On the inheritance of direction and degree of asymmetry. In *Cerebral Lateralization in Nonhuman Species*, ed. S.D. Glick, pp. 41–72. New York: Academic Press.

Cowell, P.E. (2002). Size differences in the corpus callosum: Beyond the main effects. In *The Parallel Brain: Cognitive Neuroscience of the Corpus Callosum*, eds E. Zaidel & M. Iacoboni. Cambridge, MA: MIT Press, in press.

Cowell, P.E. & Hugdahl, K. (2000). Individual differences in neurobehavioural measures of laterality and interhemispheric function as measured by dichotic listening. *Developmental Neuropsychology*, 18, 95–112.

Cowell, P.E., Allen, L.S., Zalatimo, N.S. & Denenberg, V.H. (1992). A developmental study of sex and age interactions in the human corpus callosum. *Developmental Brain Research*, 66, 187–192.

Cowell, P.E., Fitch, R.H. & Denenberg, V.H. (1999). Laterality in animals: Relevance to schizophrenia. *Schizophrenia Bulletin*, 25, 41–62.

Cowell, P.E., Kertesz, A. & Denenberg, V.H. (1993). Multiple dimensions of handedness and the human corpus callosum. *Neurology*, 43, 2353–2357.

Cowell, P.E., Turetsky, B.I., Gur, R.C., Shtasel, D.L., Grossman, R.I. & Gur, R.E. (1994). Sex differences in aging of the human frontal and temporal lobes. *Journal of Neuroscience*, 14, 4748–4755.

Cowell, P.E., Waters, N.S. & Denenberg, V.H. (1997). Effects of early environment on the development of functional laterality in Morris maze performance. *Laterality*, 2(3/4), 221–232.

De Lacoste-Utamsing, C. & Holloway, R.L. (1982). Sexual dimorphism in the human corpus callosum. *Science*, 216, 1431–1432.

Denenberg, V.H. (1970). Experimental programming of life histories and the creation of individual differences. A review. In *Effects of Early Experience, 1968, Miami Symposium on the Prediction of Behavior*, ed. M.R. Jones, pp. 61–91. Coral Gables, FL: University of Miami Press.

Denenberg, V.H. (1977). Assessing the effects of early experience. In *Methods in Psychobiology*, ed. R.D. Myers, pp. 269–281. New York: Oxford University Press.

Denenberg, V.H. (1980). General systems theory, brain organization, and early experiences. *American Journal of Physiology: Regul. Intergrat. Comp. Physiol.*, 238, R3–R13.

Denenberg, V.H. (1981). Hemispheric laterality in animals and the effects of early experience. *Behavioral and Brain Sciences*, 4, 1–49.

Denenberg, V.H. (1988). Handedness hangups and species snobbery. *Behavioral and Brain Sciences*, 11, 721–722.

Denenberg, V.H. & Yutzey, D.A. (1985). Hemispheric laterality, behavioral asymmetry, and the effects of early environment in rats. In *Cerebral Lateralization in Nonhuman Species*, ed. S.D. Glick, pp. 109–133. Orlando, FL: Academic Press.

Denenberg, V.H., Kertesz, A. & Cowell, P.E. (1991). A factor analysis of the human's corpus callosum. *Brain Research*, 548, 126–132.

Denenberg, V.H., Garbanati, J. Sherman, G., Yutzey, D.A. & Kaplan, R. (1978). Infantile stimulation induces brain lateralization in rats. *Science*, 201, 1150–1152.

Denenberg, V.H., Hofmann, M., Garbanati, J.A., Sherman, G.F., Rosen, G.D. & Yutzey, D.A. (1980). Handling in infancy, taste aversion and brain laterality. *Brain Research*, 200, 123–133.

Denenberg, V.H., Fitch, R.H., Schrott, L.M., Cowell, P.E. & Waters, N.S. (1991a). Corpus callosum: interactive effects of infantile handling and testosterone in the rat. *Behavioral Neuroscience*, 105, 562–566.

Denenberg, V.H., Mobraaten, L.E., Sherman, G.F., Morrison, L. Schrott, L.M., Waters, N.S., Rosen, G.D., Behan, P.O. & Galaburda, A.M. (1991b). Effects of the autoimmune uterine/maternal environment upon cortical ectopias, behavior, and autoimmunity. *Brain Research*, 563, 114–122.

Denenberg, V.H., Sherman, G., Schrott, L.M., Rosen, G.D. & Galaburda, A.M. (1991c). Spatial learning, discrimination learning, paw preference and neocortical ectopias in two autoimmune strains of mice. *Brain Research*, 562, 98–104.

Denenberg, V.H., Sherman, G., Schrott, L.M., Waters, N.S., Boehm, G.W., Galaburda, A.M. & Mobraaten, L.E. (1996). Effects of embryo transfer and cortical ectopias upon the behavior of BXSB-Yaa and BXSB-Yaa+mice. *Developmental Brain Research*, 93, 100–108.

Diamond, M.C. (1991). Hormonal effects on the development of cerebral lateralization. *Psychoneuroendocrinology*, 16, 121–129.

Diamond, M.C., Dowling, G.A. & Johnson, R.E. (1981). Morphologic cerebral cortical asymmetry in male and female rats. *Experimental Neurology*, 71, 261–268.

Diamond, M.C., Johnson, R.E., Young, D. & Singh, S.S. (1983). Age-related morphologic differences in rat cerebral cortex and hippocampus: Male–female; right–left. *Experimental Neurology*, 81, 1–13.

Diaz, E., Pinto-Hamuy, T. & Fernandez, V. (1994). Interhemispheric structural asymmetry induced by a lateralized reaching task in the rat motor cortex. *European Journal of Neuroscience*, 6, 1235–1238.

Fitch, R. H., Berrebi, A.S., Cowell, P.E., Schrott, L.M. & Denenberg, V.H. (1990a). Corpus callosum: Effects of neonatal hormones on sexual dimorphism in the rat. *Brain Research*, 515, 111–116.

Fitch, R.H., Cowell, P.E., Schrott, L.M. & Denenberg, V.H. (1990b). Corpus callosum: Neonatal hormones and development. Paper presented at the International Society for Developmental Psychobiology Annual Meeting. King's College, Cambridge, England.

Fitch, R.H., Cowell, P.E., Schrott, L.M. & Denenberg, V.H. (1991). Corpus callosum: Ovarian hormones and feminization. *Brain Research*, 542, 313–317.

Fitch, R.H., Brown, C.P. & Tallal, P. (1993). Functional lateralization for auditory temporal processing in male and female rats. *Behavioral Neuroscience*, 107, 844–850.

Fitch, R.H., Tallal, P., Brown, C.P., Galaburda, A.M. & Rosen, G.D. (1994). Induced microgyria and temporal auditory processing in rats: A model for language impairment? *Cerebral Cortex*, 4, 260–270.

Fitch, R.H., Brown, C.P., Tallal, P. & Rosen, G.D. (1997). Effects of sex and MK-801 on auditory-processing deficits associated with developmental microgyric lesions in rats. *Behavioral Neuroscience*, 111, 404–412.

Fitch, R.H., Miller, S. & Tallal, P. (1997). Neurobiology of speech perception. *Annual Reviews in Neuroscience*, 20, 331–353.

Fitch, R.H., Cowell, P.E. & Denenberg, V.H. (1998). The female phenotype: Nature's default. *Developmental Neuropsychology*, 14, 213–231.

Fleming, D.E., Anderson, R.H., Rhees, R.W., Kinghorn, E. & Bakaitis, J. (1986). Effects of prenatal stress on sexually dimorphic asymmetries in the cerebral cortex of the male rat. *Brain Research Bulletin*, 16, 395–398.

Friedman, H. & Davis, M. (1938). 'Left-handedness' in parrots. *Auk*, 80, 478–480.

Galaburda, A.M. (1993). Neurology of developmental dyslexia. *Current Opinion in Neurobiology*, 3, 237–242.

Galaburda, A.M., Sherman, G.F., Rosen, G.D., Aboitiz, F. & Geschwind, N. (1985). Developmental dyslexia: Four consecutive cases with cortical anomalies. *Annals of Neurology*, 18, 222–233.

Galaburda, A.M., Menard, M.T., Rosen, G.D. & Livingstone, M.S. (1994). Evidence for aberrant auditory anatomy in developmental dyslexia. *Proceedings of the National Academy of Sciences*, 91, 8010–8013.

Garbanati, J.A., Sherman, G.F., Rosen, G.D., Hoffman, M., Yutzey, D.A. & Denenberg, V.H. (1983). Handling in infancy, brain laterality and muricide in rats. *Behavioural Brain Research*, 7, 351–359.

Geffen, G.M., Jones, D.L. & Geffen, L.B. (1994). Interhemispheric control of manual motor activity. *Behavioural Brain Research*, 64, 131–140.

Geschwind, N. & Behan, P. (1982). Left-handedness: Association with immune disease, migraine, and developmental learning disorder. *Proceedings of the National Academy of Sciences USA*, 79, 5097–5100.

Geschwind, N. & Galaburda, A.M. (1986). Cerebral lateralization. Biological mechanisms, associations, and pathology: I. A hypothesis and a program for research. *Archives of Neurology*, 42, 428–459.

Goldman, P.S., Crawford, H.T., Stokes, L.P., Galkin, T.W. & Rosvold, H.E. (1974). Sex-dependent behavioural effects of cerebral cortical lesions in the developing rhesus monkey. *Science*, 186, 540–542.

Gur, R.E., Cowell, P.E., Turetsky, B.I., Gallacher, F.P., Cannon, T., Bilker, W. & Gur, R.C. (1998). A follow-up magnetic resonance imaging study of schizophrenia: Relationship of neuroanatomical changes to clinical and behavioral measures. *Archives of General Psychiatry*, 55, 145–152.

Habib, M., Gayraud, D., Oliva, A., Regis, J., Salamo, G. & Khalil, R. (1991). Effects of handedness and sex on the morphology of the corpus callosum: A study with brain magnetic resonance imaging. *Brain and Cognition*, 16, 41–61.

Hayhow, W.R., Sefton, A. & Webb, C. (1962). Primary optic centers of the rat in relation to the terminal distribution of the crossed and uncrossed optic nerve fibers. *Journal of Comparative Neurology*, 118, 295–321.

Herman, A., Galaburda, A., Fitch, R.H., Carter, A.R. & Rosen, G. (1997). Cerebral microgyria, thalamic cell size and auditory temporal processing in male and female rats. *Cerebral Cortex*, 7, 453–464.

Hernandez-Mesa, N. & Bures, J. (1985). Lateralized rewarding brain stimulation affects forepaw preference in rats. *Physiology and Behavior*, 34, 495–499.

Hiscock, M. (1998). Brain lateralization across the life span. In *Handbook of Neurolinguistics*, eds B. Stemmer & H.A. Whitaker, pp. 357–368. San Diego, CA: Academic Press.

Hiscock, M., Inch, R., Jacek, C., Hiscock-Kalil, C. & Kalil, K.M. (1994). Is there a sex difference in human laterality? I. An exhaustive survey of auditory laterality studies from six neuropsychology journals. *Journal of Clinical and Experimental Neuropsychology*, 16, 423–435.

Holman, S.D. & Hutchison, J.B. (1991). Lateralized action of androgen on development of behavior and brain sex differences. *Brain Research Bulletin*, 27, 261–265.

Hugdahl, K. (1995). Dichotic listening: Probing temporal lobe functional integrity. In *Brain Asymmetry*, eds R.J. Davidson & K. Hugdahl, K., pp. 123–156. Cambridge, MA: MIT Press.

Hugdahl, K. (2002). Attentional modulation of interhemispheric transfer of auditory information. A 'two-channel threshold' model. In *The Parallel Brain: Cognitive Neuroscience of the Corpus Callosum,* eds E. Zaidel & M. Iacoboni. Cambridge, MA: MIT Press, in press.

Isasa Chueca, J.A., Lahoz Gimeno, M. & Redondo Marco, J.A. (1992). Morphometric study of the interhemispheric asymmetries in the Wistar rat and the effects of early experience. *Histology and Histopathology*, 7, 489–492.

Jancke, L., Staiger, J.F., Schlaug, G., Huang, Y. & Steinmetz, H. (1996). The relationship between corpus callosum size and forebrain volume. *Cerebral Cortex*, 7, 48–59.

Juraska, J.M. & Kopcik, J.R. (1988). Sex and environmental influences on the size and ultreastructure of the rat corpus callosum. *Brain Research*, 450, 1–8.

Krech, D., Rosenzweig, M.R. & Bennett, E.L. (1962). Relations between brain chemistry and problem-solving among rats raised in enriched or impoverished environments. *Journal of Comparative Physiological Psychology*, 55, 801–807.

Kulynych, J.J., Vladar, K., Jones, D.W. & Weinberger, D.R. (1994). Gender differences in the normal lateralization of the supratemporal cortex: MRI surface-rendering morphometry of Heschl's gyrus and the planum temporale. *Cerebral Cortex*, 4, 107–118.

LaMantia, A.-S. & Rakic, P. (1990). Cytological and quantitative characteristics of four cerebral commissures in the rhesus monkey. *Journal of Comparative Neurology*, 291, 520–537.

Lemon, R. E. (1973). Nervous control of the syrinx in white-throated sparrows (*Zonotrichia albicollis*). *Journal of Zoology*, 171, 131–140.

Lipp, H.-P., Collins, R.L. & Nauta, W.J.H. (1984). Structural asymmetries in brains of mice selected for strong lateralization. *Brain Research*, 310, 393–396.

Livingstone, M.S., Rosen, G.D., Drislane, F.W. & Galaburda, A.M. (1991). Physiological and anatomical evidence for a magnocellular defect in developmental dyslexia. *Proceedings of the National Academy of Sciences USA*, 88, 7943–7947.

Lund, R.D. (1965). Uncrossed visual pathways of hooded and albino rats. *Science*, 149, 1506–1507.

Mack, C.M., Fitch, R.H., Cowell, P.E., Schrott, L.M. & Denenberg, V.H. (1993). Ovarian estrogen acts to feminize the female rat's corpus callosum. *Developmental Brain Research*, 71, 115–119.

Maier, S.E. & Crowne, D.P. (1993). Early experience modifies the lateralization of emotionality in parietally lesioned rats. *Behavioural Brain Research*, 56, 31–42.

Mittelman, G., Wishaw, I.Q. & Robbins, T.W. (1988). Cortical lateralization of function in a visual reaction time task. *Behavioural Brain Research*, 31, 29–36.

Moulden, J.A. & Persinger, M.A. (2000). Delayed left ear accuracy during childhood and early adolescence as indicated by Roberts' dichotic word listening test. *Perceptual and Motor Skills*, 90, 893–898.

Nakahara, D. & Ikeda, T. (1984). Differential behavioral responsiveness to ipsilateral and contralateral visual stimuli produced by unilateral rewarding hypothalamic stimulation in the rat. *Physiology and Behavior*, 32, 1005–1010.

Nottebohm, F. (1970). Ontogeny of bird song. *Science*, 167, 950–956.

Nottebohm, F. (1977). Asymmetries in neural control of vocalization in the canary. In *Lateralization of the Nervous System*, eds S. Harnad, R.W. Doty, L. Goldstein, J. Jaynes & G. Krauthamer, pp 23–44. New York: Academic Press.

Nottebohm, F. (1979). Origins and mechanisms in the establishment of cerebral dominance. In *Handbook of Behavioral Neurobiology*, Vol. 2, ed. M.S. Gazzaniga, pp. 295–344. New York: Plenum Publishing.

Nottebohm, F. & Nottebohm, M.E. (1976). Left hypoglossal dominance in the control of canary and white-crowned sparrow song. *Journal of Comparative Physiology*, 108, 171–192.

Nottebohm, F., Stokes, T. M. & Leonard, C. M. (1976). Central control of song in the canary, *Serinus canarius*. *Journal of Comparative Neurology*, 165, 456–486.

Nunez, J.L. & Juraska, J.M. (1998). The size of the splenium of the rat corpus callosum: Influence of hormones, sex ratio and neonatal cryoanesthesia. *Developmental Psychobiology*, 33, 295–303.

Okada, T., Kato, I., Watanabe, S. & Takeyama, I. (1991). Retinal ganglion cells projecting to the nucleus of the optic tract in the rat. *Acta Oto-Laryngologica Supplement*, 481, 227–229.

Orbzut, J.E., Horgesheimer, J. & Boliek, C. (1999). A 'threshold effect' of selective attention on the dichotic REA with children. *Developmental Neuropsychology*, 16, 127–137.

Petersen, M.R. Beecher, M.D., Zoluth, S.R., Moody, D.B. & Stebbins, W.C. (1978). Neural lateralization of species-specific vocalization by Japanese macaques (*Macaca fuscata*). *Science*, 202, 324–327.

Preis, S., Jancke, L., Schmitz-Hillebrecht, J. & Steinmetz, H. (1999). Child age and planum temporale asymmetry. *Brain and Cognition*, 40, 441–452.

Riddle, D.R. & Purves, D. (1995). Individual variation and lateral asymmetry of rat primary somatosensory cortex. *Journal of Neuroscience*, 15, 4184–4195.

Rodriguez, M., Santana, C. & Alfonso, D. (1994). Maternal ingestion of tyrosine during rat pregnancy modifies the offspring behavioral lateralization. *Physiology and Behavior*, 55, 607–613.

Rogers, L.J. (1980). Lateralization in the avian brain. *Bird Behaviour*, 2, 1–12.

Rogers, L.J. (1982). Light experience and asymmetry of brain function in chickens. *Nature*, 291, 223–225.

Rogers, L.J. (1990). Light input and the reversal of functional lateralization in the chicken brain. *Behavioural Brain Research*, 38, 211–221.

Rogers, L. J. & Anson, J. M. (1979). Lateralization of function in the chicken fore-brain. *Pharmacology Biochemistry Behavior*, 10, 679–686.

Rosen, G.D., Sherman, G.F. & Galaburda, A.M. (1989). Interhemispheric connections differ between symmetrical and asymmetrical brain regions. *Neuroscience*, 33, 525–533.

Rosen, G.D., Sherman, G.F., Mehler, C., Emsbo, K. & Galaburda, A.M. (1989). The effect of developmental neuropathology on neocortical asymmetry in New Zealand Black mice. *International Journal of Neuroscience*, 45, 247–254.

Rosen, G.D., Sherman, G.F. & Galaburda, A.M. (1992). Biological substrates of anatomic asymmetry. *Progress in Neurobiology*, 39, 507–515.

Rosen, G.D., Waters, N.S., Galaburda, A.M. & Denenberg, V.H. (1995). Behavioral consequences of neonatal injury of the neocortex. *Brain Research*, 681, 177–189.

Rosenzweig, M.R., Bennett, E.L. & Diamond, M.C. (1972). Brain changes in response to experience. *Scientific American*, 226, 22–29.

Sanders, G. & Wenmoth, D. (1998). Verbal and music dichotic listening tasks reveal variations in functional cerebral asymmetry across the menstrual cycle that are phase and task dependent. *Neuropsychologia*, 36, 869–874.

Savage-Rumbaugh, E.S., Murphy, J., Sevcik, R.A., Brakke, K.E., Williams, S.L. & Rumbaugh, D.M. (1993). Language comprehension in ape and child. *Monographs of the Society for Research in Child Development*, Vol. 58, University of Chicago Press.

Schrott, L.M., Denenberg, V.H., Sherman, G.F., Waters, N.S., Rosen, G.D. & Galaburda, A.M. (1992). Environmental enrichment, neocortical ectopias, and behavior in the autoimmune NZB mouse. *Developmental Brain Research*, 67, 85–93.

Sherman, G.F., Garbanati, J.A., Rosen, G.D., Yutzey, D.A. & Denenberg, V.H. (1980). Brain and behavioral asymmetries for spatial preference in rats. *Brain Research*, 192, 61–67.

Stewart, J. & Kolb, B. (1988). The effects of neonatal gonadectomy and prenatal stress on cortical thickness and asymmetry in rats. *Behavioral and Neural Biology*, 49, 344–360.

Tan, U. & Kutlu, N. (1993). The end point of the Sylvian fissure is higher on the right than the left in cat brain as in human brain. *International Journal of Neuroscience*, 68, 11–17.

Uylings, H.B.M., Malofeeva, L.I., Bogolepova, I.N., Amunts, K. & Zilles, K. (1999). Broca's language area from a neuroanatomical and developmental perspective. In *The Neurocognition of Language*, eds C.M. Brown & P. Hagoort, pp. 319–336. New York: Oxford University Press.

Wada, J.A., Clarke, R. & Hamm, A. (1975). Cerebral hemispheric asymmetry in humans: Cortical speech zones in 100 adult and 100 infant brains. *Archives of Neurology*, 32, 239–246.

Waters, N.S. & Denenberg, V.H. (1994). Analysis of two measures of paw preference in a large population of inbred mice. *Behavioural Brain Research*, 63, 195–204.

Waters, N.S., Sherman, G.F., Galaburda, A.M. & Denenberg, V.H. (1997). Effects of cortical ectopias on spatial delayed-matching-to-sample performance in BXSB mice. *Brain Research*, 84, 23–29.

Whimbey, A. E. & Denenberg, V. H. (1967). Experimental programming of life histories: The factor structure underlying experimentally created individual differences. *Behaviour*, 29, 296–314.

Witelson, S.F. (1991). Sex differences in neuroanatomical changes with aging. *The New England Journal of Medicine*, 325, 211–212.

Witelson, S.F. (1989). Hand and sex differences in the isthmus and genu of the human corpus callosum. A postmortem morphological study. *Brain*, 112, 799–835.

Witelson, S.F. & Kigar, D.L. (1992). Sylvian fissure morphology and asymmetry in men and women: Bilateral differences in relation to handedness in men. *Journal of Comparative Neurology*, 323, 326–340.

Witelson, S.F. & Pallie, W. (1973). Left hemisphere specialization for language in the newborn: Neuroanatomical evidence of asymmetry. *Brain*, 96, 641–646.

Zimmerberg, B. & Mickus, L.A. (1990). Sex differences in corpus callosum: Influence of prenatal alcohol and maternal undernutrition. *Brain Research*, 537, 115–122.

Zimmerberg, B. & Riley, E.P. (1988). Prenatal alcohol exposure alters behavioral laterality of adult offspring in rats. *Alcoholism: Clinical and Experimental Research*, 12, 259–263.

9

Posture and Laterality in Human and Non-human Primates: Asymmetries in Maternal Handling and the Infant's Early Motor Asymmetries

E. DAMEROSE AND J. VAUCLAIR

9.1. Introduction

This chapter is concerned with the question of the relations and possible influences of environmental factors on the establishment of patterns of manual lateralization in human and non-human primates. More specifically, we are interested in the relation between maternal postures and laterality in non-human primates (e.g. bias in cradling behaviour and hand preference of the mother) and the development of patterns of manual preferences in infants. In order to understand fully the many ways in which these variables could interact, we first review the evidence of postural biases in human adults when cradling and carrying their offspring. Next, we examine the divergent hypotheses advanced to explain the observed biases. The same is then done for non-human primates. A second part of our chapter (see Section 9.3) describes the different asymmetric patterns observed during the development of the infant concerning head turning, nipple preference, etc. in both human and non-human primates.

In a third part (see Section 9.4), we describe and compare the methods and definitions used by the different authors in their work. Then, we present the descriptions and definitions of behaviours that we are using in an ongoing study of Olive baboons (*Papio anubis*).

Our main goal is to contribute to understanding the phylogenetic origins of hand laterality in humans by examining some of its possible determinants in non-human primates. Despite numerous efforts to propose models and explanatory schemata of hand laterality in non-human primates (e.g. MacNeilage, Studdert-Kennedy and Lindblom, 1987; Fagot and Vauclair, 1991; Ward and Hopkins, 1993), this question is still largely unresolved. On the other hand, views of determinants of hand laterality in humans still fluctuate between genetically (e.g. Annett, 1985, 1995; Hopkins and

Rönnqvist, 1998) and environmentally based determination (e.g. Provins, 1997). A primate (non-human) model of hand preference is likely to advance our understanding greatly of human lateralization and its evolution. The interest in having a primate model is obvious. For example, if we were able to demonstrate that non-human primates exhibit the same cradling asymmetries as humans, we could better understand the respective roles of biological versus experiential/cultural factors in the causation of this phe-nomenon. Moreover, by attempting to relate postural biases of the mother to the hand biases of infants (in terms of manual preference), we should also be able to understand the importance of the maternal environment on the determination of the infant's biases and preferences.

Given that the main external agent acting on the foetus and then on the infant is the mother, it is necessary to consider in what ways the mother may determine the newborn's immediate environment and how her behaviour may affect the neonate's subsequent manual choices.

9.2. Asymmetries in Cradling

The first behaviour of interest is cradling. Cradling most often refers to the hand (and the arm) used by the mother to hold her infant. We will examine the available evidence in relation to asymmetrical biases in cradling first in human and then in non-human primate mothers.

9.2.1. Human Studies

The first published investigation examining the possible association between maternal behaviour and the infant's developing laterality was carried out by Salk (1960). This author studied cradling behaviour of the human mother. Casual observations of a mother rhesus monkey carried out by Salk (1960) revealed that this female held her newborn on the left side 40 times out of a total of 42 observations, frequently with the newborn's ear pressed against her heart. This finding prompted Salk to conduct observations on human mother–infant pairs. Salk found a left-side cradling bias at a population level in 80% of the observations, regardless of mother's handedness. This bias was spontaneous and without awareness on the part of the mother. Salk (1960) proposed that the heartbeat of the mother constituted a comforting stimulus to which the foetus was imprinted. In subsequent studies, Salk (1962, 1973) tried to establish whether the observed bias was an instinctive response evolved from a need on the part of the infant to continue to experience the maternal heartbeat rhythm or, alternatively, whether it was based on learning a familiar sensation during intrauterine life. In order to investigate what was the effect of the heartbeat sound on the infant, two groups of infant newborns were formed. In the experimental group, the infants were exposed to sounds identical to those produced by their mother's heartbeat, whereas in the control group, newborns were not exposed to these sounds. The results of the study showed an increase in body weight, without any change of food intake and a decrease of injuries and stress level; these effects were significantly greater in the experimental group than in the control group.

Following Salk's pioneering work, numerous studies have reported the same leftward maternal cradling in human populations (see references in Saling and Kaplan-Solms, 1989). Interestingly, this bias is in marked contrast to the usual rightward skew that seems to govern the actions directed by *Homo sapiens* on the environment. Moreover, this bias appears to represent a universal pattern in the human female. Indian as well as African mothers are reported to hold their infant with the left hand, even though some cultural variations during the transport of the infant have been described (Saling and Cooke, 1984). Contrary to the opinion advanced by Salk (1970), according to which early postpartum experience was critical for the emergence of the left-sided cradling preference, Saling and Tyson (1981) showed that almost 90% of nulliparous women spontaneously cradled a doll (which served as the 'infant', see Section 9.5.1.3) in the left arm. This asymmetrical pattern initially reported for infant holding was also present when mothers held their children (Richards and Finger, 1975). Indeed, left-side infant cradling appears to be characteristic of female behaviour. Thus, an analysis of 'family album' photographs revealed that it is primarily a behaviour of human females and not of males (Manning, 1991; Manning and Denman, 1994), although some 'contamination' between wives and husbands in holding habits cannot be excluded (Dagenbach, Harris and Fitzgerald, 1988).

9.2.1.1. *Functions of Left Cradling*

Several hypotheses have been proposed in order to explain the left bias of human mothers in cradling, and showing related behaviours (e.g. transport) towards their infants and children.

(a) **The heartbeat hypothesis**. One hypothesis is derived from Salk's (1962; and above) experiments with newborns exposed to a recorded normal adult heartbeat sound. The positive effect of the heartbeat on the decrease of crying and weight gain prompted Salk (1973) to conclude that the sound of the normal adult heartbeat had a soothing effect on the newborn. As a consequence, when the baby is held on the mother's left side, he/she receives soothing sensations from the mother's heartbeat. The mother's heartbeat is thus interpreted as an imprinting stimulus creating a low-anxiety situation that the mother will tend to repeat when holding her baby. This imprinted response to an innately determined releasing stimulus was, for Salk (1962), also present in the mother who tried to increase awareness of her own heartbeat. A similar explanation was suggested by Weiland (1964), who emphasized the role of the mother's heartbeat in relieving anxiety of both the infant and the adult.

(b) Lower threshold of the left side for tactile perception. An assumption of psychophysics is that the left side of the body has a lower threshold for tactile perception compared to the right side. This view was revived with a study by Weinstein (1963) who measured left and right female breast sensitivity and reported a greater sensitivity on the left that was apparently independent of handedness. Kaplan-Solmes and Saling (1988) tried to replicate this reported sensory asymmetry but failed to support Weinstein's finding of a lower pressure threshold for the left than for the right breast of females.

(c) The maternal emotion hypothesis. Weiland and Sperber (1970) asked females to hold a pillow against their chest and found no side preference. When the women were told to imagine that they were holding a threatened infant, most participants held the pillow on the left (see also Manning and Chamberlain, 1991). The authors concluded from their study that maternal emotions might influence cradling preferences. It is possible that the observed left lateral placement in response to anxiety serves to intensify heartbeat sensation. In this respect the present hypothesis is thus formally equivalent to that described above (under Section 9.2.1.1.a).

A view based on the predominant role of the right hemisphere in the control of emotional behaviour has been proposed by Manning and Chamberlain (1991). The hypothesis goes as follows: when a female participant cradles an infant on the left side of her body, the infant's face is located on the extreme left of the participant's visual field. Because of the organization of the visual pathway connecting the retina to the brain, the left part of the visual field connects the nasal retina to the right hemisphere (Corballis and Beale, 1976). Manning and Chamberlain (1991) directly tested the hypothesis that perception of an infant (or a life-sized doll) via the left eye would stimulate left-side cradling. In order to test this hypothesis directly two experiments were conducted, one with mothers and one with young females (from 6 to 16 years of age). Mothers had to pick up and carry their infants, while non-mothers had to pick up and carry a life-sized doll. Three groups were made within each population: (1) a left-eye occluded group (an opaque

Table 9.1. *Percentages of left cradling as a function of the population and group in Manning and Chamberlain's (1991) experiment*

	Mother's cradling	Non-mother's cradling
Left eye occluded	40	60
Right eye occluded	69	72
Control	66	80

patch was placed over the participant's left eye); (2) a right-eye occluded group (similar procedure with the right eye); (3) a both eyes occluded group; and (4) a control group. Percentages of left or right cradling were recorded (see Table 9.1).

The results show clearly that occluding the right eye had no significant effect on left cradling biases. The results were similar in this condition and in the control condition (two eyes available), whereas occlusion of the left eye led to a frequency of left-side cradling which was significantly lower than that of the control group but not different from an expectation of 50% left side cradling.

From such differences in left cradling distributions, Manning and Chamberlain (1991) concluded that this bias had a twofold function: (1) it allows the mother to monitor her infant's well-being with her left visual field (thus facilitating the communication with her right cerebral hemisphere); and (2) it allows the infant to monitor the mother's emotional condition, given that the most expressive side (i.e. the left) of the mother is visible to him/her.

This appealing explanation of left cradling has, however, not been confirmed in the investigations conducted by Lucas, Turnbull and Kaplan-Solms (1993). No relation was found between cradling and the perception and expression of emotions. However, the methodology they chose (use of a population of non-mothers for carrying a doll and separate tests for evaluating asymmetries in expressing and producing emotions) precludes any direct comparison between the two studies.

A final approach needs to be mentioned. Harris, Almerigi and Kirsch (2000) asked university undergraduates to imagine holding a young infant in their arms. This investigation was an extension of one of Weiland and Sperber's (1970; see above) experiments with mothers. The researchers found that holding biases were correlated with handedness and sex of the participants. The main results of this study indicate a significant left-hand preference in holding for both right-handed men and right-handed women. However left-side preference for holding was weaker for left-handed women, and absent in left-handed men. Clearly, a left bias when holding an infant is a robust phenomenon that does not require a real stimulus in order to be shown. In addition, these results are consistent with involvement of right hemisphere arousal and leftward direction of attention in the phenomenon; this seems to be more pronounced in females than in males.

9.2.2. *Non-human Primate Studies*

It is interesting to note that the first study on cradling in humans was stimulated by Salk's (1960) observations of a rhesus macaque mother who showed a clear left bias in cradling her infant. The first systematic study with non-human primates awaited the publication by Manning and Chamberlain (1990), who investigated left-side cradling in chimpanzees, gorillas and orang-utans. The three species exhibited strong left-side cradling (85% for chimpanzees, 82% for gorillas, and for three out of the four orang-utans).

To explain the observed bias in the great apes, Manning and Chamberlain (1990, 1991) and Manning, Heaton and Chamberlain (1994) relied on a hypothesis already proposed for humans (see above) according to which a left-side cradling would make easier the monitoring of emotions and could have a twofold function. Firstly, with a left-side cradling, the mother would be more able to monitor the emotional level of her infant with her left visual field (and maybe with her left hearing field). Such a bias would activate, in a direct way, the right hemisphere, which is reputed to be the most efficient in the processing of emotions. Secondly, with this postural bias, the infant would be better able to interpret the emotional condition of his/her mother while the left hemiface of the mother is presenting to him/her; this is the more expressive side of the face at an emotional level. It is worth mentioning that Hauser (1993) reported in rhesus monkeys an identical asymmetrical pattern for emotional expressions, namely more expressiveness for left side (see Chapter 13 by Weiss et al.). Other studies have revealed differentiated patterns of expressions depending on the nature of the emotions. Thus, marmosets were shown to exhibit right hemisphere specialization for the production of negative emotional expressions and vocalizations, and left hemisphere specialization for the production of social contact communication (Hook-Costigan and Rogers, 1998). Left carrying (but not cradling) was also found in 41 rhesus macaque mothers (Tomaszycki et al., 1998). However, the presence of left-side cradling or holding biases has not been unequivocally reported in non-human primates. Thus, Tanaka (1989) found no population bias in a group of eight Japanese macaque mothers. In the great apes, Manning, Heaton and Chamberlain (1994) found a left bias in cradling among chimpanzees. But such a bias was not observed in a small group of wild chimpanzees (Nishida, 1993), whereas in the same species a slight right bias was reported (Dienske, Hopkins and Reid, 1995) or no bias at all (Hopkins et al., 1993a). Finally, two out of four rehabilitated orang-utan mothers (Rogers and Kaplan, 1996) exhibited a significant right-side preference. Moreover, data collected by these same authors (pers. comm., June 2000) on wild orang-utans have confirmed the right-side bias for carrying the babies.

9.3. Early Motor Asymmetries

This section reports published evidence concerning early motor asymmetries in newborns and young infants. As in the preceding section, we start by reporting and summarizing human data and then move to literature on non-human primates.

9.3.1. *Human Studies*

Some motor asymmetries in human neonates need to be mentioned because of their possible relation with handedness. In addition to the available evidence of prenatal behavioural asymmetries (e.g. in thumb-sucking; Hepper, Shahidullah and White, 1991), several studies have demonstrated that human newborns, while supine, predominantly turn the head into a right-sided position (Gesell and Ames, 1947). For example, Coryell and Michel (1978) found that, during the first 3 months, an average of 75% of the infants ($n = 35$) laid in a supine position maintained a significant right-side head posture (see also Thompson and Smart, 1993; Rönnqvist and Hopkins, 1998). Moreover, this phenomenon is also apparent in preterm infants (Konishi et al., 1987).

Several authors have postulated and found a relation between head-turn bias and later manual preferences. Thus, Coryell and Michel (1978) observed that the asymmetrical head position preference is a reliable elicitor of an asymmetrical tonic neck reflex (ATNR). This ATNR has the consequence of providing the infants with more visual experience of their right hand than of their left hand. Shown by the latter authors, at 12 weeks of age, infants begin to exhibit a difference between their right- and left-hand's response to a visually presented stimulus. In addition, Michel (1981) observed that head-turn bias was present until 2 months of age, and that it was a predictor of preferential hand use in pre-reaching tasks at both 16 and 22 weeks. This association between head orientation preference and hand preference was explained by the differential visuomotor experience of the two hands, the right hand receiving more visually guided contacts than the left hand. Because of this opportunity to observe their right hand, 'infants may develop better eye–hand coordination with that hand, thereby giving it an advantage over the left in visually guided reaching' (Michel, 1981, p. 687). A similar relation was obtained in a study by Konishi et al. (1987): 80% ($n = 82$) of their supine infants turned their head to the right and more than 88% of these participants preferentially used their right hand when tested in a reaching task at 9 and 18 months of age.

9.3.2. Studies with Monkeys and Apes

Two kinds of asymmetries in non-human primate infants are worth mentioning: one concerns nipple preference, and the other is related to asymmetries in head turning (see the above section reporting data on human infants).

Nipple preference was recorded among a group of 40 Japanese macaque infants by Tanaka (1989). After 1 month of age, all infants used only one of the two nipples during more than 80% of contact time; however, the choice of nipple was symmetrical at the group level (21 preferences for the right and 19 for the left nipple). Some evidence of nipple suckling preference was described by Nishida (1993) among wild chimpanzee mother–infant pairs. The infants displayed preference for the left nipple (as measured by first contact and duration). Moreover, an explicit relation is made by the author between the left nipple preference he observed and cradling by the mother (see also the asymmetries reported in the same species by Manning and Chamberlain, 1990) by stating that 'cradling patterns of neonates by their mothers are likely to be related to the left nipple preference in chimpanzees' (Nishida, 1993, p. 50). Rogers and Kaplan (1998) examined the relation between nipple preference, carrying by the mother and hand preferences in a sample of common marmosets during the first 60 days postpartum. The authors found no relation between nipple preference and hand preference, and no relation between cradling and hand preference. However, their results indicate a significant relation between nipple preference and side of the mother on which the infant was carried (incidentally, it should be noted that this relation disappeared when cradling by the father was included in the analyses). This latter result is thus consistent with Nishida's (1993) findings concerning chimpanzees.

A final study is available on nipple preference in monkeys. Tomaszycki et al. (1998) observed 41 rhesus macaque mother–infant dyads during the first 6 weeks of life. A significant left-side population bias emerged at the group level for nipple preference. This preference was evident during the first 3 weeks but vanished in the later weeks. Since the authors did not find a lateral bias in maternal cradling (see above), they hypothesized that the cradling bias (when observed) might reflect a bias in the infant's nipple preference rather than a bias in the mother's behaviour. If so, asymmetries in cradling might result from adjustment on the part of the mother to allow the infant to reach its preferred nipple.

Asymmetries in head orientation in infants have also been studied in some monkeys and apes. Thus, Hopkins and Bard (1995) observed a sample of 43 nursery-reared chimpanzees from birth to 3 months of age in both prone and supine positions. While the chimpanzees in the prone position did not exhibit any side bias, 36 of them expressed a consistent bias when in a supine position: 30 oriented toward the right and six toward the left. Recording of head orientation during the first two postnatal weeks was performed in tufted capuchins (Westergaard, Byrne and Suomi, 1998). More specifically, the author noted the length of time for which an infant maintained its head on the right or left side of its mother's back as she carried the infant dorsally. Although 12 infants showed longer durations on the left side and only four on the right, statistical comparisons of the two biases did not reach significance.

9.4. Comparative Approach of the Methods

This variability in the results for maternal behaviours could be caused by differences in the methodologies used among authors. Some (e.g. Hopkins et al., 1993a) have focused on newborn chimpanzees, while others (e.g. Manning and Denman, 1994) have observed human infants from birth until 2 years of age. Moreover, some researchers have chosen the position of the head as a measure (e.g. chimpanzees: Manning and Chamberlain, 1990; humans: Manning and Chamberlain, 1991), while others (e.g. macaques: Hopkins et al., 1993a; Tomaszycki et al., 1998) recorded the hand used by the mother to hold her infant when she was seated (cradling) or when she was walking (carrying). It seems likely that this tendency for a left-side cradling is related to the nature of the object that is held or, for humans, the manner in which the experimenter asked the mother to hold the object could have an effect on the laterality of the cradling side. It is thus important to take into consideration the nature of the object (inanimate, lifesize newborn dolls or real newborns) in the interpretation of the data.

As shown in the preceding sections, asymmetrical cradling of their infant by human mothers, in particular during non-feeding interactions, appears to be a robust phenomenon. However, there is still a paucity of detailed naturalistic work in this area of human (Wind, 1982) and non-human ethology (Saling and Cooke, 1984). For example, most of the human research was carried out under experimental conditions or consisted of surveys of published photographic material.

Furthermore, an important aspect of this left-side cradling has not been examined sufficiently: namely, the role of the infant or its substitute in determining all or part of the observed bias in the mother. Thus, the nature of the object that is held as well as the instruction provided by the experimenter to the participant about to hold the object could have an effect on the laterality of the cradling side. In fact, in some studies (e.g. Souza-Godeli, 1996), human participants did not express the expected bias when they were requested to hold inanimate objects. Moreover, most of the studies refer to cradling behaviour by recording the position of the infant in mother's arms or on the mother's hip (Salk, 1960; Saling and Cooke, 1984). Only some authors also took into consideration the hand (left or right) used by the mother to hold the infant (Saling and Tyson, 1981; Saling and Bonert, 1983).

The presence of a left-side preference for maternal cradling has not always been reported for non-human primates. This variability in the results for maternal behaviours may be due to differences in the methodologies used among authors. Thus, Manning and Chamberlain (1990) did not record manual laterality of the mother, but they considered the position of the head of the infant in relation to the midline of the mother's body. Nevertheless, these authors interpreted their results in terms of cradling side preference. In other words, with the same terminology (i.e. 'cradling'), some researchers have chosen the infant head position as a measure of lateral bias (e.g. Manning and Chamberlain, 1990), while other researchers have recorded the hand (or arm) used by the mother to hold the infant when she was seated (e.g. Tomaszycki et al., 1998; Damerose and Vauclair, 1999). Further, some studies distinguished between maternal cradling and maternal carrying. The latter term usually implied that the mother was walking (e.g. Hopkins et al., 1993b; Tomaszycki et al., 1998). However, Nishida (1993) used the term of cradling, but the behaviour recorded corresponded more to carrying, according to the above definition.

It is thus useful to describe and to compare the different methods and definitions used by the authors in their work. For that purpose, we will distinguish four basic behavioural categories (i.e. holding side, head side, nipple preference and grasping side).

9.5. Cradling, Carrying or Holding Patterns

9.5.1. *Asymmetries in Human Participants*

We first review the evidence concerning patterns of asymmetries for infant cradling, carrying and holding, from the human literature.

9.5.1.1. *Free Observation*

We report in this paragraph some studies concerned with the different behaviours listed above which did not use any specific sampling technique. Thus, Salk (1960) conducted the first study on the laterality of maternal cradling and for that purpose observed 287 human mothers with their newborn babies. To assess a preference for one side of infant holding by the mother, Salk relied on free observation sessions of the mother–infant human dyads at a hospital. He found (Salk, 1960) that both right- and left-handed mothers had a significant tendency to hold their babies on the left side (see Table 9.2), close to the heart. This effect appeared to be automatic and without awareness on the part of the mother. This original finding opened the way to other investigations in order to find out the factors responsible for this observed bias in cradling. Thus, Salk (1973) hypothesized that a postpartum separation could have an effect on the holding pattern. In order to investigate this hypothesis, Salk selected, as an experimental group, 115 mothers who had experienced prolonged separation (at least 24 h) from their infant after birth. In addition, as a control group, 286 mothers were selected at random from mothers attending the clinic who had experienced any prolonged postpartum separation from their infant. During the experimental session, the experimenter took the baby and presented the infant directly to the midline of the mother's body. Next, the experimenter noted on which side the mother held the baby and then asked a series of questions to obtain background data, including handedness of the mother. The control group showed a marked preference for holding the baby on the left side, while the experimental group did not show such a side preference (see Table 9.2).

In order to assess whether the head orientation could influence the side of cradling, Thompson and Smart (1993) unobtrusively observed a total sample of 150 mother–infant pairs during a first test session. Between the first head-turning test (see Section 9.6.1, Table 9.4) and grasping tests (see Section 9.8.1, Table 9.7) the mother was asked to pick up her baby from the crib to give it a cuddle. Observations were made of the side on which each mother held her baby and the position she laid her baby back down in the crib after the grasping tests: right or left lateral, supine or prone. Thompson and Smart found a close to significant ($p = 0.074$) relationship between head-turning classification and maternal holding, with a higher proportion of babies held to the left (Table 9.2) turning their head to the right (see Table 9.4). Here an inverse relation between the side of the cradling and the head orientation of the infant was observed. However, other studies (Bundy, 1979; Saling and Tyson, 1981), which used a doll resembling a lifesize real infant and controlled for the head orientation, did not confirm these findings (see Section 9.5.1.3; Table 9.2).

Most studies bore out Salk's assumption of a left-side tendency in infant holding (Table 9.2). The question of lateral preferences has to be addressed when an object is held instead of a human or a substitute. In a recent study, Hopkins and Parr (unpublished observations) employed a naturalistic observational method with 1505 adult human participants (746 males, 859 females). The experimenters observed individuals while they were carrying either objects or infants in natural social settings. Carrying was here defined as walking upright for at least 5 s, while holding an infant or an object ventrally between the waist and shoulder. Left-sided carrying was defined as the lateral displacement of the object or infant to the left of the participant's midline. Right-sided carrying was similarly defined as a lateral displacement to the right. Objects or infants positioned at the midline were omitted from analysis. The gender of the participant, but not the sex or age of the infant, was recorded. Overall, a significant interaction was found between gender and carrying bias (χ^2 (1,1505) = 17.70, $p < 0.001$) with females showing more left-sided carrying than males (Table 9.2), but no significant interaction was found between the lateral bias and the type of item being carried by the participant.

Here carrying an infant or an object produced similar results. However, Weiland and Sperber (1970) did not find a left-side tendency when women held a pillow (see Table 9.2). Nevertheless, in the first part of their study, Weiland and Sperber (1970) observed 48 patients (28 females, 20 males) in an anxiety situation (they were receiving dental therapy). The dentist asked patients to hold a rubber ball firmly against their chest and the observer noted with which hand the ball was held and the side position of the ball on the chest. Participants showed a tendency to hold the ball on the left side, but most of them used the right hand (Table 9.2). In addition, the observed left-side preference for holding was significantly stronger for women than for men ($p < 0.05$). Thus, for the same behaviour, Weiland and Sperber (1970) found opposite results when they recorded the side position of the ball instead of the hand used to hold the ball. These findings reinforce the need to describe clearly and define what behavioural scores the observer is collecting while recording cradling behaviour.

Table 9.2. *Summary of studies on cradling, holding and carrying in humans.*

Study	Participant	Procedure	Measure	Individual criterion	Condition	N	Right	Left	No preference	Bias	Statistics
Free observation											
Salk (1960)	Mother-infant pair	Obs.	Infant holding side	1 bout	Left-hander	32	21.9%	78.1%	0%	L	?
					Right-hander	255	16.9%	83.1%	0%	L	?
Weiland and Sperber (1970)	Patient of dental therapy	Obs. in anxiety situation	Hand used to hold ball	1 bout	Female	28	20	5	3	R	?
					Male	20	14	2	4	R	?
			Ball position on chest	1 bout	Female	28	1	19	8	L	?
					Male	20	1	7	12	L	?
Salk (1973)	Mother-infant pair	Obs.	Infant holding side	1 bout	Control group	286	23%	77%	0%	L	?
					Prolonged separation group	115	47%	53%	?	No	?
Thompson and Smart (1993)	Mother-infant pair	Obs.	Mother cradling side	1 bout		132	28	104	0	L	?
Hopkins and Parr (submitted)	Adult female	Naturalistic obs.	Carrying position	1 bout	Infant	381	148	233	?	L	?
					Non-infant	378	126	252	?	L	?
	Adult male	Naturalistic obs.	Carrying position	1 bout	Infant	371	177	194	?	L	?
					Non-infant	375	172	203	?	L	?
Sequential sampling											
Saling and Cooke (1984)	Mother-infant pair	Sequential sampling	Infant cradling position in arm	1 bout	All the three groups	134	13	110	11	L	?
			Infant cradling position on hip	1 bout	All the three groups	16	3	13		L	?
			Infant cradling position	1 bout	Black group	50	4	40	6	L	$\chi^2 = 49.11$ $p < 0.001$
					Colored group	50	6	41	3	L	$\chi^2 = 53.55$ $p < 0.001$
					Indian group	50	6	42	2	L	$\chi^2 = 58.23$ $p < 0.001$
			Infant transport position in arm	1 bout	All the three groups	112	14	81	17	L	?
			Infant transport position on hip	1 bout	All the three groups	38	6	32		L	?
			Infant transport position	1 bout	Black group	50	3	30	17	L	$\chi^2 = 21.88$ $p < 0.001$
					Colored group	50	9	41	0	L	$\chi^2 = 55.71$ $p < 0.001$
					Indian group	50	8	42	0	L	$\chi^2 = 59.67$ $p < 0.001$

Study	Sample	Test	Cradling position	Bouts	Group	N	Left	Right	Other	Dir.	χ²	p
Manning and Chamberlain (1991)	Mother–infant pair	Sequential test	Infant cradling position	1 bout	Control group	50	38%	62%	0%	L	?	?
					REO[a]	50	24%	64%	12%	R	$\chi^2 = 1.62$	$p > 0.05$
					LEO[a]	50	60%	40%	0%	No	?	?
					Left-hander	20	10	9	1	No	?	?
					Right-hander	130	51	74	5	L	?	?
Imagined situation												
Weiland and Sperber (1970)	Women	Hold pillow in anxiety situation	Pillow holding side	1 bout	Pillow holding	21		20		No	?	?
		Imagine real infant in anxiety situation	Pillow holding	1 bout	Pillow as infant	21	10	10		10 shifts No -> L		
Bundy (1979)	Student male and female	Imagine real infant situation	Infant baby doll position	1 bout	Doll's head in midline	47		78%		L	?	?
					Doll's head on the left	47		68%		L	?	?
					Doll's head on the right	47		85%		L	?	?
Saling and Tyson (1981)	Nulliparous female	Imagine real infant situation	Arm used to hold the doll	1 bout	All the three groups	120	13	107	0	L	$\chi^2 = 73.63$	$p < 0.001$
					Doll's head in midline	40	7	33	0	L	$\chi^2 = 32.4$	$p < 0.001$
					Doll's head on the left	40	4	36	0	L	$\chi^2 = 25.6$	$p < 0.001$
					Doll's head on the right	40	2	38	0	L	$\chi^2 = 16.9$	$p < 0.001$
Saling and Bonert (1983)	Preschooler girl	Imagine real infant situation	Arm used to hold the doll	1 bout		53	18	35	0	L	$\chi^2 = 5.45$	$p < 0.02$
Bogren (1984)	Couple with pregnant female	Imagine real infant situation	Child position	Position shown by participant	Female	81	20%	80%	0%	L	?	?
					Male	81	17%	83%	0%	L	?	?
Manning and Chamberlain (1991)	Girls	Imagine real infant situation	Infant baby doll position	1 bout	Control group	95	20%	80%	0%	L	?	?
					REO[a]	98	21%	79%	0%	L	?	?
					LEO[a]	99	38%	61%	1%	L	$\chi^2 = 4.5$	$p < 0.05$
					BEO[a]	96	33%	66%	1%	L	$\chi^2 = 9.4$	$p < 0.05$

Table 9.2. (*cont.*)

Study	Participant	Procedure	Measure	Individual criterion	Condition	N	Right	Left	No preference	Bias	Statistics	
Turnbull and Lucas (1991)	Male student	Imagine real infant situation	Infant baby doll cradling preference	1 bout	Infant baby doll	67		46%		No	$\chi^2 = 0.19$	$p > 0.05$
Lucas, Turnbull and Kaplan-Solms (1993)	Nulliparous female student	Repeated imagine real infant situation	Infant baby doll holding position	1 bout	Infant baby doll	86	24	62	0	L	$\chi^2 = 8.40$	$p < 0.01$
Souza-Godeli (1996)	Brazilian child	Imagine real infant situation	Infant baby doll holding position	1 bout	Infant baby doll	520		Left		L	$\chi^2 = 198.58$	$p < 0.001$
Turnbull and Matheson (1996)	Adult	Cradle infant doll	Infant doll cradling position	1 bout	Package	520	35%	14%	51%	No	$\chi^2 = 102.63$	$p < 0.001$
					Congenitally deaf	12		12		No	?	?
					Congenitally blind	12			No pref.	L	?	?
		Imagine real infant situation	Infant doll cradling position	1 bout	Congenitally deaf	12		10		No	?	?
					Congenitally blind	12			No pref.	L	?	?
Harris, Almergi and Kirsch (2000)	Female student	Imagine real infant situation	Holding infant's head side imagined	Written response	Right-hander	351		74%		L	?	$p < 0.001$
					Left-hander	38		55%		No	?	ns
	Male student	Imagine real infant situation	Holding infant's head side imagined	Written response	Right-hander	150		68%		L	?	$p < 0.001$
					Left-hander	15		47%		No	?	ns
Photographs survey												
Salk (1973)	Child holding work of art	Paintings and sculptures survey	Child holding position	1 shot		466	93	373	0	L	?	?
Richards and Finger (1975)	Photograph of women holding	Categorized photographs	Infant holding position	1 shot	Western culture			65%		L	Bin	$p < 0.01$
					Eastern culture			74%		L	Bin	$p < 0.001$
					American Indian culture			76%		L	Bin	$p < 0.001$
Harris and Fitzgerald (1985)	Photograph of adult holding	Categorized photographs	Infant head position	1 shot	Female	123	45	78		L	$\chi^2 = 8.85$	$p < 0.005$
					Male	48	18	30		L	$\chi^2 = 3.0$	$p < 0.10$

Study	Stimulus	Laterality measure	Method	Analysis	Sex	N	%	Bias	Statistics
Manning (1991)	Photograph of adult-infant pair	Infant holding position	Categorized photographs	1 shot	Female	1119	61%	L	?
					Male	557	47%	No	?
									?
Thompson and Smart (1993)	Photograph of mother-infant pair	Mother cradling side	Photography	1 shot		86	67	L	?
						19	0		
Manning and Denman (1994)	Photograph of adult holding	Infant holding position	Categorized photographs	Mean LCFs[b]	Female	167	96[c]	L	$\chi^2 = 9.94$, $p = 0.002$
						57[c]	14		
					Male	67	24[c]	R	$\chi^2 = 2.77$, $p = 0.09$
						37[c]	6		

Obs, observation; N, number of participants; L, left bias; R, right bias; No, no preference; ?, information not provided in the original paper; ns, non-significant. See details in text: [a]REO, right eye occluded; LEO, left eye occluded; BEO, both eye occluded; [b]LCFs = left cradling frequencies. [c]Right, LCF of 0–49%; Left, LCF of 51–100%; No, 50%.
The abbreviation 'Bin' refers to a binomial test.

In a second part of the study by Hopkins and Parr (unpublished observations), these authors also assessed handedness, carrying an infant, as well as different kinds of objects, and footedness as well as participant's variables including age, gender and parity, with a 26-item questionnaire. Participants were 761 human adults (380 males, 381 females). The qualitative response to the 22 laterality questions (five-point scale) were coded numerically and each response was assigned a weighted value (−2 to +2) corresponding to the participants response (always left to always right). A Pearson product moment correlation revealed a significant association between each of the four laterality indices. A significant proportion of variance was accounted for in this analysis ($R = 0.57$, $F(6,719) = 56.68$, $p < 0.001$). Handedness, followed by footedness, age and sex, all significantly predicted laterality in carrying biases.

It is worth reporting that Jenni and Jenni (1976) showed sex differences in carrying books by recording the spontaneous book-carrying methods of 2256 students from kindergarten to old age (distinguishing ten levels). In carrying methods used by females, arms and books partially covered the front of the body, while those used by males employed 'open' positions, with the body unobstructed. Jenni and Jenni divided positions into two basic types. In type I, one or both arms wrapped around the books, whilst the forearm, on the outside of the books, supported them. In type II, one arm and hand at the side of the body supported books. A single record of each individual's carrying method was made: no significant difference in carrying between males and females ($p > 0.50$) was found up to grade 1. Both displayed a pattern equivalent to the college males and usually used some variant of the type II. From grade 2, significant differences emerged between the sexes: females were more likely to use type I, and males more likely to use type II carrying.

9.5.1.2. Sequential Sampling

Saling and Cooke (1984) used 50 human mother–infant pairs in each of the population groups (black, coloured and Indian) to investigate any cultural effects on the holding pattern. Women were observed whilst in the waiting room of clinics for healthy babies. A sequential sampling procedure was used. Holding positions were coded in terms of the placement of the infant against the mother's body (i.e., infant held in one arm, supported on one hip, or held against the ventral surface of the mother's body in a midline position) and the laterality of the holding position (right, left or midline) was measured. After each observation, the mother was questioned about her own age and that of her infant. For descriptive purposes, the holding positions observed were referred to as cradling behaviour. The majority of women in each group cradled their infants in a lateralized fashion, and only 7.33% of mothers exhibited a midline cradling position. Of the two lateralized cradling positions (arm and hip), the arm position (88.5%) was the more frequently used in each group. The distribution of the cradling position was significantly biased towards the left side in the three cultural groups. There was no evidence of cross-cultural variation in the direction and degree of lateral bias ($\chi^2(4) = 2.89$, $p > 0.50$), with the infant being cradled in the arm and on the hip more often on the left side (Table 9.2).

With another group in a second part of their study, Saling and Cooke (1984) used 50 human mother–infant pairs in each of the three population groups (black, coloured and Indian). Mothers were observed while carrying their infants towards the clinic. A dorsal midline position was included in coding the holding position. The second part was procedurally similar to the first one. For descriptive purposes, the holding positions observed were referred to as transport behaviour. The majority of women in each group transported their infants in a lateralized fashion and only 11.33% of mothers exhibited a midline transport position. The arm position (71.4%) was preferred to the hip position. A majority of infants were on the left side (see Table 9.2).

Using the same sampling procedure, Manning and Chamberlain (1991) assessed the importance of the visual field used during cradling. They assigned 150 mothers with their infants (ranging in age from 12 days to 11 months) to one of three groups: control, left eye occluded (LEO), and right eye occluded (REO). The mother was asked to walk and place her infant on a cot and return to her seat. She was then requested to walk to the end of the cot, pick up the infant and return (carrying the infant) to the seat, and sit down. The position of cradling (i.e. left, midline or right) was noted when the mother had completed the task. The hand that the mother normally used for writing was recorded. The control and REO groups exhibited a tendency to hold on the left side, which was absent in LEO (Table 9.2). Manning and Chamberlain (1991) also noticed an effect of infant age, namely that left-side cradling decreased with increasing infant age, and suggested that their findings revealed some evidence of an association between cradling tendency and handedness. Of the 150 participants, 20 reported a predominant use of the left hand for writing (Table 9.2). However, the differences between the left- and right-handed participants were not significant ($\chi^2 = 0.97$, $p > 0.05$).

9.5.1.3. Imagined Situation

Several studies have examined the effect on cradling a doll of instructions to imagine that it was real or had some other status. In a vast majority of those studies, participants were required to imagine that the doll was a real baby and had to hold it. Thus, to examine the possibility that the mother's preference for holding infants on the left side was based upon the normal asymmetry of the infant's head orientation, Bundy (1979) divided 141 students into three groups (15 males, 32 females in each group). Participants were then required to hold a lifelike infant doll as if it was a real baby. The head position of the doll was midline, or turned to the left or the right side. Handedness and prior experience with infants were evaluated by questionnaire. Between 68% and 85% of participants held the doll on the left side, irrespective of the head orientation of the doll (Table 9.2). The more experience mothers had had with infants, the stronger was the leftward holding bias. Males (73%) had an equivalent left holding tendency to females (79%). Subjects with little experience of infants were more influenced by the doll's head orientation.

In a study that examined whether experience of infants affected cradling bias (Bundy's hypothesis; Bundy, 1979), Turnbull and Lucas (1991) observed 67 male students (all non-fathers). Cradling preferences were established with a lifesize neonatal doll, and participants also completed a handedness and infant experience questionnaire. No significant leftward cradling was found for male participants (Table 9.2). Moreover, cradling preference was not related to handedness. No relation appeared between direction of cradling and the participant's experience of or attitude towards infants (obtained by the questionnaire). Saling and Tyson (1981) tested 120 nulliparous female students. Participants were required to cradle a lifelike doll in the preferred arm. The head position of the doll was rotated either to the left or the right of its body midline, or fixed in the midline position. Each condition contained 40 participants. Saling and Tyson (1981) found that the majority of participants (i.e., 89%) spontaneously cradled the doll in the left arm (Table 9.2). These findings were close to those obtained by Bundy (1979).

These different studies seem to show that the head position of the doll does not disrupt the leftward preference in cradling position, contrary to the assumptions made by Thompson and Smart (1993).

Also using an imagined situation, Saling and Bonert (1983) examined cradling behaviour and its relation to handedness in 53 pre-school girls. The experimenter stood opposite to the girl, presented a doll (lifelike infant size) and then asked the girl to hold it so that it could fall asleep. Handedness was established via the use of ten familiar unimanual actions. In this condition, the leftward cradling bias (Table 9.2) appeared to be independent of total handedness score. With the same aim, the psychiatrist, Bogren (1984), conducted semistructured interviews in 81 parents. All women and men were interviewed during the 13th to 14th week of the mother's pregnancy, and again during the week following delivery. During the first interview, questions were asked about social and psychological background, and feelings during pregnancy. During the second interview, participants were questioned about pregnancy and parturition. The women and men were asked to imagine that they were holding their infant, and then to show how they did it. All subjects were asked if they used the right or left hand when writing. Women and men showed a tendency to hold their infant to the left (see Table 9.2) and their lateral preference was independent of their handedness.

Several explanations have been provided to explain the leftward tendency for cradling. Thus, in order to investigate Manning and Chamberlain's (1990) suggestion that a relationship exists between the right hemisphere specialization for emotional processing and leftward cradling (see above), Lucas, Turnbull and Kaplan-Solms (1993) tested 86 nulliparous female students. They divided the testing procedure into three stages. In the first stage, they assessed all participants for their preferred cradling bias using a lifesize doll. Each participant was requested to stand in front of a centrally placed doll and was asked to imagine that the doll was a real baby, and to pick up and rock it to sleep. This procedure was then repeated to ensure that the direction of cradling first chosen was stable for that participant. In the second stage, the authors used a tachistoscopic test to measure perception of facial emotion (Campbell, 1978). The final stage involved an assessment of facial expression (asymmetry of smile, Bennett et al., 1987): 16 leftward and 16 rightward cradling participants were randomly recruited from the total group and individually photographed when smiling. Findings revealed a significant leftward side for holding position of an infant baby doll, as if it were a real infant, in an imagined situation (Table 9.2). However, no significant difference between cradling groups was noticed for perception or expression of facial emotions. Thus, perception and expression of facial affect appeared to be independent.

Nevertheless, Manning and Chamberlain (1991) reported that the visual field could have an important effect (see Section 9.2.1.1.c). They assigned 388 girls to four groups: control, LEO, REO, and both eyes occluded (BEO). Each participant was asked to hold the doll as if it were a small baby. The position of the cradling (i.e. left, midline or right) was recorded. Girls in the REO group held the doll on the left side in the same proportion as displayed by controls; however, LEO and BEO groups differed from controls (Table 9.2). All participants were asked which hand they usually used for writing. No association between handedness and lateral cradling preferences was found.

To investigate whether the role of auditory information could also be as critical as the visual information (Sieratzki and Woll, 1996), Turnbull and

Matheson (1996) hypothesized that the left-cradling bias should be reduced in deaf mothers. For that purpose, 12 congenitally deaf people and 12 congenitally blind individuals were requested to cradle a lifesize infant doll, or to imagine cradling an infant. No clear lateral cradling bias was found for deaf people. In contrast, blind individuals displayed a left-side preference, both in the 'doll condition' and in the 'imaginary condition' (Table 9.2). However, the size of the sample was too small to establish the importance of auditory information.

In a second part of the study by Weiland and Sperber (1970), 21 adult women were asked to hold a pillow against them. Then, the same participants were offered the pillow a second time after being placed in an anxiety-oriented situation. The participants were asked to hold the pillow as if it were a real baby and to comfort it. A significant number of individuals that first held the pillow in the midline shifted the pillow to the left. In the anxiety-oriented situation (Table 9.2), females as well as males tended to hold a ball on the left side of the chest. In contrast, using a naturalistic condition, Hopkins and Parr (submitted) failed to find any difference in carrying between a real infant and a non-infant object (see above and Table 9.2).

Furthermore, Souza-Godeli (1996) investigated what were the side preferences for cradling a doll and for holding a package of the same size and weight. The participants were 520 Brazilian children. The doll as well as the package were presented (on different trials, in a randomized sequence) in the vertical position at the midline of the child's body. The observer recorded each child's behaviour. Souza-Godeli (1996) reported a left-sided preference for holding the doll but not for the package (Table 9.2). In order to assess whether the reaching-hand preference in an emergency situation was a good indicator of the mothers' handedness, Hatta and Koike (1991) monitored the hand used by 43 adult humans (28 right-handers, 10 left-handers and 5 ambidextrous participants) to take up an object. Handedness was scored. Participants were required to take up an important object as fast as possible in an experimentally simulated emergency situation. Participants were first informed that the object they were to take up and bring back was a very important material. Two experimenters recorded the hand used to take up the object. A total of 12 trials were conducted for each participant. No results are given concerning the asymmetry of the manner of taking up a baby, but the correlation coefficient between hand used and handedness was 0.898.

Harris, Almerigi and Kirsch (2000) proposed a completely imagined simulation study where 554 university undergraduates were used as participants: 501 right-handers (150 males, 351 females) and 53 left-handers (15 males, 38 females). Participants, with closed eyes, were asked to imagine holding a young infant in their arms; they were later asked to write on which side (left or right) they were holding the infant's head. In addition, they were requested to complete a short questionnaire in order to rate: their experience with young babies; their comfort level while imagining themselves holding the infant; their sex and age; and, finally, their handedness on a five-point scale. Right-handed women and men exhibited significant left-side holding biases that did not appear for left-handed subjects (Table 9.2).

9.5.1.4. Surveys of Photographs

Salk (1973) collected paintings and sculptures involving an infant or a child being held by an adult. Of 466 such works of art that were examined in art museums and galleries, 80% depicted the young being held on the left side and 20% on the right side (Table 9.2). Thompson and Smart (1993) photographed live mothers when they were holding their infants and also found a left-side preference (Table 9.2). In an attempt to assess the validity and universality of Salk's (1960) contentions, Richards and Finger (1975) analysed 268 photographs of men and women from Western (100), Eastern (112) and American Indian (56) cultures. A significant trend for women to hold children on the left side of the body was also observed in each culture. Men, however, did not show such a preference. Each photograph was categorized by the sex of the adult holding the child, whether the child was positioned on the left or the right side, and whether the child was nursing. However, the authors presented their results in a figure that did not allow one to know exactly the distribution of individuals in each condition. Nonetheless, findings indicated that women in all three cultural divisions were more inclined to hold their children on their left side than on the right (Table 9.2).

In the assessment of whether a similar left-side cradling preference might occur in men as well as in women, Manning (1991) sampled 1696 'family album' photographs (1119 females, 577 males). The pictures were then categorized with respect to sex and age of the children, the sex of the adult, the kinship relationship of the adult with the child, and whether the child was positioned to the left or right side. Findings demonstrated a highly significant difference between women and men ($\chi^2 = 27.7$, $p < 0.001$). Earlier, Harris and Fitzgerald (1985) investigated lateral holding preferences, in the same conditions, with a sample of 216 photographs of adults (52 males, 164 females) holding infants. Women showed a significant left-side preference for holding infants in non-feeding and feeding situations (63%). Men also (65%) exhibited a significant left-side preference in holding behaviour (Table 9.2). The breast-feeding exhibited a leftward bias (56%), but was not significant compared to the non-feeding (63%) and the bottle-feeding (75%) conditions.

Manning and Denman (1994) also classified 3297 photographs in which infants were cradled on one side (left or right) of adults (2361 females, 936 males). From this sample, the authors identified 167 women and 67 men, represented by at least six photographs. Photographs were categorized by whether the child was positioned to the left or right side. Photographs for which the infant was in a midline position or being fed were excluded. For holding, women showed a left-side bias (mean 60.1%, Table 9.2) and men a right side (mean 54.1%, Table 9.2); this difference between the means of men and women was found to be significant ($t(231) = 3.57$, $p = 0.0004$).

In brief, for the three categories of cradling, holding and carrying, the majority of the studies confirmed the original findings of Salk (1960; Table 9.2) that about 80% of individuals cradled, held or carried infants or children on the left side.

9.5.2. Asymmetries in Non-human Primates

We now examine the relevant literature pertaining to the existence of cradling, carrying and holding behaviours in non-human primate species. As in the preceding section, we have categorized the available studies by considering the different methods used to assess these lateral biases.

9.5.2.1. Free Observation

While observing the behaviour of a mother rhesus monkey and her newborn, Salk (1960) noticed for the first time that the mother had a marked tendency to hold the newborn on her left side, frequently with the newborn's ear pressed against her heart. In an attempt to systematize this observation, Salk described the behaviour of one mother rhesus monkey and of her newborn, and in particular the side of cradling. Salk (1960) observed that the newborn was held on the left side 40 times and on the right side only twice (Table 9.3).

Hatta and Koike (1991) investigated monkeys' hand preference in an emergency situation, caused either by an alarming voice or a moving mirror that frightened all the individuals in the cage. The subjects were 8 monkey mother–infant dyads (3 Japanese macaques, 4 Taiwan monkeys, 1 Bonnet monkey). The experimenter checked the hand with which the mother took her baby up when she was in the experimentally provoked emergency situation. Recording was conducted between 17 and 32 times with an inter-trial interval of 40 min over 3 days. Of the 8 mothers, 7 significantly used their left-hand more frequently than the right-hand (Table 9.3). No mother showed a right-hand preference in this experimentally induced emergency situation.

9.5.2.2. *Continuous Recording*

Nishida (1993) investigated the effect of maternal cradling on nipple preference; for that purpose, hand support of five mother–infant dyad chimpanzees was recorded continuously when the mother was supporting the infant with one hand while walking and carrying it. Unfortunately, it is not clear how the data were analysed. Except for one pair, involving a male infant cradled more with the mother's left hand (43 bouts with the left hand and 15 with the right hand), the number of bouts was too small to reach any conclusion about individual biases in the other four mother–infant pairs (Table 9.3).

Dienske, Hopkins and Reid (1995) observed each of nine chimpanzee mother–infant pairs for 1 h a day, and recorded whether or not the mother had her hand or arm around the infant (i.e. holding). No left bias was found in these ape groups (Table 9.3).

Some studies, however, showed a leftward tendency in cradling or carrying patterns. Thus, Hopkins et al. (1993b) videotaped 11 bonobos (five males and six females) during their morning feeding for approximately 2 days each week. Seven behavioural units were observed and scored for laterality. Among these behaviours, the carrying was coded according to whether an animal used one or both hands to grasp an object, an infant, or a food item and physically carried it to a different location in the cage, with at least three strides. For all measures, the unit of analysis was a bout rather than a single act. The mean percentage of right-hand use for carrying was 42.7%, which indicated a significant left-side bias (Table 9.3). The effect of quadrupedal and bipedal postures on lateral bias in carrying was also analysed. For eight subjects, data were available in both postural conditions. The bipedal condition had the effect of increasing the use of the left hand in the carrying measure. Hopkins and De Waal (1995) studied ten additional bonobos (five males and five females). Videotaped observation periods were 60 min in duration. Binomial z-scores and the percentages of right-hand use were also calculated in the same condition as in Hopkins et al. (1993b). Hopkins and De Waal (1995) found a significant deviation from 50% right-hand use, with a mean percentage right usage of 38% for carrying. When the data of Hopkins et al. (1993b) were combined with those of Hopkins and De Waal (1995) in order to establish the presence or absence of asymmetries at a population level for each of the behavioural measures, carrying patterns were significantly biased towards the left hand (Table 9.3). A similar finding

Table 9.3. *Summary of studies on cradling, holding and carrying in non-human primates*

Study	Species	Subject age	Procedure	Measure	Individual criterion	N	Right b	Right s (%)	Left b	Left s (%)	No preference s (%)	Bias	Statistics
Free observation													
Salk (1960)	*Macaca mulatta*	D (Nw)	Obs.	Side-cradling bouts	1 bout	1	2 bouts		40 bouts		?	L	?
Hatta and Koike (1991)	*Macaca fuscata*	D (Y)	Experimentally simulated emergency situation	Picking up the baby bouts	?, 1%	3	0		100	100		L	?
	Taiwan monkey	D (Y)				4	0		100	100		L	?
	Macaca radiata	D (Y)				1	0		100	100		L	?
Continuous recording													
Nishida (1993)	*Pan troglodytes*	D (Nw)	Continuous *ad libitum*	Arm-cradling bouts	?	5	40		60	20	0	?	?
Hopkins, Bales and Bennett (1993)	*Pan paniscus*	Y	Continuous videotape	Arm-carrying bouts	Bin z-score > |1.96|, 5%	11	27.3		72.7	45.5	0	L	$t = -1.82$ $p < 0.05$
Hopkins and De Waal (1995)	*Pan paniscus*	Y	Continuous videotape	Arm-carrying bouts	Bin z-score > |1.96|, 5%	10	20		80	40	0	L	$t = -2.54$ $p < 0.05$
	Pan paniscus	Y	Combined data of Hopkins et al. (1993) and Hopkins and De Waal (1995)			21	23.8		76.2	42.9	0	L	$t = -2.99$ $p < 0.01$
Dienske, Hopkins and Reid (1995)	*Pan troglodytes*	D (Y)	Continuous recording	Holding bouts	Runs test, 5%	9	55.6	33.3	44.4		0	No	?
Damerose and Hopkins (in press)	*Papio anubis*	D (Nw)	15-min focal-dyad sampling	Arm-cradling duration	LB	10	30		70		0	L	$z = 1.27$ ns
Instantaneous sampling													

Reference	Species	Type	Method	Bouts	N	b	% of right hand	s		Test	p
Tanaka (1989)	*Macaca fuscata*	D (Nw)	1-min scan sampling		16	43.7	50	6.3	No	?	?
Manning and Chamberlain (1990)	*Pan troglodytes*	D (Y)	15-s scan sampling	Infant head position bouts	10	16	84	0	L	Bin	p = 0.004
	Gorilla gorilla	D (Y)			4	18	82	0	L	Bin	p = 0.003
	Pongo pygmaeus	D (Y)			4	38	62	0	No	Bin	ns
Manning, Heaton and Chamberlain (1994)	*Pan troglodytes*	D (Y)	15-s scan sampling	Infant head position bouts	20		80	0	L	Bin	p = 0.004
	Gorilla gorilla	D (Y)			15		87		L	Bin	p = 0.003
	Pongo pygmaeus	D (Y)			8		62.5		No	Bin	ns
	Hylobates sp.	D (Y)			9		77.8		L	Bin	p = 0.18
Rogers and Kaplan (1998)	*Callithrix jacchus jacchus*	A (Y)	1-min sampling	Father-carrying bouts	8	25	75		R	z-score	?
				Mother carrying bouts	8	62.5	37.5		L		?
Tomaszycki et al. (1998)	*Macaca mulatta*	D (Nw)	1-min scan sampling	Arm-cradling bouts	41		LB[a]		L	t-test	ns
				Arm-carrying bouts	41		LB[a]		L	t-test, t = 2.38	p < 0.01
				Arm-retrieval bouts	41		LB[a]		L	t-test	ns
Damerose and Vauclair (1999)	*Papio anubis*	D (Nw)	1-min scan sampling	Arm-cradling bouts	10	40	60	30	L	LB[a], t-test, 5%; z = 0.63	ns

Obs., observation; D, mother–infant dyad; A, adult; Nw, newborn; Y, young; N, number of subjects; b, number (or percentage) of biased subjects; s, percentage of significant biased subjects; ?, information not provided in the original paper; No, no preference; R, right bias; L, left bias; Bin refers to a binomial test. See details in text: [a]LB (lateral bias), $(\#R - \#L)/(\#R + \#L)$. Bin refers to a binomial test. A percentage was calculated whenever possible from the available data. All results are given by reporting percentages of the side preference.

is provided by Damerose and Hopkins (2001) who recorded continuously ten mother–infant dyads of baboons: a tendency for the left arm to be used by mothers to cradle the infant was found (Table 9.3).

9.5.2.3. *Instantaneous Sampling*

The studies described in this section investigated whether left-side cradling is widespread in apes. First, Manning and Chamberlain (1990) published preliminary observations on 18 captive apes. Each primate mother–infant pair was observed for at least 1 h and observation sessions were uninterrupted whenever possible. Cradling behaviour was defined as ventral holding of the infant when the infant's head was touching the mother's torso. A scan and instantaneous sampling with an interval of 15 s was used to record the infant's head position. The infant's head was recorded as being on the mother's left side, on the right side or in the midline. The position of the infant's head was the only variable recorded. For example, suppose the infant's head to be on the left and the torso and limbs to the right of the mother's midline; this would then be recorded as a left-side position. Observations of suckling periods were excluded from the analyses. Fourteen animals exhibited an appreciable excess of left-side cradling (Table 9.3). Only one animal, an orang-utan mother, showed a marked preference for cradling her infant on the right side (92%). Considering the overall average lateralization ($N = 18$), a significant ($\chi^2 = 11.5$, df $= 1$, $p < 0.005$) lateral bias appeared towards the left side (79%) versus the right side (21%). In each species, an average of 80% mothers cradled on the left side.

In a new study, Manning, Heaton and Chamberlain (1994) presented data on a sample of 52 mother–infant pairs of captive apes. The total sample of captive mother–infant ape pairs was 20 chimpanzees, 15 gorillas, 8 orang-utans and 9 gibbons. For this study the authors used the same method adopted by Manning and Chamberlain (1990; see above). Chimpanzees, gorillas and gibbons showed a left-side cradling bias that did not appear in orang-utans (Table 9.3).

Several works using a scan and instantaneous sampling are available in investigations with monkeys. For example, Rogers and Kaplan (1998) scored behaviour in eight common marmoset infants by direct observation for half an hour with 1-min interval sampling, four times daily and for the first 60 days of life. Carrying was recorded separately for mother, father and siblings, and concerned four locations on the body: in front or on the back (middle positions), on the left and on the right. Scores for carrying did not include suckling. Hand preferences during the first 2 months after birth were also recorded in a range of situations, including holding an object, holding food and touching others. Results show that mothers carried their infants more often than fathers (53.3% and 44.7%, respectively) with a left bias for fathers and a right bias for mothers (Table 9.3). Tanaka (1989), also using an instantaneous sampling with a 1-min sample interval, recorded the hand used by the mother to embrace her newborn in 40 mother–infant Japanese macaque dyads. This study failed to find a significant bias (Table 9.3).

Tomaszycki et al. (1998) followed longitudinally 41 rhesus macaque mother–infant dyads (25 females, 16 males) living in social groups. Their goal was to assess whether rhesus monkeys show evidence of a left-side cradling bias. They measured the relative contribution of mother and infant to lateral cradling, as well as the effect of several maternal and infant characteristics on lateral cradling. Observation sessions began on either the first or second day of the infant's life and continued until the infant reached 6 weeks of age. Scan and instantaneous sampling procedures with 60-s intervals were used in data collection for lateral biases in nipple preferences (see Section 9.7.2.) and maternal cradling. An *ad libitum* sampling procedure was also used throughout the observation sessions to record infant carrying and retrieval. The observer coded the hand used to perform maternal cradling, infant carrying and retrieval. Mothers exhibited a significant left-arm bias for carrying their infant, but no population bias for maternal cradling or retrieval were observed. However, the cradling bias became more left-sided with increasing maternal experience (Table 9.3). Using the same procedure, Damerose and Vauclair (1999) also found a significant left-side cradling preference in ten baboon mother–infant dyads (Table 9.3).

In short, asymmetry patterns concerning cradling, holding and carrying of the infant by the mother are not as obvious for monkeys as they are for apes. All together, most of the studies with monkeys and apes showed either a left side or no preference for cradling as well as for carrying. This pattern of biases is remarkably similar to the pattern we have described for humans in the preceding section.

9.6. Head Orientation and Position

We examine in this section a number of studies concerned with asymmetries in head orientation and head position of newborns, both human and non-human, as these patterns might be related to postural asymmetries of the mother.

9.6.1. Asymmetries in Human Participants

9.6.1.1. Free Observation

Thompson and Smart (1993) conducted an observational study with a total sample of 150 mother–infant pairs. The direction of 35 baby's lateral head movement at birth was recorded as the first observable lateral movement. It was noted whether the baby rotated to its right or left or not at all: a score of right, left or zero was given for each baby. Most babies turned their head significantly more to the right than to the left (Table 9.4).

9.6.1.2. Experimental Studies

The studies described below recorded the head-turning response in human infants, following head rotation of the infant by the experimenter.

Michel and Goodwin (1979) tested 109 newborn infants (50 females and 59 males). They set three criteria for a participant's inclusion in the study: (1) a full-term vaginal delivery birth; (2) a birth record; and (3) the attending obstetrician must have filled out the researcher's forms indicating the position of the infant's head during descent. These forms also gave the experimenters permission to visit the mother to obtain her agreement for observing her baby. The infant's birth position was determined by occiput position relative to the mother at four phases during delivery. The newborn's posture was observed during 16–50 h postpartum. The infant was placed in a supine position, the head was held gently in midline position for 1 min. Then, the head was released and state, head and digit positions were recorded on a checklist every 6 s for 1 min. After this observation, the head was held gently with the left ear flat on the mattress for 1 min. Again, postures and state were recorded every 6 s for the following minute. Finally, the head was held in the same way to the right. The order of starting head orientations (midline, left, right) was counterbalanced between infants.

Immediately following these procedures, the infant was held in a prone position, and the whole set of procedures was repeated. The infant's head was placed in midline and turned to the left and then also turned to the right. Head orientation while supine was recorded in three categories: head-right turn, defined as nose/chin to right of right nipple with right ear touching the mattress; head-left turn, defined similarly for left nipple; midline position, defined as nose/chin position between right and left nipples. Head orientation while prone was recorded as: head-right, occiput oriented over or beyond left scapula; head-left, occiput oriented over or beyond right scapula; midline position, occiput between scapulae. Head orientation data were analysed separately for the supine and prone conditions. Significantly more infants oriented right and had significant head-right preferences while supine (Table 9.4). There were no significant differences while prone (Table 9.4).

Later, using the same procedure, the direction of supine head orientation was determined by Michel (1981) for 150 normal full-term neonates (81 males and 69 females), by two separate assessments during the 16–48 h after birth. Each assessment consisted of three 2-min trials. The distribution of preferences was biased significantly to the right (Table 9.4), and sex differences were not significant.

Konishi, Mikawa and Suzuki (1986) selected 44 infants (25 males and 19 females) from 82 relatively low-risk preterm infants: all were born before the 37th week of pregnancy. An additional sample of 53 healthy, mature infants were used as controls. The infant was placed in the supine position and the experimenter held the head, at the temple, between thumb and forefingers until no resistance was felt. The head was then released and the direction of the first head turn was recorded. This test was repeated five times. Infants turning their heads toward the same side four times or more were considered to have asymmetrical head-turning preferences; those turning their heads to one side three times or less were regarded as having bilateral head-turning preferences. Hand preference was examined at the age of nine and 18 months and was defined in terms of reaching. The infant sat on the mother's lap and the experimenter placed a toy directly in front of the infant at chest level for 60 s. Preference in using the right or left hand was established when an infant reached for the toy with the same hand three times or more. Until the age of 6 months, more than half of the preterm infants turned their heads to the right side preferentially, while most full-term infants turned their heads bilaterally at 3 months (Table 9.4). At 9 months, a significant ($p < 0.02$) number of preterm and full-term infants who had preferred to turn their heads to the right at 1 month, used their right hand more than their left. Similarly, most of the preterm and full-term infants with head preference to the left used their left hand more often than their right at 9 months.

With a total sample of 150 mother–infant pairs, Thompson and Smart (1993) recorded lateral head turning when the infant was supine. Neonatal head turning behaviour was tested on two consecutive days following birth. The baby was laid in a central, supine position with its head held in the midline for 15 s, then released. The baby's first lateral head turn, after release, in excess of 20° was recorded. After a period of 1 min the baby's head was returned to the midline. The baby was tested for five trials with intertrial intervals of 1 min and then a further five trials were given a day later. Of the 144 infants tested for head-turning preference, 96 showed a predominant lateral head-turning bias, but no sex differences were found (Table 9.4).

Table 9.4. Summary of studies on infant head orientation in human.

Study	Participant	Procedure	Measure	Individual criterion	Condition	N	Right b	Right s (%)	Left b	Left s (%)	No preference (%)	Bias	Statistics
Free observation													
Thompson and Smart (1993)	*Newborn infant*	Obs. following birth	Head-turn bouts	1 bout		35	24		11		0	R	$\chi^2 = 4.83$ $p < 0.03$
Experimental studies													
Michel and Goodwin (1979)	*Full-term infant*	3 trials (2-min)	6-s interval head position bouts	z-score >1.96I	Supine	109	75%	43	25%	9		R	Bin $p < 0.0001$
					Prone	109	54%	14	45%	8		No	Bin $p < 0.2236$
Michel (1981)	*Full-term infant*	3 trials (2-min)	6-s interval head orientation bouts	z-score>1.8I	Male	81	46		13		22	R	$\chi^2 = 36.9$ $p < 0.005$
Konishi, Mikawa and Suzuki (1986)	*Preterm infant*	5 trials	Head-turn bouts	Same side at least 4 times	Female	69	51		9		9	R	$\chi^2 = 59.7$ $p < 0.001$
					1 month	44	79.5%		11.4%		9.1%	R	?
					3 months	44	77.3%		9.1%		13.6%	R	?
					6 months	44	59.1%		4.5%		36.4%	R	?
					9 months	44	27.3%		4.5%		68.2%	No	?
	Full-term infant	5 trials	Head-turn bouts	Same side at least 4 times	1 month	53	52.9%		28.3%		18.9%	R	?
					3 months	53	11.3%		11.3%		77.4%	No	?
					6 months	53	3.8%		1.9%		94.3%	No	?
					9 months	53	0%		0%		100%	No	?
Thompson and Smart (1993)	*Newborn infant*	5-trial tests (2 times)	Head-turn bouts	Head-sided turn		144	72		24		48	R	$\chi^2 = 24.26$ $p < 0.0001$
Rönnqvist and Hopkins (1998)	*Newborn infant*	3 trials with apparatus	Head-turning side	Same side at least 2 times	Male	8	6		2		0	R	?
					Female	10	6		3		1	R	?
			Head-turning maintenance	Same side at least 2 times	Male	8	6		1		1	R	?
					Female	10	6		3		1	R	?
Rönnqvist et al. (1998)	*Newborn infant*	3 trials with apparatus	Head-turning side	Same side at least 2 times	Male	5	4		1			R	?
					Female	10	6		3		1	R	?
													?

Obs., observation; *N*, number of participants; *b*, number (or percentage) of biased participants; *s*, percentage of significant biased participants; L, left bias; R, right bias; No, no preference; ?, information not provided in the original paper; ns, non significant. Other symbols as for Table 9.3.

Rönnqvist and Hopkins (1998) carried out a study on head orientation using an apparatus with 18 newborn infants (10 females and 8 males). The supine newborn was secured on a custom-built platform by a belt covering the trunk. The head was placed in a holder that, when unlocked, allowed the head to move from side to side through a range of 140°. A standard pro-tractor mounted on the head end of the platform together with a pointer attached to the holder's axis were used to measure the number of degrees that the head turned from the midline position. At the start of a trial, the head holder was locked in the midline position. It was unlocked once the infant's head was judged to be in the body midline with the eyes centered. This procedure was repeated three times. For each trial, the movement of the head, as indicated by the pointer traversing the protractor in 5° units, was scored second by second from the video recordings up to a maximum of 20 s. After the third repetition, the maintenance of a head position was recorded for 5 min. The mean number of turns to the right was 2.0, compared to 0.94 for left-sided and only 0.1 for the midline position. The percentage of time involving a right-sided position ($M = 60.6\%$) was significantly greater than for left-sided position ($M = 27.9\%$), and midline ($M = 11.4\%$). There was a significant difference between the two positions, $t(17) = 2.1$, $p < 0.05$, such that the angle to the right ($M = 35.4°$) was larger than to the left ($M = 27.3°$). In short, the majority of infants turned maximally more to the right than to the left for both assumption and maintenance (Table 9.4). The difference between boys and girls was not significant for either assump-tion or maintenance. In a study using the same procedure as that described by Rönnqvist and Hopkins (1998), Rönnqvist et al. (1998) sampled 15 new-born infants (10 females and 5 males) and found a similar pattern (Table 9.4): namely, the mean number of turns to the right was 2.0 compared to 0.93 for those to the left.

To summarize, most of the human studies reported a right-side bias for head turning of the infant while it was supine, but not when it was in prone (e.g. Michel, 1981). This rightward tendency seems to develop with the age of the infant (Konishi, Mikawa and Suzuki, 1986).

9.6.2. *Asymmetries in Non-human Primates*

9.6.2.1. *Free Observation*

All studies of head orientation or position in non-human primates (Table 9.5) were conducted with different observational sampling procedures. Thus, Hopkins and Bard (1995) observed the behaviour of 43 chimpanzees (18

Table 9.5. *Summary of studies on infant head orientation and position in non-human primates.*

Study	Species	Subject age	Procedure	Measure	Individual criterion	Condition	N	Right		Left		No preference (%)	Bias	Statistics	
								b	s (%)	b	s (%)				
Free observation															
Hopkins and Bard (1995)	*Pan troglodytes*	Y	Obs. prior to NBAS test	Head-orientation bouts	LB[a]	Prone	40	40		60		0	No	$\chi^2 = 1.60$	$p > 0.10$
						Supine	36	83.3		16.7		0	R	$\chi^2 = 16.00$	$p = .001$
Continuous recording															
Dienske, Hopkins and Reid (1995)	*Pan troglodytes*	Y	Continuous recording	Head-position bouts	Runs test, 1%		9	66.7	33.3	33.3	11.1	0	R	?	?
Westergaard, Byrne and Suomi (1998)	*Cebus apella*	Y	10-min (6 sessions)	Head-orientation duration	LI[a], t-test, 5%		16	25		75		0	No	$t(15) = -1.87$	ns
Damerose and Hopkins (in press)	*Papio anubis*	Nw	15-min focal-dyad sampling	Head-position duration	LB[a]		10	80		20		0	R	$z = 1.90$	$p = 0.10$
Scan sampling															
Damerose and Vauclair (1999)	*Papio anubis*	Nw	1-min scan sampling	Head-position bouts	LB[a], t-test, 5%		10	90	40	10	10	0	R	$z = 2.53$	$p = 0.05$

Obs., observation; Nw, newborn; Y, young; N, number of subjects; b, number (or percentage) of biased subjects; s, percentage of significant biased subjects; L, left bias; R, right bias; No, no preference; ?, information not provided in the original paper; ns, non-significant; [b]NBAS, the Neonatal Behavioral Assessment Scale. See details in text: [a]LB (lateral bias) = LI (lateral index), (#R − #L) / (#R + #L).
All results are given by reporting percentages of the side preference. A percentage was calculated whenever possible from the available data.

338

females and 25 males) for the first 3 months of life. Head orientation data were collected prior to the administration of the NBAS test (the Neonatal Behavioral Assessment Scale; Brazelton, 1984) by recording which side of the chimpanzee's face was resting on the sleeping mat. If the chimpanzees were lying prone, then a left-face contact was scored as a right-orientation bias, and a right-face contact was scored as a left-orientation bias. In the event that the chimpanzees were supine (sleeping on their back), a right-face contact was scored as a right-orientation bias, whereas a left-face contact was scored as a left-side bias. At the end of the 3-month observation period, the total frequency of observations for right- or left-sided biases that were cross-classified with prone or supine posture was determined for each subject. An overall significant right-side lateral bias was found for head orientation in the supine posture while not in the prone posture (Table 9.5). A trend toward greater right-side bias in females compared with males was observed but failed to reach statistical significance.

These data suggest that asymmetries in head orientation are present early in life in chimpanzees, and that they may be correlated with functional asymmetries observed in adulthood. For both sleeping postures, no significant sex differences were found in the direction of lateral bias; however, in the supine posture, right-side biases in head orientation were higher in females (93%) compared to males (76%). The average LB (that is, right occurrences minus left occurrences divided by right plus left occurrences) score of subjects sleeping in supine posture ($M = 0.34$) was significantly ($t(37) = -4.45$, $p < 0.01$) shifted to the right side in contrast to the average LB scores of subjects in a prone posture ($M = -0.05$) which was not different from zero ($t(41) = -9.07$, $p > 0.10$).

9.6.2.2. Continuous Recording

Dienske, Hopkins and Reid (1995) observed nine mother–infant pairs of chimpanzees for 1 h a day. All types of contacts between the infant and the mother's ventral trunk were recorded as on-mother (see above). A distinction was made between head positions being left, midline and right on the mother's chest. Of the 69 testable observation periods, on-mother was eight times significantly left-sided and 21 times significantly right-sided. However, more subjects had a right-side head position than left (Table 9.5). Also, with another small sample of olive baboons (ten infants), Damerose and Hopkins (2001) using a focal-dyad sampling with continuous recording found a right-ward head position (Table 9.5).

By contrast, in another study, Westergaard, Byrne and Suomi (1998) collected data from videotape recordings during six 10-min sessions in 16 mother-reared, tufted capuchin infants (11 males and 5 females) housed in social groups. These authors adapted a measure of head orientation for humans and chimpanzees (humans: Michel, 1981; apes: Hopkins and Bard, 1995) for use in monkeys that carry the infants dorsally. The observers noted the duration for which an infant maintained its head on the right or the left side of its mother's back as she carried the full weight of the infant (dorsal riding). Most of the subjects were left-side turned (Table 9.5), and no significant differences appeared between males and females.

9.6.2.3. Instantaneous Sampling

Damerose and Vauclair (1999) followed the behaviours of ten mother–infant dyad baboons using a scan and instantaneous sampling procedure. They found a significant right-side head position, with significant individual biases toward the left or the right (Table 9.5).

In brief, similarly to human studies, non-human primate investigations reported most often a right turn for the head of the infant, with the exception of two studies that revealed no preference (in prone posture: Hopkins and Bard, 1995; Westergaard et al., 1998).

9.7. Nipple Preference

Preferences expressed by infants may have an important effect on the postural behaviour of the mother (i.e. cradling). We believe it useful to examine the published data in relation to this behaviour in non-human primates (to our knowledge, no data on this possible bias in humans are available).

9.7.1. Continuous Recording

Tanaka (1989) used focal-animal sampling with continuous recording on 40 mother–infant Japanese macaque dyads. The point when the infant's mouth touched and left its mother's nipple, and which of the two nipples the infant used, right or left, were recorded. The infant mostly used only one nipple, an average of 90.9% of nipple contact time. After 1 month of age, all the infants used only one of the two nipples during more than 80% of nipple contact time. The same tendency, albeit weak, was also detected in the first month, but it did not appear to be biased at the population level (Table 9.6). In another study, Dienske et al. (1995) recorded nine chimpanzee mother–infant pairs, but failed to find any left or right bias (Table 9.6).

On an *ad libitum* basis, Nishida (1993) observed 34 mother–infant chimpanzee dyads with a focal-animal sampling procedure implying continuous

Table 9.6. *Summary of studies on nipple preference in non-human primates*

Study	Species	Subject age	Procedure	Measure	Individual criterion	N	Right b	Right s(%)	Left b	Left s(%)	No preference	Bias	Statistics
Continuous recording													
Tanaka (1989)	*Macaca fuscata*	D (Nw)	Continuous focal-animal	Nipple contact duration	% of right nipple	40	52.5	?	47.5		0	No	? -
Nishida (1993)	*Pan troglodytes*	D (Nw)	Continuous *ad libitum*	Nipple contact duration	Bin, 5%	32	40.6		59.4	21.9	0	L	?
				First sucking	Bin, 5%	34	38.2		55.9	23.5	5.9	L	Bin, $p < 0.05$
Dienske, Hopkins and Reid (1995)	*Pan troglodytes*	D (Y)	Continuous recording	Nipple contact bouts	Runs test, 5%	9	44.4	22.2	55.6	11.1	0	No	?
Damerose and Hopkins (in press)	*Papio anubis*	D (Nw)	15-min focal-dyad sampling	Nipple contact duration	LB[a]	9	22.2	22.2	77.8		0	L	$z = 1.67$, $p < 0.10$
Scan sampling													
Rogers and Kaplan (1998)	*Callithrix jacchus jacchus*	D (Y)	1-minute sampling	Teat preference	z-score > 12.54I, 1%	15	60	40	40	33.3	0	No	?
Tomaszycki et al. (1998)	*Macaca mulata*	D (Nw)	1-min scan sampling	Nipple contact bouts	LB[a]	41			Left			L	$t(38) = 2.66$, $p < 0.01$
Damerose and Vauclair (1999)	*Papio anubis*	D (Nw)	1-min scan sampling	Nipple contact bouts	LB[a], *t*-test, 5%	9	33.3	22.2	66.7	66.7	0	No	$z = 1$, ns
Test sessions													
Erwin and Anderson (1975)	*Macaca nemestrina*	D (Nw)	15-min focal-dyad sampling	Distinct nipple contact	Bin, 5%	56	Right					R	$\chi^2 = 2.17$, $p < 0.10$

Obs., observation; D, mother–infant dyad; A, adult; Nw, newborn; Y, young; N, number of subjects; b, number (or percentage) of biased subjects; s, percentage of significant biased subjects; L, left bias; R, right bias; No, no preference; ?, information not provided in the original paper; ns, non significant. See details in text: [a]LB (lateral bias) = (#R − #L) / (#R + #L). A percentage was calculated whenever possible from the available data. Other symbols as for Table 9.3.

recording. A suckling bout was defined as a period with nipple contact during which no interruption to nipple contact lasted for more than 1 min. Nipple contact renewed after interruptions longer than 1 min was considered to constitute the start of a new bout. Only nipple contacts lasting more than 10 s continuously were treated as a suckling bout, since brief contacts appeared to be 'reassurance nipple contacts'. The number of occurrences of such bouts was 736. Nipple preferences were determined in terms of the nipple sucked first and by measuring the duration of suckling. All of the infants who showed a statistically significant laterality were biased to the left, for both first nipple and contact duration of nipple (Table 9.6). It was likely that the left nipple preference was facilitated by the mothers' tendency to support neonates with their left arm. Damerose and Hopkins (2001) found also a leftward bias in baboons (Table 9.6).

Tanaka (1997) undertook a study with 150 mother–infant dyads in free-ranging Japanese macaques (primiparous and multiparous). The aim of the study was to examine how parity-related physiological differences in lactation affect suckling behaviour with an estimate of milk secretion rates. For that purpose, Tanaka used a focal-animal sampling method for mother–infant dyads to record when the infant began to suck its mother's nipple and when it stopped, and computed the duration of a nutritive sucking bout. The findings showed that the infants of primiparous mothers used the preferred nipple less during nutritive sucking than did the infants of multiparous mothers. In nutritive sucking, Japanese macaque infants preferred one single nipple. The infant of primiparous mothers appeared to supplement the physiological drawbacks in their mothers by behavioural means (i.e. by use of the supplementary nipple).

9.7.2. *Instantaneous Sampling*

Rogers and Kaplan (1998) scored the behaviour of a total of 15 common marmosets infants obtained from three different parents. Behaviour was scored daily by direct observation for half an hour of interval sampling every minute. Scoring took into account the side on which each infant suckled, how often it attached to the teat and which infant suckled. The results reveal that the infants had teat preferences with a significant right or left bias at the individual level, and that twins preferred opposite teats. However, teat preferences were not an artifact of twins suckling together, that is, due to only one infant having a preference and thus imposing a preference on the other infant. Their correlation was also not due to counting suckling simultaneously as carrying. 'Carrying' was scored as a non-suckling activity. Comparing infants raised by their father and mother with all infants, including those raised with their father as well as those raised without their father but in the presence of siblings, the authors found a weak but significant correlation between suckling and carrying. According to them, these findings suggest the possibility that the presence of the father had an influence on the side of infant suckling and carrying by the mother. The scores for hand preferences in the first 2-month period of testing showed no significant biases (i.e. hand preferences had not developed). Also, in the first 2 months of life, there was no significant relationship between hand preference and teat preference ($r = 0.03$, $p = 0.9$, Spearman correlation). Hand preferences for holding food developed later in all offspring and stabilized by 10–12 months of age. No significant relationship was found between teat preference and hand preference at any age. Among the 15 individuals tested, six infants exhibited a right hand preference and nine had a left-hand preference.

Following the procedure described in Section 9.5.2.3, Tomaszycki et al. (1998) collected data on lateral biases concerning nipple preferences among 41 rhesus macaque mother–infant dyads. Infants showed a significant left-side nipple preference (Table 9.6) in the first weeks of life, and this decreased and became not significant at 3 weeks. Damerose and Vauclair (1999) in nine baboon mother–infant dyads also found, in the same conditions, a trend toward left-side nipple preference (Table 9.6; see also Damerose and Hopkins, 2001).

9.7.3. *Experimental Studies*

The nursing behaviour of 56 (27 males and 29 females) singleton infant pig-tail monkeys (*Macaca nemestrina*) was monitored by Erwin and Anderson (1975). Each infant was observed on two occasions for as long as was necessary to observe ten distinct nipple contacts. Thus, each session was composed of ten two-choice trials. Observers scored each nipple contact only after a period of non-contact had intervened after a previous nipple contact. To allow assessment of the consistency of the phenomenon, a second observation session was conducted on a different day. Infants over 1 month of age showed strong preferences for one of their mother's nipples, while few infants less than 1 month old did so. A trend toward lateral preference (for right nipple) was identical for male and female infants (Table 9.6). Of the 56 infants observed, 46 (82%) showed significant preferences for one nipple over the other. Twenty-four infants exhibited exclusive contact with the preferred nipple. The intersession reliability was remarkably high, 0.94. More subjects preferred the right to the left nipple, and no sex differences were found.

In summary, a tendency for a left preference for the nipple suckled seems to emerge from the preceding studies, although a few of them reported opposite biases (see Table 9.6).

9.8. Grasping

Asymmetry in the grasping reflex may allow prediction of the future hand-edness of the infant. It is, however, important to observe that grasping is not a behaviour that involves mother and infant, but is to be considered a motor act similar to reaching. We decided to include this behaviour in our review because spontaneous reaching preferences might affect later hand preferences both in human and in non-human primates.

9.8.1. Human Studies

For studying the asymmetry in grasping (Table 9.7), authors have used experimental paradigms to determine which hand first grasps a small rattle (Caplan and Kinsbourne, 1976) or asymmetry in grasping duration (Caplan and Kinsbourne, 1976; Thompson and Smart, 1993). Thus, Caplan and Kinsbourne (1976) tested 21 infants (2 months of age), none of whom had a familial left-hander, and five babies who had familial left-handers. In this study, each participant was placed in a sitting position, leaning against the knees of the reclining examiner. At the beginning of each trial, the partici-pant's trunk, limbs and head were centered. There were two parts in the experiment. In the first part, a small rattle was placed in one of the baby's hands and the number of seconds before it was dropped was recorded. Eight trials were run, alternating hands. In the second part, two identical rattles were used, and one was placed in each hand simultaneously. The first hand to drop the rattle and the duration for which the grasp persisted were noted; a total of four trials were run. Infants displayed a significant right-hand grasp-ing when one rattle was presented, but no bias when two rattles were simul-taneously presented in each hand (Table 9.7); no sex differences emerged for the five familial left-handed participants.

Thompson and Smart (1993) carried out a study with a total of 150 new-born infants. In addition to measuring the infant's head-turning biases (see Section 9.6.1.2), and the maternal cradling-side (see Section 9.5.1), the authors tested the palmar grasp reflex while the baby was sitting centrally on its mother's lap. A rod was placed alternately in the open palm of the

Table 9.7. *Summary of studies on infant grasping in humans.*

Study	Participant	Procedure	Measure	Condition	N	Right (%)	Left (%)	No preference (%)	Bias	Statistics
Experimental studies										
Caplan and Kinsbourne (1976)	*Familial right-hander infant*	8 trials	Grasping duration	One rattle	21	57.1	14.3	28.6	R	$\chi^2 = 6.00$ $p < 0.05$
		4 trials	Grasping duration and first hand	Two rattles	20	50	15	35	No	χ^2 ns
	Familial left-hander infant	Part one and two combined	Grasping duration and first hand		5	40	20	40	No	
Thompson and Smart (1993)	*Newborn infant*	5-trial (2 times) with 15-sec interval	Rod grasp-sided duration	Palmar	149	51.7	47.6	0.7	No	χ^2 ns
			Disc grasp-sided duration	Plantar	147	53.1	45.6	1.3	No	χ^2 ns

N, number of participants; R, right bias; No, no preference; ns, non-significant. A percentage was calculated whenever possible from the available data. All results are given by reporting percentages of the side preference.

baby's right and left hands. The palmar grasp was defined as occurring when all four fingers closed around the rod and terminated when the rod was released from the grasp of all four fingers. The plantar grasp response was elicited by placing a plastic disc alternately under the baby's right or left toes. The plantar grasp was defined as all four smaller toes gripping the disc and terminated when all four toes extended. The duration of grasp was timed, and two periods of five trials were performed. Ten such trials were given each day, with a 15-s intertrial interval. No significant lateral differences were found in palmar or plantar grasp (Table 9.7). As many babies held for the same duration with their right hand as with the left, and the same held for the plantar grasp response.

9.8.2. *Non-human Primate Studies*

Studies involving non-human primates (Table 9.8) most often utilized experimental test sessions, just as for human studies, with the exception of Westergaard, Byrne and Suomi (1998), who used observations to determine grasp side bouts in 16 tufted capuchin infants (11 males and 5 females). After the head orientation was recorded (see Section 9.6.2.2), the frequency with which an infant used its right or left hand to grasp and manipulate objects, during the time that the infant was not in contact with its mother, was noted. The observer recorded only the initial grasp of an object during each succession of grasping actions in order to maintain independence between data points (McGrew and Marchant, 1997). At 23–24 weeks, subjects showed a significant grasping to the left that disappeared at 47–48 weeks (Table 9.8), and no significant differences were found between males and females.

Other studies used test sessions to determine grasping asymmetries. Thus, Bard et al. (1990) administered the NBAS every other day from 2 days after birth to 6 weeks after birth and then once a week through 12 weeks of life in 12 (7 males and 5 females) chimpanzees. The NBAS is a test designed for neonatal humans that investigates the neurobehavioural integrity and consists of 28 behavioural items and 18 reflexes. Among these behavioural items, we will consider in particular the defensive grasps and for the reflexes, the plantar (foot) grasp, the hand grasp and the tonic neck reflex (TNR; an asymmetric reflex elicited by turning the head to the side when supine). No clear bias was revealed (Table 9.8) in this study.

To explore whether or not newborn chimpanzees exhibit an asymmetry in grasping similar to that observed in human infants, Fagot and Bard (1995) used as subjects 13 neonate chimpanzees. Grasping bouts were recorded using an apparatus made of a gripometer equipped with a strain gauge.

Table 9.8. *Summary of studies on infant grasping in non-human primates*

Study	Species	Subject age	Procedure	Measure	Individual criterion	Condition	N	Right		Left		No preference (%)	Statistics
								b	s(%)	b	s(%)		
Free observation													
Westergaard, Byrne and Suomi (1998)	Cebus apella	Y	Obs.	Initial grasp sides	LI[a], t-test, 6%	23-24 weeks	16	18.8		75		6.2	$t(15) = -4.05$ $p < 0.01$
						47-48 weeks	16	25		68.8		6.2	$t(15) = -2.18$ $p > 0.05$
Experimental studies													
Bard, Hopkins and Fort (1990)	Pan troglodytes	Nw	NBAS[b] test	Defensive grasp bouts	DI[a], z-score, 5%	?	12	58.3	16.7	33.3	8.3	8.3	?
				Plantar grasp bouts	DI[a], z-score, 5%	?	7	28.6		57.1		14.3	?
				Hand grasp bouts	DI[a], z-score, 5%	?	5	60		20		20	?
Fagot and Bard (1995)	Pan troglodytes	Nw	Gripometer	TNR[c] bouts	z-score, 5%	?	12	58.3	8.3	41.7		0	?
				Duration and strength of grasping		?	12	75		25		0	Bin $p = 0.07$

Obs., observation; Nw, newborn; Y, young; N, number of subjects; b, number (or percentage) of significant biased subjects; s, percentage of significant biased subjects; ?, information not provided in the original paper; [b]NBAS, the Neonatal Behavioral Assessment Scale; [c]TNR, tonic neck reflex. See details in text: [a]LI (lateral index), DI (dominance index), $(\#R - \#L) / (\#R + \#L)$;

All results are given by reporting percentages of the side preference. A percentage was calculated whenever possible from the available data. Other symbols as for Table 9.3.

The gripometer consisted of a Lexan stick longitudinally split along two-thirds of its length. Grasping applied forces to the strain gauge fixed at the end opposite to the split. During test sessions, data were sampled at 15 Hz. Grasping responses of the hand and foot were recorded. Tests were carried out at least twice a week. The chimpanzee was placed supine on a blanket on the floor. The experimenter recorded one grasping response per hand and foot, following a pseudo-random predetermined order. Grasps were recorded only if the chimpanzees were: (1) in a symmetrical posture; (2) awake but quiet; and (3) not sucking thumb or hand. For both duration and strength, right-hand (and right-foot) responses were longer and stronger than left-hand (and left-foot), respectively, but, except for the strength of the hand, these differences were not significant (Table 9.8).

All together, these studies, which are more numerous for non-human primates than for humans, reveal a wide variability in the asymmetric patterns of grasping.

9.9. A Methodological Proposal to Study the Relationships between Cradling and Other Motor Biases

The above survey of the different methods used to study maternal behaviours shows how important it is to describe clearly and define the procedures, as well as the behavioural units of interest. In fact, we have shown that the bias, for example, for infant holding, can be different depending on the unit of analysis chosen. Thus, Manning and Chamberlain (1990) recorded only the head position of the infant and concluded that there is a cradling bias, whereas Hopkins et al. (1993a) recorded the arm used by the mother to measure infant cradling. We suggest that researchers systematically verify the relation between the units selected before providing any firm conclusion. It is also necessary to specify clearly the sampling procedures used as well as the kind of data analysis performed. In several human studies, there is even an absence of statistical analyses.

Before proposing our descriptions and definitions of measures of lateral biases as we use them in the investigations of laterality in olive baboons, we discuss briefly the advantage of having a non-human primate model to study these questions. The evaluation of human handedness depends on the cultural background and on the criteria used to determine the preferred hand (Porac and Coren, 1981). Moreover, from birth, the world of objects (both social and inanimate) influences humans in an asymmetrical manner and this action expresses itself in a prominent way in maternal behaviour. Provins (1997) assumes that cultural biases in human populations tend to favour the use of the right hand. As a result, individuals who begin in infancy by using their right hand in some unimanual tasks would have this tendency strengthened. The more common the functional neuromuscular activity, the more it will facilitate the acquisition or learning of other motor skills. An example is provided by the mastery of the manipulative action pattern involved in holding a crayon or a paintbrush with the finger of one particular hand.

During the first year of life, the human infant is placed in an oriented environment. As the development of the manual asymmetry in humans starts very early in life (perhaps even during the prenatal period; see Hepper, Shahidullah and White, 1991), the most likely source of influence would come from acts and attitudes of the parents and the teachers. Thus, the study of the effect of maternal behaviour on infants' hand laterality in humans is made difficult by the fact that we are not able to distinguish exactly this environmental contribution from the cultural ones.

Compared to humans, we can assume that non-human primates do not have the same kind of cultural pressures. Firstly, this situation allows us to remove a factor that is hard to determine and to control. The investigation of the development of handedness in non-human primates, in particular in monkeys, should permit us to isolate the effect of maternal behaviour on these biases. Secondly, unlike apes, monkeys and notably cercopithecids, such as baboons, do not have a long breast-feeding period. The young baboon is usually independent from the mother after 5–6 months of age (Altmann, 1980), whereas in chimpanzees the infant can stay with the mother until 4–5 years of age (Lawick-Goodall, 1968). This long breast-feeding period in chimpanzees can induce overlap of two offspring, which in turn might influence infant behaviour and thus affect the postural biases expressed by the mother. Conducting such a study with baboons should prevent overlaps between siblings and thus control better for the influence of maternal behaviour.

9.9.1. *Procedure*

We shall now report the techniques we used for studying lateral biases of behavioural units that involve the mothers' as well as the infants' preferences and postures, in an ongoing study with olive baboons (*Papio anubis*). The main objective of our study is the effects of postures and hand preferences of the mother on her infant's manual laterality. For that purpose, our investigations involve both observations and experiments. We rely on observations of (1) the asymmetrical nursing behaviours of the mother as well as the postural behaviours of the infant in relation to the mother, and (2) the exploratory activities of infants (e.g. carrying of food and non-food items). Moreover, we have devised a set of tasks (both unimanual and bimanual) in order to investigate manual preferences of the mother. It is also planned that, when the infants reach 6 months of age, their manual preferences will be evaluated with the same sets of tasks used for the mothers.

The subjects are 43 mother olive baboons (*Papio anubis*) belonging to the colony ($n = 90$ baboons) housed in two large outdoor compounds ($20\,\mathrm{m} \times 25\,\mathrm{m}$) at the field station of the Station de Primatologie of the CNRS (Rousset-sur-Arc, France). The group is composed of 2 adult males, 43 adult females, subadults, juveniles and infants. The adults are wild-born and their ages are not known. A tunnel connects the two outside enclosures. Each outdoor compound has an attached indoor building $7\,\mathrm{m} \times 6\,\mathrm{m}$. The baboons can move freely between the indoor and outdoor enclosures, except during the observation periods when they are locked in the outdoor enclosure. The study includes all the pregnant females and their respective infants over a 24-month period. A small experimental enclosure ($3\,\mathrm{m} \times 2\,\mathrm{m}$) connected to one of the outdoor compounds permits temporary isolation of one or more individual(s) from the rest of the group so that the different tests of hand preference can be carried out. All animals involved in the study are identified by a collar.

9.9.1.1. Nursing Behavioural Units

Nursing is evaluated through five basic measures in mother–infant dyads, from birth to 6 weeks of age. Behavioural observation sessions of mother–infant dyads begin on either the first or second day after delivery, and continue until the infant is 6 weeks old. Throughout the first 6 weeks, all mother–infant dyads are observed twice a week using two sampling procedures: (1) scan sampling, in 1-h observation sessions; and (2) focal-dyad sampling, in 15-min observation sessions. The five basic measures of lateral bias that are recorded from either the mother or the infant are maternal cradling, maternal carrying, infant retrieval, infant head position and infant nipple preference. *Cradling* behaviour is recorded when the mother holds the newborn with one hand (left- or right-cradling arm) or with both hands (both cradling).

This use of cradling is congruent with that proposed by Tomaszycki et al. (1998). Cradling is only recorded when the mother is seated and when she holds the infant ventrally. During the breast-feeding periods, a *nipple preference* is recorded whenever the infant is suckling the left or right nipple. No distinctions are made between the different suckling phases (Tanaka, 1997; feeding or non-feeding period). Thus, all nipple contacts of the infant are counted as suckling. The infant *head position* is assessed in relation to the midline of the body of the mother (Manning and Chamberlain, 1990). Nonetheless, it could happen that the infant has the head on the right while the mother cradles it with the left arm. We have consequently defined this behaviour clearly. If more than a half of the infant's head is in contact on one side of the mother's torso, we consider it to be a lateralized behaviour (i.e. left or right head position). However, if it is difficult or impossible to determine this lateral bias, then we record it as a 'middle head position'. *Infant retrieval* occurs when the infant is apart from the mother and the mother reaches to retrieve the infant for any reason. The hand used to retrieve the infant is recorded as left, right or both. Finally, when the mother is walking with the infant held ventrally with one arm (left or right carrying), an event of *maternal carrying* is defined. Frequently, during maternal carrying, the infant is held with one hand and the balance of the mother when she is walking in a tripedal position is kept with the other hand. When the mother walks without holding the infant, this behaviour is coded as 'no carrying'. In order to complete this ethogram, we include a 'no behaviour' category (i.e. 'no cradling', 'no nipple', 'no head'). This 'no category' is used when behaviours do not correspond precisely to these above defined behavioural units.

Likewise, when it is difficult to specify these behaviours or when they are not clearly visible (e.g. the mother is huddled under other baboons), the behaviour is coded as 'non-visible'.

9.9.1.2. Scan and Instantaneous Sampling

The mother–infant dyad is sampled using a scan and instantaneous sampling with an observation session lasting 1 h, with recording at 60-s intervals (Altmann, 1974; Martin and Bateson, 1993). Consequently, a total of 60 scans are obtained for each observation session. A check sheet is used, and, at each scan, the behavioural measures of lateral biases described above are recorded. The observer scans the compound, and changes of behaviour in the mother–infant dyad target at each scan interval are noted, together with a note of whether the infant is suckling or not. Thus, according to the number of dyad targets observed during this 1-h observation session, the observer notes at each interval (the duration of which in seconds = 60/number of observed dyads) the behaviour of a new dyad. For that purpose, a stopwatch with a timer, adjustable to the time of 1 s and with an automatic zero-replay, is utilized. When the same behaviour continues during several 60-s intervals, each scan is considered as a separate instance.

9.9.1.3. 15-min Focal-Dyad Sampling

Conjointly, in the same mother–infant dyads, another observational technique is used, namely the 15-min focal-dyad sampling with continuous recording (Altmann, 1974; Martin and Bateson, 1993). The same behavioural measures of lateral biases (see above, Section 9.9.1.2) are also recorded. For this method, a portable computer (with specific software written in QuickBasic that we have developed) is used in order to record, in the same session, the mothers' and infants' behaviours. By contrast to the scan and instantaneous samplings, this method focuses exclusively on one mother–infant dyad, which allows the observer to follow it with a continuous recording (Lehner, 1996; Damerose and Hopkins, 2001). As our goal is to study the laterality of behaviours, and not their occurrences, we have decided to begin an observation session only if any two of the three main behaviours are present (i.e. maternal cradling, infant head in contact with the mother's torso, nipple contact). If this criterion is not met, the session is delayed by one day for this specific dyad.

9.9.1.4. Hand Preference

Individuals are tested in an experimental enclosure. Pieces of corn and sunflower seeds are used as food items. Preference data are recorded on an audiotape. A trial is recorded only when the baboon is seated at one food-reaching apparatus (screen with a small food box) facing the food box in the midsagittal plane, so that the recorded reach corresponds to a real choice between the hands. A handedness index (HI; see Hopkins, 2001) is calculated with the total number (a minimum of 100 grasps per subject) of right and left reaching, using the same formula as the lateral bias (see Section 9.9.2.1).

To determine the hand preference of each subject, a different apparatus is used, and it provides information concerning the hand used in a one-choice hand situation. Furthermore, other tasks require the baboon to perform a cooperation as well as a coordination of both hands. For example, we use a vertical sliding panel (see Fagot and Vauclair, 1988a, 1988b), a haptic discrimination task (Lacreuse, Fagot and Vauclair, 1992), a sloping plane and a tube task developed by Hopkins (1995).

9.9.1.5. Exploratory Behaviours of the Infant

A focal-dyad sampling method is used to record different categories of exploratory behaviour expressed by the infant (e.g. carrying of and/or reaching for food, and non-food objects; one-arm suspensions on a tyre and touching).

9.9.2. Data Analysis

9.9.2.1. At the Individual Level

To assess the presence of a population bias in nipple preference, maternal cradling (arm and/or hand used), infant head position, maternal carrying and infant retrieval, the scores for each subject are summed for the 12 observation sessions. Based on these totals, lateral bias ($LB = R - L/R + L$) scores as well as absolute value ($ABS - LB$) scores are calculated. Significant individual biases are assessed using a z-score for each LB, with a $z \geqslant 1.96$ ($p < 0.05$) in absolute value indicating an individual with a significant bias and a positive value reflecting a right-side bias and a negative value reflecting a left-side bias.

9.9.2.2. At the Population Level

Population biases are assessed using one-sample t-tests for each measure and the LB scores are compared to a normal distribution with a mean of zero.

9.9.3. Results and Discussion

In this part, we only report results obtained using the 15-min focal-dyad sampling; during analysis the number of bouts for each measure collected with 19 mother–infant dyads of olive baboons were summed across the 12 observation sessions. Depicted in Table 9.9 are the percentages of subjects classified as left- or right-sided bias based on the sign of their LB score as well as of their z-score for each behavioural measure. Subjects with negative values were classified as left-sided and subjects with positive scores were classified as right-sided. Population-level biases for each behavioural measure were assessed using one-sample t-tests (see Table 9.9). For nipple preference and head position of the infant, the mean LB was significant (0.29 and 0.15, respectively), indicating a preference for the infant to suckle on the right and to have the head on the right as well. A Spearman correlation coefficient [$r(19) = 0.79$, $p < 0.0001$], indicates a significant correlation between these two measures. However, for maternal cradling, infant retrieval; and maternal carrying, the mean LB were not significantly biased (0.06, 0.16, and −0.04, respectively).

One-sample t-tests for the $ABS - LB$ measures reveal consistent asymmetries for all of the measures, with a value of 0.16 for cradling, 0.55 for nipple preference, 0.21 for head position, 0.45 for retrieval and 0.36 for carrying. Asymmetries in strength of bias were evident at a high significance level ($p < 0.0001$) for each behavioural measure: maternal cradling bias,

Table 9.9. *Results of lateral biases in olive baboons*

Study	Species	Subject age	Procedure	Measure	Individual criterion	N	Right		Left		No preference (%)	Bias	Statistics
							b	s (%)	b	s (%)			
Continuous recording													
This study	*Papio anubis*	D (Nw)	15-min focal-dyad sampling number of bouts	Arm-cradling bouts	LB[a], bin *z*-score > |1.96|, 5%	19	52.6	21.1	42.1	10.5	5.3	No	$t(18) = 1.11$ ns
				Nipple-contact bouts		19	68.4	68.4	31.6	26.3	0	R	$t(18) = 2.12$ $p < 0.05$
				Head-position bouts		19	73.7	52.6	26.3	5.3	0	R	$t(18) = 3.00$ $p < 0.05$
				Infant-retrieval bouts		18	61.1	5.6	38.9	5.6	0	No	$t(17) = 1.21$ ns
				Arm-carrying bouts		19	42.1	15.8	52.6	31.6	5.3	No	$t(18) = -0.48$ ns

D, mother–infant dyad; Nw, newborn; N, number of subjects; b, percentage of biased subjects; s, percentage of significant biased subjects; R, right bias; No, no preference; ns, non-significant. See details in text: [a]LB (lateral bias), (#R − #L) / (#R + #L). Other symbols as Table 9.3

$t(18) = 5.42$, nipple preference, $t(18) = 12.96$, infant head position, $t(18) = 8.24$, infant retrieval, $t(17) = 7.17$, and maternal carrying bias $t(18) = 5.83$. These significant values of t-tests indicate that the behaviours were not normally distributed with respect to lateral biases.

In other words, the fact that we did not find any significant mean for the LB score but significant means for the ABS − LB score reveals that even if, at the population level, non-human primates do not necessary exhibit a strong bias, most of the individuals are biased toward the right or the left side.

Our survey of the postural asymmetry patterns in human and non-human primates also indicates that it is not clear whether there are some relations between different behaviours such as maternal cradling, nipple suckling and the position of the head of the infant. Exploring the relation between, for example, head position of the infant and the hand used by the mother during cradling is possible by using a focal-dyad sampling procedure. While studies on cradling in humans mostly use instantaneous methods, Damerose and Hopkins (2001) reported a strong comparability of data from the number of bouts (scan and instantaneous) and those from the total duration and/or number of bouts (focal dyad with continuous recording). The use of a single technique should allow us to determine whether or not cradling, infant head position and nipple suckling are independent behavioural units. In particular, there was no significant positive correlation between maternal cradling and infant's head position [Spearman coefficient $r(19) = 0.11$, $p = 0.66$] as the conclusion of Manning and Chamberlain (1990, 1991) would have predicted (see above).

Our procedures make it possible to combine tests with a set of different apparatuses and unprompted observations of free actions used to determine the handedness of each individual (e.g., a food-reaching task using an apparatus that forces choice with one hand; spontaneous hand preference when carrying food and other items; one-arm suspension by a tyre to establish which is the stronger arm). Moreover, the use of different types of devices allows us to record hand performance in seminatural conditions. Examples are (1) the haptic discrimination task, which establishes the more sensitive hand for tactile exploration, and (2) the sloping plane which shows which is the faster and more skilful hand.

In short, the advantage of our approach is to record, in the same subjects, both spontaneous reaching in naturalistic conditions and experimentally induced reaching in a test. Very few studies have chosen this approach in the handedness literature (but see Marchant and McGrew, 1991). Furthermore, McGrew and Marchant (1997) argue that most studies on the laterality of hand function focus on one-handed tasks and ignore what the other hand is (or is not) doing. In a meta-analysis of methods performed to compare studies of the laterality of functions in apes, Marchant and McGrew (1991) reported that only seven studies have tested bimanuality, using a sequential task, such as sliding a panel to align two openings, through which a food reward could then be reached (e.g. Fagot and Vauclair, 1988a, 1988b). Few studies explicitly included tasks that require the two hands to be used simultaneously and complementary with a single object (Byrne and Byrne, 1991). Moreover, the typical task for assessing laterality of functions is one-handed, non-sequential and required global rather than fine movements.

Our techniques and devices allow us to experimentally and spontaneously determine hand preference of adult females as well as that of very young infants (from the age of 3 months), and also hand performance and hand collaboration. The kind of studies in which we are engaged should lead to a better integration of research on environmental (e.g. maternal cradling, infant posture) and biological (physiological, neurological, hormonal, genetic, etc.) factors that affect the development of hand preference of the primate infant.

9.10. Conclusion

Most of the studies cited above are concerned with determining, at the population level, hand preference in human and non-human primates. In general, authors sum or average collected data distinguishing only infant and adult groups. None of the studies with prosimians, New World primates, Old World primates and great apes have yielded clear conclusions as to the effects of age on the development of manual preferences in non-human primates (for reviews, see Hook-Costigan and Rogers, 1996).

Concerning the origin of handedness, Matoba, Masataka and Tanioka (1991) propose, from their studies with marmosets, that hand preference of the infant may be genetically determined or may develop with experience, for example, through imitation of the mother's hand use. In a recent study, Dellatolas et al. (1997) suggest that manual asymmetry in humans could in fact be under the influence of: (1) biological (i.e. genetic and prenatal) factors; (2) a 'right-biased' environment; and (3) a learning phenomenon. Nonetheless, it is still not known whether and how maternal influences may affect the hand preference of an infant (Hook-Costigan and Rogers, 1996).

Thus, with respect to maternal cradling biases, Hopkins et al. (1993a) reported that maternal cradling bias correlated inversely with the infant's hand preference for simple reaching when tested at 3 years of age. In addition, another study with chimpanzees revealed the existence of a relation between mother's hand preference and the hand preference of her infant (Hopkins, Bales and Bennett, 1993). These latter authors compared the strength and the direction of hand preference between generations, parents and siblings. Their results were in favour of the existence of a hereditary component for the expression of handedness. Using the coordinated bimanual tube task to test parent and offspring chimpanzees, Hopkins (1999) suggests that the direction of hand preference is heritable, although it is unlikely that the mechanism of transmission is genetic. It could be instead that it is behavioural, with infant handedness being determined by the behaviour of the mother. Explanations of this sort include effects of: (1) maternal cradling bias (Provins, 1997); (2) intrauterine fetal position (Previc, 1991); or (3) prenatal hormonal environment (Geschwind and Galaburda, 1985). But for the time being, there are no strong data that can be used to support or to challenge any of these views. Further studies of the type we are conducting, which measure a number of different parameters, are needed to assess the effects of each of the abovementioned factors.

References

Altmann, J. (1974). Observational study of behavior: sampling methods. *Behaviour*, 49, 227–267.

Altmann, J. (1980). *Baboon Mothers and Infants*. Cambridge, MA: Harvard University Press.

Annett, M. (1985). *Left, Right, Hand and Brain: The Right Shift Theory*. Hillsdale, NJ: Lawrence Erlbaum.

Annett, M. (1995). The right shift theory of a genetic balanced polymorphism for cerebral dominance and cognitive processing. *Current Psychology of Cognition*, 14, 427–480.

Bard, K.A., Hopkins, W.D. & Fort, C.L. (1990). Lateral bias in infant chimpanzees (*Pan troglodytes*). *Journal of Comparative Psychology*, 104, 309–321.

Bennett, H.L., Delmonico, R.L. & Bond, C.F. (1987). Expressive and perceptual asymmetries of the resting face. *Neuropsychologia*, 25, 681–687.

Bogren, L. (1984). Side preference in women and men when holding their newborn child: psychological background. *Acta Psychiatrica Scandinavia*, 69, 13–23.

Brazelton, T.B. (1984). *Neonatal Behavioral Assessment Scale*, 2nd edn. Philadelphia: Spastics International Medical Publications.

Bundy, R.S. (1979). Effects of infant head position on sides preference in adult handling. *Infant Behavior and Development*, 2, 355–358.

Byrne, R.W. & Byrne, J.M. (1991). Hand preferences in the skilled gathering tasks of mountain gorillas (*Gorilla gorilla berengei*). *Cortex*, 27, 521–536.

Campbell, R. (1978). Asymmetries in interpreting and expressing a posed facial expression. *Cortex*, 14, 327–342.

Caplan, P.J. & Kinsbourne, M. (1976). Baby drops the rattle: Asymmetry of duration of grasp by infants. *Child Development*, 47, 532–534.

Corballis, M.C. & Beale, I.L. (1976). *The Psychology of Left and Right*. Hillsdale, NJ: Lawrence Erlbaum.

Coryell, J. & Michel, G.P. (1978). How supine postural preferences of infants can contribute towards the development of handedness. *Infant Behavior and Development*, 1, 245–257.

Dagenbach, D., Harris, L.G. & Fitzgerald, H.E. (1988). A longitudinal study of lateral biases in parents' cradling and holding of infants. *Infant Mental Health Journal*, 9, 219–233.

Damerose, E. & Hopkins, W.D. (2001). A comparison of scan and focal sampling procedures in the assessment of laterality for maternal cradling and infant nipple preferences in olive baboons (*Papio anubis*). *Animal Behaviour*, in press.

Damerose, E. & Vauclair, J. (1999). Development of hand asymmetry in Olive baboons (*Papio anubis*): Effects of the mothers' posture and manual laterality. *Folia Primatologica*, 70, 208–209.

Dellatolas, G., Tubert-Bitter, P., Curt, F. & De Agostini, M. (1997). Evolution of degree and direction of hand preference in children: methodological and theoretical issues. *Neuropsychological Rehabilitation*, 7, 387–399.

Dienske, H., Hopkins, B. & Reid, A. K. (1995). Lateralization of infant holding in chimpanzees: New data do not confirm previous findings. *Behaviour*, 132, 801–809.

Erwin, J. & Anderson, B. (1975). Nursing behavior of infant pigtail monkeys (*Macaca nemestrina*): Preferences for nipples. *Perceptual and Motor Skills*, 592–594.

Fagot, J. & Bard, K. A. (1995). Asymmetric grasping response in neonate chimpanzees (*Pan troglodytes*). *Infant Behavior and Development*, 18, 253–255.

Fagot, J. & Vauclair, J. (1988a). Handedness and bimanual coordination in the lowland Gorilla. *Brain, Behavior and Evolution*, 32, 89–95.

Fagot, J. & Vauclair, J. (1988b). Handedness and manual specialization in the baboon. *Neuropsychologia*, 26, 795–804.

Fagot, J. & Vauclair, J. (1991). Manual laterality in nonhuman primates: A distinction between handedness and manual specialization. *Psychological Bulletin*, 109, 76–89.

Geschwind, N. & Galaburda, A.M. (1985). Cerebral lateralization: Biological mechanisms, associations and pathology: I. A hypothesis and program of research. *Archives of Neurology*, 42, 428–459.

Gesell, A. & Ames, L.B. (1947). The development of handedness. *Journal of Genetic Psychology*, 70, 155–175.

Harris, L.J. & Fitzgerald, H.E. (1985). Lateral cradling preferences in men and women: Results from a photographic study. *The Journal of General Psychology*, 112, 185–189.

Harris, L.J., Almerigi, J.B. & Kirsch, E.A. (2000). Side-preference in adults for holding infants: Contributions of sex and handedness in a test of imagination. *Brain and Cognition*, 43, 246–252.

Hatta, T. & Koike, M. (1991). Left-hand preference in frightened mother monkeys in taking up their babies. *Neuropsychologia*, 29, 207–209.

Hauser, M.D. (1993). Right hemisphere dominance for the production of facial expression in monkeys. *Science*, 261, 475–477.

Hepper, P.G., Shahidullah, S. & White, R. (1991). Handedness in the human fetus. *Neuropsychologia*, 29, 1107–1111.

Hook-Costigan, M.A. & Rogers, L.J. (1996). Hand preferences in new world primates. *International Journal of Comparative Psychology*, 9, 173–207.

Hook-Costigan, M.A. & Rogers, L.J. (1998). Lateralized use of the mouth in production of vocalizations by marmosets. *Neuropsychologia*, 36, 1265–1273.

Hopkins, B. & Rönnqvist, L. (1998). Human handedness: Developmental and evolution perspectives. In *The Development of Sensory, Motor and Cognitive Capacities in Early Infancy: From Perception to Cognition,* ed. F. Simion & G. Butterworth, pp. 191–236. London: Psychology Press.

Hopkins, W.D. (1995). Hand preferences for a coordinated bimanual task in 110 chimpanzees (*Pan troglodytes*): Cross-sectional analysis. *Journal of Comparative Psychology*, 109, 291–297.

Hopkins, W.D. (1999). Heritability of hand preference in chimpanzees (*Pan troglodytes*): Evidence from a partial interspecies cross-fostering study. *Journal of Comparative Psychology*, 113, 307–313.

Hopkins, W.D. (2001). On the other hand: Statistical issues in the assessment and interpretation of hand preference data in nonhuman primates. *International Journal of Primatology*, in press.

Hopkins, W.D. & Bard, K.A. (1995). Asymmetries in spontaneous head orientation in infant chimpanzees (*Pan troglodytes*). *Behavioral Neuroscience*, 109, 808–812.

Hopkins, W.D. & De Waal, F.B.M. (1995). Behavioral laterality in captive bonobos (*Pan paniscus*): Replication and extension. *International Journal of Primatology*, 16, 261–276.

Hopkins, W.D. & Parr, L.A. A multimethod assessment of laterality in carrying in humans (submitted).

Hopkins, W.D., Bales, S.A. & Bennett A.J. (1993). Heritability of hand preference in chimpanzees (*Pan*). *International Journal of Neuroscience*, 74, 17–26.

Hopkins, W.D., Bard, K.A., Jones, A. & Bales, S.L. (1993a). Chimpanzee hand preference in throwing and infant cradling: Implications for the origin of human handedness. *Current Anthropology*, 34, 786–790.

Hopkins, W.D., Bennett, A.J., Bales, S.L., Lee, J. & Ward, J.P. (1993b). Behavioral laterality in captive bonobos (*Pan paniscus*). *Journal of Comparative Psychology*, 107, 403–410.

Jenni, D.A. & Jenni, M.A. (1976). Carrying behavior in humans: Analysis of sex differences. *Science*, 194, 859–860.

Kaplan-Solms, K.L. & Saling, M.M. (1988). Lateral asymmetry and tactile sensitivity. *Perceptual and Motor Skills*, 67, 55–62.

Konishi, Y., Mikawa, H. & Suzuki, J. (1986). Asymmetrical head-turning of preterm infants: Some effects on later postural and functional lateralities. *Developmental Medicine and Child Neurology*, 28, 450–457.

Konishi, Y., Kuruyama, M., Mikawa, H. & Suzuki, J. (1987). Effect of body position on later postural and functional lateralities of preterm infants. *Developmental Medicine and Child Neurology*, 29, 751–757.

Lacreuse, A., Fagot, J. & Vauclair, J. (1992). Manual preferences for tactile discrimination in the baboon (*Papio papio*). Poster presented at the *XIVth Congress of the International Primatological Society*, Strasbourg, France.

Lawick-Goodall, J. van (1968). Behaviour of free-ranging chimpanzees in the Gombe Stream Reserve (Tanzania). *Animal Behavior Monographs*, 1, 161–311.

Lehner, P.N. (1996). *Handbook of Ethological Methods*, 2nd edn. Cambridge: Cambridge University Press.

Lucas, M.D., Turnbull, O.H. & Kaplan-Solms, K.L. (1993). Laterality of cradling in relation to perception and expression of facial affect. *Journal of Genetic Psychology*, 154, 347–352.

MacNeilage, P.F., Studdert-Kennedy, M.G. & Lindblom, B. (1987). Primate handedness reconsidereted. *Behavioral and Brain Sciences*, 10, 247–263.

Manning, J.T. (1991). Sex differences in left-side infant holding: results from 'family album' photographs. *Ethology and Sociobiology*, 12, 337–343.

Manning, J.T. & Chamberlain, A.T. (1990). The left-side cradling preference in great apes. *Animal Behaviour*, 39, 1224–1227.

Manning, J.T. & Chamberlain, A.T. (1991). Left-side cradling and brain lateralization. *Ethology and Sociobiology*, 12, 237–244.

Manning, J.T. & Denman, J. (1994). Lateral cradling preferences in humans (*Homo sapiens*): Similarities within families. *Journal of Comparative Psychology*, 108, 262–265.

Manning, J.T., Heaton, R. & Chamberlain, A.T. (1994). Left-side cradling: Similarities and differences between apes and humans. *Journal of Human Evolution*, 26, 77–83.

Marchant, L.F. & McGrew, W.C. (1991). Laterality of function in apes: A meta-analysis of methods. *Journal of Human Evolution*, 21, 425–438.

Martin, P. & Bateson, P. (1993). *Measuring Behaviour: An Introductory Guide*, 2nd edn. Cambridge: Cambridge University Press.

Matoba, M., Masataka, N. & Tanioka, Y. (1991). Cross-generational continuity of hand-use preferences in marmosets. *Behaviour*, 117, 281–286.

McGrew, W.C. & Marchant, L.F. (1997). On the other hand: Current issues in, and meta-analysis of the behavioral laterality of hand function in nonhuman primates. *Yearbook of Physical Anthropology*, 40, 201–232.

Michel, G.F. (1981). Right-handedness: A consequence of infant supine head-orientation preference? *Science*, 212, 685–687.

Michel, G.F. & Goodwin, R. (1979). Intrauterine birth position predicts newborn supine head position preferences. *Infant Behavior and Development*, 2, 29–38.

Nishida, T. (1993). Left nipple suckling preference in wild chimpanzees. *Ethology and Sociobiology*, 14, 45–52.

Porac, C. & Coren, S. (1981). *Lateral Preferences and Human Behavior*. New York: Springer.

Previc, F.H. (1991). A general theory concerning the prenatal origins of cerebral lateralization in humans. *Psychological Review*, 98, 299–334.

Provins, K.A. (1997). Handedness and speech: A critical reappraisal of the role of genetic and environmental factors in the cerebral lateralization of function. *Psychological Review*, 104, 554–571.

Richards, J.L. & Finger, S. (1975). Mother–child holding patterns: A cross-cultural photographic survey. *Child Development*, 46, 1001–1004.

Rogers, L.J. & Kaplan, G. (1996). Hand preferences and other lateral biases in rehabilitated orang-utans (*Pongo pygmeus pygmeus*). *Animal Behaviour*, 51, 13–25.

Rogers, L.J. & Kaplan, G. (1998). Teat preference for suckling in common marmosets: Relationship to side of being carried and hand preference. *Laterality*, 3, 269–281.

Rönnqvist, L. & Hopkins, B. (1998). Head position preference in the human newborn: A new look. *Child Development*, 69, 13–23.

Rönnqvist, L., Hopkins, B., Van Emmerik, R. & De Groot, L. (1998). Lateral biases in head turning and the Moro response in the human newborn: are they both vestibular in origin? *Developmental Psychobiology*, 339–349.

Saling, M.M. & Bonert, R. (1983). Lateral cradling preferences in female preschoolers. *Journal of Genetic Psychology*, 142, 149–150.

Saling, M.M. & Cooke, W.L. (1984). Cradling and transport of infants by South African mothers: A cross-cultural study. *Current Anthropology*, 25, 333–335.

Saling, M.M. & Kaplan-Solms, K. (1989). On lateral asymmetry in cradling behaviour and breast sensitivity. *Current Anthropology*, 30, 210–211.

Saling, M.M. & Tyson, G. (1981). Lateral cradling preferences in nulliparous females. *Journal of Genetic Psychology*, 139, 309–310.

Salk, L. (1960). The effects of the normal heartbeat sound on the behavior of the new-born infant: Implications for mental health. *World Mental Health*, 12, 168–175.

Salk, L. (1962). Mother's heartbeat as an imprinting stimulus. *Transactions of the New York Academy of Sciences*, 24, 753–763.

Salk, L. (1970). The critical nature of the post-partum period in the human for the establishment of the mother–infant bond: A controlled study. *Nervous System*, 31, 110–113.

Salk, L. (1973). The role of the heartbeat in the relations between mother and infant. *Scientific American*, 228, 24–29.

Sieratzki, J.S. & Woll, B. (1996). Why do mothers cradle babies on their left? *Lancet*, 347, 1746–1748.

Souza-Godeli, M.R. (1996). Lateral cradling preferences in children. *Perceptual and Motor Skills*, 83, 1421–1422.

Tanaka, I. (1989). Change of nipple preference between successive offspring in Japanese macaques. *American Journal of Primatology*, 18, 321–325.

Tanaka, I. (1997). Parity-related differences in suckling behavior and nipple preference among free-ranging Japanese macaques. *American Journal of Primatology*, 42, 331–339.

Thompson, A.M. & Smart, J.L. (1993). A prospective study of the development of laterality: Neonatal laterality in relation to perinatal factors and maternal behavior. *Cortex*, 29, 649–659.

Tomaszycki, M., Cline, C., Griffin, B., Maestripieri, D. & Hopkins, W.D. (1998). Maternal cradling and infant nipple preferences in rhesus monkeys (*Macaca mulatta*). *Developmental Psychobiology*, 32, 305–312.

Turnbull, O.H. & Lucas, M.D. (1991). Lateral cradling preferences in males: The relationship to infant experience. *Journal of Genetic Psychology*, 152, 375–376.

Turnbull, O.H. & Matheson, E.A. (1996). Left-sided cradling. *Lancet*, 348, 691–692.

Ward, J.P. & Hopkins, W.D. (eds) (1993). *Primate Laterality: Current Behavioral Evidence of Primate Asymmetries*. New York: Springer Verlag.

Weiland, H. (1964). Heartbeat rhythm and maternal behavior. *American Academy of Child Psychiatry*, 3, 161–164.

Weiland, H. & Sperber, Z. (1970). Patterns of mother–infant contact: The significance of lateral preference. *Journal of Genetic Psychology*, 117, 157–165.

Weinstein, S. (1963). The relationship of laterality and cutaneous area to breast sensitivity in sinistrals and dextrals. *American Journal of Psychology*, 76, 475–479.

Westergaard, G.C., Byrne, G. & Suomi, S.J. (1998). Early lateral bias in tufted capuchins (*Cebus apella*). *Developmental Psychobiology*, 32, 45–50.

Wind, J. (1982). On child transport in Sri Lanka. *Current Anthropology*, 23, 333.

Part Three

Cognition and Lateralization

10

Evidence for Cerebral Lateralization from Senses Other than Vision

RICHARD. J. ANDREW AND J.A.S. WATKINS

A fundamental issue concerning the nature of vertebrate lateralization is the extent to which such lateralization is generated by mechanisms common to all senses. At the lowest level, there is linkage by simple physical constraints: if a bird turns its head to look at an object with the left lateral visual field, it cannot avoid also turning the left ear towards the object. Attentional mechanisms are bound also to be important: any target of interest identified by any sense will tend to become a target for all. However, there is also evidence that there is left–right specialization of perceptual processing for hearing, olfaction and touch, which resembles that which is more fully established for vision. Briefly, this specialization involves the use of mechanisms of the left side of the forebrain in the control of response to an identified target, and the use of mechanisms of the right side in analysis of a wide range of cues, including spatial relations.

10.1. Hearing

In the case of hearing, unlike that of vision, there is no possibility of inputs being initially routed exclusively to one side of the central nervous system. Sounds may reach the ears at markedly different amplitudes or times, but they nevertheless reach both ears. However, the properties of an auditory stimulus do affect lateralized processes of perception: it will be shown that perceived position of a sound affects which hemisphere is responsible for its processing. This sort of effect also occurs in visual perception (Chapter 3 by Andrew and Rogers), in addition to the effects of differences between what left and right eyes actually see.

In the human dichotic test, attention can be directed to either right or left ear input (e.g. by instructions); when this is done, the input to the ear, to which attention is directed, receives priority in processing (Hugdahl, 1995).

365

This then causes the contralateral hemisphere to take the lead in processing, revealing right ear advantage for verbal, and left ear advantage for emotional content of speech. Such direction of attention is equivalent to attending to sounds in left or in right hemispace: perceived position is known to be important in determining whether right hemisphere (RHem) or left hemisphere (LHem) abilities are shown in processing a sound (Pierson, Bradshaw and Nettleton, 1983).

There are now a number of studies of effects in animals that appear to depend on the perceived position of sounds. In the Japanese macaque, a procedure comparable with the human dichotic test (Petersen et al., 1978, 1984) has revealed a right ear advantage for discrimination of two variants of a species specific vocalization.

Comparable effects can be obtained by a simpler procedure: Ehret (1987) showed that, when either the right or left external meatus of a lactating mouse was blocked, she would go to her calling pups only when the right ear was open. McKenzie (1991) used the same procedure in the domestic chick. The chick undergoes a series of shifts in bias to control by one or other hemisphere during development (Chapter 5 by Andrew). When the meatus contralateral to the controlling hemisphere (LHem day 2; RHem day 4) is plugged, chicks tend to remain still and not approach the sound of a cluck, whereas they approach readily when it is the other meatus that is plugged.

Such effects of the hemispace into which a sound appears to fall may explain why both birds and mammals turn the head to put a sound source on the right or left, according to its properties. This has been clearly shown for rhesus monkeys (Hauser and Andersson, 1994; Hauser, Agnetta and Perez, 1998; Chapter 13 by Weiss et al.), where the right ear is turned towards conspecific vocalizations of several types. Such use of the right ear disappears when the vocalization is so transformed as (apparently) to cease to be treated as a conspecific vocalization.

The hypothesis that right ear use under these circumstances results in LHem involvement in the analysis of conspecific vocalizations has in general been confirmed by the results of unilateral lesions of auditory structures. Some limitations of the available evidence are noted by Hopkins and Carriba (Chapter 12).

In the domestic chick, the right ear is turned towards a source of hen clucks after 2 h prior exposure to clucks (Miklósi, Andrew and Dharmaretnam, 1996); it is shown below that the same right ear use is present in naive chicks. The tests were conducted at low intensities of infrared light in an entirely visually homogeneous environment, so that head positions were not affected by the selection of possible visual targets. This right ear use was progressively replaced by left ear use as prior exposure increased. Significant bias to left ear use appeared after 6 h. Watkins (1999) showed that use of the right ear is especially marked in the last few seconds before approach (Figure 10.1), which is consistent with the hypothesis that it is associated with taking a decision to approach. This would also explain why the right ear is used early in attachment, before approach becomes automatic.

A direct test of this hypothesis was provided by experiments in which a novel feature (a flute note) was added to a familiar cluck (Watkins, 1999). The addition of the note caused left ear use to be replaced by right ear use, the difference being clearest just before approach (Figure 10.1), exactly when the decision to approach the novel sound was being taken.

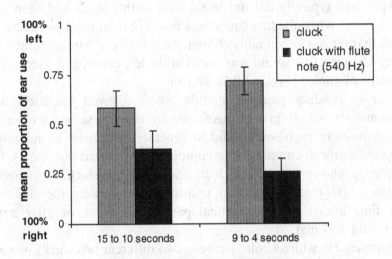

Figure 10.1. Female chicks were exposed to a particular cluck overnight in the hatcher, before test on the second day of life. The test followed immediately on removal of each chick from the hatcher, so that not only was the long-term memory trace available to recall, but so also were short-term traces (see text). Ear use became clear just before approach. The head turned to point towards the sound source as the chick moved off: this sometimes occurred a few seconds before the move, so that the last 3 s were treated separately. Data are for the period from 15 to 4 s before movement, divided into two 6-s blocks. Ear use was significantly different from random for both left ear turned to the novel sound and right ear turned to the familiar sound ($p < 0.05$). Ear use differed between the two sounds: $F(1,91) = 10.21$, $p = 0.003$.

These findings closely parallel eye use when presented with a silent imprinting object: naive chicks use the right eye when deciding whether to approach the imprinting object. Such chicks shift towards left eye use after exposure of one or more hours (McKenzie, Andrew and Jones, 1998). In the case of visual lateralization in the chick (Chapter 3 by Andrew and Rogers), right eye use is often caused by the need to control a response visually, and shift to left eye use occurs when the response is so practised or simple that there is little need for close control; instead, the main priority is to check that the object or surroundings are indeed entirely familiar.

Striking confirmation that hearing a sound in the right hemispace initiates response to the sound, presumably under LHem control, in a way which does not hold for sounds in the left hemispace, is provided by Watkins (1999). Two different sounds, both more or less equally acceptable (below), were played in short bouts alternatively on the left and on the right of the chick. Approach typically did not begin until both sounds had been heard more than once. When the first sound was heard in right hemispace, that was the sound that tended to be finally chosen, irrespective of which sound it was. However, when the first sound was heard in the left hemispace, there was no clear pattern of choice (Figure 10.2a and b).

In order to produce pairs of sounds, which allowed the effect to be clearly demonstrated, it proved necessary to use the same cluck as the basic component of each sound, and to generate differences by manipulating an added artificial component (a computer-synthesized flute note). Data are shown for choice between a cluck and the same cluck with the flute added after a brief interval (600 ms), and for choice between the same cluck with the flute added either as a final part of the cluck or after a brief interval (again 600 ms).

In experiments in which choice between two different hen clucks was measured, one cluck proved more likely to be chosen than the other. A hemispace effect was still present: if the 'attractive' cluck were heard first, it was chosen eventually, given that its perceived position lay in the right hemispace. In contrast, when the first sound was heard in left hemispace, there was no clear choice. If the first sound was the 'unattractive' cluck, no clear choice followed, even if it had been heard in the right hemispace. The explanation is probably that some chicks eventually adopted the more attractive cluck as the one to approach, even if it had not been the first to be heard. This did not happen when the same cluck with an artificial added component was used for both sounds, presumably because the two sounds did not differ in intrinsic attractiveness. The flute note acted only as a cue that distinguished the two sounds, one from the other.

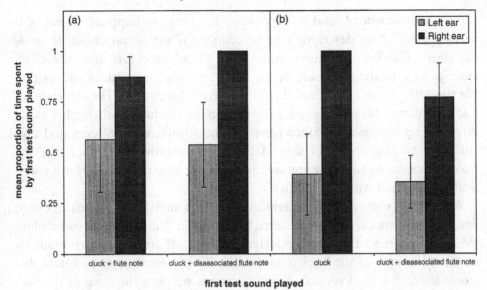

Figure 10.2.   (a) Female chicks that had never heard the sound of a cluck were placed in the test arena with one ear turned towards one sound source and the other ear towards the other source, the position being counterbalanced between chicks. The two sounds were both made up of the same cluck and flute note, but differed in the gap between cluck and flute. The flute note was 195 ms long, with the main energy in the fundamental at 540 Hz, but with first and second harmonics also present. The note either followed immediately after the cluck, with a 60 ms gap, or after a gap of 600 ms. The ultimate choice (see text) differed according to which side of the head (which hemispace; see text) was turned towards the first sound heard [$F(1,11) = 5.37$, $p = 0.049$]. (b) Conditions were as for (a), except that the choice was between a cluck and the same cluck with the separated flute note. The choice again differed according to the ear turned towards the first sound heard [$F(1,14) = 8.59$, $p = 0.014$].

This technique offers an interesting new approach to the lateralization of processes based on very short-term memory traces: the effect with two different clucks that is mentioned above was obtained with chicks that had never heard clucks before.

However, the point that we wish to stress here is that hearing a cluck in the right hemispace at once makes it a candidate for approach, and that this is maintained despite distraction by other types of cluck elsewhere in the environment. The resemblance to the use of right visual hemispace when sustaining a visually controlled response is clear. In both cases, it is likely that the behaviour requires a very recently elaborated record to be maintained to specify the stimulus to be approached. However, in the case of visual stimuli, the chick clearly sees the stimulus of interest throughout fixation. Auditory stimuli are transitory, and if they move location (as happened here) it is necessary for some description to be retained if the initial choice is to be sustained. For both auditory and visual stimuli, the fact that the chick chooses to turn the right side of the head towards the stimulus means that the stimulus is placed and sustained in the right hemispace. The evidence for auditory stimuli is that this will directly promote holding that stimulus as the target to be approached. Since hens present attractive visual cues and also cluck, by turning the right side of the head towards the mother, a chick makes it easier to sustain approach to both the kinds of cue that are specifically associated with the mother.

What may be comparable lateralization of the analysis of sounds has been described for another bird, the zebra finch (Cynx, Williams and Nottebohm, 1992). Either the LHem or the RHem was cut off from auditory input by unilateral lesions of the nucleus ovoidalis. Birds were trained to sound discriminations for food reward. Discriminations between the song of the bird itself and a comparable song of a familiar neighbour were carried out better by the LHem than by the RHem. Discriminations that required the detection of a single inconspicuous feature (the change in the harmonic profile of one syllable) were more difficult; they were also carried out better by the RHem. These findings clearly show greater ability of the RHem to detect inconspicuous change in the details of a familiar sound, much as has been postulated above for the chick RHem.

The nature of the LHem advantage is less obvious at first glance, but here too the chick condition may give a clue. A key feature of familiar songs of conspecifics is that they call for decision, including what song to use in reply. In some songbirds (e.g. *Melospiza melodia*; Beecher et al., 2000), the rules governing reply are complex. They take into account the degree of threat implied by the position and nature of the song. A familiar song from a position which is usual for the singer may appropriately be ignored, whereas a song which implies approach to a mate, or other resource, calls for an appropriate reply, perhaps backed up by approach and preparations for fighting. Thus, the LHem may be involved because of the need to take a decision.

Early in the evolution of vocalization, the recognition of the call of a conspecific required little elaborate neural machinery, being instead heavily dependent on peripheral tuning (Capranica and Moffat, 1981). However, the control of response must have made much the same demands as in more advanced tetrapods: a call recognized as that of a conspecific had to evoke reply and/or approach (or avoidance). Left–right asymmetries in the control of calling were probably already present: lesions on the left side of the hindbrain are more effective in *Rana pipiens* in reducing species-specific vocalizations than are corresponding ones on the right (Bauer, 1993). It is thus possible that the lateralization of the control of response to vocalizations to the left side of the brain is an extremely ancient feature of auditory lateralization.

Finally, it is worth noting that MacNeilage, Studdert-Kennedy and Lindblom (1987) argued that control of posture by the LHem (see above) explains the acquisition of control of vocalization by the LHem, since the assumption of a characteristic posture is usual when calling. A wider evolutionary perspective makes this unlikely. Glottally produced sound almost certainly preceded the origin of the tetrapods. It, therefore, preceded the evolution of limbs, which could be used in support or grasping. The hypothesis that control of response to the sound of a conspecific call was the earliest cause of the lateralization of the control of vocalization is as applicable to calling before the evolution of limbs, as after tetrapods had evolved.

The usual formulation of the relation of the LHem to vocalization, namely that it is responsible for performance and recognition of species-specific vocalizations, should probably be recast. The LHem is responsible for appropriate response to such sounds, just as it controls visually guided response (and perhaps response mediated by other senses, see below). Since the response to hearing a conspecific call is commonly to reply, the control of vocalization is an integral part of such functioning.

The lateralized control of song in birds provides much the most extensive comparative data. As is well known, Nottebohm demonstrated that left hypoglossal control of the syrinx is predominant in the canary and *Zonotrichia* spp. (Nottebohm, 1970, 1971; Nottebohm and Nottebohm, 1976). Further, lesion of left forebrain song areas disturbs song in the canary far more profoundly than lesion of corresponding areas on the right (Nottebohm, Stokes and Leonard, 1976). This finding was the most important single step in the rekindling of interest in lateralization in non-mammalian vertebrates. The resemblance to the LHem control of speech led to the assumption that it was the control of learned vocalizations that called for LHem involvement. However, caveats, such as the importance of bilateral control of sounds in some birds, were clearly set out in the initial papers.

The interpretation of evidence from the control of left and right sides of the syrinx is complicated. Bilateral control of syringeal function is essential for the generation of sound of precisely controlled pitch as a result of beats between differing frequency of vibration of the two sides of the syrinx. This is important in some song birds (e.g. calls of *Parus* spp.; Nowicki, 1989). Bilateral control of syringeal sounds in parrots (Nottebohm, 1976) extends the evidence for such control to birds other than songbirds. At the same time, predominantly left-sided control is present in birds in which there is little evidence for learned vocalizations, such as the domestic fowl (Youngren, Peek and Phillips, 1974; Nottebohm, 1976). Here, it is unlikely that the needs of controlling complex learned patterns of vocalization are important.

Predominantly LHem control of vocalization may occur when answering conspecific song (or calls). Most songbirds that have been tested show left hypoglossal dominance (chaffinch, canary, white-crowned sparrow and Java sparrow; Nottebohm, 1981), and couple this with well-developed counter-singing (with the probable exception of the Java sparrow). Countersinging may depend on complex rules with the appropriate song sometimes being different from that heard (Beecher et al., 1996); difficult decisions thus may be called for. The zebra finch is unusual in showing right hypoglossal dominance (and RHem control; Williams et al., 1992).

The use of countersinging may predispose to LHem control, as already argued, whilst the need to use RHem abilities to learn the details of complex single sounds may push towards RHem predominance. The possibility that the zebra finch has an unusual need to learn details of song components is raised by the fact that, in this species, many such components are derived from complex conspecific calls that are learned before singing begins (Zann, 1990).

LHem control of vocalizations thus may be best viewed not as a special and unique feature, but as a consequence of hemispheric specializations that affect other senses as well. Note that control of response to sounds other than calls used in communication between conspecifics may sometimes be important. Cobb (1964) showed, in an owl (*Aegornis*) and the oilbird (*Steatornis*), that the left auditory area of the midbrain was the larger. This suggests that there is asymmetric involvement of hearing in the control of (respectively) the location of prey by sound (followed by the response of prey catching), and the use of emitted sounds in echolocation to guide locomotion.

It is possible that one feature in which hearing, in general, differs from vision has been influential in shaping the involvement of the RHem in hearing. Valent visual features can usually be fixated sustainedly, whereas sounds are usually only temporarily available to perception. It is thus particularly important for hearing mechanisms to hold a short-term record of the position and properties of a valent sound, so that it can be recognized as being probably the same stimulus at a second occurrence. The use of information about both position and detailed properties is likely to require the RHem to be involved as well as the LHem.

10.2. Touch

In humans, asymmetries of perception based on tactile information from the right and left hand have often been reported. The best known is the dichaptic test, in which both hands simultaneously palpate two different nonsense shapes, and the subject has to judge whether either has been felt before. Witelson (1976) reported that in males, but not females, the left hand showed marked advantage in this task. This was confirmed by Nilsson and Geffen (1987) in a study that found that the male asymmetry was entirely due to poorer performance by males using the right hand; female performance with right and left hand was the same as that of males using the left hand.

However, in an extensive review of the literature, Summers and Ledermann (1990) showed that most studies fail to replicate Witelson's study. They were unable to establish why this might be. Thus, in experiments of their own, which varied methods of reporting identification (either by verbal statement or by matching with pictures), there was no effect of method of reporting on the failure to find left–right differences. Smith, Chu and Edmonston (1977) reported that, in purely haptic tests requiring matching a Braille letter with the appropriate one in a sample series, poor right-hand performance became good when the left ear heard music. They interpreted this as removing the RHem interference in the Braille task with LHem performance. It is thus possible that the task is sensitive to environmental effects that are not usually considered or reported.

Other hand differences confirm that engagement of the hemisphere contralateral to the hand in use is important. Thus, the same painful stimulus (ice water) is tolerated better with the right than the left hand (Schiff and Gagliese, 1994); chronic shoulder pain in the left shoulder leads to more hypochondria than does right shoulder pain. This is presumably because of the association of the RHem with intense and negative emotion (see Chapter 3 by Andrew and Rogers).

The main evidence of which we are aware, for lateralization of touch in vertebrates other than humans, is a remarkable study by LaMendola and Bever (1997). This used rats tested in an eight-arm radial maze, the same five arms of which were always baited. Intramaze cues included a large cue unique to each arm that was suspended above the entry point to the arm. Extramaze cues included eccentrically placed room lighting. Errors were scored as returns to a baited arm, which had already been visited, or entry of one of the three arms that were never baited. Fewer errors were made when left whiskers were anaesthetized (and so only right whiskers were in use) than when only left whiskers were in use. The dependence of this effect on LHem involvement in the analysis of right whisker input was confirmed by unilateral spreading depression of the left or right cortex (in rats with both sets of whiskers in use): LHem depression resulted in more errors.

Most of the errors shown by left-whisker rats were due to re-entries of baited arms. LaMendola and Bever argue that this is because left-whisker rats have access only to a record of a fixed route through the maze coded as a series of choices. As a result, once a baited arm has been missed, they have to return to the beginning and repeat the route (so entering again previously visited baited arms). Right-whisker rats, on the other hand, are argued to have access to a map, which allows them to correct such mistakes. Further evidence was provided by experiments in which the maze was rotated so that intramaze and extramaze cues were no longer in their usual relationship. The result was a reversal of the relative performance of right- and left-whisker rats, which LaMendola and Bever interpreted as showing that the strategy used by right-whisker rats was disrupted by maze rotation. However, right-whisker performance was actually little affected by maze rotation. Whatever the strategy in use, right-whisker use allowed adequate performance under both conditions, but at levels well below those shown by normal rats with all whiskers in use. It was the left-whisker performance that changed: in the maze in its familiar condition, with intramaze and extramaze cues in normal register, performance was very poor, whereas after rotation it was very good, indeed as good as that of normal rats.

Whatever the explanation of the differences between right- and left-whisker rats, it seems likely that reliance on whiskers of only one side tends to put the contralateral hemisphere in charge; this is strongly suggested by the results of the spreading depression experiment. Evidence from other vertebrates (Chapter 3 by Andrew and Rogers) suggests that RHem use will give advantage in the use of environmental layout. This is true for the rat as well. Although Bianki (1988) is cited by LaMendola and Bever as presenting no clear evidence for right–left asymmetry in spatial ability in the rat, extensive evidence showing RHem superiority in spatial tests is presented in Bianki's book and in his other studies (see Chapter 3 by Andrew and Rogers).

This provides an alternative hypothesis: the problem for left-whisker/ RHem rats in the unrotated maze may be caused by the salience of the intramaze cues, and the improved performance after rotation may reflect increased reliance on extramaze cues. The intramaze cues provide a conspicuous and unique label for each arm. Predominant use of these cues would allow the task to be solved by a record based on association of the labelling cue of each arm with the fact that it had recently been visited. The insensitivity of right-whisker/LHem rats to maze rotation would be well explained, if they tended to rely on such a labelling strategy. This is entirely consistent with evidence from birds (domestic chick and *Parus* spp.) that LHem control causes reliance on cues specific to the local site, and is followed by better recording of the consequences of response (Chapter 3 by Andrew and Rogers; Chapter 15 by Andrew; Chapter 11 by Vallortigara and Regolin). Use of extramaze cues would in contrast favour dependence on a record based on the position of visited arms in the overall layout of the maze.

The explanation of the poor performance of the left-whisker/RHem rats in the unrotated maze may then be as follows: the use of the intramaze cues may tend to impose a labelling strategy during arm entry. However, successful use of a labelling strategy may be difficult without right whiskers. Whisker information is apparently of great importance for interaction with food. It was observed that rats with both right and left whiskers anaesthetized moved through the maze, but 'had great difficulty in the last stages of locating the food rewards' (LaMendola and Bever, 1997, note 13, p. 485). Right-whisker information may be particularly important in recording successful discovery of food.

Whatever the interpretation of these fascinating findings, it is clear, as LaMendola and Bever point out, that they prove that consistent lateralization for the use of whisker information exists in the rat. A most intriguing question for the future is whether there is any asymmetry of whisker use in exploring the environment comparable to the use of right or left eye or ear.

10.3. Olfaction

Olfaction, like vision and touch, potentially allows inputs to be analysed by only one side of the brain. Since each nostril supplies the ipsilateral olfactory bulb, this input is uncrossed, unlike that from the other senses.

Once again, the bulk of the evidence for lateralization is human. Olfactory lateralization resembles that for other senses in a number of ways. RHem analysis (right nostril application) tends to judge unpleasant odours as more unpleasant than does LHem analysis (Ehrlichman, 1986). The same holds for visual stimuli (Chapter 3 by Andrew and Rogers). It is not clear how far analysis of purely olfactory properties differs between LHem and RHem. Gilbert, Greenberg and Beauchamp (1989) found that there were no differences between right and left nostril judgements of odour similarities plotted on two-dimensional maps of similarity. Abraham and Matthai (1983) found that right temporal lobectomy disturbed olfactory memory much more than left. However, this may have been a consequence of unilateral neglect, which has been demonstrated for RHem damage: it involves both ignoring odorants presented to the right nostril, and reporting left nostril odorants as being given to the right nostril (Bellas et al., 1988).

Somewhat unexpectedly, human evidence suggests that mechanisms exist that can bias olfactory input to one or other hemisphere. Commonly, one nostril is congested by vasodilation due to dominantly sympathetic control, whilst the other is decongested by predominantly parasympathetic control (Werntz, Bickford and Shannahoff-Khalsa, 1987). This nasal cycle has been shown to be coupled to shifts in relative electroencephalogram (EEG) amplitudes in RHem and LHem, with relatively variable periodicity (25–200 min; Werntz et al., 1983). The most important point here is that such regulation of airflow would potentially allow biasing of olfactory input to one or other hemisphere, comparable with that which is so important for visual inputs.

Olfactory lateralization appears to have been described outside humans only for the domestic chick. Vallortigara and Andrew (1994) found that chicks, which had been reared with an imprinting object that was scented, would choose by smell, when presented with objects identical visually with the familiar object but differing in smell, but only when using the right nostril. This result parallels choice when presented with the familiar object and a visual transformation of it, which is shown only when the left eye is in use (male chicks; Vallortigara and Andrew, 1991); in both cases sensory input to the RHem is required for choice.

A second study (Rogers, Andrew and Burne, 1998) showed that, when a chick is presented with a stimulus that is visually conspicuous and novel, the presence of this stimulus tended to inhibit response to a scent presented via the left nostril. Perceptual analysis of such a scent would be by the LHem, which is likely to be strongly engaged in deciding how to respond to the visual stimulus. There was no such interference when the visual stimulus was less conspicuous or when input from the scent was via the right nostril. It is thus likely that, with both nostrils normally in use, it will be the RHem that is responsible for processing scents associated with a conspicuous visual stimulus.

10.4. Attention and response

All of the senses are likely to be affected by the organization of attentional mechanisms. Geschwind (1980) suggested that the earliest lateralized functions might indeed have been attentional, with the RHem picking out possible targets to which the focus of attention might shift. There is extensive human evidence that the RHem sustains all round attention, so that right hemispace is to some extent also covered by the RHem. However, in humans, left hemispace is only dealt with by the RHem, so that RHem depression (e.g. Wada test) or lesioning the RHem results in left-side neglect (Heilmann, 1995). Another correlate of these attentional asymmetries is that the RHem is better able to sustain diffuse attention, whereas the focused attention characteristic of the LHem makes this much more difficult (Dimond, 1979). Consistent with this, in normal subjects, distractors are more likely to be noticed in the left hemispace (Heilmann, 1995).

Diffuse attention should be accompanied by a wider recording of properties and spatial context than focused attention would allow. Goldberg and Costa (1981, p. 148) provide what is still an excellent formulation of human lateralization: the RHem is better able to 'deal with informational complexity' and 'to process many modes of representation within a single cognitive task'. The LHem, in contrast, excels 'in tasks which require fixation upon a single mode of representation or execution'. The key word in the last phrase is execution: the most obvious function for such focused attention is to allow decision and to sustain a course of action, once it is decided on. Such differences in RHem and LHem function have been described here, and in other chapters, for vision, hearing and olfaction, in a variety of vertebrates; it may also be shown by touch.

Important anatomical correlates of features of lateralization that affect all senses are likely to be the asymmetries of noradrenergic and dopaminergic supply to the forebrain (Liotti and Tucker, 1995). In humans, the RHem receives the greater noradrenergic supply and the LHem the greater dopaminergic. Asymmetries of noradrenergic and dopaminergic supplies are also present in rats. In the case of the noradrenergic supply, there is apparently an effective positive feedback on this supply from the RHem but not the LHem in the rat. Pearlson and Robinson (1981) showed that frontal lesions (infarct or suction), which caused marked depletion of noradrenaline in both hemispheres when induced in the RHem, had no effect in the LHem.

The noradrenergic asymmetry is likely to be related to left–right differences in attention. One function commonly ascribed to the dorsal bundle component of the ascending noradrenergic system is that of sustaining selective attention on relevant cues and cue dimensions (Mason and Iversen, 1979). Fallon and Loughlin (1987) note that units in the locus coeruleus (the source of the ascending noradrenergic supply) fire less when a rat is engaged in activities like feeding and grooming. This agrees well with the usual control of such units by medullar nuclei driven by painful or striking polymodal inputs (Aston-Jones et al., 1986). A key property of the RHem noradrenergic supply is thus that it is likely to shift between promoting capture of attention by unexpected but potent stimuli, and sustaining attention on a stimulus selected by the RHem due to feedback control from the RHem itself. Such alternation is probably what is needed to allow analysis of and learning about a very wide range of properties of an interesting object.

The prefrontal cortex is of central importance in asymmetries of dopaminergic action. Slopsema, van der Gugten and De Bruin (1982) showed for the prefrontal area (which receives the densest dopamine supply) that the left medial prefrontal has higher dopamine than the right. In the case of the rat, the prefrontal cortex is identified (Divac and Öberg, 1990) as the frontal area that receives input from the mediodorsal thalamic nucleus, and has the densest dopamine supply. In the rat, as in other mammals, lesions here (mesial frontal) impair delayed alternation. Vargo et al. (1988) showed that unilateral lesions of the dorsomedial prefrontal area produced unilateral neglect in rats, which affected responses to visual, somatosensory and auditory stimuli simultaneously and in identical ways. This confirms the hypothesis that asymmetries in prefrontal functioning are likely to be a major cause of features of lateralization common to all senses.

One surprising feature of the findings was that, whilst LHem damage produced the expected contralateral neglect, corresponding RHem damage either produced ipsilateral neglect (i.e. again neglect of the right hemispace), or neglect which shifted between contralateral and then ipsilateral neglect. Vargo et al. (1988) note that, if RHem lesions were to affect dopaminergic function on both sides of the forebrain, as is the case for noradrenergic supply (see above), this might provide an explanation.

Fuster (1991) reviewed largely primate evidence, and argued that the most basic function of the prefrontal cortex was to mediate 'cross-temporal contingencies'. This requires the 'integration' of sensory information received at separate times, but for which there was evidence of 'contingency'. The evidence for contingency might be simply that one input follows another after a typical interval. It almost certainly also includes information about the stimulus to which the response was directed, and the outcome of response. Direct involvement of dopaminergic input in signalling the consequence of response is suggested by the firing of neurones, in the nuclei from which the dopaminergic supply originates, to both reward and conditioned stimuli for reward (Ljungberg, Apicella and Schultz, 1991). Dopaminergic modulation of the ability of prefrontal cortex to sustain coding of correct delay has been demonstrated in primates by Sawaguchi and Goldman-Rakic (1991).

All of this evidence is consistent with the role of the LHem in sustaining attention on a stimulus to which a motor response is planned. This has been argued in this book (Chapter 2 by Andrew; Chapter 3 by Andrew and Rogers) to be a very early and basic property of vertebrate lateralization.

In its simplest form it sustains readiness to respond during approach. The ability to delay response to a stimulus (which is already accessible) until a precise interval has elapsed could easily be derived from the more primitive condition (e.g. via the use during approach of estimates of time to reach the target). During this period, attention should be sustained on the target. It is, therefore, striking that dopamine agonists may cause pathological locking of attention to particular stimuli in humans (Matthysse, 1978).

For the particular purposes of this chapter, it is important to note that it is unlikely that the effects of the monoaminergic systems (including asymmetries of action) are confined to any one sense. These systems themselves are typically driven polymodally, and the dopaminergic system in particular acts on structures that are not associated with any one sense. It is likely that many of the resemblances between the lateralization of each of the senses stem from asymmetric functioning of systems controlling attention and response.

No doubt differential learning (if nothing else) leads to left–right differences, which are specific to particular senses. Analysis of the spatial relations between a number of cues, for example, is likely to result in records that include details of such relations, which differ between the senses involved. This in turn would itself generate left–right differences in future ability to analyse spatial relations. However, one conclusion to be drawn from the arguments set out in this chapter is that it is necessary to consider the possible effects of lateralized differences in attentional strategies before concluding that, for any particular sense, there are differences in perceptual analysis that are peculiar to that sense.

References

Abraham, A. & Matthai, K.V. (1983). The effect of right temporal lobe lesions on matching of smells. *Neuropsychologia*, 21, 277–281.

Aston-Jones, G., Ennis, M., Pieribone, V.A., Nickell, W.T. & Shipley, M.T. (1986). The brain nucleus locus coeruleus: Restricted afferent control of a broad efferent network. *Science*, 234, 734–737.

Bauer, R.H. (1993). Lateralization of neural control of vocalisation by the frog (*Rana pipiens*). *Psychobiology*, 21, 243–248.

Beecher, M.D., Stoddard, P.K., Campbell, S.E. & Horning, C.L. (1996). Repertoire matching between neighbouring song sparrows. *Animal Behaviour*, 51, 917–923.

Beecher, M.D., Campbell, S.E., Burt, J.M., Hill, C.E. & Nordby, J.C. (2000). Song-type matching between neighbouring song sparrows. *Animal Behaviour*, 59, 21–27.

Bellas, D.N., Novelly, R.A., Eskenazi, B. & Wasserstein, J. (1988). Unilateral displacement in the olfactory sense: A manifestation of the unilateral neglect syndrome. *Cortex*, 24, 267–275.

Bianki, V.L. (1988). The Right and Left Hemispheres of the Animal Brain: Cerebral Lateralization of Function. Monographs in Neuroscience, Vol. 3. New York: Gordon & Breach.

Capranica, R.R. & Moffat, A.J.M. (1981). Neurobehavioural correlates of sound communication in anurans. In *Advances in Vertebrate Neuroethology*, ed. J.-P. Ewert, R.R. Capranica & D.J. Ingle, pp. 701–730. New York: Plenum Press.

Cobb, S. (1964). A comparison of the size of an auditory nucleus (*n. mesencephalicus lateralis, pars dorsalis*), with the size of the optic lobes in twenty seven species of birds. *Journal of Comparative Neurology*, 122, 271–280.

Cynx, J., Williams, H. & Nottebohm, F. (1992). Hemispheric differences in avian song discrimination. *Proceedings of the National Academy of Sciences USA*, 89, 1372–1375.

Dimond, S.J. (1979). Disconnection and psychopathology. In *Hemisphere Asymmetries of Function in Psychopathology*, ed. P. Flor-Henry, pp. 35–46. New York: Elsevier.

Divac, I. & Öberg, R.G.E. (1990). Prefrontal cortex: The name and the thing. In *The Forebrain in Nonmammals*, ed. W.K. Schwerdtfeger & P. Germroth, pp. 213–220. Berlin: Springer-Verlag.

Ehret, G. (1987). Left hemisphere advantage in the mouse brain for recognising ultrasonic communication calls. *Nature*, 325, 249–251.

Ehrlichman, H. (1986). Hemispheric asymmetry and positive-negative affect. In *Duality and Unity of the Brain*, ed. D. Ottoson, pp. 194–206. Dordrecht, The Netherlands: Kluwer.

Fallon, J.H. & Loughlin, S.E. (1987). Monoamine innervation of cerebral cortex and a theory of the role of monoamines in cerebral cortex and basal ganglia. In *Cerebral Cortex*, Vol. 6, ed. E.G. Jones & A. Peters, pp. 41–127. New York: Plenum Press.

Fuster, J.M. (1991). The prefrontal cortex and its relation to behaviour. *Progress in Brain Research*, 87, 201–211.

Geschwind, N. (1980). Some comments on the neurology of language. In *Biological Studies of Mental Processes*, ed. D. Caplan, pp. 301–319. Cambridge, MA: MIT Press.

Gilbert, A.N., Greenberg, M.S. & Beauchamp, G.K. (1989). Sex, handedness and side of nose modulate human odour perception. *Neuropsychologia*, 27, 505–511.

Goldberg, E. & Costa, L.D. (1981). Hemisphere differences in the acquisition and use of descriptive systems. *Brain and Language*, 14, 144–173.

Hauser, M.D. & Andersson, K. (1994). Left hemisphere dominance for processing vocalisations in adult, but not infant rhesus monkeys: field experiments. *Proceedings of the National Academy of Sciences USA*, 91, 3946–3948.

Hauser, M.D., Agnetta, B. & Perez, C. (1998). Orienting asymmetries in rhesus monkeys: The effect of time-domain changes on acoustic perception. *Animal Behaviour*, 56, 41–47.

Heilmann, K.M. (1995). Attentional asymmetries. In *Brain Asymmetry*, ed. R.J. Davidson & K. Hugdahl, pp. 217–234. Cambridge, MA: MIT Press.

Hugdahl, K. (1995). Dichotic listening: probing temporal lobe functional integrity. In *Brain Asymmetry*. eds R.J. Davidson & K. Hugdahl, pp. 123–156. Cambridge, MA: MIT Press.

LaMendola, N.P. & Bever, T.G. (1997). Peripheral and cerebral asymmetries in the rat. *Science*, 278, 483–486.

Liotti, M. & Tucker, D.M. (1995). Emotion in asymmetric corticolimbic networks. In *Brain Asymmetry*, ed. R.J. Davidson & K. Hugdahl, pp. 389–423. Cambridge, MA: MIT Press.

Ljungberg, T., Apicella, P. & Schultz, W. (1991). Response of monkey midbrain dopamine neurones during delayed alternation performance. *Brain Research*, 567, 337–341.

McKenzie, R. (1991). Lateralization of auditory and visual learning in the domestic chick. D.Phil. thesis, University of Sussex.

McKenzie, R., Andrew, R.J. & Jones, R.B. (1998). Lateralization in chicks and hens: New evidence for control of response by the right ear system. *Laterality*, 1, 215–224.

MacNeilage, P.F., Studdert-Kennedy, M.G. & Lindblom, B. (1987). Primate handedness reconsidered. *Behavioural Brain Sciences*, 10, 247–263.

Mason, S.T. & Iversen, S.D. (1979). Theories of the dorsal bundle extinction effect. *Brain Research Reviews*, 1, 107–137.

Matthysse, S. (1978). A theory of the relation between dopamine and attention. In *The Nature of Schizophrenia*, eds L.C. Wynne, R.L. Cromwell & S. Matthysse, pp. 148–150; 307–310. London: John Wiley.

Miklósi, A., Andrew, R.J. & Dharmaretnam, M. (1996). Auditory lateralization; shifts in ear use during attachment in the domestic chick. *Laterality*, 1, 215–224.

Nilsson, J. & Geffen, G. (1987). Perception of similarity and laterality effects in tactile shape recognition. *Cortex*, 23, 599–614.

Nottebohm, F. (1970). Ontogeny of birdsong. *Science*, 167, 950–956.

Nottebohm, F. (1971). Neural lateralization of vocal control in a passerine bird. I. Song. *Journal of Experimental Zoology*, 179, 35–50.

Nottebohm, F. (1976). Phonation in the Orange-winged Amazon Parrot, *Amazona amazona*. *Journal of Comparative Physiology A*, 108, 157–170.

Nottebohm, F. (1981). Laterality, seasons and space govern the learning of a motor skill. *Trends in Neurosciences*, 4, 104–106.

Nottebohm, F. & Nottebohm, M.E. (1976). Left hypoglossal dominance in the control of canary and white-crowned sparrow song. *Journal of Comparative Physiology A*, 108, 171–192.

Nottebohm, F., Stokes, T.M. & Leonard, C.M. (1976). Central control of song in the canary, *Serinus serinus*. *Journal of Comparative Neurology*, 165, 457–486.

Nowicki, S. (1989). Vocal plasticity in captive black-capped chickadees: The acoustic basis and rate of call convergence. *Animal Behaviour*, 37, 64–73.

Pearlson, G.D. & Robinson, R.G. (1981). Suction lesions of the frontal cerebral cortex in the rat induce asymmetrical behavioural and catecholaminergic responses. *Brain Research*, 218, 233–242.

Petersen, M.R., Beecher, M.D., Zoloth, S.R., Moody, D.B. & Stebbins, W.C. (1978). Neural lateralization of species-specific vocalisations in Japanese macaques (*Macaca fuscata*). *Science*, 202, 324–327.

Petersen, M.R., Beecher, M.D., Zoloth, S.R., Green, S., Marler, P.R., Moody, D.B. & Stebbins, W.C. (1984). Neural lateralization of vocalisations by Japanese macaques: Communicative significance is more important than acoustic structure. *Behavioural Neuroscience*, 98, 779–790.

Pierson, J.M., Bradshaw, J.L. & Nettleton, N.C. (1983). Head and body space to left and right, front and rear. 1. Unidirectional competitive auditory stimulation. *Neuropsychologia*, 21, 463–473.

Rogers, L.J., Andrew, R.J. & Burne, T.H.J. (1998). Light exposure of the embryo and development of behavioural lateralization in chicks, I: Olfactory responses. *Behavioural Brain Research*, 97, 195–200.

Sawaguchi, T. & Goldman-Rakic, P.S. (1991). D1 dopamine receptors in prefrontal cortex: Involvement in working memory. *Science*, 251, 947–950.

Schiff, B.B. & Gagliese, L. (1994). The consequences of experimentally induced and chronic unilateral pain: Reflections of hemispheric lateralization of emotion. *Cortex*, 30, 255–267.

Slopsema, J.S., van der Gugten, J. & De Bruin, J.P.C. (1982). Regional concentrations of noradrenaline and dopamine in the frontal cortex of the rat: Dopaminergic innervation of the frontal subareas and lateralization of prefrontal dopamine. *Brain Research*, 250, 197–200.

Smith, M.O., Chu, J. & Edmonston, W.E. (1977). Cerebral lateralization of haptic perception: Interaction of responses to Braille and music reveals a functional basis. *Science*, 197, 689–690.

Summers, D.C. & Lederman, S.J. (1990). Perceptual asymmetries in the somatosensory system: A dichaptic experiment and critical review of the literature from 1929 to 1986. *Cortex*, 26, 201–26.

Vallortigara, G. & Andrew, R.J. (1991). Lateralization of response by chicks to change in a model partner. *Animal Behaviour*, 41, 187–194.

Vallortigara, G. & Andrew, R.J. (1994). Olfactory lateralization in the chick. *Neuropsychologia*, 32, 417–423.

Vargo, J.M., Corwin, J.V., King, V. & Reep, R.L. (1988). Hemispheric asymmetry in neglect produced by unilateral lesions of dorsomedial prefrontal cortex in rats. *Experimental Neurology*, 102, 199–209.

Watkins, J.A.S. (1999). Lateralization of auditory learning and processing in the domestic chick (*Gallus gallus domesticus*). D.Phil. thesis, University of Sussex.

Werntz, D.A., Bickford, R.G., Bloom, F.E. & Shannahoff-Khalsa, D.S. (1983). Alternating cerebral hemispheric activity and the lateralization of autonomic nervous function. *Human Neurobiology*, 2, 39–43.

Werntz, D.A., Bickford, R.G. & Shannahoff-Khalsa, D.S. (1987). Selective hemispheric stimulation by unilateral forced nostril breathing. *Human Neurobiology*, 6, 165–171.

Williams, H., Crane, L.A., Hale, T.K., Esposito, M.A. & Nottebohm, F. (1992). Right-side dominance for song control in the zebra finch. *Journal of Neurobiology*, 23, 1006–1020.

Witelson, S.F. (1976). Sex and the single hemisphere: Specialisation of the right hemisphere for spatial processing. *Science*, 193, 425–427.

Youngren, O.M., Peek, F.W. & Phillips, R.E. (1974). Repetitive vocalisations evoked by local electrical stimulation of avian brain: III. Evoked activity in the tracheal muscles of the chicken (*Gallus gallus*). *Brain, Behaviour and Evolution*, 9, 393–421.

Zann, R. (1990). Calls and song learned in wild zebra finch. *Animal Behaviour*, 40, 811–828.

11

Facing an Obstacle: Lateralization of Object and Spatial Cognition

GIORGIO VALLORTIGARA AND
LUCIA REGOLIN

11.1. Introduction

We might not be surprised by the fact that a predator may wait for the reappearance of a prey that has, momentarily, disappeared from view, or by the fact that the predator's prey remains motionless even when the predator has vanished temporarily from view or is only partly visible because it partly occluded by some obstacles. There could be (in some species, at least) several different abilities operating in these situations and, although such sorts of behaviours can be observed rather commonly in a natural environment, a scientific investigation on these aspects of higher cognition has begun only recently.

As human adults, we perceive our surroundings as a layout of continuous surfaces furnished with physical objects. According to the traditional Piagetian perspective, the ability to do this is actually the final outcome of a progression of stages. For instance, at a certain age, an infant will be able to grasp a partly hidden object, whereas a fully hidden one will not be retrieved. This sort of 'object permanence' for stimuli that are completely (not just partially) out of sight is, according to Piaget (1953), not developed in humans before the age of 7–8 months. More recent research, however, has provided evidence for object permanence abilities in human infants as young as 2 months old (Baillargeon, 1994), suggesting that the classical Piagetian phenomena do not reflect, in infants, a different concept of 'object' from that of adults, and that they have more to do with the development of 'action' than with the development of object representation (Carey and Spelke, 1996; Spelke, 1998a, 1998b). Such a change in perspective is particularly encouraging from a comparative point of view because it (hopefully) allows cognitive scientists to take into account the results obtained with species that are characterized by a precocial pattern of development, such as the young domestic chick (*Gallus gallus*), which is the protagonist of this chapter.

A major problem remains, however, with respect to the kind of analyses that have been conducted so far concerning the abilities of biological organisms to deal with objects in their environment. In fact, the notion of 'representation' is frequently used to 'explain away' the problem of identifying the actual cognitive mechanisms that may underlie animal behaviour. A striking example of this is provided by the detour behaviour, which has been associated with the presence (or the lack) of '*insight*' and representational abilities. We shall show that, in the domestic chick, relatively simple mechanisms seem to be necessary to deal with detour problems, and quite *specific* representations (e.g. those of position-specific and/or object-specific characteristics) are needed in different detour tasks. Moreover, the left and right cerebral hemispheres of the chick seem to be differently involved in dealing with the various aspects of these abilities, and this can provide a useful instrument to dissect the underlying mechanisms.

11.2. Perceiving Partly Occluded Objects

Two types of abilities, of a perceptual and of a cognitive nature, can be distinguished with respect to an animal's reaction towards objects surrounding it. Let us first consider some perceptual issues. Objects are typically represented as internally connected and externally bounded, with surfaces that continue behind nearer, occluding objects (Spelke, 1998a, 1998b). In our visual experience, when an object is partially concealed by an obstacle, we do not perceive just pieces or fragments of that object: the parts that are directly visible usually suffice for recognition of the whole object. Thus, in humans, visual perception is little impaired when objects are partly hidden. This is such a pervasive experience that makes it hard to appreciate the existence of a genuine scientific problem behind it. Although previous knowledge and memory of how objects are formed may sometimes play a part in this recognition, it is widely accepted that they are secondary to a more fundamental perceptual process of '*amodal*' completion (see Kanizsa, 1979; Michotte, 1963), which generates a genuine phenomenal presence of the non-visible parts, and which depends on detection of certain configurational relationships in visual scenes, such as the alignment of visible parts and similarities in their colours and textures. Perception of object unity in certain partial occlusion displays has been demonstrated in human infants as young as 2 months of age (Kellman and Spelke, 1983; Johnson and Aslin, 1995). Do other species also show recognition of partly occluded objects?

Among mammals, some evidence has been collected for mice (Kanizsa et al., 1993) and monkeys (Osada and Schiller, 1994). Among birds, the evidence is quite equivocal. Some studies would suggest that perception of object unity occurs in pigeons (Towe, 1954; Hamme, Wasserman and Biederman, 1992; White, Alsop and Williams, 1993), but others would not (Cerella, 1980; Sekuler, Lee and Shuttleworth, 1996). The possibility that avian species, which have usually very sophisticated visual abilities, could not perceive amodal completion would be somewhat surprising because, as a result of a bird's movements, the visible parts of the objects that are partly occluded undergo continuous changes in size and shape. So, in the absence of amodal completion, the animal would experience a weird, mutating, visual world. Note, however, that all these studies using birds have made use of conditioning procedures. We reasoned that filial imprinting, the learning process through which the young of some animals (usually of precocial species) come to recognize an object by being exposed to it, could be a very interesting and ecologically reliable situation to study perception of partly occluded objects.

We reared chicks singly with a freely moving triangle made of red cardboard (see Regolin and Vallortigara, 1995) (Figure 11.1). The triangle was suspended by a fine thread at about head height for the chick. At the test, on day 3, separate groups of chicks were presented with pairs of stimuli located at the opposite ends of a test cage. When faced with a choice between a complete and a triangle with a section removed, chicks clearly preferred to associate with a complete triangle, the stimulus with which they had been reared (Figure 11.2a). The choice did not seem to be due to a generic preference for figures with more extended red areas: when the part of the triangle that was removed was dislocated so as to produce a 'scrambled' triangle, chicks still preferred the complete one (Figure 11.2b). When faced with a partly occluded and a triangle with a section removed (both stimuli with exactly the same amount of red and black areas), chicks clearly chose the partly occluded triangle (Figure 11.2c).

An alternative explanation to amodal completion would suppose that chicks have a preference for visually 'compact' objects as opposed to 'fragmented' ones. We tried to test this by devising a situation in which the occluding bar was split and the two halves juxtaposed on the left and right side of the central missing part of the triangle. In this case, the background in the middle region of the triangle was clearly visible (much more clearly than in the schematic representation shown in Figure 11.2d because of the thickness of the cardboard figure), and so was the 'fragmented' nature of the triangle, but there was compactness of the overall figure, particularly with

(a)

(b)

Figure 11.1.   Schematic representation of the experimental design to investigate perception of partly occluded objects in chicks. Newly hatched chicks were imprinted on a red triangular stimulus (a) and then tested for choice (b) using various transformations of the original imprinting stimulus (see Figure 11.2).

respect to the continuity of the external boundary. Even in this case, however, chicks showed a preference for the partially occluded triangle (Figure 11.2d).

When we reared chicks with a partly occluded shape, the opposite outcome could be observed: chicks exposed to a partly occluded triangle preferred a complete triangle to a fragmented one (Figure 11.3a). This choice cannot be due to a generic preference for the stimulus with a larger coloured area, because chicks reared with a fragmented triangle did indeed prefer the frag-

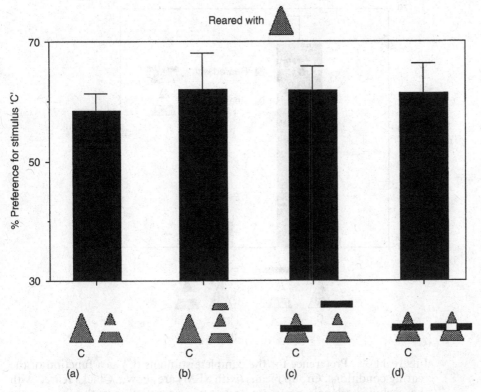

Figure 11.2. Preference for the complete (or amodally completed) stimulus (C). Group means [with standard error of the mean (SEM)] are shown. When faced with a choice between the complete triangle and the triangle with a section removed (a), chicks chose the complete triangle [$t(44) = 2.543$, $p \leqslant 0.01$], the same occurred using a 'scrambled' triangle as opposed to a complete one (b) [$t(34) = 2.115$, $p \leqslant 0.05$]. When faced with a partly occluded triangle and the one with a section removed, chicks chose the partly occluded triangle (c) [$t(43) = 2.667$, $p \leqslant 0.01$]. The same occurred when the occluding bar was split in half and juxtaposed to the missing part of the triangle with a section removed (d) [$t(28) = 2.306$, $p \leqslant 0.05$]. After Regolin and Vallortigara (1995).

mented triangle (smaller red area) to the complete (larger red area) one (Figure 11.3b).

Our results have been duplicated by Lea, Slater and Ryan (1996). In order to provide comparisons with data obtained in human infants, they used the same sort of stimuli that are employed by developmental psychologists (see Kellman and Spelke, 1983). Chicks were imprinted to: (1) two rod pieces moving above and below a central occluder; (2) a complete rod, in the absence of the occluder; and (3) two rod pieces also in the absence of the occluder. Chicks reared with the two rod pieces preferred at test this stimulus

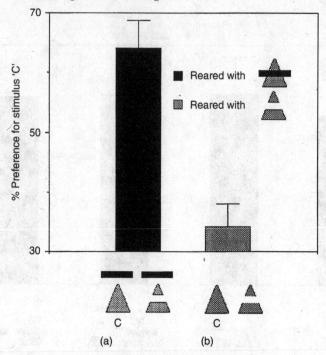

Figure 11.3.   Preference for the complete stimulus (C) as a function of the rearing conditions. Group means (with SEM) are shown. Chicks reared with the occluded triangle preferred at test the complete over the fragmented triangle (a) [$t(28) = 2.967$, $p \leqslant 0.01$]; chicks reared with the fragmented triangle preferred at test the fragmented over the complete triangle (b) [$t(29) = -4.033$, $p \leqslant 0.001$]. After Regolin and Vallortigara (1995).

to a complete rod, whereas chicks imprinted with an occluded rod preferred at test the solid rod over the two rod pieces.

Finally, indirect evidence for perception of object unity has been recently obtained by Forkman (1998) using adult hens. Using a touchscreen procedure, hens were trained to peck at the one of two stimuli (a square and a circle) that was the highest up on a screen with a grid providing pictorial cues of depth (i.e. at the stimulus that to a human observer appeared as being further away). When, during subsequent probe trials, hens were presented with either the circle overlapping the square or vice versa, they pecked significantly more at the stimulus that was occluded, thus showing that they completed or continued occluded objects.

Even though alternative explanations could be possible for each single experimental situation described in these studies, it seems to us that the overall evidence favours the idea that chicks (and hens) do possess visual abilities very similar to those found in human visual perception of partly occluded objects. This is not surprising. The visual system of vertebrates has probably evolved only once, so that its basic operating principles are likely to be common to all classes. Although some mechanisms (e.g. colour vision) evolved independently in different species in relation to specific ecological pressures, the problem of negotiating with a world of mostly opaque objects and therefore with occlusion phenomena is common to all ecological niches. It is also little surprising that chicks, a precocial species, show perception of object unity soon after hatching, whereas human infants, probably the most altricial of the species, require several months for its development. It would be simply useless for a newborn human (that cannot move autonomously to reach the mother) to develop very early such visual abilities.

11.3. Perceiving Partly Occluded Objects and the Cerebral Hemispheres

Lateralization of brain functions is well attested in the domestic chick. A variety of behavioural (Andrew, 1991), pharmacological (Rogers, 1982, 1989, 1991, 1995a, 1995b; Rogers and Anson, 1979) and neurobiological (Horn, 1990; Rose, 1992) techniques have revealed that the two hemispheres differ in fundamental ways in modes of analysis and storage of perceptual information (see Chapter 3 by Andrew and Rogers).

Procedures that restrict direct sensory input to one or other hemisphere have proved to be particularly valuable. Chicks using their left eye tend to choose between objects to which they are socially attached on the basis of small changes in their appearance. These same changes tend instead to be ignored by chicks using their right eye (Vallortigara and Andrew, 1991; Vallortigara, 1992a). [Following convention, the term 'hemisphere' will be used here for brevity to stand for the structures contralateral to the seeing eye, which in birds receive the direct visual input, and whose specializations are assumed to be responsible for the differences between right- and left-eyed chicks (see Andrew, 1991; but also see Chapter 6 by Deng and Rogers).] Chicks using their right nostril (and so predominantly their right hemisphere) show a similar pattern in choice based on olfactory changes (Vallortigara and Andrew, 1994a; see also Chapter 10 by Andrew and Watkins). Chicks using their left eye also have a striking advantage in topographical orientation based on visual cues (Rashid and Andrew, 1989). Overall, these findings suggest a special competence of the right hemisphere in spatial analysis and in response to novelty (see Andrew, Mench and Rainey, 1982; Andrew, 1991; Vallortigara and Andrew, 1994b). The left hemisphere, in contrast, appears to be concerned with select cues allowing stimuli to be assigned to categories that accommodate a range of different exemplars, despite variation between stimuli in a variety of other properties (see Andrew, 1991; Andrew, Mench and Rainey, 1982).

Differences in modes of analysis between the cerebral hemispheres provide a unique opportunity to look at the way in which the various aspects of visual representations of objects are organized in a bird's brain. To start with perception of partly occluded objects: are amodal completion abilities available to both hemispheres? It would seem reasonable to assume that these are such basic visual abilities that there is little chance of perceptual asymmetry. However, Corballis et al. (1998) recently reported that, following callosotomy, only the human patient's right hemisphere could benefit from amodal completion. Whether this reflects different abilities of the two hemispheres in early visual processing or rather in attentional mechanisms, however, remains to be established. There are no direct studies on this issue in birds, but there is some interesting evidence from a closely related phenomenon, perception of subjective contours (see Kanizsa, 1979). In a study on subjective contours with Kanizsa's triangles and squares, Prior and Güntürkün (1999) were able to demonstrate that some (but not all) of the pigeons that they tested reacted to the test stimuli as if they were seeing subjective contours. This is perhaps not particularly surprising, given that perception of subjective contours in birds has been demonstrated using behavioural methods in the domestic chick (Zanforlin, 1981) and in the barn owl (Nieder and Wagner, 1999); moreover, in the latter species, neurones have been found in the visual Wulst, the discharge rate of which is selectively modulated by subjective contours. It is interesting, however, that only a minority of pigeons responded to subjective contours. As indicated by control tests, pigeons responding to subjective contours were attending to the 'global' pattern of stimuli, whereas pigeons not responding to subjective contours were attending to extracted elements of the stimuli. Perception of subjective contours is closely linked to amodal completion. In natural situations in which objects occlude one another, boundaries may vanish and interpolation mechanisms to reconstruct contours absent from retinal images are sometimes needed. The fact that only pigeons attending to the more 'global' aspects of the stimulation respond to subjective contours suggests that such individual variability in attending 'globally' or 'locally' to visual scenes can explain why pigeons sometimes fail in amodal completion tests, which are effective in mammals (see Sekuler, Lee and Shettleworth, 1996). Although the pattern emerging in monocular left and right stimulation in the Prior and Güntürkün's study was similar, suggesting that both the left and the right hemispheres of pigeons that responded to subjective contours were capable of 'filling in' processes, it could be that a basic difference between the hemispheres still exists. This difference between the hemispheres may be revealed by the difference between pigeons that respond and pigeons that do not respond to subjective contours. For instance, it could be that dominance of one or other hemisphere favours a 'global' or 'local' strategy of analysis of visual stimuli.

We are currently testing these ideas in chicks, using the imprinting paradigm applied to recognition of partly occluded objects in birds tested monocularly. Soon after hatching, chicks were imprinted on a red square partly covered by a black rectangular bar, freely suspended at about head height in the animals' homecage. At the test, on day 3, separate groups of chicks were presented with pairs of stimuli located at the opposite ends of a test cage. Chicks could approach either a complete square or one with a section missing, with the black bar juxtaposed without covering the stimuli. Our preliminary results suggest that differences between the two eye systems indeed exist. When tested with only the left eye in use, 3-day-old chicks behaved very much like binocular chicks, choosing the complete stimulus (the square); when tested with only the right eye in use, in contrast, chicks chose the square with a section missing. In principle, these results could be explained by differences in approach–avoidance tendencies, but this seems unlikely because a great deal of previous evidence suggests that, as a rule, it is the left eye (right hemisphere) that shows stronger tendencies to explore novelty (see above). Thus, although further experiments are certainly needed, we would claim that the left eye system (the right hemisphere) is in chicks more inclined to a 'global' analysis of visual scenes, whereas the right eye system (left hemisphere) seems to be more inclined to a 'featural' analysis. This is, of course, a very broad and quite imprecise way to characterize cerebral functions in the two hemispheres. We hope, however, to become progressively more accurate in the following sections. Following the same scheme used for perception of partly occluded objects, in each section we shall first consider the evidence for a particular cognitive ability in the chick and then the role played in this ability by the left and right hemisphere.

11.4. Retrieving Completely Occluded Objects

Sometimes objects are not just partly hidden but are totally concealed behind other objects. Yet, even when an object disappears completely from sight because of an obstacle, we usually do not believe that it is gone 'out of existence'. Here perception meets memory and cognition, for we are able both *to perceive* certain kinetic displays as 'the hiding' of an object (Michotte, 1963) and *to remember* that the object is hidden behind an occluder in a certain spatial location.

The cognitive capacity functioning to attribute continued existence to objects that have disappeared and to localize physical and social objects in space appears to be of high adaptive value (Etienne, 1974): it is so basic a concept for successful interaction with the external world that its ecological relevance has rarely been questioned (Dumas and Doré, 1989, 1991). Only after Piaget (1937, 1953) suggested that this capacity is not immediately available, but develops with age and can be acquired to varying degrees, did a number of studies investigate these issues in animal cognition (apes: Mathieu et al., 1976; Natale et al., 1986; Redshaw, 1978; monkeys: Mathieu et al., 1976; cats: Gruber, Girgus and Banuazizi, 1971; Triana and Pasnak, 1981; Thinus-Blanc et al., 1982; Doré, 1986, 1990; Dumas and Doré, 1989; dogs: Triana and Pasnak, 1981; Gagnon and Doré, 1992). These studies demonstrated that the capacity of at least some non-human mammals [e.g. chimpanzees (*Pan troglodytes*), gorillas (*Gorilla gorilla*) and dogs (*Canis familiaris*)] for solving 'object permanence' problems is comparable to that of humans (for a review, see Etienne, 1974; Dumas, 1992).

Detour behaviour, i.e. the development of itineraries that allow for obstacles between subject and goal, is of particular interest with respect to the issues of object representation and object permanence. Typically, in the detour test, the animal is required to abandon a clear view of a desired goal-object in order subsequently to achieve that goal. Thus, in the absence of local orienting cues emanating from the goal, detour performance suggests the maintenance of a mental representation of at least some of the characteristics of the object that has disappeared, of the type that developmental psychologists have investigated extensively in human infants (Piaget, 1953).

It is clear that a proper understanding of these aspects of the animal perceptual and cognitive world is a *sine qua non* condition for facing even the more basic ethological questions (e.g. prey–predator interaction, social and individual recognition, and so forth). In the following sections we shall describe in great detail an experimental analysis aimed at penetrating the perceptual and cognitive characteristics of the domestic chick through an analysis of its behaviour while dealing with an obstacle. However, we first summarize the present state of knowledge about the cognitive mechanisms implicated in detour behaviour.

11.5. The Detour Problem: Turning Around Obstacles to Reach either Visible or Completely Occluded Objects

The detour problem consists of a situation in which an animal has to develop an itinerary to go around an obstacle, i.e. it has to move away from a goal in order eventually to reach it. Köhler (1925) first introduced this problem as a test of insight learning. He stated that the detour problem can be solved in one trial by *insight* whenever the overall problem situation is perceptually recognizable by the animal. Köhler insisted on an immediate perception of functional relationships and stressed that the insight is independent of previous experiences. The idea of insight is implied also in Lewin's (1933, 1935) and Tolman's (1932) approaches, which attributed failures in the detour task to an incomplete representation of the situation. They maintained that a certain level of representational integration is necessary in order to consider moving away from the goal as a required step toward a possible solution.

Even small changes in the experimental situation can affect detour behaviour. Factors that have been shown to be relevant are length, number, complexity and angular deviation of the available paths leading to the goal, distance of the goal itself, and nature of the cues whereby it can be located (see Poucet, Thinus-Blanc and Chapuis, 1983; Chapuis, 1987; Rashotte, 1987).

Blachenteau and Le Lorec (1972) dissociated two of these factors in a detour situation in which two routes were available at the choice point: the shorter route formed a wider angle with the straight line to the goal than the longer one. The goal position was clearly marked by a light just above it. During the early trials, rats chose the path with the smaller deviation with respect to the direction of the goal, although it was the longest. In the successive trials, rats rapidly learned to choose the shorter (but more divergent) path.

Chapuis, Thinus-Blanc and Poucet (1983) analysed the role of length and angular deviation of the paths leading to the goal according to whether or not the goal could be seen from the starting position. The animals (dogs) had to choose between two routes leading to the goal, differing in their respective lengths and/or angular deviations at the choice point with respect to the straight-line direction to the goal. In one condition, the goal was visible from the choice point (transparent screen), whereas in the other it was not (opaque screen). The animals preferred the shortest and less divergent path when one of the two dimensions (angular deviation and length of the path) was relevant or when both were relevant, and there was no contrast between them. When, on the other hand, the shorter route formed a larger angle than the longer route, dogs had more difficulties in their choice and the visibility of the goal seemed to be of great importance. Using an opaque screen, the length of the path became the main variable, whilst the angular deviation had less importance. As a result, the dogs tended to choose the shorter route more frequently than the less divergent one. Using a transparent screen, dogs had greater difficulty in discriminating the shortest path, and no preference for either path was shown. In this case, choices depended almost entirely on the angular deviation, the less divergent route being preferred independently of its length.

Poucet, Thinus-Blanc and Chapuis (1983) obtained very similar results with cats. The only difference was that, in the opaque condition, when the angular deviation was the only relevant factor, cats tended to prefer the minimum angular deviation route. Interestingly, this agrees with Hull's (1938) proposal that, in the case of an asymmetrical barrier, the less divergent path would be preferred even though it is longer, since, from the starting point, it would be the more attractive and successful one. Beside this confirmation, the results of Poucet et al. are unexpected because adding information about the location of the goal should have made the task easier (in fact the only information to be taken into account and processed in this case would have been the length of the route). The sight of the goal seemed to act as a 'perceptual anchor' inhibiting the integration of previous experiences (in particular eliminating the use of the distance cue) and forcing the animals to the most immediate strategy, i.e. reducing the distance between themselves and the goal. It appeared that a perceptual dominance of the sight of the goal prevented the animals from taking into account the path lengths.

These results could be accounted for in terms of the O'Keefe and Nadel's hypothesis (1987) concerning two classes of spatial mechanisms: mapping and route-following. Spatial behaviour could be guided by direct perception of a target to be approached or avoided ('guidance mechanism') or by a rigid sequence of landmarks defining a route with reference to which the animal may reorient its body axis ('orientation mechanism'). These two mechanisms would belong to a 'taxon system'. Besides this, a so-called 'locale system' (see O'Keefe and Conway, 1980) could also be available and provide a spatial map of the environment that would allow animals to locate and orient themselves. From this hypothesis, it appears possible to differentiate the mechanisms involved in spatial orientation according to whether the goal is visible or not. Maps would in fact be characterized by their flexibility, high information content and relative invulnerability to loss of information together with low speed of operation; in contrast, the properties of routes would be high speed of operation but also rigidity, low information content and high sensitivity to loss of information. This last set of properties seems exactly to match the characteristics that emerged in dogs' and cats' behaviour when dealing with a visible goal (Chapuis, Thinus-Blanc and Poucet, 1983; Poucet, Thinus-Blanc and Chapuis, 1983).

There is other evidence consistent with a negative effect of goal visibility on detour performance. Poucet (1985) trained cats in two spatial learning tasks differing in the location of one single available visual cue that could be either remote from the goal ('mapping situation') or just above the goal ('guidance situation'). Cats using a mapping mechanism performed better at delayed-reaction tests than did those using a guidance mechanism. Similarly, Schiller (1949) trained minnows to find roundabout pathways. He noticed that animals mastered the task better if the goal was not visible while approaching it; while if the goal was visible, the detour was usually interrupted as soon as it became visually closer. Transparent partitions required almost twice as long as partly opaque barriers, and Schiller observed that most of the solving time was actually spent circling in front of the visible goal. Studies with human infants also indicate that detours around opaque barriers are easier and emerge at an earlier stage of development than detours around transparent barriers (Lockman, 1984).

Goal distance is another critical variable in detour problems. Köhler (1925) first stressed that the distance of the goal affects its degree of attractiveness. In fact, a lower detour time would be expected with a more distant goal since the ratio between the lengths of the two paths (the direct and the indirect one) decreases as the distance from the goal increases (see also Lewin, 1933, 1935; Tolman, 1932). Hull (1938) also attributed an excitatory potential to the goal with decreasing gradient at increasing distance. Interestingly, however, there has been very little empirical support for these claims, and failure to observe any effects of goal distance on detour behaviour has sometimes been reported (e.g. Sandler et al., 1968).

11.6. The Chick in Front of a Barrier

Traditionally birds have been regarded as cognitively inferior to mammals, a view which has been challenged in recent years on both neuroanatomical (Karten and Shimizu, 1989) and behavioural grounds (Emmerton and Delius, 1993; Rogers, 1995b, 1997). The chicken, in particular, has always had the reputation of being a 'poorly intelligent' creature. Folk-ethology stereotypes contributed to this view and, on the scientific side, the detour problem has long been considered as evidence of the chicken's alleged cognitive limitation.

Little is known about detour learning in birds. Köhler (1925) reported that chickens had difficulty in detour problems, and this has been interpreted traditionally as indicating poor abilities to form cognitive maps in this species (e.g. Pearce, 1987, p. 224). Krushinsky (1990), in contrast, reported that chickens have a certain capacity of what he called 'extrapolation reflex'. He performed tunnel experiments in which a feeder was moved on a rail so that the animal could follow it and obtain some food. After travelling some distance, the feeder entered a tunnel, so that after having watched the initial trajectory of the feeder, the animal had to extrapolate the direction of movement. Pigeons made no attempts to look for the food in the direction of its disappearance. Chickens, in contrast, were searching for food when it disappeared and some animals moved in the direction in which the food was moving before it disappeared; in several cases they followed alongside the tunnel for about 50 cm. Chickens (and ducks) were, however, not as good at doing this as were magpies and crows, which searched actively for the food all along the length of the tunnel.

Some studies have been performed on detour learning in chicks. Etienne (1973) presented 6-day-old chicks with a mealworm that disappeared behind one of two screens opposite each other. At the first trial, chicks did not necessarily choose the screen behind which the mealworm had disappeared. After repeated testing, however, all chicks developed searching behaviour behind either screen and a minority of the birds (24%) learned spontaneously to orient their delayed response directly towards the correct screen. Similar results were obtained with 3-, 6- and 14-day-old chicks, although in the intermediate age group (6 days) learning was more difficult than in the other two groups (Etienne, 1974). Scholes (1965) and Scholes and Wheaton (1966) also showed that young chickens less than 2 weeks old could, after repeated testing, learn to solve a problem requiring them to leave a clear view of a desired goal (a group of conspecifics) in order to reach that goal.

These studies seemed to indicate that chicks can learn the correct route to the goal after repeated trials (*detour learning*), but have difficulties in solving the problem on the first attempt, in the absence of previous experience of the correct spatial route (*detour behaviour*). Nonetheless, several results show quite good spatial learning abilities in this species. Chicks can easily learn to orient in an environment to find food using both distant and nearby cues (Vallortigara and Zanforlin, 1988; Rashid and Andrew, 1989) as well as the geometrical properties of the spatial arrangement of the surfaces (Vallortigara, Zanforlin and Pasti, 1990; Tommasi, Vallortigara and Zanforlin, 1997). If the chicks are quite able to orient in space, why would they have difficulties at solving the detour task on their first attempt? Is it that they simply need some experience of the spatial arrangement of the test situation, or do they lack any abilities at all to form a representation of the objects that have disappeared? After all, we know chicks can build up internal representations and, in certain circumstances, this occurs at once and does not require repeated experiences (e.g. in the passive avoidance test; see Rose, 1992). Also, in delayed matching to sample paradigms, hens perform similarly to other bird species (Foster et al., 1995) and in tasks requiring recognition of a conspecific individual they even surpass species such as the pigeon (Ryan and Lea, 1994). Krushinsky (1990) also claimed that there is a certain capacity for 'extrapolation' in adult chickens. Thus, before attributing substantial cognitive limitation to young chicks, alternative explanations should be considered carefully.

11.7. Perceiving an Obstacle (and a Goal)

Köhler (1925) thought that in order to solve the detour problem, the overall problem situation has to be recognizable *perceptually* by the animal. The barrier used by Scholes (1965), for instance, was a clear glass plate behind which cagemates were visible. Transparent obstacles are infrequent and unnatural in the animal world: the chicks' persistence in attempting to pass through this sort of barrier may therefore simply be due to their difficulty in perceiving it as an obstacle (and there is evidence of a similar difficulty being experienced by human infants; Lockman, 1984; see also above).

Moreover, all major theoretical analyses of detour behaviour (Tolman, 1932; Lewin, 1933; Hull, 1938) maintain that the degree of attractiveness of the goal is the main determinant of detour performance of naive animals. Thus, goal visibility should contribute to the difficulty of the task: the more visible a goal, the more attractive and stronger the tendency to choose the direct rather than the indirect route to reach it. The sight of the goal seems to act as a 'perceptual anchor', forcing the animal to adopt the most immediate strategy in order to reduce the distance between itself and the goal. (The use of transparent barriers might thus have been particularly unfortunate because complete visibility of the goal is combined with a very reduced visibility of the obstacle.)

Also, even among opaque barriers, there could be some that are perceived by the animals as less (or more) of an obstacle than others. Miller (1973) reported that, when confronted with low-density mesh barriers, ring doves and pigeons displayed very long detour times. The birds seemed to perceive the low-density mesh barrier as lesser obstacles than the higher density mesh barriers, inasmuch as they tended to spend more time trying unsuccessfully to penetrate these barriers rather than detour around of them.

To investigate the role of the perceived characteristics of the obstacle and the goal, we used cagemates as the goal, placed at different distances behind barriers concealing them to different degrees (Figure 11.4). Six different barriers were used: a transparent barrier (T), a horizontal-bar barrier (H), a vertical-bar barrier occluding the same amount of area as the horizontal-bar barrier (V), a small-grid barrier obtained by superimposing the vertical-bar barrier on the horizontal-bar one (SG), a large-grid barrier with the same degree of occlusion as the H (or V alone) barrier (LG), and an opaque barrier (O). Percentages of physical occlusion of the goal were: 0% (T), 20% (H, V, LG), 40% (SG), 100% (O). There was also another type of barrier (S), studied only at short distance, which consisted of a translucent screen. A lamp was placed beyond the transparent cage where the goal was located, so that the silhouettes of these chicks projected on to the translucent barrier and were visible to the test chick.

Chicks were reared socially in groups of three of the same sex (trios) and tested on day 2. One of the chicks was selected randomly as the 'test animal' and placed behind the barrier. The other two chicks were used as the 'goal' and put together in the transparent cage in front of the barrier at either a short (0 cm) or long (30 cm) distance. The time needed to solve the task was measured. The same procedure was then repeated for the other two animals of each trio, so that each chick became, in turn, either a 'goal' (together with

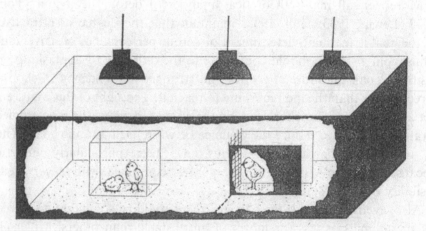

Figure 11.4.   The experimental apparatus used to investigate the effects of different types of barriers in the detour problem. The task was considered accomplished when the entire body of the test chick crossed the dotted line. After Regolin, Vallortigara and Zanforlin (1994b).

another chick) or a 'test' chick. The means of the data for each trio were used as individual data for statistical purposes (for further details, see Regolin, Vallortigara and Zanforlin, 1994).

Figure 11.5 shows the results. The transparent and vertical-bar barriers appeared to be the most difficult ones, the opaque barrier the easiest. The shadow barrier also made the task easier, but solving times were longer than with the opaque barrier. This means that chicks were not simply orienting using acoustic stimuli (in which case no difference would be expected between the opaque and the shadow barriers), but were responding, to a certain degree, to the shadows themselves. In general, the task became easier with increasing goal distance, an effect that has long been postulated in the literature (see above), although seldom proved experimentally.

The degree of visual occlusion had a crucial effect on detour behaviour. When visibility of the goal was reduced by occlusion, the task became easier. With the small grid barrier, the task was easier than with the vertical- or horizontal-bar barriers. However, with the small-grid barrier (obtained by superimposing the vertical- and horizontal-bar barriers) the amount of physical occlusion was doubled. Our results suggest that facilitation depended on the larger degree of occlusion produced by this barrier. With the large-grid barrier, which produced the same degree of occlusion as the vertical- or horizontal-bar barriers alone, the task was in fact more difficult than when the small-grid barrier was used.

Bar orientation also affected the results. The vertical-bar barrier took longer to detour than the horizontal-bar barrier, even though the physically occluded area was identical. There are two possible explanations for this asymmetry. One hypothesis concerns the different ways adaptation to particular environments shapes perceptual abilities of animals with identical sense organs. Lorenz (1971) speculated that organisms whose habitats were characterized by little structured differentiation would show less sophisticated and less differentiated orienting behaviour. The extreme example would be certain free-swimming organisms dwelling in the open sea, e.g. jelly fish, completely lacking spatially oriented responses to external stimuli. Lorenz also observed that, among steppe-living organisms, there are some that are unable to comprehend and master (even after specific training) a vertical-structured obstacle and, interestingly enough, this peculiarity appears to be specific of certain behavioural patterns. For instance, Lorenz believed that while flying partridges can take vertical obstacles into account, running partridges cannot (flying birds need to be able to cope with woods, vertical faces and the like). Our chicks had never seen grass blades, twigs or similar structures in their short lifetime before the experiments. Nonetheless, it could be

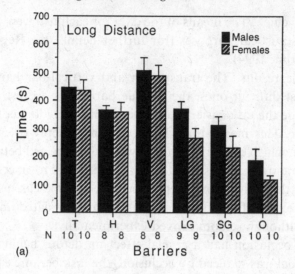

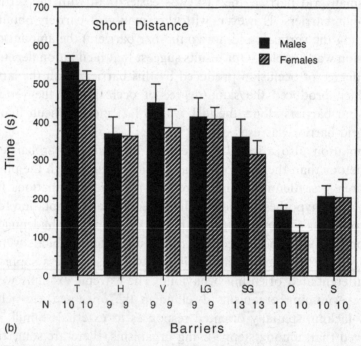

Figure 11.5.   Mean times (± SEM) needed to solve the detour problem as a function of sex and type of barrier with the goal placed at a long (a) or a short (b) distance from the test chick. N, number of trios; T, transparent barrier; H, horizontal-bar barrier; V, vertical-bar barrier; LG, large-grid barrier; SG, small-grid barrier; O, opaque barrier; S, shadow barrier. Redrawn from Regolin, Vallortigara and Zanforlin (1994b).

that vertical bars are not perceived as natural obstacles by running galliformes. In a natural environment, patterns of vertically elongated fine bars (such as grass and twigs) offer little resistance to an animal whose movements can easily bend such structures as it walks through them.

The second hypothesis is related to the fact that horizontal and vertical occluders are equivalent only from the perspective of a completely motionless observer. Anyhow, a walking animal experiences occlusion and disocclusion of different parts of the goal when the occluders are vertical, whereas there are no optical effects in the case of horizontal occluders (assuming, of course, that movements of the head along its vertical axis are negligible with respect to movements of the entire body along the horizontal axis). As a result, although both horizontal and vertical bar gratings occlude the same physical surface, they would be optically different for a moving organism: in a given time, vertical-bar barriers allow the animal to see 'more' of the goal than do horizontal-bar barriers. Note, also, that the same objections apply even to a stationary organism with binocular vision, where the binocular field of one eye can compensate for the occlusion experienced by the other eye and vice versa. Since chicks have a certain (though reduced) degree of binocular overlap, this mechanism could have been operating at least when the goal object was located at a close distance. If enhanced visibility of the goal increases the difficulty of the task, then it is understandable that chicks take longer with vertical-bar than with horizontal-bar barriers.

Further experimental work allowed us to decide between the two hypothesis. In our original design, goal visibility and visual characteristics of the barriers were linked. Did highly occlusive barriers facilitate the task only because they concealed the goal to a larger extent or was it that because they were also recognized as 'natural obstacles' more easily than other kinds of barriers? We tried to disentangle these two factors by devising an experimental situation in which the use of different types of barriers did not affect the visibility of the goal. This was achieved by reducing the height of the barriers until they were below the chick's eye level (11 cm) and by raising the transparent cage containing the goal to the same level. In this way, regardless of the chick's position in the corridor, the barrier could never occlude any portion of the goal. (To prevent chicks from jumping over the barrier, a glass screen was placed 1 cm beyond the barrier.)

Figure 11.6 shows the results. Vertical-bar, horizontal-bar and grid barriers appeared to be equivalent, while the transparent barrier continued to present the greatest difficulty. This suggests that chicks do indeed have problems in considering transparent barriers as 'obstacles', and that this effect is not simply due to the complete lack of goal concealment.

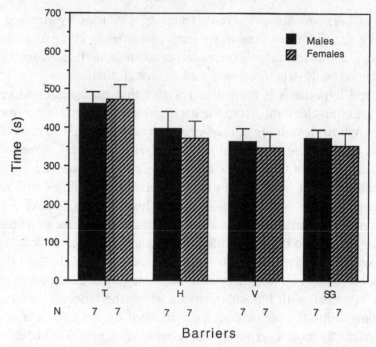

Figure 11.6. Mean times (± SEM) needed to solve the detour problem with lower barriers as a function of sex and type of barrier. N = number of trios; T, transparent barrier; H = horizontal-bar barrier; V, vertical-bar barrier; SG = small-grid barrier. Re-drawn from Regolin, Vallortigara and Zanforlin (1994b).

One might argue that the absence of differences between horizontal-bar and grid-barriers demonstrates that concealment of the goal rather than of the rest of the environment is the crucial variable. Yet, it might simply be that, with lower barriers, only strong differences between them could be taken into account by the chick (e.g. those resulting from comparisons between completely transparent and half-transparent barriers), whereas slight differences remained unnoticed (e.g. comparisons between different bar-patterns in barriers below eye level).

With lower barriers, in the absence of goal occlusion, one would expect an increase in the times needed to solve the task as compared to higher barriers. In particular, if this increase is caused by the degree of concealment produced by the different barriers, one would expect only slight differences with transparent and vertical-bar barriers, a slight increase with the more occlusive horizontal-bar barriers and a strong increase with the very strongly occlusive small-grid barriers. These predictions proved to be correct (see Figure 11.6). A puzzling result, however, occurred with vertical-bar barriers, with which times were shorter when lower barriers were used. Clearly, the degree of physical occlusion cannot explain this result. Lower vertical barriers did not occlude the goal at all, hence one would expect times to be longer or at least equal to those shown with goal-occluding barriers of identical orientation. The conclusion seems to be that vertical-bar barriers occluding the goal (or at least high enough for the animal to take into account the orientation of the bars) produced an increment in difficulty that cannot be explained in terms of disocclusion mediated by the animal's movement.

Some of the perceptual and motivational factors affecting chicks' performance in the detour task appeared to be reasonably clear at this point, but they were not the entire story. Perception provides organisms with information about the *social* as well as the physical environment. In our experiments test chicks were faced with pairs of animals that moved and interacted freely with each other and with the environment. Most of the information about the nature of this interaction was visually and acoustically available to the test animal.

During the experiments, we noticed that the two goal chicks placed together in the transparent compartment interacted with each other (i.e. they spent most of the time in reciprocal contact, gently pecking at each other and emitting soft peeping) and with the environment (looking for small fragments to peck at and showing the usual social enhancement effects associated with pecking behaviour). They did not emit distress calls, and they seemed to pay very little attention to the isolated test chick placed on the other side of the barrier. The behaviour of the test chick was, in contrast, quite different: it frequently emitted distress calls, and appeared to be constantly looking for tactile, visual and acoustic interaction with the companions placed behind the barrier. This outcome is probably little surprising considering that the three chicks had been reared socially in a cage with no limits to free interaction, and also considering the reassuring effects of the presence of a social companion when a chick is placed in a novel environment (Vallortigara and Zanforlin, 1988). It is interesting, however, to investigate how much the behaviour of the goal chicks could have affected detour performance. It seems reasonable that the tendency to reinstate social contact along the direct rather than the indirect route depends on how attractive the goal is, which in turn depends not only on the mere visibility of the goal, but also on its behaviour. To what extent, then, does the goal itself actively attract the chick's attention by means of visual and acoustic signals? To provide an answer, we proceeded by placing an opaque partition inside the transparent compartment so as to split it sagittally into two halves. As a result, the two 'exposure' chicks were visually separated from each other, whereas from the test chick's perspective the number of goal conspecifics remained the same.

When the goal chicks could interact, the task was much easier than when interactions were limited to the ones between the 'goal' and the two separated 'test' chicks (Figure 11.7). Since the number of companions visible to the test chick remained identical in the two conditions (i.e. with or without the sagittal partition), it seems reasonable to conclude that it was the different behaviour of the goal chicks (either isolated or non-isolated) that affected detour performance of the test chick.

This experimental outcome does not require attribution of sophisticated cognitive abilities to the chick. The simplest interpretation is that the chick would respond to the social signals emitted by an isolated conspecific. There is independent evidence for chicks being able to do this (and even something more than just this): socially reared chicks prefer unfamiliar socially reared chicks to unfamiliar chicks reared in isolation (Sigman, Lovern and Schulman, 1978). This suggests that 48-h-old chicks respond to conspecifics on the basis of behavioural characteristics other than morphology. Interestingly, even if there is evidence that chicks reared with conspecifics

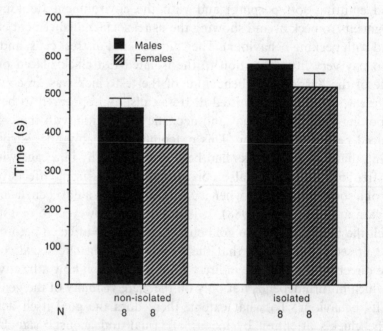

Figure 11.7. Mean times (± SEM) needed to solve the detour problem using pairs of chicks visually isolated by a sagittal partition or using pairs of non-isolated chicks as a goal. N, number of trios. Redrawn from Regolin, Vallortigara and Zanforlin (1994b).

are sensitive to reductions in group size (Jones and Harvey, 1987), manipulations involving the number of goal chicks visible behind the barrier with respect to the number of rearing cagemates had no significant influence on chicks' detour performance (for further details, see Regolin, Vallortigara and Zanforlin, 1994).

To summarize, our results suggest that the chick's difficulties in detour problems should mostly be ascribed to the motivational overtones and perceptual ambiguities of the experimental situation. In fact, the task difficulty can be changed dramatically simply by modifying the characteristics of the obstacle or of the goal, i.e. the less visible the goal, the easier the task. Moreover, apart from the attractiveness of the goal itself, there seem to be barriers that are perceptually less of an 'obstacle' than others, either because they are not true obstacles in a natural environment for that species (e.g. vertical barriers) or because they are somewhat special and not normally encountered in a natural environment (e.g. transparent barriers). Social signals emitted by the goal appear to be a crucial factor too.

It is true, of course, that, even in the most favourable conditions, the solution of the detour problem still requires a few minutes (see Figure 11.5). Yet, since animals were separated abruptly from their conspecifics and placed in a novel environment, the necessity of some time to become familiarized with the testing situation should certainly be considered. As we shall see later, it is in fact possible that the shortening of the time needed to solve the problem, occurring with repeated experiences in the test situation (see e.g. Scholes, 1965), would reflect adaptation to the novel environment and reduced emotional responses rather than some form of increased spatial learning *per se*. The evidence we collected does indeed suggest that the detour learning could be an all or nothing form of learning.

11.8. Representing and Locating a Goal that Has Disappeared

In the aforementioned experiments, our results might be accounted for in terms of a purely sensory guidance system; in fact, cagemates used as goals provided both visual and acoustic stimuli, and chicks' spatial behaviour could have been entirely under the control of local sensory cues emanating from the goal (the same objection could apply to other studies employing cagemates as the goal; e.g. Scholes, 1965). It may be that the chicks, after several frustrated attempts to pass 'through' the obstacle, moved about quite randomly and exited the U-shaped barrier losing sight of the goal. Then acoustic stimuli from cagemates (and perhaps asymmetries of the test arena such as closed and open ends) allowed them to orient correctly towards the goal. Motivational overtones and ambiguous perceptual information could thus contribute to the difficulty of the task but, nonetheless, chicks might, in fact, lack the ability to *represent* the goal and its spatial location in the absence of locally orienting cues.

The issue could be addressed by looking at the chick's detour behaviour at its first trial after disappearance of the goal and in the complete absence of sensory cues to orient towards the goal and of previous experiences. If chicks move randomly in the environment when the goal is no longer available to direct perception, then no straightforward conclusion can be drawn, because chicks may represent the object but lack any ability to discover its position (admittedly, a more economic and conservative tenet would deny possession of both capacities). If, on the other hand, chicks move non-randomly and show an ability to orient towards the goal that has disappeared, then some sort of mental representation of the 'remaining in existence' of the goal in a certain spatial location can be ascribed to the animals (of course, leaving open the issue of what the nature of the mechanism to localize the goal would be).

Chicks rapidly develop very strong social attachment to small artificial objects with which they have been reared for even a short time (e.g. Salzen, Lily and McKeown, 1971; Johnson and Bolhuis, 1991; Vallortigara and Andrew, 1991). We used such artificial social partners, i.e. small red cylindrical objects (see Figure 11.8), as goals, thus limiting local orienting stimuli to the visual ones.

The apparatus we used is shown in Figure 11.8. The chick placed in the starting position, within the inner corridor, could see the goal (a cylinder identical to that used during rearing) through the grid of the small square window. The goal was placed 50 cm away from the barrier. In order to reach the goal the chick could choose between four compartments (incorrect compartments: A and B; correct compartments: C and D). From the chick's perspective, the walls and openings of the apparatus were symmetrical. The chick's first choice to enter one of the compartments and the time taken were recorded (diagonal partitions prevented the sight of the closed walls immediately after moving out of the corridor).

Chicks were tested on day 2. Five (all males) of the 25 tested chicks failed to go outside the corridor within 600 s; of the remaining 20 animals, 18 chose the correct compartments, C and D [χ^2 (1) = 12.80, $p < 0.001$; males 8 versus 0, χ^2 (1) = 8.00, $p = 0.005$; females 10 versus 2, χ^2 (1) = 5.33, $p = 0.02$]. Times needed to solve the problem were lower in females than in males [males 423 ± 47 s; females 261 ± 52 s; $F(1, 18) = 5.261$, $p < 0.05$]. There were no significant differences between choices for C and D (12 versus 6,

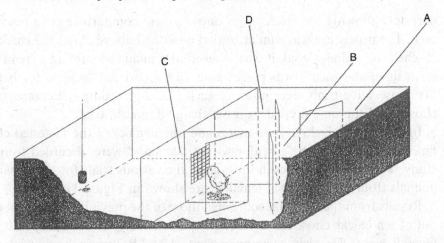

Figure 11.8. Schematic representation of the experimental apparatus used to study the representation of objects that have disappeared in the chick. The imprinting stimulus is visible behind the small window-grid barrier. Two symmetrical apertures placed in the midline of the corridor allowed the chick to adopt routes passing around the barrier. After entering the apertures, the chick is faced with a choice between a correct and an incorrect compartment (A, B are incorrect compartments; C, D are correct compartments). After Regolin, Vallortigara and Zanforlin (1995a).

$p > 0.10$) and A and B (2 versus 0, $p > 0.10$). Results showed clearly that chicks were able to turn correctly towards the goal in the absence of any locally orienting cues. It seemed that they could maintain some sort of representation of the location (and thus of the presence) of a social partner even when this was no longer available to direct perception.

Previous studies of detour learning seem to assume implicitly that the reduction in the time needed to solve the problem after repeated trials reflects spatial learning of the correct route (cf. Scholes, 1965). Our results, however, suggest that chicks may have very little to learn about the spatial localization of the goal, since they could turn correctly in the absence of previous experience of the correct route and of reinforcement contingencies associated with the following of that route. Thus, time improvement during detour learning may be attributed to lowered emotional responses to the novel environment and a better appreciation of the impossibility of following the straight route. However, because the procedure we adopted differed in several respects from that used by previous investigators (e.g. Scholes, 1965; Etienne, 1973, 1974), we performed a repeated-trials experiment to check whether a reduction in the times needed to reach the goal could be observed with our paradigm too.

In the same apparatus employed in the previous experiment a small opening was made in compartments C and D, near the external walls of the corridor, allowing the chick to go outside these compartments to reach the goal. The procedure was similar to that described above. After the chick had reached the cylinder/goal, it was allowed to remain there for 15 s (reinforcement time) and then it was placed back in the corridor, beyond the barrier. Six consecutive trials were given to each chick, recording the compartment chosen and the time needed to reach the goal in each trial.

In the first trial, three (two males and one female) of the 13 tested chicks failed to go outside the corridor within 600 s and were discarded from the study. Times needed to reach the goal in the various trials for the remaining animals (four males and six females) are shown in Figure 11.9.

Results from the first trial confirmed those of the previous experiment [nine out of ten chicks chose the correct compartments; χ^2 (1) = 6.40, $p = 0.011$]. Overall, errors (i.e. visits to compartments A and B) during the six trials were rare (8 out of 60 visits). The majority of birds alternated their correct choices between compartments C and D. Four birds were given another two trials after an interval of 2 h: they all chose the correct compartments (C, D) both times without errors.

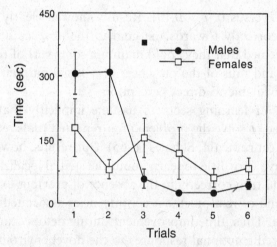

Figure 11.9. Mean times ($\pm$ SEM) needed to reach the goal over six successive trials using the apparatus shown in Figure 11.8. The analysis of variance revealed a significant main effect of trials [$F(5,40) = 5.776$, $p < 0.001$] and a significant sex $\times$ trials interaction [$F(5,40) = 3.983$, $p = 0.005$]. The main effect of sex was not significant [$F(1,8) = 0.209$]. In the first two trials females took less time to solve the task than males ($F(1,8) = 5.970$, $p = 0.039$], whereas in the last four trials, males took less time than females [$F(1,8) = 8.789$, $p = 0.017$]. Redrawn from Regolin, Vallortigara and Zanforlin (1995a).

In the single-trial experiment, the spatial localization of the goal was likely to be based on an egocentered frame of reference. This seems to require quite a long memory of the 'last-viewed side' on egocentric coordinates (a point to which we shall return later). In fact, considering the distance of the goal and the small size of the window through which it was visible, the goal was quite likely to be already out of sight after the chick, still in the corridor, had taken a few steps away from the barrier. Dead reckoning (see Gallistel, 1990) could thus be another possibility: chicks may be able to update their position with respect to the goal in a represented space moment by moment. However, whatever the orienting mechanism actually used, what is relevant in the present context is that it seems to imply the presence of some form of the idea that the goal has not gone 'out of existence' even when it can no longer be seen. Whether this should be considered as an instance of an ability to represent an 'object concept' is an issue to which we shall return.

11.9. Searching for Different Types of Objects that Have Disappeared from View

As mentioned above, Etienne (1973) devised a test in which chicks were presented with a mealworm that disappeared behind one of two screens. She found that, on being presented with the task for the first time, the chicks remained for a few seconds oriented towards the place where the mealworm had disappeared and pecked at it, and then moved randomly in the test arena. In cases in which they did go behind one screen, they did not necessarily choose the correct one (i.e. the one behind which the worm had disappeared).

How could the results obtained with the imprinted stimulus in the U-shaped detour test be reconciled with those of Etienne in the double-screen test? Several possibilities should be considered. It may be that the sample of animals used by Etienne was not large enough to reveal a preferential choice for the correct screen. (However, there was no trend for such a response bias in the data, Etienne, pers. comm.) Another possibility could be related to the use of a very brief cutoff value for each trial, i.e. 1 min after disappearance of the worm. It is likely that very few chicks actually circled around any of the screens within this very short observation period.

Another important issue concerns the type of goal-object used in these studies (i.e. food vs. social partners). It may be that, in a natural environment, different goals elicit different searching behaviour strategies, and that these strategies affect the probability of a chick solving detour problems in the laboratory. Alternatively, different goals may be associated with the evocation of different emotional and/or motivational responses, some of which could interfere with the execution of the task and mask the true cognitive abilities of the animal.

It thus seemed worthwhile attempting to duplicate Etienne's (1973) original experiments employing longer observation times, and using either a social (an imprinted object) or a non-social (a mealworm) goal.

The apparatus we used (Figure 11.10) reproduced the one originally devised by Etienne, and consisted of a white rectangular cage inside of which were placed two large parallel opaque screens (40×30 cm) spaced 40 cm apart. The task required the chick to move around either one of the two screens, behind one of which the goal (either a social partner or a palatable prey) could be found.

Chicks tested with the social partner as a goal had been reared solely with a small red plastic ball that was suspended by a fine thread in the middle of the rearing cage at about head height. At testing, on day 2 or day 6, each chick was placed into the testing cage, between the two screens, together with its own rearing red ball. The ball was suspended by a thread held by the experimenter who, after 10 s, smoothly moved it towards one of the two screens

Figure 11.10.   Schematic representation of the 'double-screen test' apparatus. For simplicity, both goal-objects (and two chicks) are shown simultaneously. After Regolin, Vallortigara and Zanforlin (1995b).

(randomly chosen) until the ball had disappeared behind the upper edge of the screen (see Figure 11.10). If the chick did not show following behaviour, nor orientation towards the moving ball, after three successive attempts it was removed from the apparatus and its performance was assigned to the category 'No ball following'. In cases in which the chick that initially followed the ball but did not subsequently walk around either screen within 5 min, the red ball was replaced between the two screens and the above procedure was repeated entirely, except that this time the ball was hidden behind the other screen. The chick was then given another 5-min trial, after which, if it still did not move around either screen, it was removed from the apparatus and assigned to the category 'No Detour'.

Chicks tested with the palatable prey as a goal were reared individually and were given, twice a day from the first day on, 2–3 mealworms, placed on the floor of their rearing cage. The procedure with the double-screen test was virtually identical to that described above for the imprinting ball, and identical to that described by Etienne (1973). At test, on day 2 (or on day 6), the chick was placed into the testing cage between the two screens. A glass tube was located on the floor, perpendicular to the two screens so that each end of it passed under one screen at its middle (see Figure 11.10). A mealworm tied with a fine thread was inserted inside the glass tube and, during the test, was slowly dragged by the experimenter through the tube until it disappeared completely under one of the screens (chosen randomly). Chicks that showed no interest in the mealworm after three attempts (each time the mealworm was placed back in the centre of the tube and dragged again towards one of the screens) were removed and assigned to the 'no prey chasing' category. Chicks that did not move around either screen after 10 min (with a new presentation of the goal after the first 5 min; see above) were assigned to the 'No detour' category.

Using the imprinted object as a goal (see Table 11.1), 61 animals chose the correct screen and 36 the incorrect one [χ^2 (1) = 6.44, $p = 0.01$]. Although there was no statistically significant age-related heterogeneity [χ^2 (1) = 0.03], an analysis limited to the two age groups revealed a significant preference for the correct screen in 2-day-old chicks [χ^2 (1) = 4.09, $p < 0.05$] but not in 6-day-old chicks [χ^2 (1) = 2.38, $p = 0.119$]. This was largely due to a sample reduction, because of a significant increase in the 6-day-old group in the number of animals that failed to choose any screen [3 versus 11; χ^2 (1) = 4.57, $p = 0.03$; 'no detour' category].

Using the mealworm as a goal (Table 11.1), no significant preferences for the correct screen appeared [84 animals chose the correct screen and 64 animals the incorrect one; χ^2 (1) = 2.70, not significant]. [Two-day-old chicks

Table 11.1.

	2-day-old chicks	6-day-old chicks	Total
Using the imprinted ball as a goal			
Correct screen	35	26	61
Incorrect screen	20	16	36
No detour	3	11	14
No ball following	1	0	1
Using the mealworm as a goal			
Chicks tested alone			
Correct screen	52	32	84
Incorrect screen	35	29	64
No detour	4	9	13
No prey chasing	15	36	51
Chicks tested in the presence of an artificial social companion			
Correct screen	9		
Incorrect screen	9		
No detour	25		
No prey chasing	11		
Chicks tested in the presence of a natural social companion			
Correct screen	32	15	47
Incorrect screen	16	7	23
No detour	1	9	10
No prey chasing	1	27	28

'Correct screen' indicates a choice for the compartment behind which the goal has disappeared; 'Incorrect screen' indicates a choice for the other screen; 'No detour' indicates the animals that failed to detour around any screens within the overall test period; 'No ball following' and 'No prey chasing' indicate the animals that did not follow the ball or the moving prey. 'Artificial social companion' and 'Natural social companion' indicate, respectively, the use of an imprinted red ball or of a conspecific-cagemate as a reassuring partner.

showed a tendency to choose the correct screen, but the difference did not reach significance; χ^2 (1) $= 3.32$, $p = 0.065$.] There was, however, no significant heterogeneity related to age [χ^2 (1) $= 0.002$]. There was a significant increase in the number of older chicks that failed to follow the goal [2-day-old: 15 chicks; 6-day-old: 36 chicks; χ^2 (1) $= 8.64$, $p = 0.003$; 'no prey chasing'.category] but not in the chicks that failed to detour any screens [4 versus 9; χ^2 (1) $= 1.92$, not significant; see 'no detour' category].

Thus, using a mealworm as a goal, we obtained the same results as Etienne (1973). That is, even with longer testing times (and a very large sample), chicks appeared unable to make use of the directional cue provided by the prey movement. Using an artificial social partner as a goal, however, the majority of the chicks appeared able to choose the correct screen.

Clearly, the use of such different goals results in relevant differences in motivational and emotional variables. An interesting possibility is that the emotional overtones associated with being placed in a novel environment are the main cause of the different behaviours shown with the social partner and with the prey. Socially reared chicks that are placed in a novel environment in which their social partner is made to disappear are obviously strongly motivated to reinstate social contact. In this situation, fear responses to the novel environment and social reinstatement responses to the partner's disappearance elicit a similar reaction, with both inducing searching behaviour towards the goal that has disappeared. After disappearance of a mealworm, on the other hand, fear responses to the novel environment are likely to compete and interfere with searching for the prey. Chicks are, in fact, faced with a situation (the novel environment) that probably overcomes any interest for the prey and for feeding behaviour in general. Finding the prey in no way can alleviate the stress of being in a novel environment, as would rejoining a social companion. Fear responses to the novel environment should thus produce (1) little interest in searching behaviour towards the prey that has disappeared and thus (2) random moving in the environment with respect to the two screens. This is indeed what our results have shown.

One indication that interference by fear responses could be the correct explanation for the poor performance with the mealworm arises from age-related behavioural changes. Developmental studies have revealed a complex time-course of fear responses, with a peak in the levels of fear at around day 5 that extends to about day 7 (Andrew and Brennan, 1983, 1984; Vallortigara, Regolin and Zanforlin, 1994). Interestingly, our results showed that there was a certain tendency for choosing the correct screen using the mealworm as a goal in 2-day-old but not in 6-day-old chicks. Moreover, the number of animals that did not show any interest in the worm increased strongly in the 6-day-old group (see above). Older chicks could not have been less motivated (i.e. less hungry) than 2-day-old ones (younger chicks are notoriously poorer feeders because they still have yolk reserves). Older chicks were visibly much more afraid of the novel environment than younger chicks, showing freezing, peeping and other fear responses quite likely to be in competition with searching for the moving prey. Interestingly, Etienne (1974) also observed that 6-day-old chicks were poorer at learning in the double-screen task than 3-day-old and 14-day-old chicks.

All this strongly suggests that the emotional overtones of the testing situation could account for the chick's failure in the double-screen task when using the mealworm. A simple way to reduce emotional responses is to test the animal in the presence of a social companion – a procedure that is known to be very effective in reducing fear responses in the open field test (for a discussion of this topic, see Vallortigara and Zanforlin, 1988). We used either an artificial (a red ball) or a natural (a cagemate) social companion.

The apparatus and general procedure were the same as in the previous experiment. The mealworm was used as a goal object. Chicks were reared with either a red ball (that became their 'artificial' social companion) or with a conspecific of the same sex (that became their 'natural' social companion). Both groups of chicks were given, in addition to normal food, 2–3 mealworms twice a day from the first day on. At the test, chicks were placed into the apparatus together with their social companion. In the 'natural' social-companion condition, responses were recorded only for the first chick of the pair that moved around one of the two screens (note that the other chick of the pair invariably followed the first one very closely). In the 'artificial' social-companion condition, the ball was suspended in the middle of the test cage, between the two screens, and kept motionless throughout the test. Chicks tested with the artificial companion were all 2-day-olds, whereas chicks tested with the natural companion were either 2- or 6-day-olds.

When using the imprinted ball as a social companion, no significant preferences for the correct screen were apparent, and most of the animals failed to detour around either screen (see Table 11.1). When using a conspecific as a social companion, in contrast, the majority of chicks chose the correct screen [47 versus 23, χ^2 (1) = 8.22, $p = 0.004$]. Although the proportion of animals that chose the correct screen was similar in the two age groups, the results showed that, as in the previous experiments, there was a strong increase in the number of older animals that did not show any interest in the mealworm [1 versus 27, χ^2 (1) = 24.14, $p = 0.001$; 'no prey chasing' category] and a consequent reduction in the absolute number of older chicks that chose the correct screen.

Artificial social partners probably lacked the important components of movement and mutual interaction to act effectively as companions in the double-screen test. The high number of animals that did not detour any screen in this condition suggests that chicks preferred to maintain strict contact with the motionless imprinted object.

Natural social partners, on the other hand, allowed the animals to solve the task even using a mealworm as a goal. Older chicks still tended to perform poorly, as shown by the large number of animals that did not pay any attention to the mealworm.

Overall, the results seem to confirm the emotional hypothesis: difficulties using the prey as a goal are probably due to the confounding presence of fear responses to the novel environment. These responses can be partially eliminated (at least their social reinstatement motivation component can be eliminated; see Gallup and Suarez, 1980) by testing the chicks together with a social companion, thus disclosing the real detour abilities of the chicks.

There seem to be age-related effects influencing the amount to which emotional responses to the test situation interfered with the correct execution of the detour task. Six-day-old chicks seemed to perform particularly poorly compared to 2-day-old chicks. This is in accordance with several independent findings indicating a peak in the levels of fear at that age (Andrew and Brennan, 1983, 1984; Vallortigara, Regolin and Zanforlin, 1994) and could also explain the reduced abilities of 6-day-old chicks in the double-screen task under repeated testing conditions (Etienne, 1974).

Although Etienne (1976) presented the double-screen test within the theoretical framework of the comparative study of object permanence, one may wonder to what extent the test should require a 'representation' of the object that has disappeared. A preference for the correct screen may be accounted for in terms of a chick's bias to remain oriented and to move closer to the screen where the goal has been seen to disappear (this objection is, of course, not valid in the case of the more traditional U-shaped detour test). The problem, however, is that such a bias would be expected to occur irrespective of the type of goal. Another problem with this argument is the long time that chicks took to detour the correct screen (about 2–3 min). Obviously, during this time, chicks were not completely motionless, nor did they remain always oriented towards the correct screen. If chicks simply moved around the correct screen because they happened to be close to it as a result of tracking the goal, then such a bias should manifest itself in extremely short time delays and should require the chicks to remain constantly near and orientated towards the correct screen throughout the whole latency time. The fact that chicks circled around the correct screen after a couple of minutes and after having looked and moved about the environment suggests instead that they maintained some sort of memory of the location of the desired object.

Of course, a different argument would be that the long detour latencies required by the chicks to detour the screens make their performance quite different from that of species that solve the problem at once. However, again, for a highly affiliative and social animal, as is the young chick, being placed in a novel environment must be a very stressful experience. It is simply unreasonable to expect the chick to move immediately in such a situation. Detouring any of the two screens means that the chick has to move in previously unexplored and potentially dangerous places. Not surprisingly, the chicks are much more prone to take these risks when they are strictly necessary (i.e. in order to rejoin a social partner) than when they are not (i.e. to find a prey).

As regards the issue of 'object concept', however, we believe that, in a sense, the ability to detour should be separated from the ability to maintain a representation of the goal object. We think a better explanation may be that the generic term 'representation of the goal object' should be substituted by a more precise specification of what aspects of the goal object need to be represented in the animal's nervous system in order to explain its behaviour. What is actually required in these tasks is a representation of the position of the goal-object. Nothing more can reasonably be said about other properties of the goal stimulus (such as its colour, shape and so on). This is important because it would be easy to extrapolate from these data on the basis of our mental experiences. In similar testing conditions we actually experience some form of 'mental images' of the object that has disappeared, but we cannot say, with these experimental designs, whether the same is true for the chick. This does not mean to imply, however, that these are intractable problems in non-human species: as we shall see shortly, using different and specially designed experimental tasks, exploration of the possible 'declarative' nature of animals' representations is possible.

11.10. Sex Differences in Detour Behaviour

Sex differences were ubiquitous in all our detour tests. Using a U-shaped barrier, females showed a slight but consistent bias towards shorter detour times than males (see Figures 11.5–11.7). (The effect tended to disappear in those conditions that required very long times to solve the problem; this 'ceiling' effect could be due to the use of a time limit of 600 s.) Sex differences in cognitive abilities have been described in several species of mammals (McNemar and Stone, 1932; Barrett and Ray, 1970; Maccoby and Jacklin, 1974; Gaulin and Fitzgerald, 1989), and there is recent evidence for similar phenomena occurring in the domestic chick (Vallortigara, 1996). Are sex differences in detour related to cognitive processes or are they rather the byproduct of emotional and motivational variables associated to the detour task?

Males are known to be less active and vocal than females when placed in a novel environment (Jones, 1977; Jones and Faure, 1981, 1982; Vallortigara and Zanforlin, 1988). This may be due to stronger motivation for social reinstatement in females than in males (Vallortigara and Zanforlin, 1988; Vallortigara, 1992b), which should make the detour task harder for females. Note, however, that in these tests we were studying detour behaviour (single trial) rather than detour learning (repeated trials). The likelihood of naive chicks to go outside the walls of the runway in their first trial depends on their moving about the environment in an attempt to join their cagemates, and females are more likely to move about than males. If this hypothesis is correct, then on subsequent trials the opposite outcome should be expected. When chicks have learnt the correct route to the goal, and also learnt that reinforcement occurred for following that route, stronger social motivation in females might compete with correct execution of the spatial task (i.e. females should be more attracted than males towards the direct route), thus resulting in better performance in males. Results of the repeated-trials test seem to confirm these conjectures (see Figure 11.9). Females did better in the very first trials, whereas males took less time to reach the goal in the subsequent trials.

Sex differences in fear may, however, have contributed to the phenomenon. Initial trials may be particularly difficult for males because of their strong fear reactions, whereas subsequent trials may be particularly difficult for females because of their strong social reinstatement tendencies. Interestingly, previous work with runway tests has revealed very similar ‘results. Using a straight runway with social reinforcement (cagemates), females ran faster than males, whereas using a V-shaped runway, the reverse occurred (Vallortigara, Cailotto and Zanforlin, 1990). Higher levels of social motivation in females facilitated behavioural performance when chicks had to proceed toward the goal object directly (straight runway) but inhibited it when chicks had to proceed toward the goal object in an indirect fashion (V-shaped runway), as in a detour task.

Further evidence comes from sex effects in the double screen-test. Although the two sexes did not appear to differ in their ability to solve the problem (see Table 11.1), there were intriguing sex effects when the times needed to detour the correct or the incorrect screen were considered separately. Among chicks that chose the correct screen, males took longer than females with the social goal object, whereas there were no sex differences with the non-social goal object. Among chicks that chose the incorrect screen, males took longer than females with the non-social goal object, whereas there were no sex differences when the social goal object was used (see Regolin, Vallortigara and Zanforlin, 1995b).

Results in chicks that solved the task correctly fit quite well with the social motivation hypothesis. Females, being more interested in reinstating social contact, performed better than males when the goal object was a social one, but not when it was a palatable prey. This is also in good agreement with previous results using runway tests, in which females have proved to run faster than males for social but not for non-social reinforcement (Vallortigara, Cailotto and Zanforlin, 1990).

The observed increase in times needed to detour the incorrect screen in 6-day-old chicks tested with the mealworm could be explained by the increase of fear levels at that age (Andrew and Brennan, 1983, 1984; see also the previous section). Sex differences disappeared using socially reared chicks tested in the presence of a natural companion (for a similar result, see Vallortigara and Zanforlin, 1988).

To summarize, sex effects in detour tests could be accounted for in terms of the well-known emotional–motivational differences between the two sexes; there is no evidence that male and female chicks would differ in any way in their cognitive capability to solve detour problems.

11.11. Temporal Characteristics of the Representation of an Object that Has Disappeared from View

If chicks possess some sort of representation of the location of an attractive object that has disappeared from view, the question arises of how long can this representation be maintained. There are reasons to believe that it cannot be a very long time, since informal observations consistently report how animals typically cease to look for an object that has disappeared from view after a rather brief searching time. This does not necessarily mean that decay of the memory trace occurs so rapidly: attentional mechanisms could be responsible for the phenomenon. Révész (1924) took a similar position when discussing why fowls do not usually peck at hidden food, although they can be easily trained to do so. His idea was that, after disappearance of a food object, fowls keep representations of it in the form of a quite immediate after-effect of the optical impressions of the object that has disappeared, but these '...images emerging immediately after primary perceptions are dragged along by the flood of numerous fresh impressions' (Révész, 1924, p. 390). Successive experiences, therefore, would only have the function of protecting and reinforcing these unstable representations, which would otherwise be easily overwhelmed by other sensory impressions.

We have developed a method for studying the temporal properties of the representation of an object that has disappeared from view. The experiment is reminiscent of the old-fashioned technique employed at the beginning of the century to investigate the so-called 'delayed reaction' (Hunter, 1913; Tinklepaugh, 1928). In a large arena, chicks are first trained to follow an object on which they have been imprinted (a small red ball suspended by a thread) that moves slowly towards and then behind an opaque screen. Chicks rapidly learn to run behind the screen in order to rejoin the imprinted object. At the test, the experimenter (not visible to the chick) moves the ball behind either one of two identical screens (see Figure 11.11). The chick is at first confined into a transparent cage from where it can see and track in the distance the moving ball. Immediately after the disappearance of the ball behind the screen (or with a certain delay) the chick is released and allowed to run towards one or other of the two screens. Sixteen trials are given

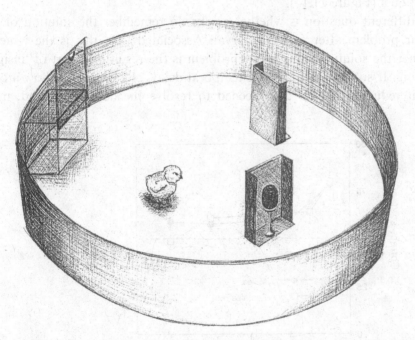

Figure 11.11. Schematic representation of the apparatus used to test retention of the location of an object that has disappeared (from Vallortigara et al., 1998). The imprinting object is made to disappear behind either one of two opaque screens, and meanwhile the chick is confined within a transparent enclosure. The chick is then released and allowed to approach the screens either immediately after disappearance of the imprinted object or after a certain time delay. (After Vallortigara et al. © 1998, with permission from Springer-Verlag GmbH & Co. KG.)

consecutively, and in each trial the screen behind which the ball is hidden is changed according to a semirandom sequence.

Figure 11.12 shows our results using time delays ranging from 0 to 240 s. It seems that, in this task, chicks can take into account the directional cue provided by the movement and concealment of the ball, at least up to a time interval of about 180 s.

Although chicks did not keep oriented in a stereotyped manner towards the correct screen during the whole delay (and actually moved around considerably in an attempt to exit from the transparent cage), we wondered whether visibility of the screens during the delay was crucial to this surprisingly long retention period. We thus tested chicks using an opaque partition, which was placed in front of the transparent cage immediately after disappearance of the goal. In this way, during the delay, neither the goal-object nor the two screens were visible. The results showed that the chicks were still capable of remembering the position of the correct screen up to a delay of about 60 s (Figure 11.13).

A different question is whether chicks do remember the solution of the detour problem after a long interval. Associated with this is the issue of whether the solution of a detour problem is the result of a sort of 'insight' learning. If so, even after a long temporal delay, chicks should show a consistent reduction in the time needed to resolve the detour test and, most

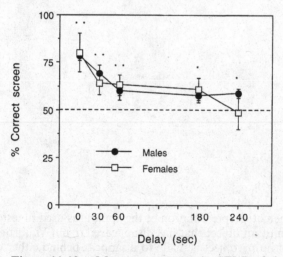

Figure 11.12. Mean percentages (± SEM) of choices for the correct screen (the one behind which the imprinted object has disappeared) as a function of the delay between object disappearance and chick's release. Significant departures from chance level (50%) are indicated by the asterisks (* $p < 0.05$; ** $p < 0.01$; two-tailed one-sample t-tests).

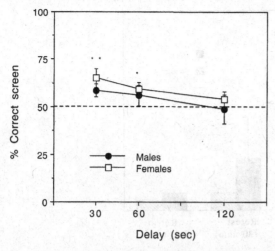

Figure 11.13. Mean percentages ($\pm$ SEM) of choices for the correct screen as a function of the delay between object disappearance and the chick's release. During this delay an opaque partition prevented the chick from seeing the two screens. Significant departures from chance level (50%) are indicated by the asterisks (* $p < 0.05$; ** $p < 0.01$; two-tailed one-sample t-tests).

importantly, there should be a clear difference in savings (in terms of trials) at retest between those chicks that have solved the problem in the first test and those that have not.

Figure 11.14 shows the results of two separate experiments (see Regolin and Rose, 1999) in which 2-day-old chicks were first tested in the U-shaped detour problem (see Figure 11.4) and then retested after 30 min and 24 h (a) or after 24 h only (b). As can be seen, chicks appear to remember the solution very well even after a 24-h interval.

In the second experiment, 10 animals out of 40 were unable to solve the problem within 600 s in the first test, but they were nevertheless retested after 24 h. It is interesting to compare the times needed to detour the barrier in the second test between chicks that had solved the task in the first test and those that had not. As can be seen from Figure 11.14b, time saving was negligible in chicks that had not solved the problem during the first test. It seems that solving the problem has mostly an all or nothing effect on successive tests. If chicks had solved the test, then they took less time to solve it again after 24 h. If chicks had not solved the first test, then they were likely to score the same times on the second test as they did on the first test. It might be argued that the effect could be related to individual differences between the chicks them- selves: chicks that did not solve the problem might have simply been 'less

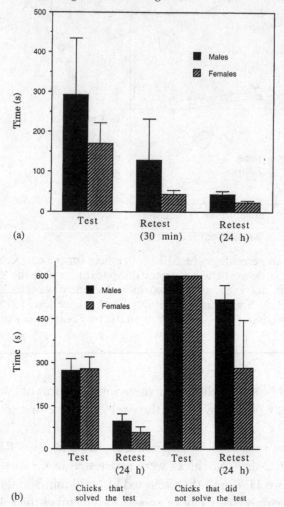

(a)

(b)

Figure 11.14. Mean times (± SEM) needed to solve the detour problem with the U-shaped barrier (Figure 11.4) at the first instance and at two retests after a delay of 30 min and 24 h (a) or at retest after a delay of 24 h (b). The first experiment (a) revealed significant main effects of sex [$F(1,8) = 6.777$, $p = 0.030$] and testing time [$F(2,16) = 7.669$, $p = 0.004$]. The interaction was not significant [$F(2,16) = 0.509$, ns]. The second experiment (b) revealed only a significant main effect of testing time [$F(1,28) = 51.817$, $p < 0.001$] although females always tended to perform better than males in the first test. There was a clear reduction in the times needed to solve the problem at the second attempt, with both the short (30 min) and long (24 h) time interval. Chicks that did not solve the problem within 600 s in the second experiment (4 females, 6 males, not included in the figure) were nevertheless retested after a 24-h interval as were the chicks that solved the problem (b, rightmost column). A comparison between times at retest of chicks that solved the task during the first test and those that did not revealed a highly significant difference between the two groups [$F(1,36) = 53.592$, $p < 0.001$].

clever' and, unsurprisingly, they remained less clever even in successive tests, and persisted in showing longer detour times. However, in the second test, females that had not solved the first test required the same time as those females that solved the problem in the first test, whereas males still required long times. This sex effect suggests that individual differences might pertain more to emotionality levels than to cognitive abilities. Moreover, in chicks that solved the problem, the correlation between times needed to solve the task during the first and the second test, though positive, was not statistically significant ($r = 0.324$, $n = 30$, $p > 0.05$). This means that, among chicks that had solved the problem, those that took longer at the first test did not necessarily take longer also at the second one.

Summing up, the evidence so far collected indicates that chicks can maintain for up to 60 or 180 s (depending on testing conditions) some sort of representation of the location of a highly attractive concealed object and that, within this temporal interval, they can update their memory trace from trial to trial. Moreover, memory of the motor strategy needed to detour an obstacle (which is likely to be of a different nature than the memory for the object itself) can be maintained in a long-term store and can be retrieved even after a 24-h interval.

11.12. The Detour Problem and the Cerebral Hemispheres

Visual lateralization can be tested even when both the bird's eyes are unobstructed, by simply considering preferential eye use in viewing of various stimuli. It has been shown, in fact, that chicks are able to bring into action the hemisphere appropriate to a particular condition and to a particular stimulus by using lateral fixation with the eye contralateral to the hemisphere to be used (Andrew and Dharmaretnam, 1993; Dharmaretnam and Andrew, 1994). Thus, for instance, McKenzie, Andrew and Jones (1998) found that, on day 1, females (but not males) fixate with the right eye rather than the left eye when naive and deciding whether to approach an imprinting object.

Informal observations during our detour tests suggested that chicks could indeed be looking at the goal using one eye preferentially. However, since we used long walls that obstructed the sight of the goal for long periods, there were apparently no clear effects of preferential eye use in the direction (right or left) of barrier detouring. We thus devised a simplified version of the test, which enabled us to study asymmetries of response (see Figure 11.15). Chicks exhibited a stereotyped behavioural strategy in this apparatus. They kept very close to the barrier with the head turned right or left so that the goal was under lateral fixation of one eye, and then moved along the barrier (i.e. in

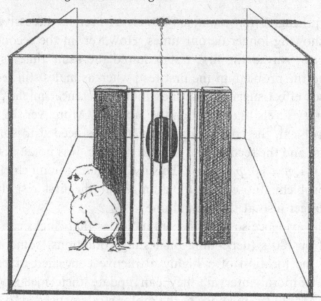

Figure 11.15.   Schematic representation of the test apparatus used to investigate asymmetries of detour responses in chicks. The chick could detour the barrier to rejoin the imprinted ball either on the right or on the left side.

the direction of their head orientation) until they circled around the obstacle. As a result of this behaviour, the right or left direction of detour around the barrier strictly reflected preferential fixation by contralateral eye during the detour (i.e. detour on the right, preferential left eye use; detour on the left, preferential right eye use). We also studied chicks wearing eye patches to check whether the two eye systems would differ in their detour-learning abilities.

Figures 11.16 and 11.17 show the results of a study in which independent groups of 2-, 3- and 4-day old chicks were tested in the apparatus using a familiar imprinted red ball as a goal (Vallortigara, Regolin and Pagni, 1999). There were separate groups tested in the binocular condition or with the right or the left eye temporarily occluded by an eyepatch. Times needed to detour the barrier (Figure 11.16) and directions of detour (Figure 11.17) were recorded.

There were two main results. Firstly, chicks using their right eye took less time to detour the barrier than chicks using their left eye (the effect was small but consistent at all ages studied and in both sexes). Secondly, binocular chicks showed a bias to detour the barrier on the left side (which is consistent with preferential right eye use). As can be seen from Figure 11.17, monocular chicks were constrained in their choice of the direction of detour by the eye in

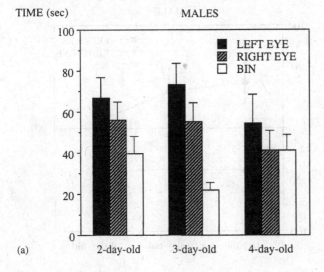

(a)

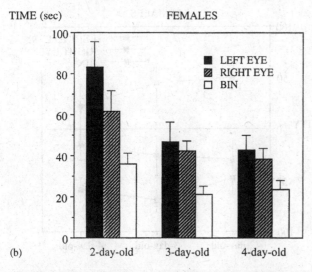

(b)

Figure 11.16. Mean times (with SEM) needed to detour the barrier in binocular (BIN) chicks and in right-eyed and left-eyed chicks using a familiar imprinted ball as a goal. (a) males; (b) females. The analysis of variance revealed significant effects of age [$F(2,153) = 8.210$, $p < 0.001$] and eye in use [$F(2,153) = 22.989$, $p < 0.001$]; there were no other statistically significant effects. Right-eyed chicks showed shorter times to detour the barrier than left-eyed chicks [$F(1,103) = 4.741$, $p = 0.029$].

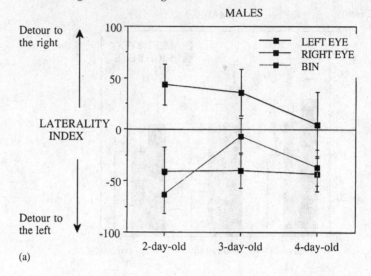

(a)

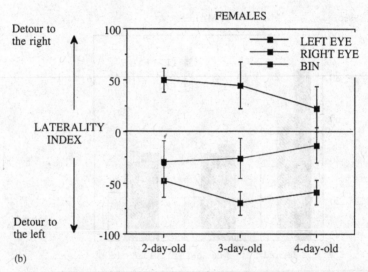

(b)

Figure 11.17.　Direction of detour responses [(detour on the right − detour on the left/(detour on the right + detour on the left) × 100; group means with SEMs are shown] in binocular (BIN) chicks and in right-eyed and left-eyed chicks using a familiar imprinted ball as a goal. (a) males; (b) females. The analysis of variance revealed only a statistically significant effect of the eye in use [$F(2,154) = 30.912$, $p < 0.001$). Left-eyed chicks showed a preference to detour to the right [$t(56) = 3.953$, $p < 0.001$]; binocular chicks showed a preference to detour to the left [$t(55) = -3.745$, $p < 0.001$] as did right-eyed chicks [$t(57) = -7.599$, $p < 0.001$].

use: those using the left eye made the detour on the right, those using the right eye made the detour on the left. Responses of binocular chicks were, however, not random but more similar to those of chicks that used the right eye.

It seems very unlikely that these asymmetries were due to a motor bias, because the chicks' direction of turn could be reversed by simply changing the visual characteristics of the red imprinting ball. Figure 11.18 shows biases in direction of detour responses in 2-day-old chicks tested with a visual transformation of the imprinted ball (i.e. a yellow, a blue, and a half-yellow–half-red ball). As can be seen, there was a shift to left eye use with some of these novel stimuli (see also Vallortigara and Andrew, 1994a, 1994b). The shift in eye use seemed to depend on an estimation of the degree of novelty of the unfamiliar ball and, interestingly, the judgement seemed to differ between males and females. Assuming that there is an increase in the degree of change ranking from the red–yellow ball (slight novelty), the yellow ball (novelty) to

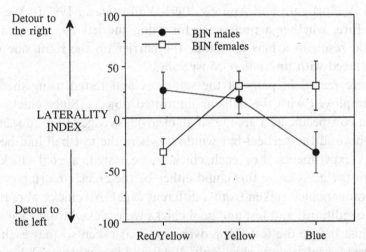

Figure 11.18. Direction of detour responses [(detour on the right − detour on the left/(detour on the right + detour on the left) × 100; group means with SEMs are shown] in binocular (BIN) male and female chicks reared with a red ball and tested with a novel ball as a goal. The ANOVA revealed a significant sex × stimulus interaction $[F(2,49) = 6.73, p < 0.002]$. There were no other statistically significant effects. Females tended to turn on the right side (left eye use) when using the yellow and the blue balls $[F(17) = 2.533, p = 0.0211]$ whereas males did the same when using the red/yellow and the yellow balls $[F(17) = 2.182, p = 0.0403]$. Detour on the left side (right eye use) was in contrast predominant in males tested with the blue ball and in females tested with the red/yellow ball $[t(19) = −3.104, p = 0.005]$.

the blue ball (strong novelty), then results for females suggested a corresponding shift from right to left eye use with increasing degree of novelty. Results for males were, however, puzzling in that they came back to right eye use when faced with the blue ball. An entirely speculative but interesting possibility would be that chicks use the right eye to minimize fear due to a very large transformation, that is, males use the right eye (and the less 'emotional' left hemisphere; see Andrew, 1991; see also Chapter 3 by Andrew and Rogers) in order 'to ignore' the change involved because the blue colour may be frightening (see Andrew and Brennan, 1983; Clifton and Andrew, 1983).

Lateral asymmetries could occur even using different goals. A study (Regolin and Andrew, unpublished observations) carried out using mealworms as goals has revealed preferential right eye use on chicks tested first on day 4 and then on day 5 (in this last day the effect was, however, less pronounced). Whereas the left eye involvement in response to novelty seems to be well established, preferences in the use of the right eye appear to be somewhat more puzzling. Previous work had shown, in monocularly tested chicks, that the left eye is mainly involved in social discrimination between conspecifics (Vallortigara and Andrew, 1991; Vallortigara, 1992a). We wondered, therefore, whether a preference for using the left eye in estimating novelty could result in a bias to detour the barrier on the right side when chicks were faced with unfamiliar conspecifics.

Chicks were reared in pairs of the same sex and tested using the same procedure employed with the balls as imprinted objects. Since chicks were tested with a conspecific as a goal, a small clear-glass cage ($10 \times 20 \times 20$ cm) was placed beyond the vertical-bar window, where the test ball had been in the previous experiments. For each chick to be tested, a goal chick was positioned in the glass cage; this could either be the chick's rearing companion or a stranger chick (taken from a different cage). All chicks were reared in identical conditions, and test and goal chicks were picked from their cages at random just before the test. Moreover, care was taken so that each goal chick was used for testing one chick only. We found that males took longer to detour the barrier when the goal was a stranger, females did the same when it was a companion. Similar sex differences have been reported previously in social recognition tests (Vallortigara, 1992b), which fit in well with our previous discussion about stronger social reinstatement tendencies in females than in males (see section above on sex differences). As to lateral preferences, as expected, chicks showed a significant bias to detour the barrier to the right side (left eye use) when tested with a stranger as a goal, but there was no significant bias when they were tested with a companion as a goal. Results thus confirm that the left eye is chiefly involved when the chick is faced with slight changes in the visual aspect of an otherwise highly familiar stimulus. Interestingly, however, no bias for the use of the right eye was apparent using the cagemate as a goal. This contrasts with the results obtained with the familiar artificial imprinting object. The more likely explanation is that chicks use the left eye to scrutinize stimuli of which the degree of familiarity is uncertain: when faced with a simple homogeneous stimulus, such as the red ball, estimation of its familiarity, even by the limited vision available through the barrier, is probably an easy task, but this is not true for a natural conspecific. It could be that, in the latter case, chicks have to alternate the use of the two eyes in order to check for confirmation of the identity of the social partner and, therefore, for this reason, no clear bias is present.

11.13. Representing Object-specific and Position-specific Cues in the Left and Right Hemisphere

In the delayed response task mentioned earlier, chicks had to remember the location of the correct screen. Previous work has revealed that the right hemisphere in the chick is mainly involved in topographical learning (Rashid and Andrew, 1989). Moreover, it has been shown that, during simultaneous discrimination learning, position cues seem to engage the right hemisphere and object-specific cues the left hemisphere (Vallortigara et al., 1996). Specializations of the right eye–left hemisphere for object-specific cues have been reported for the pigeon (Güntürkün, 1985; von Fersen and Güntürkün, 1990; for a review, see also Güntürkün, 1997) and complementary specializations of the left eye for position-specific cues and of the right eye for object-specific cues have been documented in tits, jays and jackdaws (Clayton and Krebs, 1994). We wondered therefore whether the two eyes–hemispheres systems could be differently involved in our delayed response task when object- and position-specific cues were available simultaneously. We used two screens differing in colour and patterning. In each trial, the imprinting object was hidden behind one of the two screens, as usual. However, during the delay (when an opaque partition prevented the chicks having any sight of the screens) the right–left position of the two screens was swapped rapidly. After the delay, which lasted 30 s, the chick was thus faced with the problem of choosing either the correct screen in the wrong position or the incorrect screen in the correct position. Our results (see Vallortigara, 2000; Vallortigara and Garzotto, in preparation) are shown in Figure 11.19 and indeed suggest that chicks using the left eye relied on the position of the screen ignoring its colour and patterning, whereas chicks using the right

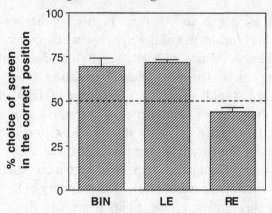

Figure 11.19. Percentages of choice for the screen in the correct position (and with the wrong colour and patterning) in chicks tested binocularly (BIN) or with only the left (LE) or the right (RE) eye in use in the delayed-response task. Each chick performed 16 trials, and the screen behind which the ball disappeared was changed according to a semirandom sequence. The screen first chosen and searched by the chick was recorded. Separate groups of chicks were tested in binocular conditions ($N = 6$) and in monocular conditions (temporarily eyepatched: right-eyed chicks, $N = 6$; left-eyed chicks, $N = 8$). The analysis of variance revealed a significant main effect of the eye in use [$F (2,30) = 19.396$, $p = 0.0001$]. *Post-hoc* comparisons (Tukey tests) showed that there were significant differences between right-eyed and left-eyed chicks ($p = 0.0001$), and between right-eyed and binocular chicks ($p = 0.0001$), but not between left-eyed and binocular chicks. Results suggest that chicks using their left eye relied on the position of the screen, ignoring its colour and pattern, whereas chicks using their right eye relied on the colour and pattern of the screen, ignoring its position (see also Vallortigara, 2000).

eye tended to use sometimes position and sometimes local cues (pattern and colour).

Similar results were recently obtained using a reference (rather than a working memory) paradigm (see Tommasi and Vallortigara, 1997, 2001; Vallortigara, 2000). Young chickens were trained to find food by ground-scratching in the centre of a closed uniform arena; the centre was indicated by a conspicuous landmark (a red stick). After binocular training with a landmark located in the centre, the landmark was displaced to a corner, so that object-specific cues (the landmark) and position-specific cues (the central position) provided contradictory information. A striking asymmetry appeared: binocular and left-eyed chicks searched at the centre (ignoring the landmark), whereas right-eyed chicks searched at the corner (ignoring purely spatial information). It is interesting to note that there were no differences in the ability to retrieve the correct spatial location between right- and left-eyed chicks after binocular training. If, after binocular training, the central landmark is removed, both eye systems appeared to be able to search in the correct spatial (central) position. Similarly, in the delayed response task, chicks tested with two identical screens or with two screens of different colours that maintained fixed spatial position during the various trials showed no asymmetry when tested with only the right or the left eye in use. It seems that the prevalent use of spatial cues by the left eye and of object cues by the right eye emerges only when these cues are available simultaneously and provide conflicting information. This could be explained by supposing that, even when only one eye is in use in the chick, information stored in the ipsilateral hemisphere could be nonetheless assessed and employed to control behaviour (for instance, to retrieve spatial information stored in the right hemisphere even when vision is confined to the right eye only) providing that other cues are not engaging the directly stimulated eye and hemisphere. When position- and object-specific cues are simultaneously available, however, the hemisphere that is directly stimulated takes charge of control of behaviour relying on its 'preferred' cues, thus giving rise to choice of position-specific cues for the left eye and object-specific cues for the right eye. This would be consistent with recent neuroanatomical evidence showing that, although primary visual projections ascend mainly contralaterally in the chick's brain, when laterality is revealed by monocular testing only, one can never be sure that it is exclusively the contralateral hemisphere involved. This is because ipsilateral and contralateral projections are both present in the thalamofugal as well as in the tectofugal pathways (Deng and Rogers, 1997, 1998a, 1998b; see also Chapter 6 by Deng and Rogers).

11.14. Representing 'Where' and Representing 'What'

We hope that the previous sections have provided ample evidence about the chick's ability to represent the spatial location of hidden goals. It should be also clear that this sort of global spatial information is mainly managed by structures located somewhere in the right side of the brain, whereas orientation based on local, object-specific cues seems to be managed by structures located in the left hemisphere. So, this is about the 'where', but what about representing the 'what'? Are chicks able to know *what* sort of goal is located (hidden) in a certain position? These sort of cognitive representations are usually indicated as 'declarative' in human cognition – because they can be 'named' by verbal language – and kept distinct from procedural representations, which could be considered as 'behavioural dispositions', i.e. as a set of instructions that are initiated by a given stimulus (see Ryle, 1949; Dickinson,

1980). However, several authors maintain that declarative representations could be formed even without a linguistic medium (for a review, see Griffiths, Dickinson and Clayton, 1999). In food-storing birds, evidence that they are able to remember the contents (other than the positions) of food caches has been obtained using procedures that exploit the observation that pre-feeding on one type of food selectively reduces the value of that food (see Clayton and Dickinson, 1999). A similar procedure has been recently used by Forkman (2000) with domestic hens. Individual hens were fed in an enclosure with two containers, each with a different type of food. Prefeeding on one of the two food types was performed before testing the hens with empty food containers. Hens chose the location previously occupied by the non-devalued type of food, showing that they have a representation of where a particular type of food was located. We recently used this devaluation procedure with young chicks (Cozzutti and Vallortigara, 1999, 2001). Chicks underwent a learning session of 10 min twice a day, consecutively for 3 days, from day 2 to day 4. During the learning trials, chicks were placed into a large experimental enclosure with two identical food plates, filled with different types of seeds. The food plates were positioned in two adjacent corners on the same side of the arena. The location of the plates remained fixed throughout the learning trials, so that a certain kind of seed was always found in the same position. On day 5 chicks underwent a procedure of devaluation in their home cages: for 30 min food-deprived animals were fed *ad libitum* exclusively one kind of seed in their homecages and, in this way, they were satiated selectively for that kind of seed. Each chick underwent the procedure of devaluation and was tested subsequently tested either in binocular conditions or with one eye temporarily closed by an eyepatch. At the test, binocular chicks and chicks using the right eye preferentially approached the food plates containing the seeds that had not been devalued (Figure 11.20). In contrast, chicks using the left eye chose at random. In order to check whether monocular chicks could differ in their motivation to approach the food plates, the experiment was repeated omitting any devaluation procedure. Both right-eyed and left-eyed chicks showed interest and approached the food plates.

These findings demonstrate that, as early as on day 5 of life, chicks are able to encode and store memory information concerning the particular kind of food that is located in a particular place. Such a memory seems to reside in (or perhaps could be retrieved only from) the left hemisphere. It is intriguing that it is the right, rather than the left, hemisphere that has commonly been found to be involved in spatial orientation (Rashid and Andrew, 1989; Vallortigara et al., 1996; Tommasi and Vallortigara, 1997; Vallortigara,

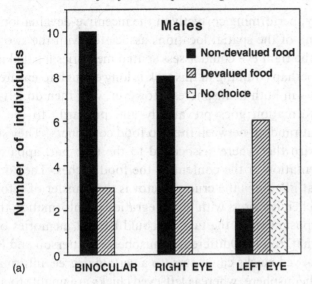

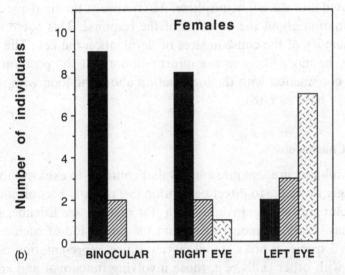

Figure 11.20.   The number of chicks choosing the devalued food, the non-devalued food or showing no choice within 5 min. (a) males; (b) females. There was significant heterogeneity associated with the three categories of choice between binocular and left-eyed chicks [$\chi^2(2) = 20.913$, $p < 0.001$] and between right-eyed and left-eyed chicks [$\chi^2(2) = 15.692$, $p < 0.001$], but not between binocular and right-eyed chicks [$\chi^2(2) = 1.172$, $p > 0.1$]. There was a significant preference for the food plate location previously occupied by non-devalued food in binocular [$\chi^2(1) = 8.16$, $p = 0.004$] and right-eyed chicks [$\chi^2(1) = 5.76$, $p = 0.0158$] but not in left-eyed chicks [$\chi^2(1) = 1.92$, $p > 0.1$]. After Cozzutti and Vallortigara (2001).

2000), and, obviously, performing correctly in the incentive-devaluation task does require retrieving of the spatial locations associated with the two food containers. Perhaps the right eye could assess spatial memories located in the right hemisphere or perhaps training in this task is long enough to ensure that spatial learning occurs in both hemispheres. However, why then do left-eyed chicks (in which information goes prevalently and primarily to the right hemisphere) choose at random between the two food containers? They surely should be able to learn the 'where' associated to the task but, apparently, they have no representation of the content of the food caches. They do not know about the 'what'. Perhaps the crucial factor is the transfer of information between the two hemispheres, with the integration of information that is crucial to the accomplishment of the task. It could be that memories of the 'where' and of the 'what' reside in different hemispheres (Patterson and Rose, 1992), and that right-eyed chicks can assess, to a certain degree, information stored into the right hemisphere, whereas left-eyed chicks are unable to assess information stored into the left hemisphere. Alternatively, the right eye could access to information about the outcome of the response. This could make access to the memory of the consequences of devaluation the key. The right eye system may be more likely to use information about the position of a certain food in conjunction with the information about the food with which the chick has last been satiated.

11.15. Conclusions

The notion that objects are separate entities that continue to exist even when they are no longer available to direct perception has recently become a focus of interest in comparative cognitive research (for reviews, see Etienne, 1976; Doré and Dumas, 1987). Demonstrating that the possession of such a concept in an animal species is not a trivial matter: object representation occurs in conjunction with other skills (e.g. those involving functional and spatial organization) that may influence or be influenced by its occurrence. Also, the ontogeny of object representation may correlate with the rhythm of ontogenetic development of a given species. For example, humans may be one of the most altricial animal species (complete maturation is achieved over a period of many years), requiring a very long time to acquire and master classical 'object permanence' tests (Piaget, 1953): deducing an object's position after its disappearance usually begins to be mastered only between the age of 15 and 18 months (Bower and Paterson, 1972; Bower, 1982). Apes should attain complete sensorimotor object permanence earlier than humans, and monkeys should reach complete (or stage six) object permanence towards the age of 6 months (at least 1 year in advance of the human child). Kittens should follow a similar time-course, but at an even accelerated rhythm, achieving at 20 weeks a level of object permanence corresponding to that which a human infant attains not earlier than at 6–9 months of age (Gruber et al., 1971). However, it should be taken into account that Piagetian tests for object permanence actually involve a constellation of abilities (motor skills, motivation, attention, memory and socialization) that may correlate to various degrees to this concept. In fact, young children might encounter some problems with the standard tasks because these require a certain level of motor skills. As noted previously (see Section 11.1) with the use of testing procedures especially devised, such as looking time and habituation/dishabituation, even babies as young as 2 months old show the possession of some notion of object permanence (Baillargeon, 1994).

Behaviours such as recovery of cached food or cavity nesting exhibited by some avian species in seminatural conditions and in the wild (e.g. nutcrackers: Kamil and Balda, 1985; marsh tits: Sherry, 1982; Shettleworth and Krebs, 1982) appear to require a concept of object permanence. However, direct evidence of object permanence from laboratory studies of avian behaviour has, until now, been limited to corvids (Krushinskii, 1970; Etienne, 1976) and parrots (Pepperberg and Kozak, 1986; Pepperberg and Funk, 1990).

Chicks as young as 2 days of age do possess the cognitive abilities required by a detour behaviour task, and some sort of representation of the goal object and its spatial location seems to be necessary to account for the chicks' performance. At the present state of knowlege, however, it would be hard to say precisely what the nature of a chick's internal representation is. For instance, although chicks do have an object concept that maintains a representation of the object in the absence of direct sensory cues, it seems that they are not able to predict the resting position of an imprinted ball from its direction of movement prior to occlusion (Freire and Nicol, 1997, 1999). It is not clear whether this reflects a basic cognitive limitation or adaptation to ecological demands; for instance, when prey or other interesting objects hide themselves behind an occluder, it is more likely that they would reappear, after some time, in the same location where they have been seen to disappear rather than at the other side of the occluder (see e.g. Haskell and Forkman, 1997). Further behavioural investigations are needed to cast more light on this appealing issue and progressively contribute to revealing all of the features of the chick's mind.

There is recent evidence that a species of jumping spiders (*Portia fimbriata*) appears able to solve successfully a detour task requiring it to move away from a prey in order to subsequently reach it (Tarsitano and Jackson, 1994). Interestingly, jumping spiders do this even in the absence of locally orienting cues emanating from the prey (Tarsitano and Jackson, 1997). Also, choice of a complete route rather than an alternative one with a gap could be done on the basis of visual scanning from the starting point (Tarsitano and Andrew, 1999). When the spider set off, it headed at once towards the beginning of the correct route (this is probably a consequence of the rules governing scanning). Scans trace out the horizontal and vertical arms of possible routes, and stop and reverse when a discontinuity is encountered. As a result, incomplete routes receive progressively less scanning; the spider solves its problem by setting up a vector specifying the direction in which it has scanned most. Holding such a vector is, therefore, equivalent to a 'representation' of one key feature of the goal (its position), but it does not amount to evidence of anything more. Similarly, we need to know possible mechanisms underlying the ill-defined concept of 'representation' in higher vertebrates like chicks (and also humans). Comparative work on route planning using animal models with a very different organization of the nervous system may prove extremely important to develop (possibly simple) mechanisms, which may underlie these (apparently very complex) behavioural performances.

It should be remembered also that, as a nidifugous bird, the chick acquires much information about its environment within the first days posthatching (i.e. when it becomes imprinted on its mother and learns to recognize food). The data discussed here would suggest that, at least in this extremely precocial species, some rudiments of the concept of object permanence may already be acquired at a very early stage of development (and are possibly inborn). It is likely that a very different developmental pattern would characterize less precocial species.

Apart from work on comparative cognition, the detour test looks to be a promising technique for the investigation of brain lateralization and brain mechanisms of memory formation (for early neurobiological work on detour behaviour, see also Scholes and Wheaton, 1966; Miller and Tallarico, 1974). As to lateralization, preferences in eye use are apparent in detour tests. They seem to be associated with the target (goal-object) under fixation in order to provide an estimation of the degree of novelty involved; moreover, in more demanding tasks, such as the delayed response task, differential involvement of the two eyes (and hemispheres) seems to be associated with the use of local and global (spatial) cues. Finally, integration of information concerning the 'where' and the 'what' of the goal seems critically to involve neural structures in the left hemisphere.

References

Andrew, R.J. (1991). The nature of behavioural lateralization in the chick. In *Neural and Behavioural Plasticity. The Use of the Chick as a Model*, ed. R.J. Andrew, pp. 536–554. Oxford: University Press, Oxford.

Andrew, R.J. & Brennan, A. (1983). The lateralization of fear behaviour in the male domestic chick: A developmental study. *Animal Behaviour*, 31, 1166–1176.

Andrew, R.J. & Brennan, A. (1984). Sex differences in lateralization in the domestic chick. *Neuropsychologia*, 22, 503–509.

Andrew, R.J. & Dharmaretnam, M. (1993). Lateralization and strategies of viewing in the domestic chick. In *Vision, Brain, and Behavior in Birds*, eds H.P. Zeigler & H.-J. Bishof, pp. 319–332. Cambridge, MA: MIT Press.

Andrew, R.J., Mench, J. & Rainey, C. (1982). Right–left asymmetry of response to visual stimuli in the domestic chick. In *Analysis of Visual Behaviour*, eds D.J. Ingle, M.A. Goodale & R.J. Mansfield, pp. 225–236. Cambridge, MA: MIT Press.

Baillargeon, R. (1994). Physical reasoning in infancy. In *The Cognitive Neurosciences*, ed. M.S. Gazzaniga, pp. 181–203. Cambridge, MA: MIT Press.

Barrett, R.J. & Ray, O.S. (1970). Behavior in the open field, Lashley III maze, shuttle box and Sidman avoidance as a function of strain, sex, and age. *Developmental Psychology*, 3, 73–77.

Blancheteau, M. & Le Lorec, A. (1972). Raccourci et detour chez le rat: Durée, vitesse et longuer des parcours. *L'Année Psychologique*, 72, 7–16.

Bower, T.G.R. (1982) *Development in Infancy*, 2nd edn. San Francisco: Freeman.

Bower, T.G.R. & Paterson, J.G. (1972). Stages in the development of the object concept. *Cognition*, 1, 47–55.

Cailotto, M., Vallortigara, G. & Zanforlin M. (1989). Sex differences in the response to social stimuli in young chicks. *Ethology, Ecology, and Evolution*, 1, 323–327.

Carey, S. & Spelke, E. (1996). Science and core knowledge. *Philosophy of Science*, 63, 515–533.

Cerella, J. (1980). The pigeon's analysis of pictures. *Pattern Recognition*, 12, 1–6.

Chapuis, N. (1987). Detour and shortcut abilities in several species of mammals. In *Cognitive Processes and Spatial Orientation in Animal and Man*, eds P. Ellen & C. Thinus-Blanc, pp. 97–106. Dordrecht: Martinus Nijhoff Publishers.

Chapuis, N., Thinus-Blanc, C. & Poucet, B. (1983). Dissociation of mechanisms involved in dogs' oriented displacements. *Quarterly Journal of Experimental Psychology*, 35B, 213–219.

Clayton, N.S. & Dickinson, A. (1999). Memory for the content of caches by Scrub Jays (*Aphelocoma coerulescens*). *Journal of Experimental Psychology: Animal Behavior Processes*, 25, 82–91.

Clayton, N.S. & Krebs, J.R. (1994). Memory for spatial and object-specific-cues in food-storing and non-storing birds. *Journal of Comparative Physiology A*, 171, 807–815.

Clifton, P.G. & Andrew, R.J. (1983). The role of stimulus size and colour in the elicitation of testosterone-facilitated aggressive and sexual responses in the domestic chick. *Animal Behaviour*, 31, 878–886.

Corballis, P.M., Fendrich, R., Shapley, R. & Gazzaniga, M. (1998). A dissociation between illusory contour perception and amodal boundary completion following callosotomy. *Journal of Cognitive Neuroscience*, Suppl. 18, p. 18.

Cozzutti, C. & Vallortigara, G. (1999). Lateralization of a declarative memory in the chick as revealed by an incentive–devaluation procedure. *Annual Meeting of the European Brain and Behaviour Society*, Rome, 29 September–2 October 1999.

Cozzutti, C. & Vallortigara, G. (2001). Hemispheric memories for the content and position of food caches in the domestic chick. *Behavioral Neuroscience*, 115, 305–313.

Deng, C. & Rogers, L.J. (1997). Differential contributions of the two visual pathways to functional lateralization in chicks. *Behavioural Brain Research*, 87, 173–182.

Deng, C. & Rogers, L.J. (1998a). Organisation of the tectorotundal and SP/IPS-rotundal projections in the chick. *Journal of Comparative Neurology*, 394, 171–185.

Deng, C. & Rogers, L.J. (1998b). Bilaterally projecting neurons in the two visual pathways of chicks. *Brain Research*, 794, 281–290.

Dharmaretnam, M. & Andrew, R.J. (1994). Age- and stimulus-specific effects on the use of right and left eyes by the domestic chick. *Animal Behaviour*, 48, 1395–1406.

Dickinson, A. (1980). *Contemporary Animal Learning Theory*. Cambridge: Cambridge University Press.

Doré, F.Y. (1986). Object permanence in adult cats (*Felix catus*). *Journal of Comparative Psychology*, 100, 340–347.

Doré, F.Y. (1990). Search behaviour in cats (*Felix catus*) in an invisible displacement test: Cognition and experience. *Canadian Journal of Psychology*, 44, 359–370.

Doré, F.Y. & Dumas, C. (1987). Psychology of animal cognition: Piagetian studies. *Psychological Bulletin*, 102, 219–233.

Dumas, C. (1992). Object permanence in cats (*Felix catus*): An ecological approach to the study of invisible displacements. *Journal of Comparative Psychology*, 106, 404–410.

Dumas, C. & Doré, F.Y. (1989). Cognitive development in kittens (*Felix catus*): A cross-sectional study of object permanence. *Journal of Comparative Psychology*, 103, 191–200.

Dumas, C. & Doré, F.Y. (1991). Cognitive development in kittens (*Felix catus*): An observational study of object permanence and sensorimotor intelligence. *Journal of Comparative Psychology*, 105, 357–365.

Emmerton, J. & Delius, J.D. (1993). Beyond sensation: Visual cognition in pigeons. In *Vision, Brain and Behavior in Birds*, eds H.P. Zeigler & H.J. Bishof, pp. 377–390. Cambridge, MA: MIT Press.

Etienne, S.A. (1973). Searching behaviour towards a disappearing prey in the domestic chick as affected by preliminary experience. *Animal Behaviour*, 21, 749–761.

Etienne, S.A. (1974). Age variability shown by domestic chicks in selected spatial tasks. *Behaviour*, 50, 52–76.

Etienne, S.A. (1976). L'Étude comparative de la permanence d'object chez l'animal. *Bulletin de Psychologie*, 327, 187–197.

Forkman, B. (1998). Hens use occlusion to judge depth in a two-dimensional picture. *Perception*, 27, 861–867.

Forkman, B. (2000). Domestic hens have declarative representations. *Animal Cognition*, 3, 135–137.

Foster, T.M., Temple, W., MacKenzie, C., Demello, L.R. & Poling, A. (1995). Delayed matching-to-sample performance of hens. Effects of sample duration and response requirements during the sample. *Journal of the Experimental Analysis of Behaviour*, 64, 19–31.

Freire, F. & Nicol, C.J. (1997). Object permanence in chicks: Predicting the position of an occluded imprinted object. *Abstracts of the ASAB Meeting 'Biological Aspects of Learning'*, St Andrews, Scotland, 1–4th July 1997, p. 21.

Freire, F. & Nicol, C. J. (1999). Effect of experience of occlusion events on the domestic chicks's strategy for locating a concealed imprinting object. *Animal Behaviour*, 58, 593–599.

Gagnon, S. & Doré, F.Y. (1992). Search behaviour in various breeds of adult dogs (*Canis familiaris*): Object permanence and olfactory cues. *Journal of Comparative Psychology*, 106, 58–68.

Gallistel, C.R. (1990). *The Organization of Learning*. Cambridge, MA: MIT Press.

Gallup, G.G., Jr. & Suarez, S.D. (1980). An ethological analysis of open-field behaviour in chickens. *Animal Behaviour*, 28, 368–378.

Gaulin, S.J.C. & Fitzgerald, R.W. (1989). Sexual selection for spatial-learning abilities. *Animal Behaviour*, 37, 322–331.

Griffiths, D., Dickinson, A. & Clayton, N. (1999). Episodic memory: What can animals remember about their past? *Trends in Cognitive Sciences*, 3, 74–80.

Gruber, H.E., Girgus, J.S. & Banuazizi, A. (1971). The development of object permanence in the cat. *Developmental Psychology*, 4, 9–15.

Güntürkün, O. (1985). Lateralization of visually controlled behaviour in pigeons. *Physiology and Behavior*, 34, 575–577.

Güntürkün, O. (1997). Avian visual lateralization – a review. *Neuro Report*, 6, iii–xi.

Hamme, L.J. van, Wasserman, E.A. & Biederman, I. (1992). Discrimination of contour-deleted images by pigeons. *Journal of Experimental Psychology: Animal Behavior Processes*, 18, 387–399.

Haskell, M. & Forkman, B. (1997). An investigation into object permanence in the domestic hen. *Abstracts of the ASAB Meeting 'Biological Aspects of Learning'*, St Andrews, Scotland, 1–4 July 1997, p. 22.

Horn, G. (1990). Neural basis of recognition memory investigate through an analysis of imprinting. *Philosophical Transactions of the Royal Society London B.*, 329, 133–142.

Hull, C. (1938). The goal-gradient hypothesis applied to some 'field-force' problems in the behaviour of young children. *Psychological Review*, 45, 271–299.

Hunter, W.S. (1913). The delayed reaction in animals and children. *Behaviour Monograph*, II, 1.

Johnson, M.H. & Bolhuis, J.J. (1991). Imprinting, predispositions, and filial preference in the chick. In *Neural and Behavioural Plasticity: The Use of the Domestic Chick as a Model*, ed. R.J. Andrew, pp. 133–156. Oxford: Oxford University Press.

Johnson, S.P. & Aslin, R.N. (1995). Perception of object unity in 2-month-old infants. *Developmental Psychology*, 31, 739–745.

Jones, R.B. (1977). Sex and strain differences in the open-field responses of the domestic chick. *Applied Animal Ethology*, 3, 255–261.

Jones, R.B. & Faure, J.M. (1981). Sex effects on open-field behaviour in the domestic chick as a function of age. *Biology of Behaviour*, 6, 265–272.

Jones, R.B. & Faure, J.M. (1982). Open-field behaviour of male and female domestic chicks as a function of housing conditions, test situations and novelty. *Biology and Behaviour*, 7, 17–25.

Jones, R.B. & Harvey, S. (1987). Behavioural and adrenocortical responses of domestic chicks to systematic reductions in group size and to sequential disturbance of companions by the experimenter. *Behavioural Processes*, 14, 291–303.

Kamil, A.C. & Balda, R.P. (1985). Cache recovery and spatial memory in Clark's nutcrackers (*Nucifraga columbiana*). *Journal of Experimental Psychology: Animal Behavior Processes*, 11, 95–111.

Kanizsa, G. (1979). *Organization in Vision*. New York: Praeger.

Kanizsa, G., Renzi, P., Conte, S. Compostela, C. & Guerani, L. (1993). Amodal completion in mouse vision. *Perception*, 22, 713–722.

Karten, H.J. & Shimizu, T. (1989). The origins of neocortex: Connections and lamination as distinct events in evolution. *Journal of Cognitive Neuroscience*, 1, 291–301.

Kellman, P.J. & Spelke, E.S. (1983). Perception of partly occluded objects in infancy. *Cognitive Psychology*, 15, 483–524.

Köhler, W. (1925). *The Mentality of Apes*. New York: Harcourt Brace.

Kovacs, G., Vogels, R. & Orban, G.A. (1995). Selectivity of macaque inferior temporal neurons for partially occluded shapes. *Journal of Neuroscience*, 15, 1984–1997.

Krushinsky, L.V. (1990). *Experimental Studies of Elementary Reasoning*. New Delhi: Oxonian Press.

Lea, S.E.G., Slater, A.M. & Ryan, C.M.E. (1996). Perception of object unity in chicks: A comparison with the human infant. *Infant Behaviour and Development*, 19, 501–504.

Lewin, K. (1933). Vectors, cognitive processes, and Mr Tolman's criticism. *Journal of General Psychology*, 8, 318–345.

Lewin, K. (1935). *A Dynamic Theory of Personality: Selected Papers*. New York: McGraw-Hill.

Lockman, J.J. (1984). The development of detour ability during infancy. *Child Development*, 55, 482–491.

Lorenz, K. (1971). *Studies in Animal and Human Behavior. I*. London: Butler & Tanner.

Maccoby, E. & Jacklin, C.N. (1974). *The Psychology of Sex Differences*. Stanford, CA: Stanford University Press.

McKenzie, R., Andrew, R.J. & Jones, R.B. (1998). Lateralization in chicks and hens: New evidence for control of response by the right eye system. *Neuropsychologia*, 36, 51–58.

McNemar, Q. & Stone, C.P. (1932). The sex difference in rats on three learning tasks. *Journal of Comparative Psychology*, 14, 171–180.

Mathieu, M., Bouchard, M.-A., Granger, L. & Herscovitch, J. (1976). Piagetian object-permanence in *Cebus capucinus*, *Lagothrica flavicauda* and *Pan troglodytes*. *Animal Behaviour*, 24, 585–588.

Michotte, A. (1963). *The Perception of Causality*. New York: Basic Books.

Miller, D.B. (1973). A comparative analysis of detour behaviour in ring doves (*Streptopelia risoria*) and rock doves (*Columba livia*) as a function of barrier mesh density. Ph.D. dissertation, University of Miami.

Miller, D.B. & Tallarico, R.B. (1974). On the correlation of brain size and problem-solving behavior of ring doves and pigeons. *Brain, Behavior and Evolution*, 10, 265–273.

Natale, F., Antinucci, F., Spinozzi, G. & Poti, P. (1986). Stage 6 object concept in non-human primates cognition: A comparison between gorilla (*Gorilla gorilla gorilla*) and Japanese macaque (*Macaca fuscata*). *Journal of Comparative Psychology*, 100, 335–339.

Nieder, A. & Wagner, H. (1999). Perception and neuronal coding of subjective contours in the owl. *Nature Neuroscience*, 2, 660–663.

O'Keefe, J. & Conway, D.H. (1980). On the trail of the hippocampal engram. *Physiological Psychology*, 8, 229–238.

O'Keefe, J. & Nadel, L. (1978). *The Hippocampus as a Cognitive Map*. Oxford: Clarendon Press.

Osada, Y. & Schiller, P.H. (1994). Can monkeys see objects under conditions of transparency and occlusion? *Investigative Ophthalmology and Visual Science*, Suppl. 35, 1664.

Patterson, T.A. & Rose, S.P.R. (1992). Memory in the chick: multiple cues distinct brain locations. *Behavioral Neuroscience*, 106, 465–470.

Pearce, J.M. (1987). *An Introduction to Animal Cognition*. Hillsdale, NJ: Lawrence Erlbaum.

Pepperberg, I.M. & Funk, M.S. (1990). Object permanence in four species of psittacine birds: An African Grey parrot (*Psittacus erithacus*), an Illiger mini macaw (*Ara maracana*), a parakeet (*Melopsittacus undulatus*), and a cockatiel (*Nymphicus hollandicus*). *Animal Learning and Behaviour*, 18, 97–108.

Pepperberg, I.M. & Kozak, F.A. (1986). Object permanence in the African Grey parrot (*Psittacus erithacus*). *Animal Learning and Behaviour*, 14, 322–330.

Piaget, J. (1937). *La construction du réel*. Neuchâtel: Delachaux et Niesté.

Piaget, J. (1953). *Origin of Intelligence in the Child*. London: Routledge & Kegan Paul.

Poucet, B. (1985). Spatial behaviour of cats in cue-controlled environments. *Quarterly Journal of Experimental Psychology*, 37, 155–179.

Poucet, B., Thinus-Blanc, C. & Chapuis, N. (1983). Route planning in cats in relation to the visibility of the goal. *Animal Behaviour*, 31, 594–599.

Prior, H. & Güntürkün, O. (1999). Patterns of visual lateralization in pigeons: Seeing what is there and beyond. *Perception*, Suppl 28, 22.

Rashotte, M.E. (1987). Behaviour in relation to objects in space: Some historical perspectives. In *Cognitive Processes and Spatial Orientation in Animal and Man* (I), eds P. Ellen & C. Thinus-Blanc, pp. 39–54, Dordrecht: Martinus Nijhoff Publishers.

Rashid, N. & Andrew, R.J. (1989). Right hemisphere advantages for topographical orientation in the domestic chick. *Neuropsychologia*, 27, 937–948.

Redshaw, M. (1978) Cognitive development in human and gorilla infants. *Journal of Human Evolution*, 7, 133–141.

Regolin, L. & Rose, S.P.R. (1999). Long-term memory for a spatial task in young chicks. *Animal Behaviour*, 57, 1185–1191.

Regolin, L. & Vallortigara, G. (1995). Perception of partly occluded objects by young chicks. *Perception and Psychophysics*, 57, 971–976.

Regolin, L., Vallortigara, G. & Zanforlin M. (1994a). Detour behaviour in the chick: A review and re-interpretation. *Atti e Memorie dell'Accademia Patavina di Scienze, Lettere ed Arti. Classe di Scienze Matematiche e Naturali*, Vol. CV, pp. 105–126.

Regolin, L., Vallortigara, G. & Zanforlin, M. (1994b). Perceptual and motivational aspects of detour behaviour in young chicks. *Animal Behaviour*, 47, 123–131.

Regolin, L., Vallortigara, G. & Zanforlin, M. (1995a). Object and spatial representations in detour problems by chicks. *Animal Behaviour*, 49, 195–199.

Regolin, L., Vallortigara, G. & Zanforlin, M. (1995b). Detour behaviour in the domestic chick: Searching for a disappearing prey or a disappearing social partner. *Animal Behaviour*, 50, 203–211.

Révész, G. (1924). Experiments on animal space perception. *British Journal of Psychology*, 14, 387–414.

Rogers, L.J. (1982). Light experience and asymmetry of brain function in chickens. *Nature*, 297, 223–225.

Rogers, L.J. (1989). Laterality in animals. *International Journal of Comparative Psychology*, 3, 6–25.

Rogers, L.J. (1991). Development of lateralization. In *Neural and Behavioural Plasticity*, ed. R.J. Andrew, pp. 507–535. Oxford: Oxford University Press.

Rogers, L.J. (1995a). *The Development of Brain and Behaviour in the Chicken*. Wallingford: CAB International.

Rogers, L.J. (1995b). Evolution and development of brain asymmetry, and its relevance to language, tool use and consciousness. *International Journal of Comparative Psychology*, 8, 1–15.

Rogers, L.J. (1997). *Minds of Their Own*. St Leonards: Allen & Unwin.

Rogers, L.J. & Anson, J.M. (1979). Lateralization of function in the chicken forebrain. *Pharmacology, Biochemistry and Behaviour*, 10, 679–686.

Rose, S.P.R. (1992). *The Making of Memory: From Molecules to Mind*. London: Bantam Press.

Ryan, C.M.E. & Lea, S.E.G. (1994). Images of conspecifics as categories to be discriminated by pigeons and chickens: Slides, video tapes, stuffed birds and live birds. *Behavioural Processes*, 33, 155–176.

Ryle, G. (1949). *The Concept of Mind*. London: Hutchinson.

Salzen, E.A., Lily, R.E. & McKeown, J.R. (1971). Colour preference and imprinting in domestic chicks. *Animal Behaviour*, 19, 542–547.

Sandler, B.E., Van Gelder, G.A., Buck, W.B. & Karas, G.G. (1968). Effect of dieldrin exposure on detour behavior in sheep. *Psychological Report*, 23, 451–455.

Schiller, P.H. (1949). Analysis of detour behavior. I. Learning of roundabout pathways in fish. *Journal of Comparative and Physiological Psychology*, 42, 463–475.

Scholes, N.W. (1965). Detour learning and development in the domestic chick. *Journal of Comparative and Physiological Psychology*, 60, 114–116.

Scholes, N.W. & Wheaton, L.G. (1966). Critical period for detour learning in developing chicks. *Life Sciences*, 5, 1859–1865.

Sekuler, A.B., Lee, J.A.J. & Shuttleworth, S.J. (1996). Pigeons do not complete partly occluded figures. *Perception*, 25, 1109–1120.

Sherry, D.F. (1982). Food storage, memory, and marsh tits. *Animal Behaviour*, 30, 631–633.

Shettleworth, S. & Krebs, J. (1982). How marsh tits find their hoards: The roles of sites preference and spatial memory. *Journal of Experimental Psychology: Animal Behavior Processes*, 8, 354–375.

Sigman, S.E., Lovern, D.R. & Schulman, A.H. (1978). Preferential approach to conspecifics as a function of different rearing conditions. *Animal Learning and Behavior*, 6, 231–234.

Spelke, E.S. (1998a). Nativism, empiricism, and the origins of knowledge. *Infant Behavior & Development*, 21, 181–200.

Spelke, E.S. (1998b). Nature, nurture and development. In *Perception and Cognition at Century's End*, ed. J. Hochberg, pp. 333–371. San Diego: Academic Press.

Tarsitano, M.S. & Andrew, R.J. (1999). Scanning and route selection in the jumping spider *Portia labiata*. *Animal Behaviour*, 58, 255–265.

Tarsitano, M.S. & Jackson, R.R. (1994). Jumping spiders make predatory detours requiring movement away from prey. *Behaviour*, 131, 65–73.

Tarsitano, M.S. & Jackson, R.R. (1997). Araneophagic jumping spiders discriminate between detour routes that do and do not lead to prey. *Animal Behaviour*, 53, 257–266.

Thinus-Blanc, C., Poucet, B. & Chapuis, N. (1982). Object permanence in cats: Analysis in locomotor space. *Behavioural Processes*, 7, 81–82.

Tinklepaugh, O.L. (1928). An experimental study of representative factors in monkeys. *Journal of Comparative Psychology*, 8, 197–236.

Tolman, E.C. (1932). *Purposive Behavior in Animals and Men*. New York: Appleton-Century-Crofts.

Tommasi, L. & Vallortigara, G. (1997). Lateralization of spatial memory tasks in the domestic chick (*Gallus gallus*). *Experimental Brain Research*, 117, S43.

Tommasi, L. & Vallortigara, G. (2001). Encoding of geometric and landmark information in the left and right hemisphere of the avian brain. *Behavioural Neuroscience*, 115, in press.

Tommasi, L., Vallortigara, G. & Zanforlin, M. (1997). Young chickens learn to localize the centre of a spatial environment. *Journal of Comparative Physiology A*, 180, 567–572.

Towe, A.L. (1954). A study of figural equivalence in the pigeon. *Journal of Comparative and Physiological Psychology*, 47, 283–287.

Triana, E. & Pasnak, R. (1981). Object permanence in cats and dogs. *Animal Learning and Behaviour*, 9, 135–139.

Vallortigara, G. (1992a). Right hemisphere advantage for social recognition in the chick. *Neuropsychologia*, 30, 761–768.

Vallortigara, G. (1992b). Affiliation and aggression as related to gender in domestic chicks (*Gallus gallus*). *Journal of Comparative Psychology*, 106, 53–57.

Vallortigara, G. (1996). Learning of colour and position cues in domestic chicks: Males are better at position, females at colour. *Behavioural Processes*, 36, 289–296.

Vallortigara, G. (2000). Comparative neuropsychology of the dual brain: A stroll through animals' left and right perceptual worlds. *Brain and Language*, 73, 189–219.

Vallortigara, G. & Andrew, R.J. (1991). Lateralization of response by chicks to change in a model partner. *Animal Behaviour*, 41, 187–194.

Vallortigara, G. & Andrew, R.J. (1994a). Olfactory lateralization in the chick. *Neuropsychologia*, 32, 417–423.

Vallortigara, G. & Andrew, R.J. (1994b). Differential involvement of right and left hemisphere in individual recognition in the domestic chick. *Behavioural Processes*, 33, 41–58.

Vallortigara, G. & Garzotto. Delayed response tasks in chicks: Binocular and lateralized monocular performance (in preparation).

Vallortigara, G. & Zanforlin, M. (1988). Open-field behavior of young chicks (*Gallus gallus*): Antipredatory responses, social reinstatement motivation, and gender effects. *Animal Learning and Behaviour*, 16, 359–362.

Vallortigara, G., Cailotto, M. & Zanforlin, M. (1990). Sex differences in the social reinstatement motivation of the domestic chick (*Gallus gallus*) revealed by runway tests with social and nonsocial reinforcement. *Journal of Comparative Psychology*, 104, 361–367.

Vallortigara, G., Regolin, L. & Zanforlin, M. (1994). The development of responses to novel-coloured objects in male and female domestic chicks. *Behavioural Processes*, 31, 219–230.

Vallortigara, G., Regolin, L. & Pagni, P. (1999). Detour behaviour, imprinting, and visual lateralization in the domestic chick. *Cognitive Brain Research*, 7, 307–320.

Vallortigara, G., Zanforlin, M. & Pasti, G. (1990). Geometric modules in animal spatial representations: A test with chicks (*Gallus gallus*). *Journal of Comparative Psychology*, 104, 248–254.

Vallortigara, G., Regolin, L., Bortolomiol, G. & Tommasi, L. (1996). Lateral asymmetries due to preferences in eye use during visual discrimination learning in chicks. *Behavioural Brain Research*, 74, 135–143.

Vallortigara, G., Regolin, L., Rigoni, M. & Zanforlin, M. (1998). Delayed search for a concealed imprinted object in the domestic chick. *Animal Cognition*, 1, 17–24.

Vander Wall, S.B. (1982). An experimental analysis of cache recovery in Clark's nutcracker. *Animal Behaviour*, 30, 84–94.

von Fersen, L. and Güntürkün, O. (1990). Visual memory lateralization in pigeons. *Neuropsychologia*, 28, 1–7.

White, K.G., Alsop, B. & Williams, L. (1993). Prototype identification and categorization of incomplete figures by pigeons. *Behavioural Processes*, 30, 253–258.

Zanforlin, M. (1981). Visual perception of complex forms (anomalous surfaces) in chicks. *Italian Journal of Psychology*, 1, 1–16.

12

Laterality of Communicative Behaviours in Non-human Primates: A Critical Analysis

WILLIAM D. HOPKINS AND
SAMUEL FERNÁNDEZ CARRIBA

12.1. Introduction

In the past 12 years, there has been a resurgence of interest in the topic of laterality in non-human species (Bradshaw and Rogers, 1993). The resurgence of interest in this topic in a host of scientific disciplines was to some extent spurred by the review article written by MacNeilage, Studdert-Kennedy and Lindblom (1987) that called into question the prevailing view that population-level hemispheric specialization was all but absent in non-human species save songbirds (see Nottebohm, 1977; Warren, 1980). The theory proposed by MacNeilage, Studdert-Kennedy and Lindblom (1987) met with considerable criticism (see commentaries on the original article) but nonetheless resulted in a subsequent plethora of studies on laterality in non-human species, notably non-human primates (for reviews, see Fagot and Vauclair, 1991; Bradshaw and Rogers, 1993; Ward and Hopkins, 1993; Hopkins, 1996; McGrew and Marchant, 1997). The overwhelming focus of these studies was on hand preference and the determination of whether or not population-level biases were evident for specific kinds of sensory or motor tasks. Parallel to studies on hand preference but in substantially lower numbers were studies that focused on sensory or perceptual asymmetries (see Morris et al., 1993). As with studies of hand preference, the focus of these studies was largely the determination of population-level biases for specific tasks and the similarity in findings relative to humans. The purpose of the current chapter is to review studies that have examined perceptual asymmetries in non-human primates. Of particular focus are studies of asymmetries in communicative behaviours including: (1) discrimination of individual identity and facial expressions; (2) production of facial expressions; (3) perception and production of vocalizations and other auditory stimuli; and (4) perception and production of symbols and gestures. We have chosen to focus on

445

these four areas because they have been studied extensively in humans and allow for a comparative analysis of the findings between species. In addition, they are of heuristic interest because of their potential parallels with the evolution of higher cognitive processes, notably language. After reviewing the literature, we provide a critical analysis of the findings and offer some suggestions for future research.

12.2. Asymmetries in the Production and Perception of Auditory Stimuli and Manual Gestures in Non-human Primates

12.2.1. *Perception of Species-specific Vocalizations and Other Auditory Stimuli*

Previous studies have reported a left hemisphere advantage in the processing of auditory stimuli in macaques (Dewson, 1977); see Table 12.1. Dewson (1977) trained macaques to discriminate auditory pure tones and bursts of noise. After reaching criterion, unilateral lesions to the temporal lobe of either the left or right hemisphere were performed. The macaques with left hemisphere lesions performed significantly less well than macaques with a right hemisphere lesion. Pohl (1983) tested four baboons for laterality in auditory discrimination for four different types of stimuli, including pure tones, musical words, vowels and consonant-vowels. Significant left ear advantages (i.e. right hemisphere asymmetry) were found for the pure tones, musical tones and consonant-vowels. No ear advantages were found for the vowels. In a follow-up study, Pohl (1984) tested four new baboon subjects on an auditory discrimination task to assess whether asymmetries in the consonant-vowel stimuli were due to an asymmetry for making discriminations in the processing of temporal resolution. In this study, the baboons had to discriminate consonant-vowel pairs ('ba' and 'pa') that differed only in terms of the temporal interval of the voice onset time. The same baboons were also trained to discriminate two broad-band, white noise sounds, one of which sometimes contained a silent gap in the middle of the sound burst. Stimuli were presented unilaterally to each ear and performance was compared between ears for each stimulus type. Three of the four baboons showed a left ear advantage (i.e. right hemisphere asymmetry) for both classes of stimuli. Moreover, all four subjects showed consistent ear advantages for both types of stimuli. Pohl (1984) concluded that the basis of consonant-vowel discrimination in baboons is due to their processing of temporal resolution. More recently, Gaffan and Harrison (1991) taught six rhesus monkeys auditory–visual conditional matching tasks and subsequently performed

Table 12.1. *Studies on perception of auditory stimuli*

Author(s)	Species	Task	Results
Dewson (1997)	Rhesus monkey (*Macaca mulatta*)	Perception of pure tones	LHem advantage
Pohl (1983)	Baboon (*Papio anubis*)	Perception of four classes of auditory stimuli	RHem advantage for three classes of stimuli
Pohl (1984)	Baboon (*Papio anubis*)	Temporal sound bursts	RHem advantage
Gaffan and Harrison (1991)	Rhesus monkey (*Macaca mulatta*)	Auditory–visual matching task	LHem advantage
Petersen et al. (1978)	Japanese macaque (*Macaca fuscata*), pig-tailed macaque (*Macaca radiata*) and vervet monkey (*Cercopithecus aethiops*)	Discrimination of Japanese macaque vocalizations	LHem advantage for processing species-specific calls in Japanese macaques
Petersen et al. (1984)	Japanese macaque (*Macaca fuscata*); other macaques included (*Macaca nemestrina*)	Discrimination of Japanese macaque vocalizations	LHem advantage for processing species-specific calls in Japanese macaques
Heffner and Heffner (1984)	Japanese macaque (*Macaca fuscata*)	Discrimination of Japanese macaque vocalizations	LHem advantage
Hauser and Andersson (1994)	Rhesus monkey (*Macaca mulatta*)	Orienting response to species-specific vocalizations and a bird call	LHem advantage for species-specific calls; RHem advantage for bird calls
Hauser, Agnetta and Perez (1998)	Rhesus monkey (*Macaca mulatta*)	Discrimination of species-specific vocalizations varying in temporal patterning	LHem advantage for vocalizations in typical temporal sequence

LHem, left hemisphere; RHem, right hemisphere.

serial ablations to the left ($N = 3$) and right ($N = 3$) prefrontal cortex followed by lesions to the auditory cortex. Ablation of the left but not right auditory cortex produced a significant impairment in the intermodal equivalence task. Gaffan and Harrison (1991) interpreted these findings as evidence for a left hemisphere specialization in auditory processing in rhesus monkeys.

In one of the most cited papers in the literature, Petersen et al. (1978) reported a left hemisphere advantage in the discrimination of species-specific calls by Japanese macaques. In this study, ten monkeys including five Japanese macaques (*Macaca fuscata*), two bonnet macaques (*Macaca radiata*), two pig-tail macaques (*Macaca nemestrina*) and one vervet monkey (*Cercopithecus aethiops*) were trained to discriminate two types of 'coo' vocalizations produced by *Macaca fuscata* that differ in the temporal location of the peak frequency. During training, the stimuli were presented to either the left or right ear of the subjects and the number of correct trials per ear was tallied for each training session and subject. For each training session, the subjects could be characterized as performing better with their left ear, right ear or neither ear. Petersen et al. (1978) then determined how many subjects performed significantly better with the right or left ear, and found that all five Japanese macaques reached criterion on the task faster when the stimuli were presented to the right ear, whereas only one of the remaining five monkeys showed the same right ear advantage. Note that none of the subjects showed a significant left ear advantage.

In a subsequent study, Petersen et al. (1984) trained two Japanese monkeys and two comparison macaques (pig-tailed monkeys) to discriminate the 'coo' vocalization but, in this study, Petersen et al. tested laterality for generalization of this discrimination. The two Japanese macaques both showed a right ear advantage in acquisition and in generalization of the discrimination, whereas neither comparison macaque showed an asymmetry in acquisition or generalization of this discrimination. It is important to note that the comparison macaques did show significant generalization of the discrimination. Petersen et al. concluded that both species of macaque were using the same acoustic features to acquire and maintain this discrimination, but that the asymmetry was specific to Japanese macaques because they were the only species to show the right ear advantage (i.e. left hemisphere asymmetry).

Capitalizing on the study by Petersen and colleagues, Heffner and Heffner (1984) trained Japanese macaques similarly to perform auditory discriminations of the 'coq' vocalizations, and then performed unilateral lesions to either the left or right posterior temporal lobes in subsamples of individuals. Heffner and Heffner (1984) subsequently examined post-lesion relearning of the 'coo' vocalization in the two unilaterally ablated samples of monkeys.

Monkeys with a left hemisphere lesion showed a greater decrement in post-operative performance and took longer to relearn the discrimination than right hemisphere lesioned monkeys. Heffner and Heffner (1984) concluded that the left hemisphere was the dominant half of the brain in performing this discrimination but emphasized that, to some extent, the asymmetry was transient as all monkeys subsequently re-acquired the discrimination. In a series of subsequent studies, Heffner and Heffner (1989) replicated their previous results and demonstrated convincingly that the left hemisphere asymmetries were not due to cortical deafness or other alternative explanations.

More recently, rather than utilize lesion approaches or more traditional laboratory approaches to the assessment of auditory laterality, Hauser and Andersson (1994) examined orienting asymmetries to different auditory stimuli in rhesus monkeys living on Cayo Santiago (see also Chapter 13 by Weiss et al.). Briefly, a speaker was positioned in a field setting and the monkeys were lured to the feeding site by the experimenters. Once positioned centrally behind the speakers (which varied from 5 to 10 m from the subject), an acoustic stimulus was presented to the subject and the experimenters recorded which way the monkeys turned (left or right) to orient toward the sound. There were four types of acoustic stimuli, including three types of conspecific calls (rhesus monkey vocalizations) and one type of heterospecific call (a local bird call). A total of 80 monkeys were tested with one trial presented to each subject. Each of the four types of calls were played to 20 different monkeys and the distribution of left-orienting contrasted with right-orienting monkeys was compared across the different stimulus types. Hauser and Andersson (1994) reported that significantly more monkeys oriented to the right, compared to the left, for the three conspecific calls but not for the heterospecific call. The authors interpreted these findings as evidence that the left hemisphere is dominant in processing specific-specific calls by rhesus monkeys. More recently, Hauser, Agnetta and Perez (1998) tested for orienting asymmetries in the same format as described above but manipulated the interpulse interval for three different types of rhesus monkey vocalizations including grunts, shrills and copulation screams. Variations in the interpulse intervals were longer and shorter than the population mean pulse interval for each of the three call types. For two types of calls (grunts and shrills), extreme variation in the interpulse interval resulted in a lack of any observed orienting asymmetry while presentation with the 'normal' range resulted in a right-sided orienting bias. For the copulation scream, changes in the interpulse interval did not significantly influence the overall right-sided orienting asymmetry.

12.2.1.1. Signing and Hand Preference

With respect to hand use and gestural communication (Table 12.2), Shafer (1988) reported that the gorilla 'Koko' had a left hand bias for signing behaviour. Similarly, Miles (1990) reported that the orang-utan 'Chantek'

Table 12.2. *Hand preference in signing and gestures*

Author(s)	Species	Study	Results
Shafer (1988)	Gorilla (*Gorilla gorilla*)	Hand use in signing	RHem advantage
Miles (1990)	Orangutan (*Pongo pygmaeus*)	Hand use in signing	RHem advantage
Steiner (1990)	Chimpanzee (*Pan troglodytes*)	Hand use in signing	LHem advantage in two subjects and no bias found in three.
Morris, Hopkins and Bolser-Gilmore (1993)	Chimpanzee (*Pan troglodytes*)	Hand use in signing	LHem advantage in one subject and RHem advantage in another
Shafer (1993)	Gorilla (*Gorilla gorilla*)	Observation of hand use in gestures	LHem advantage
Shafer (1997)	Bonobo (*Pan paniscus*)	Observation of hand use in gestures	RHem advantage
Hopkins and de Waal (1995)	Bonobo (*Pan paniscus*)	Observation of hand use in gestures	LHem advantage
Krause and Fouts (1997)	Chimpanzee (*Pan troglodytes*)	Hand use in gestures	LHem advantage for indexical pointing
Leavens, Hopkins and Bard (1996)	Chimpanzee (*Pan troglodytes*)	Hand use in gestures	LHem advantage for indexical pointing and RHem for whole-hand pointing
Hopkins and Leavens (1998)	Chimpanzee (*Pan troglodytes*)	Hand use in gestures	LHem advantage
Hopkins and Wesley (2001)	Chimpanzee (*Pan troglodytes*)	Hand use in gestures	LHem advantage

LHem, left hemisphere; RHem, right hemisphere.

primarily signed with the left hand. Steiner (1990) reported two chimpanzees had a right-hand preference and three had no bias for producing signs. Morris, Hopkins, and Bolser-Gilmore (1993) reported that one chimpanzee preferred to use the right hand for pointing to symbols while the other chimpanzee preferred the left hand.

12.2.1.2. Laterality and Manual Gestures

In non-language-trained apes, several studies have reported data on hand use in gestural communication in gorillas, bonobos and chimpanzees. In an observational study, Shafer (1993) reported that gorillas gesture more with their right compared to the left hand. Hand use in gestures was noted in 32 gorillas living in zoos and it was reported that 22 preferred the right and 10 preferred the left hand. Using the same type of ethogram, Shafer (1997) studied gestures (as well as other behaviours) in a sample of 12 captive bonobos. Of the 12 bonobos that showed a bias in hand use, nine preferred the right and three preferred the left hand. The data from Shafer (1997) are consistent with previous findings by Hopkins and de Waal (1995), who studied laterality in hand use in a sample of 20 captive bonobos. Of the 18 subjects that showed a bias in hand use, 16 preferred the right hand and two preferred the left. Krause and Fouts (1997) reported hand use for non-American Sign Language (ASL) gestures in two chimpanzees trained in ASL. Overall, one subject was left-handed and the other showed no bias in for all types of gestures. However, both chimpanzees gestured significantly more with their right hand for indexical pointing compared to whole-hand pointing. The results reported by Krause and Fouts (1997) are nearly identical to those reported by Leavens, Hopkins and Bard (1996) in a sample of three non-language-trained chimpanzees. Leavens et al. (1996) reported that the subjects used their left hand more often for whole-hand pointing, whereas right hand use was more prevalent for indexical pointing.

More recently, Hopkins and Leavens (1998) have designed a series of studies explicitly to elicit manual gestures in a sample of 115 chimpanzees. With specific reference to laterality, Hopkins and Leavens (1998) reported a population-level right-hand bias in manual gestures, particularly for food begs contrasted with whole-hand pointing (for description of this distinction, see Leavens and Hopkins, 1999). Hopkins and Leavens (1998) also found that right hand use for manual gestures was observed significantly more often if it was accompanied by a vocalization than when it was not. Interestingly, hand use for gestures did not significantly correlate with hand preferences for bimanual feeding or for a coordinated bimanual task (see Hopkins, 1994, 1995). In a follow-up study, Hopkins and Wesley (2001) examined the effect of situational factors on hand use in gestures. In this study, an experimenter elicited gestures from a sample of 113 chimpanzees when positioned either to the left, right or directly in front of the subject. Hand use and gesture type was recorded for each subject. Subjects were classified as strongly left-handed (all three gestures with the left hand), mildly left-handed (use of the left hand on two of the three trials), mildly right-handed (use of the right hand on two of the three trials) or strongly right-handed (all three gestures with the right hand). Depicted in Figure 12.1 are the numbers of subjects classified into the four hand-use categories. As can be seen, there were significantly more mildly right-handed than mildly left-handed subjects. In addition, there were significantly more strongly right-handed subjects than there were strongly left-handed subjects.

12.2.1.3. Hemispheric Activation and Visual Recognition

In contrast to gestural communication, several studies have attempted to assess perceptual asymmetries in the processing of meaningful symbols contrasted with non-meaningful symbols in language-trained chimpanzees (Table 12.3). In one of the first studies, Muncer (1982) assessed conjugate lateral eye movements (CLEM) in response to specific kinds of linguistic questions posed to two chimpanzees trained in ASL. Conjugate lateral eye movements presumably reflect the activation of the cerebral hemispheres when engaged in a specific cognitive operation. Leftward movements reflect right hemisphere activation and rightward movements reflect left hemisphere

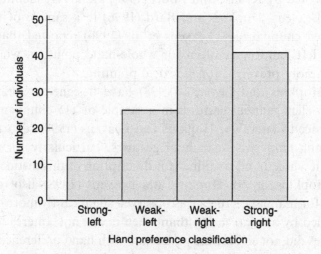

Figure 12.1. Distribution of hand use in manual gestural communication for three trials presented in different positions.

Table 12.3. *Visual recognition and hemispheric activation studies*

Author(s)	Species	Study	Results
Muncer (1982)	Chimpanzee (*Pan troglodytes*)	CLEM	RHem advantage for one subject, no bias for another
O'Neil et al. (1978)	Chimpanzee (*Pan troglodytes*)	CLEM	RHem advantage for two of eight subjects
Hopkins, Morris and Savage-Rumbaugh (1991)	Chimpanzee (*Pan troglodytes*)	Response time to meaningful and non-meaningful symbols in priming task	LHem advantage for meaningful symbols and RHem advantage for non-meaningful symbols
Hopkins et al. (1992)	Chimpanzee (*Pan troglodytes*)	Response time to symbols when semantic categories were cued in priming task	LHem advantage for cued and non-cued semantic categories; no effect for non-meaningful symbols
Hopkins (this study)	Chimpanzee (*Pan troglodytes*)	Response time to meaningful and non-meaningful symbols in recognition task	LHem advantage for 'food' and 'tool' categories and RHem advantage for 'unknown' and 'pattern' stimuli

CLEM, conjugate lateral eye movements; LHem, left hemisphere; RHem, right hemisphere.

activation. Muncer (1982) reported that question type (linguistic or spatial) had no significant influence on CLEM, although one subject did show a consistent left CLEM. The remaining subjects did not show a bias in CLEM. O'Neil et al. (1978) studied CLEM in a sample of eight chimpanzees. Rather than present subjects with particular types of questions to respond to, in this study, the experimenters made eye contact with the chimpanzee subjects. When mutual eye contact was broken by the chimpanzee, the experimenters recorded the direction of movement of the eyes. Overall, no population bias was found but two of the subjects did show a significant left eye movement bias.

In contrast to the CLEM studies, Hopkins and colleagues (Hopkins, Morris and Savage-Rumbaugh, 1991; Hopkins et al., 1992) examined asymmetries in hemispheric activation using a simple priming paradigm in three language-trained chimpanzees (Sherman, Austin and Lana). In one set of studies (Hopkins, Morris and Savage-Rumbaugh, 1991), simple reaction time was measured in response to different kinds of visual stimuli presented to either the left (LVF) or right (RVF) visual half-fields. In this paradigm, the subjects were trained to depress a key upon the appearance of a fixation stimulus presented in the centre of a computer monitor. Upon depressing the key, a foreperiod occurred ranging between 1000 and 2000 ms whereupon a warning stimulus (WS) was presented laterally of the fixation stimulus for a duration of 100 ms. On control trials, no warning stimulus was presented. After the presentation of the WS or control trial, a delay period occurred that lasted between 1000 and 2000 ms. At the end of the delay period, a response cue appeared in the centre of the screen and overlaid the fixation stimulus. The subjects were required to release the button as quickly as possible upon the appearance of the response cue. In the first study, the stimuli varied according to their familiarity to the subjects and their specific semantic meanings. The stimuli were lexigrams, or visual-graphic symbols, that comprised the lexicon used with these chimpanzees (for a description, see Rumbaugh, 1977). One class of stimuli were defined as meaningful to the subjects because they could name and comprehend these symbols under blind test conditions (Savage-Rumbaugh, 1986). Two classes of stimuli were visually familiar to the subjects but had never been paired with any exemplars in the environment or their semantic content was not well defined. A final class of stimuli was comprised of symbols that were unknown to the subjects. Hopkins, Morris and Savage-Rumbaugh (1991) found a RVF advantage in priming for meaningful symbols but no visual field advantages for the remaining stimuli. This suggests that the left hemisphere showed greater activation to meaningful than to familiar and unknown stimuli.

In a follow-up study, Hopkins and Morris (1994) tested for semantic priming within and between hemispheres using a simple reaction time test. In this study, specific semantic categories (either foods or tools) were cued by altering the fixation cue within a trial. The same paradigm that was described above was used in this study but, rather than use a neutral fixation stimulus, either the symbol for 'food' or the symbol for 'tool' was presented as the fixation stimulus. Upon pressing down the key and initiating the trial, a WS was presented and was either a lexigram representing a specific food, a lexigram representing a specific tool, or an unknown lexigram. Thus, on some trials, the WS was compatible with the cued semantic category while on other trials they were incompatible. Hopkins and Morris (1994) found RVF asymmetries for both semantic categories but did not find semantic priming. However, the cueing paradigm had a significant effect on priming with much greater priming occurring within this paradigm compared to a non-cued paradigm (for a review, see Hopkins and Morris, 1994).

One limitation of these findings was the lack of any recognition data. In other words, no direct measures of the subjects' perception of the stimuli were assessed in the priming studies but rather were inferred from reaction time data. As a follow-up to the priming studies, a recognition task was performed in which visual stimuli were presented within a divided visual half-field format (for a review, see Hopkins, Washburn, and Rumbaugh, 1990). In this study, Sherman, Austin and Lana served as subjects and the same stimuli used in the Hopkins et al. (1992) study were used. Specifically, there were two classes of meaningful symbols (foods and tools) with three exemplars from each class serving as stimuli. A third class of stimuli was symbols that were unknown to the subjects (i.e. never been seen before). A fourth class of stimuli was simple geometric patterns. The subjects were tested for laterality using an automated computerized test system that allowed for lateralized presentation of the stimuli. Specifically, after making a fixation response, a stimulus was presented for 100 ms to either the LVF or RVF. After stimulus presentation, two comparison stimuli appeared equidistant above and below the fixation point. One stimulus was identical to the sample and the other was a foil from the overall stimulus set. Subjects were required to move their cursor to the comparison that matched the sample as quickly as possible. Both accuracy and reaction time were measured. Overall, accuracy was high in this experiment and exceeded 80% for all classes of stimuli. Depicted in Figure 12.2 are the mean reaction times as a function of visual half-field and stimulus types. A RVF advantage was found for the food and tool stimuli, while a LVF advantage was found for the unknown and pattern stimuli.

12.2.2. Summary and Interpretations

Based on the findings reviewed, there are several tentative conclusions that can be put forward. One consistent finding from the review of the auditory processing or discrimination studies in monkeys is a left hemisphere advantage in processing these stimuli, except for the studies by Pohl (1983, 1984). These results are largely consistent with data from human subjects and suggest some homology in structure and function for the left auditory cortex in primates. These data are also consistent with data from other mammals and

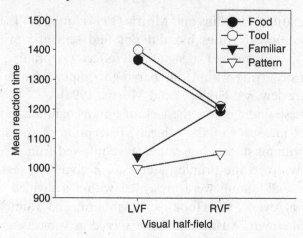

Figure 12.2.   Mean response time as a function of stimulus type in a visual half-field paradigm.

non-mammals (for a review, see Bradshaw and Rogers, 1993). More difficult to interpret from the extant literature is whether the reported left hemisphere asymmetries are selective to species-specific vocalizations or whether they represent a general capacity of the left hemisphere for processing all kinds of auditory stimuli. We believe there are three fundamental problems with the existing data that warrant a reconsideration of the general view that left hemisphere asymmetry in auditory processing is in response to species-specific calls or is dissociative from a general auditory laterality system.

First, the view that left hemisphere asymmetries in auditory processing are in response to species-specific calls comes from the original findings by Petersen et al. (1978). Upon close examination of the data presented in that paper, it is not so clear that the results from Petersen et al. (1978) are robust. For example, if one ignores whether the subjects had a *significant* ear bias or not, four of the five non-Japanese macaques *did show* a right ear advantage. These findings were not discussed in the paper but it certainly places some constraints on the interpretation of the findings and suggests that the reported differences are quantitative rather than qualitative. In fact, the entire findings of Petersen et al. (1978) rest on the simple observation that all five Japanese monkeys had a significant right ear advantage while only one of the five non-Japanese monkeys showed a significant right ear advantage. These are not large enough sample sizes to conduct appropriate non-parametric statistics and we believe this raises some questions about the interpretation of these findings.

Second, the findings by Heffner and Heffner (1984) clearly support the results of Petersen et al. (1978) and suggest a left hemisphere asymmetry in response to species-specific calls. However, Heffner and Heffner (1984) never tested any macaques other than *Macaca fuscata* and never presented any of their training data. This presents two problems. First, even though non-Japanese macaque monkeys may not manifest a left hemisphere asymmetry in the acquisition of the 'coo' vocalization discrimination, it could be expressed in the context of a lesion study. Thus, monkeys with a left hemisphere lesion may be more disrupted in post-lesion performance than right hemisphere monkeys even though no differences in performance may have been evident prior to the lesion. Second, with specific reference to Heffner and Heffner (1984), it is not clear whether the lesion induced the asymmetry effects or whether a behavioural asymmetry was present that was removed after the lesion was performed.

Third, the findings by Hauser and Andersson (1994) and Hauser, Agnetta and Perez (1998) are intriguing from the standpoint of the species specificity in processing of vocalizations and lateral bias. One strong appeal of this approach is that it is a relatively simple procedure and many subjects can be tested, an obvious limitation of more traditional experimental and lesion approaches. Notwithstanding, in our opinion, there are some conceptual and methodological issues that need to be resolved with this procedure before one can accept the findings without impudence. Specifically, the studies by Hauser and colleagues have not attempted to determine individual variation in responding to more than one stimulus. In short, individual monkeys have not been tested on more than one stimulus exemplar within a study. This is relevant because no baseline measures of turning biases were recorded and therefore it is not clear whether the biases expressed are simply based on an endogenous turning bias or specific to rhesus monkey calls. Certainly, it could be argued that because the right-turning biases were specific to rhesus monkey calls and not bird calls, the asymmetry is in response to specific classes of stimuli; however, the bottom line is that a better approach would be to test individual monkeys on all stimulus types. This would allow for a double dissociation of the laterality effects and would rule out variation due to individual turning biases. Moreover, it was not clear from these studies whether all subjects assigned to a group were tested on the same day or setting, or whether different auditory cues were presented within a given day to different subjects. Finally, it is not at all clear what neural system is mediating the turning biases or whether the turning is even governed by the contralateral hemisphere (but see Chapter 10 by Andrew and Watkins). For example, the neck muscles are controlled by the ventromedial pathway, which projects ipsilaterally in the brain. Thus, turning the head to the right would reflect ipsilateral control of the movement, in other words, the right hemisphere. This would lead to the opposite interpretation of the effects reported by Hauser and colleagues.

Regarding the remaining studies, it is obvious that the studies focusing on hand use and signing behaviour and the processing of symbolic information are plagued by small sample sizes and limited procedures (e.g. no lesion or invasive studies). It is also not clear, particularly with reference to hand use and signing, whether the signs were trained through moulding procedures that would have imposed a bias on the subjects by the experimenter in terms of hand use. The CLEM studies have yielded the most discrepant findings and this is likely to be due to some serious methodological and theoretical problems with this paradigm (Ehrlichman and Weinberger, 1978).

The evidence of a right-hand bias in gestural communication by gorillas, bonobos and chimpanzees is perhaps the best evidence for an asymmetry in the production of an intentional communicative signal by non-human primates. However, there are some serious constraints on these findings, not unlike those that apply to the results reported by Hauser and colleagues (see above and Chapter 13 by Weiss et al.). Specifically, it is not clear what cognitive or neurological mechanism is mediating this response. For example, nearly all of the reported gestures were observed in the context of the subjects alternating their gaze between the referent and the human experimenter which operationally defined them as being intentional (see Leavens and Hopkins, 1998). This raises the question of whether the intentionality and referential nature of the gesture induces the right-hand bias, but it is difficult to manipulate this variable because the chimpanzees do not gesture to the exemplars without a human present (the so-called audience effect; see Leavens and Hopkins, 1999). In other words, there are no gestures that are non-intentional, making it difficult to interpret these findings.

12.3. Brain Asymmetries in the Production and Perception of Facial Expressions in Non-human Primates

Not until recently have there been studies exploring the issue of brain laterality in the production and perception of emotions and facial expressions in primates others than humans (for reviews, see Hiscock and Kinsbourne, 1995; Hauser, 1996). Such interest is relatively new compared to the study of other behavioural asymmetries in non-human primate species, despite the fact that lateralization in the production and perception of facial expressions in humans has been a topic of discussion since the 1970s (e.g. Campbell, 1978; Sackeim, Gur and Saucy, 1978; Borod and Koff, 1984; Rinn, 1984).

Studies of emotions in both human and non-human primate species have most commonly taken certain overt motor patterns, such as facial expressions, as an index of emotional behaviour. This is not surprising, as the relationship between facial expressions and emotion has been reported for centuries, particularly in the writings of Charles Darwin (Fridlund, 1994). Thus, the study of hemispheric specialization in emotions has become the study of lateralization in the production and perception of facial expressions. However, it is important to emphasize that, at least in humans, the perception of facial stimuli is not equivalent to the perception of emotions. Depending on the nature of the task and stimuli, alternative modes of processing are feasible.

In reviewing the studies in non-human primates, it is important to bear in mind that two distinct theoretical models of emotion and laterality have been proposed in the human literature. One model proposes that the right hemisphere (RHem) is dominant or specialized for the processing of all emotions (Borod, 1992; Best, Womer and Queen, 1994; Lane et al., 1995). In contrast, the valence theory proposes that there is differential hemispheric involvement as a function of emotional valence, or pleasantness–unpleasantness (for an excellent review, see Borod, Haywood and Koff, 1997). The emotional valence distinction has been championed by Davidson (1992, 1995), who argues based on electroencephalograms (EEGs), that in human adults, the RH is more active during negative–withdrawal emotion, whereas the left hemisphere (LHem) is more dominant for positive–approach emotion. It is not clear how or whether these models apply to data from non-human primates but they offer a framework from which to interpret these data (see also Chapter 3 by Andrew and Rogers).

12.3.1. Single-cell Recording Studies

After decades of research with human subjects, attention has shifted to non-human primates but any conclusions are as yet tentative (Table 12.4). One of the most relevant lines of research in the study of neurological bases of perception of facial stimuli in non-human primates suggests the presence of 'face cells' in the inferior temporal cortex that are activated in response to such a specific stimulus as a face. The existence of such cells that are sensitive to faces seems logical given the importance of faces in communicative interactions in terrestrial primates and, for almost 20 years, single-unit recording studies of facial discrimination in rhesus monkeys have been carried out (for a review, see Tovee and Cohen-Tovee, 1993).

Table 12.4. *Studies on perception of faces and emotions*

Author(s)	Species	Study	Results
Perret, Rolls and Caan (1982)	Rhesus monkey (*Macaca mulatta*)	Single-cell recording study	RHem advantage
Perret et al. (1988)	Rhesus monkey (*Macaca mulatta*)	Single-cell recording study	LHem advantage
Hamilton (1977a, 1977b)	Rhesus monkey (*Macaca mulatta*)	Discrimination of facial cues	No bias
Overman and Doty (1982)	Pig-tail monkey (*Macaca nemestrina*)	Perception of chimeric faces	No bias
Hamilton and Vermeire (1983)	Rhesus monkey (*Macaca mulatta*) (split-brain subjects)	Perception of facial expressions	RHem advantage
Hamilton and Vermeire (1988)	Rhesus monkey (*Macaca mulatta*) (split-brain subjects)	Perception of facial expressions	RHem advantage
Vermeire and Hamilton (1988)	Rhesus monkey (*Macaca mulatta*) (split-brain subjects)	Perception of normal and inverted faces	RHem advantage for normal vs. inverted faces
Vermeire and Hamilton (1998)	Rhesus monkey (*Macaca mulatta*) (split-brain subjects)	Perception of normal and inverted faces	RHem advantage for normal faces and RHem disadvantage for inverted faces
Deruelle and Fagot (1997), Fagot and Deruelle (1997)	Baboon (*Papio papio*)	Global vs. local processing of non-facial stimuli	RHem advantage for global processing
Hopkins (1997)	Chimpanzee (*Pan troglodytes*)	Global vs. local processing of non-facial stimuli	LHem advantage for local processing
Morris and Hopkins (1993)	Chimpanzee (*Pan troglodytes*)	Perception of human chimeric facial expressions	RHem advantage
Casperd and Dunbar (1996)	Gelada baboon (*Theropithecus gelada*)	Eye preference during agonistic encounters	Left eye preference

Table 12.4. (*cont.*)

Author(s)	Species	Study	Results
Hook-Costigan and Rogers (1998a)	Marmoset (*Callithrix jacchus*)	Eye preference during the presentation of emotion-inducing stimuli	Right eye preference for food, a watch, mirror, a model beetle and the experimenter's hand
Parr and Hopkins (2001)	Chimpanzee (*Pan troglodytes*)	Tympanic membrane temperature change while viewing emotional video scenes	Increased tympanic temperature was found for the RHem negative emotional scenes
Rogers, Ward and Stafford (1994)	Small-eared bush-baby (*Otolemur garnetti*)	Eye preference during the presentation of emotion-inducing stimuli	Left eye preference for the tester and food. Right eye preference in females for their babies.

LHem, left hemisphere; RHem, right hemisphere.

Particularly relevant to this chapter is the finding of more temporal neu-rones selectively sensitive to faces in the right than in the left hemisphere (Perret, Rolls and Caan, 1982). These results, however, have not been repli-cated and a slightly higher number of face-sensitive cells were found in the left hemisphere in a later study, although any asymmetries in processing faces were eliminated when the image was rotated to an inverted position (Perret et al., 1988). Interestingly, another study found cells that were highly responsive to particular facial expressions independent from identity in the superior temporal gyrus and other cells that responded to identity independently from facial expression in the inferior temporal gyrus (Hasselmo, Rolls and Baylis, 1989).

12.3.2. Behavioural Studies of Individual Discrimination and Perception of Facial Expressions

With respect to behavioural studies, Hamilton (1977a, 1977b) was the first to investigate asymmetries extensively in the discrimination in facial expressions by monkeys and the results were equivocal. In a later study, Overman and Doty (1982) similarly reported no evidence of asymmetries in processing facial stimuli. Overman and Doty (1982) carried out a study in which chimeric photographs of the left and right halves of the faces of both humans and rhesus monkeys were created, and the subjects were required to select the one that most resembled the face in its normal appearance. Overman and Doty tested 20 human subjects and six macaques and found that the human subjects, but not the monkeys, identified right–right chimeras as being more similar to the originals than chimerics made of left half composites. Moreover, the right–right asymmetry was found only for human faces. The macaques showed no bias for either class of stimuli. This result is consistent with the hypothesis of humans having a tendency to pay more attention to their left side (i.e. right half of the observed subject) and, therefore, with the hypothesis of a right hemisphere dominance in processing facial information.

More recently, a different picture has emerged as more data on discrimination of facial expressions have been collected. Hamilton and Vermeire (1983, 1988) found a right hemisphere superiority for discriminating conspecific individuals and their facial expressions in split-brain rhesus monkeys. In 19 out of 27 split-brain monkeys (70%), the right hemisphere was more adept at discriminating two photographs of two different individuals with the same facial expression than two photographs of the same individual with two different facial expressions. In a retest carried out 6 months later, the right hemisphere continued to have advantage over the left in the previously learned facial discriminations. Hamilton and Vermeire (1988) replicated these findings using new photographs of the same subjects, demonstrating that these lateralized processes implicated the use of facial attributes rather than incidental details in the photographs. The same authors also found a superiority of the left hemisphere for discrimination of line orientation and absence of asymmetry in the discrimination of several geometric patterns (Hamilton and Vermeire, 1988). There was not, however, a significant correlation between left hemisphere specialization in line orientation discrimination and right hemisphere specialization in face processing, which implies that they are independent functions.

Vermeire and Hamilton (1988) reported more similarities between humans and monkeys when they found the expected right hemisphere dominance pattern in 16 split-brain monkeys in the processing of faces in a normal position, but not when faces were presented in the inverted position. Similarly, Vermeire and Hamilton (1998) reported a right hemisphere advantage in split-brain macaques for both learning and discriminating individual conspecific faces. When these faces were inverted, subjects showed impaired discrimination performance when the faces were presented to the right hemisphere. No deficits were reported for discriminations involving the same inverted stimuli presented to the left hemisphere. Taken together, these results suggest that face recognition is primarily a function of the right hemisphere and that the recognition of inverted faces is impaired when stimuli are laterally presented to this hemisphere.

The decrement in performance found for inverted faces is referred to as the *inversion effect* and has been widely studied in humans and is often used as a marker supporting evidence of a specialized face-processing mechanism, possibly located in the right hemisphere (Farah, Tanaka and Drain, 1995; Yin, 1969). Whether non-human primates exhibit the *inversion effect* is a matter of some debate. For example, the inversion effect has recently been demonstrated in chimpanzees for human and chimpanzee faces, but not cebus monkey faces and cars (Parr, Dove and Hopkins, 1998). Parr, Dove and Hopkins (1998) interpreted these findings to suggest that the inversion effect is found for stimuli for which subjects have developed an expertise (but see Tomonaga, Itakura and Matsuzawa, 1993). This supports the view of the inversion effect as evidence of holistic strategies developed by the visual processing system when dealing with familiar stimuli, also typically associated with right hemisphere functions (Carey and Diamond, 1977; Benton, 1980; Diamond and Carey, 1986). In contrast to chimpanzees, the data are less clear in monkeys with some studies reporting evidence of the inversion effect (Dittrich, 1990; Swartz, 1983; Tomonaga, 1994) and others not (Rosenfeld and Van Hoesen, 1979; Bruce, 1982; Parr, Winslow and Hopkins, 1999).

Although not directly relevant to issues of face perception, the issue of familiarity and holistic processing is relevant to the topic of laterality in human and non-human primates. It has been suggested that holistic processing is primarily done by the right hemisphere and the left hemisphere processes details or features (Bradshaw and Nettleton, 1981). This hypothesis has been tested in baboons and chimpanzees using comparable testing paradigms (Deruelle and Fagot, 1997; Fagot & Deruelle, 1997; Hopkins, 1997). In both of these studies, a visual half-field paradigm was employed and the subjects were required to discriminate between a sample and comparison stimuli on the basis of either their global configuration or local elements comprising the compound stimuli. Chimpanzees showed a RVF advantage for local processing and a borderline significant LVF advantage for global processing (Hopkins, 1997). Baboons showed a similar pattern of lateralization with a LVF advantage for global processing and a borderline significant

RVF advantage for local processing (Deruelle and Fagot, 1997; Fagot and Deruelle, 1997).

Only one study has aimed to explore hemispheric asymmetries in the perception of facial stimuli in apes. Morris and Hopkins (1993) found a left visual field advantage in three chimpanzees discriminating human chimeric stimuli. Using Levy's free visual discrimination paradigm (Levy et al., 1983), these authors taught the subjects to select the photograph with the happy face compared to the same individual posing with a neutral facial expression. Stimuli were made of a neutral half and a smiling half: each half could be the left or the right half of a whole expression and be placed in its original side or in the opposite one (using the mirror-reversed duplicate of the original in the last case). In 62% of the trials, the chimpanzees selected the stimuli with the smiling half on the left side, and hence in their left visual field.

12.3.2.1. Eye Dominance and Emotional Stimuli

Rather than study perceptual biases in discrimination learning, a number of recent investigators have examined eye dominance in relation to the presentation of different types of emotional inducing stimuli. Although a number of early studies assessed eye dominance in monkeys (Cole, 1957; Kruper, Boyle and Patton, 1966), it is only recently that eye dominance patterns have been linked to particular kinds of stimuli. Casperd and Dunbar (1996) recorded the asymmetries in body orientation of wild male gelada baboons (*Theropithecus gelada*) during agonistic encounters. Left-sided orientation was considered to reflect a left eye preference for viewing the other baboon, while right-sided orientation was considered to reflect a right eye preference. Casperd and Dunbar (1996) reported that typically both the aggressor and recipient baboons exhibited left-sided orientation asymmetries. Casperd and Dunbar (1996) also reported that these effects were evident for all types of levels of aggression.

Hook-Costigan and Rogers (1998a) examined eye preference in a sample of 21 marmosets. Stimuli were placed outside the subjects' enclosure and the experimenters recorded the eye used to view specific kinds of stimuli, including familiar food, a watch, mirror, model beetle, model snakes and the experimenters' hand. A right eye preference was found for all of the stimuli with the exception of the snake models. For this class of stimuli, there was a significant shift toward greater preferential use of the left eye but the shift did not reach statistical significance.

Rogers, Ward and Stafford (1994) examined eye preferences in a sample of six small-eared bush babies (*Otolemur garnetti*). In this study, the subjects were tested for eye preference when looking through a small hole for various

Table 12.5. *Studies on production of emotions*

Author(s)	Species	Study	Results
Ifune, Vermeire and Hamilton (1984)	Rhesus monkey (*Macaca mulatta*) (split-brain subjects)	Emotional responses to emotional inducing stimuli	RHem advantage
Hauser (1993)	Rhesus monkey (*Macaca mulatta*)	Timing asymmetries in facial expressions	RHem advantage
Hook-Costigan and Rogers (1998b)	Marmoset (*Callithrix jacchus*)	Intensity asymmetries in facial expressions	RHem advantage for fear and LHem advantage for contact call
Davidson, Kalin and Shelton (1992)	Rhesus monkey (*Macaca mulatta*)	Electrophysiological activation	LHem activation in a state of 'less anxiety'
Kalin et al. (1998)	Rhesus monkey (*Macaca mulatta*)	Electrophysiological activation	RHem activation for stressful circumstances
Boyce et al. (1996)	Rhesus monkey (*Macaca mulatta*)	Physiological	LHem increase in TMT for stressful individuals
Hopkins et al. (2000)	Chimpanzee (*Pan troglodytes*)	Physiological	LHem increase in TMT and right hemisphere decrease in TMT

LHem, left hemisphere; RHem, right hemisphere; TMT, tympanic membrane temperature.

types of stimuli. For viewing the tester and food, a significant left eye preference was found in the original six subjects. In a subsequent test, the experimenters measured eye preference for three females looking at their babies and they found a shift away from left eye dominance toward greater use of the right eye. Finally, in one monkey, a right eye dominance was found for looking at a toy monkey and an artificial snake.

12.3.3. *Behavioural Studies of Asymmetries in the Production of Facial Expressions*

Much less is known about the relative contribution of the right and left hemisphere in the production of facial expressions in non-human primates (Table 12.5). Ifune, Vermeire and Hamilton (1984) presented to each visual field of split-brain rhesus monkeys video sequences of human and non-human primates along with segments that contained other animals and scenes in an attempt to elicit emotional responses in these subjects. A higher number of submissive and aggressive facial expressions were elicited during right compared to left hemisphere stimulation.

Hauser (1993) recently reported that, during the production of facial expressions by neurologically intact rhesus monkeys, the left side of the face began to move first and was more expressive (as reflected in the number of skin folds and height of the corner of the mouth) than the right side. This facial asymmetry was tested for four different facial expressions, including the fear grimace, copulation grimace, open mouth threat and ear flap threat, which took place spontaneously while subjects were engaged in social inter-actions. The sample size varied from 4 to 17 subjects depending on the expression. In addition to these data, Hauser (1993) created chimerics of the fear grimace expression for three monkeys and asked 43 human subjects to rate which chimeric expression looked more like the original expression. In 41 of the subjects, they reported that the left–left chimeric was more expressive and looked more like the original than the right–right chimeric.

A study not unlike that of Hauser (1993) has been reported by Hook-Costigan and Rogers (1998b) in marmosets but with slightly different results. Hook-Costigan and Rogers (1998b) assessed asymmetries in the production of facial expressions in a sample of 11 marmosets. The marmosets were videotaped while producing three facial expressions. Two expressions had an accompanying vocalization and were referred to as the tsik (characterized as fearful) and the twitter (defined as a social contact call). The third expression was simply referred to as the silent fear expression. For each call, the experimenters recorded areas left and right of midline of the mouth to quantify laterality in the intensity of the expression. They also recorded the distance form midline to the side of the mouth as an indicator of asymmetry. For the area measure, a left side asymmetry was found for the fear and tsik expression, while a right side asymmetry was found for the twitter expression. For the distance to midline measure, a left side bias was found for the fear and tsik expression, but no effect was found for the twitter expression.

12.3.4. Asymmetries in Physiology and Emotion

Recently, some investigators have begun to attempt to link specific physiological responses to certain emotional events or behaviours. Davidson, Kalin and Shelton (1992) recorded baseline scalp EEGs from the left and right prefrontal regions of the brain in a sample of nine rhesus monkeys when they were very young. Davidson, Kalin and Shelton (1992) then used the asymmetries in EEG scalp recordings to predict freezing behaviour in stressful settings as well as the monkeys' responses to diazepam, a drug used to reduce anxiety. They found that the monkeys showed a shift towards the left hemisphere in EEG power in response to the diazepam which they inferred as a reflection of less anxiety. Davidson et al. also found that the shift in EEG asymmetry was positively correlated with the measure of freezing time. More recently, Kalin et al. (1998) examined the relation between asymmetries in EEG activity, cortisol responses to stresses and fear in rhesus monkeys. They found that baseline EGG activity is stable in monkeys and that monkeys with higher right EEG activity exhibited significantly higher cortisol responses and a greater amount of fear in stressful circumstances.

Rather than assess brain activity, Boyce et al. (1996) examined the association between behavioural temperament and tympanic membrane temperature (TMT) in a sample of 19 2-year-old rhesus monkeys. The tympanic membrane shares blood vessels with the hypothalamus and it was argued that, therefore, changes in TMT would reflect asymmetric activation of the hypothalamic–pituitary–adrenal axis induced by stress. Boyce et al. (1996) inserted a thermometer that measured TMT in each ear and recorded the magnitude of difference in TMT between the left and right ears. These data were then correlated with behavioural and physiological measures of stress in the same monkeys. They found that monkeys with greater temperature asymmetries (as reflected in larger left hemisphere temperatures) showed the most stress-related (i.e. greater agitation in novel environments and lower levels of corticotrophin and cortisol) when separated from their social groups as infants.

Similar findings were reported in humans for a group of 8-year-old children. Parents of children with greater TMT asymmetries, again with higher temperatures on the left side, reported them as less resilient and less capable of adaptive responses in novel circumstances. Children in whom these asymmetries were largest also showed more behavioural problems, were hyperaggressive, had a greater incidence of depression and social withdrawal, and exhibited schizoid behaviours (Boyce et al., 1996).

Finally, the asymmetries in TMT are not restricted to rhesus monkeys and humans, but have also been reported in chimpanzees in the context of cognitive functions (Hopkins and Fowler, 1998). Recently, we have examined the association between TMT in response to different emotional scenes (Parr and Hopkins, 2001) as well as TMT and stress as induced by cognitive challenge (Hopkins et al., submitted; Leavens et al., 2001). In terms of TMT in response to different emotion-inducing scenes,

Parr and Hopkins (2001) played video-clips ranging from 30 s to 1 min in duration of chimpanzees engaged in either positive social behaviours (such as play, tickle or wrestling), negative social behaviours (severe aggression) or neutral scenes (the sky, grass, cement, etc.). TMT was recorded in each ear at baseline (pretest), 5 min and 10 min. Overall, the right TMT increased in response to the negative social scenes. No differences between hemispheres were found for TMT in response to either of the remaining stimulus sets.

With respect to TMT and stress, the interest has been in the association between mild stress, scratching and TMT. Our interest in scratching as an indicator of stress comes from previous studies in monkeys that have reported increased self-directed behaviours, notably scratching, in the context of socially stressful circumstances (Maestripieri, 1993). Additional studies have reported that anxiolytic and anxiogenic drugs selectively increase and decrease scratching in monkeys (Schino et al., 1996). In one study (Leavens et al., 2001), scratching was examined in relation to increasing cognitive demands in a sample of eight chimpanzees. Cognitive challenge was manipulated by altering the difficulty of discrimination problems presented to the chimpanzee subjects. Of specific interest were rates of scratching as a function of increased stress. They found that scratching increased in relation to cognitive challenge. In a follow-up to this study, Hopkins et al. (submitted) examined the association between increasingly cognitive challenge, TMT and scratching. Task difficulty was manipulated in the same way as in previous studies. Seven chimpanzees were tested on an easy and hard version of a matching to sample task for 15 min. Within a test session, TMT were assessed at 5-min intervals. In addition, the frequency and hand use in scratching was noted in the same 5-min intervals. Finally, the side of the body to which the scratching was directed was noted in 5-min intervals. Depicted in Figure 12.3 are the mean changes from baseline in TMT for the hard and easy tasks within each ear.

No differences in TMT were found for the easy task, but for the hard task, the TMT for the right ear was significantly lower than the left ear, suggesting greater cooling of the right cerebral cortex. In addition, we also found that subjects tended to scratch with their right hand and hand use of scratching was negatively correlated with the side of the body to which they directed their scratch (see Table 12.6). Finally, significant positive correlations were found between TMT asymmetry and frequency of scratching ($r = 0.70$, df $= 5$, $p < 0.04$, one-tailed test) and scores for hand use in scratching with the total number of scratches ($r = 0.72$, df $= 5$, $p < 0.04$, one-tailed test).

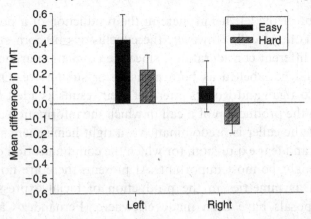

Figure 12.3. Mean change in tympanic membrane temperature (TMT) from baseline for each ear and test condition (easy and hard).

Table 12.6. *Frequency of hand use and body location of SDBs*

Subject	Hand use				Body location			
	Rubs		Scratch		Rubs		Scratch	
	Left	Right	Left	Right	Left	Right	Left	Right
Jarred	0	13	0	8	10	3	7	1
Scott	4	14	20	16	15	3	16	20
Lamar	2	13	0	14	4	11	8	6
Katrina	0	9	4	14	4	5	9	9
Carl	2	10	3	24	5	7	9	18
Winston	14	5	6	2	8	11	2	6
Clint	12	1	4	0	3	10	1	3
Mean	4.85	9.29	5.29	11.14	7.00	7.14	7.43	9.00

12.3.5. Summary and Interpretations

Overall, the general findings suggest that the right hemisphere is involved in both individual discrimination and discrimination of facial expressions in monkeys and possibly chimpanzees. The data are less clear with respect to the production of facial expressions, and particularly in relation to the valence of the emotion and facial expression. The data from Hauser (1993) and Hook-Costigan and Rogers (1998b) are not in complete agreement. Hauser showed that the right hemisphere was more involved than the left in the production of the only positive emotion analysed in his study, the copulation grimace. On the other hand, Hook-Costigan and Rogers showed a greater involvement of the left hemisphere in the production of a positive emotion, the social contact call. However, the conclusions of both studies might be subject to different considerations, since the copulation grimace of rhesus monkeys could be labelled as both pleasant or submissive. On the other hand, Hook-Costigan and Rogers interpret their results as a left hemisphere superiority in the production of a call in which the information about the social identity of the caller is predominant vs. a right hemisphere superiority in the fear call and fear expression, for which the communication of the emotional context might be most important. At present, there are no data from great apes on asymmetries in the production of facial expressions, although some proposals have been made elsewhere (Fernández-Carriba and Loeches, 1997). Findings from eye dominance studies are somewhat consistent with the discrimination studies with left eye dominance more prevalent for emotion-evoking stimuli, with the exception of the findings from bush babies (Rogers et al., 1994). Finally, the few studies that have examined physiological measures of asymmetry in relation to certain behaviours implicate the right hemisphere in the processing of stimuli that induce some degree of negative affect.

Despite the general agreement in findings between studies, there are some general issues that need to be resolved in future research. Specifically, with regard to asymmetries in the production of facial expressions, the findings by Hauser (1993) are difficult to interpret because no individual data were presented in the study. Thus, it is not clear how many subjects are contributing to the total number of observations for each facial expression. In contrast, Hook-Costigan and Rogers (1998b) did provide individual data and had repeated observations on each subject, making for a more complete interpretation of their findings. This may explain why the results were not entirely consistent between these two studies, although they did differ with respect to methodology as well. The complete lack of data from great apes is also unfortunate and warrants investigation.

The eye dominance studies rest on the premise that the emotion-inducing stimuli induce a bias in the visual fields for viewing. Unfortunately, at present, there is no strong and independent conformation that the stimuli are emotionally inducing other than the *post-hoc* explanation of the eye dominance findings. This is an easily adaptable procedure and has good promise but, conceptually, it could be improved by employing stimuli that are known to evoke specific emotional responses. A good example comes from the study by Rogers, Ward and Stafford (1994) in which they separated mothers from infants, which is a known stressor to primates and would presumably invoke a strong and pronounced viewing bias.

Finally, the EEG and TMT studies in monkeys and apes show good promise for future studies. However, it should be emphasized that, in the studies by Davidson and colleagues as well as by Boyce and colleagues, the physiological measures are being assessed at baseline level and used to predict responses under stressful conditions. Thus, the variables are considered to reflect some trait of the individual monkey rather than a state induced by the emotional context. It would be ideal to have 'on-line' data or, in other words, obtain physiological measures while the animals are experiencing different emotions or exposed to different emotional context. Obviously, this is pragmatically more difficult but seems to be a necessary step in the evolution of studies in this field.

12.4. Overall Discussion and Conclusion

Overall, there has been good progress in recent years in the study of asymmetries involved in communicative behaviours. On the whole, there still remain relatively few studies in great apes and prosimians, and the number of subjects in most of these studies are comparatively small. There is a fairly large discrepancy in studies focusing on laterality in the perception contrasted with the production of species-specific calls. This stands in sharp contrast to studies in other animals, such as birds and frogs in which the overwhelming emphasis has been on asymmetries in the production of species-specific calls. Thus, we would suggest that greater efforts be made to rectify this discrepancy.

We believe that there is also a need for some continuity in procedures across different species in order for more precise interpretations of the findings.

There remain a number of unresolved questions or issues that need to be addressed in future research. Most obvious is the apparent contradictory findings between asymmetries in the perception of vocalizations and the production of facial expressions, some of which are accompanied by vocalizations. Specifically, Hauser (1993) has reported that the right hemisphere is dominant in the production of facial expressions, which is the opposite hemisphere that is apparently dominant in the perception of vocalizations (Hauser and Andersson, 1994). If we accept these findings as valid, then it would suggest that the perception and production of vocalizations are in opposite hemispheres, which intuitively makes no evolutionary sense and is inconsistent with the lateralized organization of speech perception and production in humans. This would raise some questions about the homology of functions between monkeys and humans. This entire scenario rests on the assumption that the calls produced by rhesus monkeys (and perhaps other monkeys) have semantic functions rather than the function of communicating affect. If we accept the premise that the right hemisphere is involved in the production of vocalizations, then this raises some questions about the semanticity of these calls (see Gouzoules, Gouzoules, and Marler, 1984).

A further issue that needs to be resolved in both human and non-human primate studies is the apparent contradiction in the asymmetries in the perception and production of facial expressions. Specifically, if the right hemisphere is dominant for both perception and production, the side of the face that is the most expressive (left) falls in the visual field that is paid the least amount of attention (left visual field, i.e. right side of the face in front of the observer). The exception is the study by Hauser (1993) where human raters exhibited a right visual field bias and perceived left–left chimeras of rhesus monkeys as closer to the original faces. This finding is contradictory with the rest of the literature, which shows in both human and non-human primates a right hemisphere (left visual field) dominance in the perception of facial expressions. Indeed, if the valence theory is correct, then visual field biases should be dissociated on the basis of the type of expression. As far as we know, this issue has been tested only in a small sample of human subjects and there is no strong support for this theory. Perhaps key to this issue is the fact that faces can be processed as conveying emotion or not, depending on the nature of the task and other experimental variables. In short, it is not clear whether right hemisphere superiority in the perception of facial expressions is the result of a specific face-processing mechanism (as face-cells studies would suggest), a holistic cognitive strategy in dealing with familiar stimuli (as inversion-effect studies propose), or it is actually the manifestation of a right hemisphere specialization for emotions. These various explanations warrant further empirical investigation.

Finally, there is clearly a need to develop imaging technologies for the purposes of linking behavioural with physiological or neuroanatomical asymmetries. These studies will be extremely important for assessing whether the same or different neurological systems are involved in various types of perceptual and cognitive lateralized processes between species. Recent studies clearly indicate that structural asymmetries of the cerebral cortex can be documented in great apes and monkeys, but that certain morphological asymmetries may not be evident in all species, at least using the same anatomical landmarks (Zilles et al., 1996; Hopkins et al., 1998; Hopkins and Marino, 2000). These studies are an important first step, but morphological differences will not reflect functional differences in brain areas nor will they be very specific. Thus, the use of emerging functional imaging technologies in conjunction with solid behavioural tasks should be sought in future studies.

Acknowledgements

This research was supported by NIH grants NS29574, NS-36605, and HD-38051 to WDH, AP96.20194893 to SFC and RR00165 to the Yerkes Regional Primate Research Center. The Yerkes Center is fully accredited by the American Association for Accreditation of Laboratory Animal Care. APA guidelines for the ethical treatment of animals were adhered to during all aspects of this study. Correspondence concerning this article should be addressed to Dr William D. Hopkins, Division of Psychobiology, Yerkes Regional Primate Research Center, Emory University, Atlanta, Georgia, 30322. Email: lrcbh@rmy.emory.edu or whopkins@berry.edu

References

Benton, A.L. (1980). The neuropsychology of faces. *American Psychologist*, 35, 176–186.

Best, C.T., Womer, J.S. & Queen, H.F. (1994). Hemispheric asymmetries in adults' perception of infant emotional expressions. *Journal of Experimental Psychology Human Perception and Performance*, 20, 751–765.

Borod, J.C. (1992). Interhemispheric and intrahemispheric control of emotion: A focus on unilateral brain damage. Special Section: The emotional concomitants of brain damage. *Journal of Consulting and Clinical Psychology*, 60, 339–348.

Borod, J.C., Haywood, C.S. & Koff, E. (1997). Neuropsychological aspects of facial asymmetry during emotional expression: A review of the normal adult literature. *Neuropsychology Review*, 7, 41–60.

Borod, J.C. & Koff, E. (1984). Asymmetries in affective facial expression: Behavior and anatomy. In *Psychobiology of Affective Development*, ed. N.A. Fox and R.J. Davidson, pp. 293–323. Hillsdale, NJ: Lawrence Erlbaum Associates.

Boyce, W.T., Higley, J.D., Jemerin, J.J., Champoux, M. & Suomi, S.J. (1996). Tympanic temperature asymmetry and stress behavior in rhesus macaques and children. *Archives of Pediatric and Adolescent Medicine*, 150, 518–523.

Bradshaw, J.L. & Nettleton, N.C. (1981). The nature of hemispheric specialization in man. *Behavioral and Brain Sciences*, 4, 51–63.

Bradshaw, J.L. & Rogers, L. (eds) (1993). *The Evolution of Lateral Asymmetries, Language, Tool-use, and Intellect*. San Diego, CA: Academic Press.

Bruce, C. (1982). Face recognition by monkeys: Absence of an inversion effect. *Neuropsychologia*, 20, 515–521.

Campbell, R. (1978). Asymmetries in interpreting and expressing a posed facial expression. *Cortex*, 14, 327–342.

Carey, S. & Diamond, R. (1977). From piecemeal to configurational representation of faces. *Science*, 195, 312–314.

Casperd, J.M. & Dunbar, R.I.M. (1996). Asymmetries in the visual processing of emotional cues during agonistic interactions by gelada baboons. *Behavioural Processes*, 37, 57–65.

Cole, J. (1957). Laterality in the use of the hand, foot and eye in monkeys. *Journal of Comparative Psychology*, 50, 296–299.

Corballis, M.C. (ed.) (1992). *The Lopsided Brain: Evolution of the Generative Mind*. New York: Oxford University Press.

Davidson, R.J. (1992). Emotion and affective style: Hemispheric substrates. *Psychological Science*, 3, 39–43.

Davidson, R.J. (1995). Cerebral asymmetry, emotion and affective style. In *Brain Asymmetry*, ed. R.J. Davidson & K. Hugdahl, pp. 361–387. Cambridge, MA: MIT Press.

Diamond, R. & Carey, S. (1986). Why faces are and are not special: An effect of expertise. *Journal of Experimental Psychology: General*, 115, 107–117.

Davidson, R.J., Kalin, N.H. & Shelton, S.E. (1992). Lateralized effects of diazepam on frontal brain electrical asymmetries in rhesus monkeys. *Behavioral Neuroscience*, 32, 438–451.

Deruelle, C. & Fagot, J. (1997). Hemispheric lateralization and global precedence effects in the processing of visual stimuli by humans and baboons (*Papio papio*). *Laterality*, 2, 233–246.

Dewson, J.H. (1977). Preliminary evidence of hemispheric asymmetry of auditory function in monkeys. In *Lateralization in the Nervous System*, ed. S. Harnad, R.W. Doty, L. Goldstein, J. Jaynes & G. Krauthamer, pp. 63–74. New York: Academic Press.

Dittrich, W. (1990). Representation of faces in longtailed macaques (*Macaca fasicularis*). *Ethology*, 85, 265–278.

Ehrlichman, H. & Weinberger, A. (1978). Lateral eye movements and hemispheric asymmetries: A critical review. *Psychological Bulletin*, 85, 1080–1101.

Fagot, J. & Deruelle, C. (1997). Processing of global and local visual information and hemispheric specialization in humans (*Homo sapiens*) and baboons (*Papio papio*). *Journal of Experimental Psychology: Human Perception and Performance*, 23, 429–442.

Fagot, J. & Vauclair, J. (1991). Manual laterality in nonhuman primates: A distinction between handedness and manual specialization. *Psychological Bulletin*, 109, 76–89.

Farah, M.J., Tanaka, J.W. & Drain, H.M. (1995). What causes the face-inversion effect? *Journal of Experimental Psychology: Human Perception and Performance*, 21, 628–634.

Fernández-Carriba, S. & Loeches, A. (1997). Chimpanzee's facial expressions: A proposal for a study of lateralization of emotion. Personal communication presented at the *IV International Ethological Conference*, Vienna, August, 1997.

Fridlund, A.J. (ed.) (1994). *Human Facial Expression*. New York: Academic Press.

Gaffan, D. & Harrison, S. (1991). Auditory–visual associations, hemispheric specialization and temporal–frontal interaction in the rhesus monkey. *Brain*, 114, 2133–2144.

Gouzoules, S., Gouzoules, H. & Marler, P. (1984). Rhesus monkey (*Macaca mulatta*) screams: Representational signaling in the recruitment of agonistic aid. *Animal Behaviour*, 32, 182–193.

Hamilton, C.R. (1977a). An assessment of hemispheric specialization in monkeys. *Annals of the New York Academy of Sciences*, 299, 222–232.

Hamilton, C.R. (1977b). Investigations of perceptual and mnemonic lateralization in monkeys. In *Lateralization in the Nervous System*, eds S. Harnad, R.W. Doty, J. Jaynes, L. Goldstein, & G. Krauthamer, pp. 45–62. New York: Academic Press.

Hamilton, C.R. & Vermeire, B.A. (1983). Discrimination of monkey faces by split-brain monkeys. *Behavioural Brain Research*, 9, 263–275.

Hamilton, C.R. & Vermeire, B.A. (1988). Complimentary hemispheric specialization in monkeys. *Science*, 242, 1694–1696.

Hasselmo, M.E., Rolls, E.T. & Baylis, G.C. (1989). The role of expression and identity in the face-selective responses of neurones in the temporal visual cortex of the monkey. *Behavioural Brain Research*, 32, 203–218.

Hauser, M.C. (1993). Right hemisphere dominance in the production of facial expression in monkeys. *Science*, 261, 475–477.

Hauser, M. (ed.) (1996). *The Evolution of Communication*. Cambridge, MA: MIT Press.

Hauser, M.D. & Andersson, K. (1994). Left hemisphere dominance for processing vocalizations in adult, but not infant, rhesus monkeys: Field experiments. *Proceedings of the National Academy of Sciences USA*, 91, 3946–3948.

Hauser, M.D., Agnetta, B. & Perez, C. (1998). Orienting asymmetries in rhesus monkeys: The effect of time-domain changes on acoustic perception. *Animal Behaviour*, 56, 41–47.

Heffner, H.E. & Heffner, R.S. (1984). Temporal lobe lesions and perception of species-specific vocalizations by macaques. *Science*, 226, 75–76.

Heffner, H.E. & Heffner, R.S. (1989). Cortical deafness cannot account for in the inability of Japanese macaques to discriminate species-specific vocalizations. *Brain and Language*, 36, 275–285.

Hiscock, M. & Kinsbourne, M. (1995). Phylogeny and ontogeny of cerebral lateralization. In *Brain Asymmetry*, eds R.J. Davidson & K Hugdahl, pp. 535–578. Cambridge, MA: MIT Press.

Hook-Costigan, M.A. & Rogers, L.J. (1998a). Eye preference in common marmosets (*Callitrix jacchus*): Influence of age, stimulus and hand preference. *Laterality*, 3, 109–130.

Hook-Costigan, M.A. & Rogers, L.J. (1998b). Lateralized use of the mouth in production of vocalizations by marmosets. *Neuropsychologia*, 36, 1265–1273.

Hopkins, W.D. (1994). Hand preferences for bimanual feeding in 140 captive chimpanzees (*Pan troglodytes*): Rearing and ontogenetic factors. *Developmental Psychobiology*, 27, 395–407.

Hopkins, W.D. (1995). Hand preferences for a coordinated bimanual task in 110 chimpanzees: Cross-sectional analysis. *Journal of Comparative Psychology*, 109, 291–297.

Hopkins, W.D. (1996). Chimpanzee handedness revisited: 54 years since Finch (1941). *Psychonomic Bulletin and Review*, 3, 449–457.

Hopkins, W.D. (1997). Hemispheric specialization for local and global processing of hierarchical visual stimuli in chimpanzees (*Pan troglodytes*). *Neuropsychologia*, 35, 343–348.

Hopkins, W.D. & de Waal, F.D. (1995). Behavioral laterality in captive bonobos (*Pan paniscus*): Replication and extension. *International Journal of Primatology*, 16, 261–276.

Hopkins, W.D. & Fowler, L.A. (1998). Lateralized changes in tympanic membrane temperature in relation to different cognitive tasks in chimpanzees (*Pan troglodytes*). *Behavioral Neuroscience*, 112, 83–88.

Hopkins, W.D. & Leavens, D.A. (1998). Hand use and gestural communication in chimpanzees (*Pan troglodytes*). *Journal of Comparative Psychology*, 112, 95–99.

Hopkins, W.D. & Marino, L.M. (2000). Cerebral width asymmetries in nonhuman primates as revealed by magnetic resonance imaging (MRI). *Neuropsychologia*, 38, 493–499.

Hopkins, W.D. & Morris, R.D. (1994). Priming as a technique in the study of lateralized cognitive processes in language-trained chimpanzees: Some recent findings. In *Language and Communication: Comparative Perspectives*, eds H. Roitblat, L. Herman & P. Natigall, pp. 401–413. New York: Lawrence Erlbaum and Associates.

Hopkins, W.D. & Wesley, M.J. (2001). Gestural communication in chimpanzees (*Pan troglodytes*): The effect of situational factors on gesture type and hand use. *Laterality*, in press.

Hopkins, W.D., Washburn, D.A. & Rumbaugh, D.M. (1990). Processing of form stimuli presented unilaterally in humans, chimpanzees (*Pan troglodytes*) and monkeys (*Macaca mulatta*). *Behavioral Neuroscience*, 104, 577–582.

Hopkins, W.D., Morris, R.D. & Savage-Rumbaugh, E.S. (1991). Evidence for asymmetrical hemispheric priming using known and unknown warning stimuli in two language-trained chimpanzees. *Journal of Experimental Psychology: General*, 120, 46–56.

Hopkins, W.D., Morris, R.D., Savage-Rumbaugh, E.S. & Rumbaugh, D.M. (1992). Hemispheric priming by meaningful and nonmeaningful symbols in language-trained chimpanzees: Further evidence of a left hemisphere advantage. *Behavioral Neuroscience*, 106, 575–582.

Hopkins, W.D., Marino, L., Rilling, J. & MacGregor, L. (1998). Planum temporale asymmetries in great apes as revealed by magnetic resonance imaging (MRI). *NeuroReport*, 9, 2913–2918.

Hopkins, W.D., Fowler, L.A., Aureli, F. & Leavens, D.A. Tympanic membrane asymmetry, scratching and cognitive challenge (submitted).

Ifune, C.K., Vermeire, B.A. & Hamilton, C.R. (1984). Hemispheric differences in split-brain monkeys and responding to videotape recordings. *Behavioral and Neural Biology*, 41, 231–235.

Kalin, N.H., Larson, C., Shelton, S.E. & Davidson, R.J. (1998). Asymmetric frontal brain activity, cortisol, and behavior associated with fearful temperament in rhesus monkeys. *Behavioral Neuroscience*, 112, 286–292.

Krause, M.A. & Fouts, R.S. (1997). Chimpanzee (*Pan troglodytes*) pointing: Hand shapes, accuracy, and the role of eye gaze. *Journal of Comparative Psychology*, 111, 330–336.

Kruper, D.C., Boyle, B.E. & Patton, R.A. (1966). Eye and hand preferences in rhesus monkeys (*Macaca mulatta*). *Psychonomic Science*, 5, 277–278.

Lane, R.D., Kivley, L.S., Du-Bois, M.A., Shamasundara, P. & Schwartz, G.E. (1995). Levels of emotional awareness and the degree of right hemispheric dominance in the perception of facial emotion. *Neuropsychologia*, 33, 525–538.

Leavens, D.A. & Hopkins, W.D. (1998). Intentional communication by chimpanzees (*Pan troglodytes*): A cross-sectional study of the use of referential gestures. *Developmental Psychology*, 34, 813–822.

Leavens, D.A. & Hopkins, W.D. (1999). The whole-hand point: The structure and function of pointing from a comparative perspective. *Journal of Comparative Psychology*, 113, 417–425.

Leavens, D.A., Hopkins, W.D. & Bard, K.A. (1996). Indexical and referential pointing in chimpanzees (*Pan troglodytes*). *Journal of Comparative Psychology*, 110, 346–353.

Leavens, D.A., Aureli, F., Hopkins, W.D. & Hyatt, C.W. (2001). Lateralized scratching in response to task difficulty: Effects of cognitive challenge on self-directed behaviors by chimpanzees (*Pan troglodytes*). *American Journal of Primatology*, in press.

Levy, J., Heller, W., Banich, M.T. & Burton, L. A. (1983). Asymmetry of perception in free viewing of chimeric faces. *Brain and Cognition*, 2, 404–419.

MacNeilage, P.F., Studdert-Kennedy, M.G. & Lindblom, B. (1987). Primate handedness reconsidered. *Behavioral and Brain Sciences*, 10, 247–303.

Maestripieri, D. (1993). Maternal anxiety in rhesus macaques (*Macaca mulatta*): I. Measurement of anxiety and identification of anxiety-eliciting situations. *Ethology*, 95, 19–31.

McGrew, W.C. & Marchant, L.F. (1997). On the other hand: Current issues in and meta-analysis of the behavioral laterality of hand function in nonhuman primates. *Yearbook of Physical Anthropology*, 40, 201–232.

Miles, L. (1990). Hand use in a signing ape. Personal communication presented at the *Laterality: Evolution and Mechanism Conference*, Memphis, Tennessee.

Morris, R.D. & Hopkins, W.D. (1993). Perception of human chimeric faces by chimpanzees (*Pan troglodytes*): Evidence for a right hemisphere asymmetry. *Brain and Cognition*, 21, 111–122.

Morris, R.D., Hopkins, W.D. & Bolser-Gilmore, L. (1993). Assessment of hand preference in two language-trained chimpanzees (*Pan troglodytes*): A multimethod analysis. *Journal of Clinical and Experimental Neuropsychology*, 15, 487–502.

Morris, R.D., Hopkins, W.D., Bolser, L. & Washburn, D.A. (1993). Behavioral lateralization in language-trained chimpanzees. In *Primate Laterality: Current Behavioral Evidence of Primate Asymmetries*, eds J.P. Ward & W.D. Hopkins, pp. 215–227. New York: Springer-Verlag.

Muncer, S.J. (1982). Functional asymmetry in the chimpanzee. *Perceptual and Motor Skills*, 54, 147–152.

Nottebohm, F. (1977). Asymmetries in neural control vocalization in the canary. In *Lateralization of the Nervous System*, eds S.R. Harnad, R.W. Doty, L. Goldstein, J. Jaynes & G. Krauthamer, pp. 23–44. New York: Academic Press.

O'Neil, C.R., Stratton, H.T.R., Ingersoll, R.H. & Fouts, R.S. (1978). Conjugate lateral eye movements in *Pan troglodytes*. *Neuropsychologia*, 16, 759–762.

Overman, W.H. & Doty, R.W. (1982). Hemispheric specialization displayed by man but not macaques in the analysis of faces. *Neuropsychologia*, 20, 113–128.

Parr, L.A. & Hopkins, W.D. (2001). Brain temperature asymmetries and emotional perception in chimpanzees (*Pan troglodytes*). *Physiology and Behavior*, 71, 363–371.

Parr, L.A., Dove, T. & Hopkins, W.D. (1998). Why faces may be special: Evidence of the inversion effect in chimpanzees. *Journal of Cognitive Neuroscience*, 10, 615–622.

Parr, L.A., Winslow, J.T. & Hopkins, W.D. (1999). Is the inversion effect in rhesus monkeys face-specific? *Animal Cognition*, 2, 123–129.

Perret, D.I., Rolls, E.T. & Caan, W. (1982). Visual neurones responsive to faces in the monkey temporal cortex. *Experimental Brain Research*, 47, 329–342.

Perret, D.I., Mistlin, A.J., Chitty, A.J., Smith, P.A., Potter, D.D., Broennimann, R. & Haries, M. (1988). Specialized face processing and hemispheric asymmetry in man and monkey: Evidence from single unit and reaction time studies. *Behavioural Brain Research*, 29, 245–258.

Petersen, M.R., Beecher, M.D., Zoloth, S.R., Moody, D.B. & Stebbins, W.C. (1978). Neural lateralization of species-specific vocalizations in Japanese macaques (*Macaca fuscata*). *Science*, 202, 324–327.

Petersen, M.R., Beecher, M.D., Zoloth, S.R., Green, S., Marler, P.R., Moody, D.B. & Stebbins, W.C. (1984). Neural lateralization of vocalizations by Japanese macaques: Communicative significance is more important than acoustic structure. *Behavioral Neuroscience*, 98, 779–790.

Pohl, P. (1983). Central auditory processing: ear advantages for acoustic stimuli in baboons. *Brain and Language*, 20, 44–53.

Pohl, P. (1984). Ear advantages for temporal resolution in baboons. *Brain and Cognition*, 3, 438–444.

Rinn, W.B. (1984). The neuropsychology of facial expression: A review of the neurological and psychological mechanisms for producing facial expression. *Psychological Bulletin*, 95, 52–77.

Rogers, L.J., Ward, J.P. & Stafford, D. (1994). Eye dominance in the small-eared bushbaby, *Otolemur garnetti*. *Neuropsychologia*, 32, 257–264.

Rosenfeld, S.A. & Van Hoesen, G.W. (1979). Face recognition in the rhesus monkey. *Neuropsychologia*, 17, 503–509.

Rumbaugh, D.M. (1977). *Language Learning by a Chimpanzee: The LANA Project*. New York: Academic Press.

Sackeim, H.A., Gur, R.C. & Saucy, M.C. (1978). Emotions are expressed more intensely on the left side of the face. *Science*, 202, 434–436.

Savage-Rumbaugh, E.S. (ed.) (1986). *Ape Language: From Conditioned Response to Symbol*. New York: Columbia University Press.

Schino, G., Peretta, G., Taglioni, A.M., Monaco, V. & Troisi, A. (1996). Primate displacement activities as an ethopharmacological model of anxiety. *Anxiety*, 2, 186–191.

Shafer, D.D. (1988). Handedness in gorillas. *The Gorilla Foundation*, 8, 2–5.

Shafer, D.D. (1993). Patterns of hand preference in gorillas and children. In *Primate Laterality: Current Behavioral Evidence of Primate Asymmetries*, eds J.P. Ward & W.D. Hopkins, pp. 267–283. New York: Springer-Verlag.

Shafer, D.D. (1997). Hand preference behaviors shared by two groups of captive bonobos. *Primates*, 38, 303–313.

Steiner, S.M. (1990). Handedness in chimpanzees. *Friends of Washoe*, 9, 9–19.

Swartz, K.B. (1983). Species discrimination in infant pigtail monkeys with pictorial stimuli. *Developmental Psychobiology*, 16, 219–231.

Tomonaga, M. (1994). How laboratory-raised Japanese monkeys (*Macaca fuscata*) perceive rotated photographs of monkeys: Evidence of an inversion effect in face perception. *Primates*, 35, 155–165.

Tomonaga, M., Itakura, S. & Matsuzawa, T. (1993). Superiority of conspecific faces and reduced inversion effect in face perception by a chimpanzee. *Folia Primatologica*, 61, 110–114.

Tovee, M.J. & Cohen-Tovee, E.M. (1993). The neural substrate of face processing models: A review. *Cognitive Neuropsychology*, 10, 505–528.

Vermeire, B.A. & Hamilton, C.R. (1988). Laterality in monkeys discriminating inverted faces. *Neuroscience Abstracts*, 14, 1139.

Vermeire, B.A. & Hamilton, C.R. (1998). Inversion effect for faces in split-brain monkeys. *Neuropsychologia*, 36, 1003–1014.

Ward, J.P. & Hopkins, W.D. (eds) (1993). *Primate Laterality: Current Behavioral Evidence of Primate Asymmetries*. New York: Springer-Verlag.

Warren, J.M. (1980). Handedness and laterality in humans and other animals. *Physiological Psychology*, 8, 351–359.

Yin, R.K. (1969). Looking at upside-down faces. *Journal of Experimental Psychology*, 81, 141–145.

Zilles, K., Dabringhaus, A., Geyer, S., Amunts, K., Qu, M., Schleicher, A., Gilissen, E., Schlaug, G. & Steinmetz, H. (1996). Structural asymmetries in the human forebrain and the forebrain of non-human primates and rats. *Neuroscience and Biobehavioral Reviews*, 20, 593–605.

13

Specialized Processing of Primate Facial and Vocal Expressions: Evidence for Cerebral Asymmetries

DANIEL J. WEISS, ASIF A. GHAZANFAR,
CORY T. MILLER AND MARC D. HAUSER

13.1. Introduction

In a recent review article, Corballis (1998) states that perhaps 'the major question confronting research on cerebral asymmetry is whether it will survive into the new millennium'. This volume certainly shows that the field has a strong basis, and that the contributions emerging from studies of animals are providing an increasingly precise picture of how cerebral asymmetries have evolved. Although the field has come a long way since the days when humans were considered uniquely lateralized, there are still many gaps in our knowledge. In this chapter, we attempt to fill in a portion of this gap, focusing explicitly on the non-human primates (hereafter referred to as 'primates') and the specialized mechanisms underlying the production and perception of their facial and vocal expressions.

We begin our review by discussing the logic underlying the search for neural specializations, and then briefly discuss a selective set of problems associated with the comparative method. We then discuss current evidence for specialized processing mechanisms, focusing on the perception of faces and facial expressions, the perception of vocalizations, and the production of facial and vocal expressions. We conclude the chapter with a few comments on how future studies of hemispheric specialization must integrate behavioural studies of wild and captive animals with laboratory studies of neurophysiology.

13.1.1. Why Should We Expect Neural Specializations?

Like other species, including humans, it seems reasonable to expect primates to have a suite of specialized brain structures dedicated to processing ethologically relevant behaviours. Thus, identification of species-typical behaviours, such as the production and perception of facial and vocal expressions, can guide explorations of species-specific neural mechanisms, or *specializations*. Neural specializations can manifest themselves in many ways, none of which are mutually exclusive. At the anatomical level, one may find left–right asymmetries in the sizes of surface features (sulci and gyri) or in the size and/or neurochemical composition of brains areas. Neurophysiological and experimental lesioning approaches can elucidate the functional role of the region in question, and this knowledge can guide the comparative study of behavioural and brain evolution (Camhi, 1984; Konishi, 1985; Allman, 1999; Ghazanfar and Hauser, 1999).

Using this pluralistic approach, neuroethological research has already added much to our understanding of how natural selection shapes brain design for complex sensory behaviours (for reviews, see Hauser and Konishi, 1999), such as echolocation in bats (Simmons, 1971; Suga, 1989; Kanwal, 1999), song learning in birds (Marler, 1970; Nottebohm, 1999), and mate choice in frogs (Ryan and Rand, 1999). For example, songbirds have species typical songs, which they use to attract mates and defend territories. The neural circuitry underlying this special behaviour was identified (Nottebohm, Stokes and Leonard, 1976) and demonstrated, via lesion and tracer studies, to be dedicated to the production and perception of song (for reviews, see Ball, 1999; Doupe and Solis, 1999; Nottebohm, 1999). In some species, the neural system for song has been shown to be lateralized (Arnold and Bottjer, 1985), although the degree and direction of asymmetry varies between species, especially in peripheral structures (Suthers, 1997). Neurophysiological data reveal that neurones within song-related brain structures are extremely sensitive to the birds' own songs, but not to simple stimuli or songs from other individuals from the same species (Margoliash and Fortune, 1992; Doupe and Solis, 1999).

13.1.2. The Comparative Method: Some Issues and Problems

One of the central problems in comparative biology is to determine the evolutionary mechanisms underlying similarity between species. As evolutionary biologists have pointed out, however, there are two coarse-grained categories of similarity, and each provides deep insights into phylogenetic patterns and the history of selection pressures. On the one hand are *homologies*, characters that are shared between two species because of evolution by descent from a common ancestor. On the other hand are *homoplasies*, characters that evolved independently in different taxonomic groups due to the process of convergence.

In general, studies of primates have often aimed their comparisons at humans, and this is particularly the case in the study of hemispheric specialization. When humans and primates show the same patterns or characters, it is often assumed that such similarities represent cases of homology. It is possible, however, that the similarity represents a case of homoplasy. Many cases of homology within the primates, especially at the behavioural level, have been defended on the basis of plausibility – specifically, that it is unlikely for the character to have evolved twice, once in each lineage. Although this is a reasonable argument to make for primates as a group, each case must be considered on its own. It is certainly possible that some traits, shared in common between two species, evolved after the divergence point.

Although there are historical reasons for drawing comparisons between the patterns of hemispheric specialization obtained for humans, and those obtained for primates, there are two potential problems with this kind of focus. First, when neuroscientists look to animals for comparative data, they tend to draw classificatory boundaries with respect to higher order taxa such as 'animal', 'vertebrate', or 'monkey'. Thus, several review papers on hemispheric specialization in humans present a cursory review of lateralization in 'animals', or sometimes 'monkeys and birds'. As we will document below, there are often important differences between species, even within the same genus, and such differences are informative with respect to the selective pressures on brain organization. Second, studies of hemispheric specialization should also focus on similarities and differences between primates, independently of the patterns obtained for humans. This is important because it allows us to map patterns of primate brain evolution onto existing phylogenies that have used molecular, anatomical, behavioural and ecological characters.

In the literature reviewed below, it will be apparent to the reader that our understanding of the role of cerebral asymmetries in primate communication is restricted to only a handful of species. More specifically, although we know a great deal about asymmetries in hand use, and in coarse grained anatomy for a number of primate species, our knowledge of behavioural and neurophysiological asymmetries associated with the perception and production of communicative signals is largely restricted to *Macaca mulatta*, *Macaca fuscata* and *Pan troglodytes*. As a result, our ability to draw inferences about the patterns of evolution are minimal. Our goal, therefore, is to draw attention to what we know about the few species that have been studied, and hopefully inspire others to collect the relevant data on other species.

13.1.3. Why Look for Hemispheric Asymmetries Underlying the Production and Perception of Communicative Signals?

Much of what we know about hemispheric specialization in primates comes from the extensive studies conducted on the preferential use of the right or left hand or foot during grasping, reaching and manipulating (see Chapter 3 by Andrew and Rogers, Chapter 10 by Andrew and Watkins and Chapter 9 by Damerose and Vauclair). Considerably less is known about the extent to which primates show behavioural and neural asymmetries associated with the production and perception of species-typical communicative signals. There are, however, a number of reasons to expect primates to show cerebral specializations for communicative signals. First, and as discussed in Section 13.1.1, selection tends to favour specializations for interactions that are critical to survival and reproduction. For primates, vocalizations and facial expressions play a critical role in mediating social interactions critical to survival and mating. Thus, a wide variety of primates have vocalizations that indicate the presence of predators or the location of food, vocalizations that are used during mating and dominance-related social interactions, and facial expressions that are used to convey information about the probability of escape, attack and friendly affiliation (Cheney and Seyfarth, 1990; Marler, Evans and Hauser, 1992; Preuschoft, 1995; Preuschoft and van Hooff, 1997; Hauser, 1996). Second, given the significant lateralization of human facial and vocal expressions, it seems likely that at least some primates would show comparable asymmetries, at least for those expressions that they share in common (e.g. the 'grimace' produced by humans and all primates).

We also would like to emphasize the importance of studying both perception and production. In Corballis' (1998) review, he states that the strongest evidence of cerebral asymmetries in humans comes from studies of motor production as opposed to perception, and that there are two reasons to expect selection to have designed it in this way. First, there would be significant disadvantages to an asymmetrical perceptual system since 'a deficit on one side would leave an animal vulnerable to attack from that side, or unable to capitalize on prey that emerge on that flank' (pp. 152–153). Second, asymmetries in motor production might well be advantageous, as when a task requires bimanual coordination, with each hand playing a different role. Although these are reasonable predictions, one can easily imagine advantages to a perceptual asymmetry (e.g. the barn owl's asymmetrical ruff, which leads to advantages in sound localization and numerous other examples discussed in earlier chapters), and disadvantages to a motor asymmetry (e.g. vulnerability in defence against a competitor with opposite biases). Nonetheless, the observation that humans tend to show stronger directional biases for motor than perceptual systems is important to keep in mind in looking at the primate data (reviewed in Corballis, 1998).

Another reason why we need to consider both production and perception systems is that, while some theories of communication have argued for co-evolutionary patterns, others have argued for a decoupling of these systems. For example, Ryan's (Ryan et al., 1990; Ryan and Rand, 1995) work on the Tungara frog indicates that the female's perceptual system acts as an evolutionary pressure on the male's production system. Specifically, because the tuning of the female's auditory system is for frequencies that are lower than the male's advertisement call, females show preferences for synthetic calls that fall outside the range of species-typical variation. As such, female choice acts as a selective pressure on the evolution of the male's character. Although studies of primate communication are not yet in the position to look at the evolution of production and perception systems, by discussing what is currently known about cerebral asymmetries for communicative expression and perception, we will begin to lay the groundwork for this important problem.

13.2. Specialized Processing of Faces: Perceptual Mechanisms

Faces represent a complex class of stimuli. They have highly invariant features such as the T-configuration formed by the eyes, nose and mouth, as well as variable features such as interocular distance, nose shape and so forth. In addition, faces convey many important types of information including sex, identity and emotional state. Given that the visual identification and recognition of individuals and their facial expressions is an important component of the behavioural repertoire of at least some primate species, it is of interest to investigate whether other species besides *Homo sapiens* have specialized neural mechanisms for face processing.

13.2.1. The Inversion Effect

One source of behavioural evidence supporting the notion of specialized face processing is the inversion effect. For humans, face processing is significantly affected by orientation. When subjects are presented with an inverted face, they are impaired in recognition tasks, showing both a decrease in accuracy and slower reaction times (Yin, 1969, 1970). This phenomena may be even more robust for faces of a type little experienced by the subject (e.g. other-race faces: Valentine and Bruce, 1986). This effect seems particularly strong for faces as opposed to non-face visual objects, and has been used to argue for a specialized, even modular, processing mechanism (e.g. Farah et al., 1995; Kanwisher, Tong and Nakayama, 1998). Several researchers have concluded that subjects actually encode upright faces holistically, while inverted faces are encoded in a piecemeal fashion (Yin, 1969, 1970; Carey and Diamond 1977; Bruce and Humphreys, 1994). To that effect, an inversion effect may be found for some non-face objects, such as dot configurations, provided that they are configured to be processed holistically (Farah, Drain and Maxwell, 1995). It has been posited that the neural mechanisms for face processing in humans have been designed to handle the configuration of upright faces (e.g. Perrett et al., 1988) and these mechanisms reside primarily in the right hemisphere (see Section 13.3). Furthermore, the right hemisphere bias for face recognition disappears when an inverted face is presented (Leehey et al., 1978). Therefore, the inversion effect may provide a good behavioural test for finding evidence of specialized face processing in primates.

The majority of experiments on the inversion effect have been carried out with macaques (*Macaca*). An early study by Bruce (1982) with rhesus macaques failed to find an inversion effect. Bruce concluded that macaques lack the orientation-dependent face recognition mechanism that, in humans, matures within the right hemisphere. More specifically, he suggested that macaques may not develop a configurational mechanism for face recognition because they mature at a much quicker rate than human infants (the supposition being that early developmental experience with faces may shape the configurational mechanism). Further, he suggested that macaques have minimal hemispheric specialization for any function. However, these suggestions must be re-examined in light of more recent studies (detailed below).

In a series of well-designed behavioural and neurophysiological experiments, Perrett and colleagues (1988) re-examined the claim that monkeys process faces in a different manner from humans, focusing in particular on the failure to find an inversion effect. Initially, the macaques were trained to distinguish between faces (human and macaque) and common objects (including human and macaque non-face body parts). To ensure that subjects were actually seeing the images as faces, the experimenters trained the monkeys to generalize to novel faces (in the face versus common object discrimination condition). Following this generalization phase, the stimuli were presented upright or inverted. The subjects showed no difference in reaction time for correctly identifying the face images, thus providing no evidence for an inversion effect. Based on these results, Perrett and colleagues hypothesized that the monkeys were performing the face–non-face discrimination on the basis of distinctive facial features (e.g. the eyes alone) as opposed to the configuration of features. Thus, if a subject identified a salient facial feature then it would press the 'face' button irrespective of how many other facial features were present or how they were configured.

To address this possibility, Perrett and colleagues (1988) postulated that it might be possible for primates to learn a discrimination between faces and non-faces on the basis of configuration, and then, on a subsequent test, show an inversion effect for the face category. This is a reasonable hypothesis given that, for humans, the inversion effect appears to result from processing upright faces on the basis of configuration and inverted faces on the basis of distinctive features. A second experiment was designed to test this hypothesis with rhesus macaques.

Rhesus macaques were taught to discriminate sets of human and macaque faces from scrambled face images where all the features were present, but not in their proper position. The monkeys were then trained on horizontal stimuli, followed by training on inverted stimuli. To test for an inversion effect, Perrett and colleagues presented the original training stimuli in four orientations (0°, 90°, 180°, 270°), and then retrained the subjects on another face using the same procedure. Under these conditions, the macaques took longer to respond to both horizontal and inverted orientations compared with the upright orientation. This effect was replicated with a second set of novel face stimuli. Perrett and colleagues concluded that rhesus macaques process faces in a similar fashion to humans. Prior failures to find an inversion effect in primates may have been due to a lack of constraints on processing strategies. In particular, without training subjects to attend to configuration, alternative features may be used to solve the task, resulting in a misguided understanding of the potential mechanisms underlying face processing in primates.

Despite the findings of Perrett and colleagues, there have been more recent studies that have failed to find an inversion effect for conspecific faces in rhesus macaques. A study by Wright and Roberts (1996) showed that rhesus monkeys appear to respond differently to human faces as opposed to the faces of great apes, Old and New World monkeys, and prosimians. Their study compared human and rhesus monkey subjects, finding that both species showed inversion effects for human faces, but not for the other primate faces, and not for scenes. This finding is difficult to interpret in light of the finding by Perrett et al. (1988).

A recent study by Parr, Winslow and Hopkins (1999) also challenges the hypothesis that the inversion effect found in primates is specific to faces. They suggest that rhesus monkeys may show inversion effects for a number of classes of stimuli, not only faces. Using a match to sample procedure, Parr and colleagues tested rhesus monkeys with conspecific and heterospecific faces (including humans and primates), as well as automobiles and abstract shapes. They found evidence for an inversion effect with conspecific faces, capuchin (*Cebus apella*) faces and automobiles. In contrast to previous studies, they failed to find evidence for an inversion effect with human faces, as well as for abstract shapes. Thus, Parr and colleagues concluded that the inversion effect in rhesus monkeys is not face-specific and that previous studies of this effect do not provide valid evidence for a specialized face-processing mechanism in this species. These findings are also difficult to reconcile with previous studies.

Taken together, current evidence suggests that, under certain conditions, rhesus monkeys show an inversion effect for human and conspecific faces (see Section 13.2.2). To observe this effect, however, rhesus macaques must be constrained to use configurational cues, forcing them to process faces in the same manner as humans. In addition, it is important to have a generalization task to ensure that the subjects are actually processing face images as faces. Those studies that have failed to find an inversion effect have generally failed to provide evidence that their subjects were responding to configuration as opposed to distinctive features and thus failed to show that subjects were responding to faces as 'faces'.

While there have been numerous studies of face processing in macaques, research on the Great Apes has been limited, with all efforts focused on chimpanzees (*Pan troglodytes*). Tomonaga, Itakura and Matsuzawa (1993) studied the inversion effect in a chimpanzee with a long history of formal experimental training (Matsuzawa, 1996). The central goal of this experiment was to assess whether chimpanzees process heterospecific faces (humans) differently from the way in which they process conspecific faces. Subjects (four humans and one chimpanzee) sat at a panel with six buttons, each one corresponding to a different face (there were six faces in total: three human and three chimpanzee). The subjects were trained to press the button corresponding to the presented face. The training stimuli consisted of upright faces. Test conditions involved presentation of rotated (horizontal) and inverted faces. The chimpanzees had more difficulty in identifying the horizontal faces but did not show a significant inversion effect for faces (as measured by reaction time data). In a more recent study by Parr, Hopkins and de Waal (1996), however, the inversion effect was found in four out of five chimpanzees when they were presented with unfamiliar conspecific faces. In a control, using abstract shapes, two out of five subjects showed better performance for upright orientations. Thus, there was some evidence for an inversion effect to conspecific faces, but this effect was not restricted to faces;

this result parallels the findings of Parr, Winslow and Hopkins (1999) on rhesus macaques.

Parr, Dove and Hopkins (1998) tested five chimpanzees on their ability to discriminate human, chimpanzee and capuchin faces, as well as automobiles, in both upright and inverted orientations. All subjects performed better on upright than inverted stimuli, across all conditions. However, a statistically significantly difference in performance was only obtained on upright versus inverted presentations of chimpanzee and human faces, and not for capuchin faces and automobiles.

Chimpanzees represent the only great ape tested to date, and thus, nothing can be said about face processing in bonobos, gorillas and orang-utans. In parallel with studies of rhesus monkeys and humans, however, most studies of chimpanzees reveal evidence of an inversion effect to human and conspecific faces, providing support for a specialized processing mechanism (see Section 13.2.2).

In contrast to the other primates, relatively few studies have investigated face processing, and especially the inversion effect in New World monkeys. Phelps and Roberts (1994) studied face processing and the inversion effect in squirrel monkeys, and contrasted their performance with humans and pigeons. In the initial experiment, both humans and one squirrel monkey showed inversion effects for human faces. However, neither showed significant inversion effects for monkey faces (a category comprised of many different species). A follow-up experiment showed that a different squirrel monkey showed better reference memory for upright as opposed to inverted human and great ape faces, but not for Old and New World monkey faces, prosimian faces or scenes. The pigeons showed no inversion effect at all.

Weiss, Kralik and Hauser (2001) recently completed a study of face processing in cotton-top tamarins, using a procedure that closely matched the second experiment of Perrett et al. (1988). In the initial stages of the experiment, four subjects were trained to discriminate two human faces from two scrambled human faces (Figure 13.1). After this training, none of the subjects showed an inversion effect. As previously discussed, however, this experiment does not provide a fair test of the inversion effect because there was no evidence that the tamarins were using configuration to classify the stimuli as faces. In fact, more detailed analyses revealed that the tamarins were attending to distinctive features, such as the presence or absence of an eye in a particular location, and then using these to make an appropriate response. In a second condition, one subject was trained to focus on the configuration of the face, and was then tested on a series of generalization

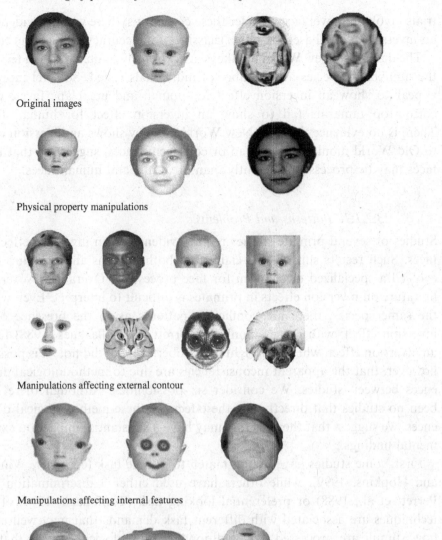

Original images

Physical property manipulations

Manipulations affecting external contour

Manipulations affecting internal features

Rotations and inversions

Figure 13.1. Sample images from Weiss, Kralik, and Hauser (2001). The top row contains the original images that the tamarins were trained to discriminate. The remaining images are samples from the different categories tested during the experiment.

trials involving novel faces. Under these conditions, there was no evidence of an inversion effect based on either classification accuracy or reaction time.

The data from New World monkeys are extremely limited, both in terms of the number of species and number of individuals tested. Squirrel monkeys appear to show an inversion effect for human and great ape faces, while cotton-top tamarins fail to show an inversion effect for human faces. There is no evidence that any New World monkey shows an inversion effect to Old World monkey, prosimian or conspecific faces, suggesting that these faces may be processed differently than great ape and human faces.

13.2.1.1. Patterns and Problems

Studies of several primate species report evidence of an inversion effect for faces. Such results support the claim that both humans and primates have evolved a specialized mechanism for face processing. Overall, however, the literature on inversion effects in primates is difficult to interpret. Even within the same species, there are conflicting patterns as to the presence of an inversion effect: with rhesus monkeys, Perrett and colleagues (1988) found an inversion effect, whereas Wright and Roberts (1996) did not. It is possible, however, that the apparent inconsistencies are due to methodological differences between studies. We consider six possibilities. Although there have been no studies that directly test the effects of these methodological differences, we suggest that these factors may have a substantial impact on experimental findings.

First, some studies have used a match to sample task (e.g. Parr, Winslow and Hopkins, 1999), while others have used either a discrimination (e.g. Perrett et al., 1988) or preferential looking task (Tomonaga, 1994). These techniques are associated with different task demands that may well affect how stimuli are processed. An additional methodological note is that the body posture of the test subjects may influence the results. Specifically, some species may use head cocking (e.g. tamarins) or other strategies in order to view faces in a different orientation than those presented by experimenters.

Second, few studies have verified that subjects were attending to the configuration of facial stimuli when selecting an appropriate response. More specifically, only two experiments (Perrett et al., 1988; Weiss, Kralik and Hauser, 2001) have provided evidence that subjects used the configuration of facial features in choosing between responses or stimuli. As studies of humans have illustrated, because the configuration of a face is critical to its categorical status as a 'face', studies of face processing in primates must first show that their subjects are attending to configuration before testing for specialized processing mechanisms.

Third, some experiments report reaction time differences (e.g. Perrett et al., 1988; Weiss, Kralik and Hauser, 2001), while others report only accuracy scores (e.g. Parr, Winslow and Hopkins, 1999). In most tests of face processing that have been conducted on humans, the inversion effect is subtle and requires statistical testing of reaction time data to find the effect. This may explain why some studies of rhesus macaques do not find inversion effects (e.g. Bruce, 1982) while other studies do (Perrett et al., 1988).

Fourth, there are significant differences in the kinds of stimuli presented across experiments. Some experiments limit the stimuli to faces while others include additional body parts, as well as inanimate objects. Further, some contrast responses to conspecific as opposed to heterospecific faces, whereas others contrast conspecific faces with non-face objects. In our opinion, tests of specialized face processing should include a generalization phase showing that subjects can respond appropriately to novel faces. In addition, comparing subjects' performance with face stimuli to performance with other stimulus types may determine whether the effects are particular to a face-processing mechanism.

Fifth, almost all studies of face processing in primates are based on small sample sizes. Rarely do studies of the face inversion effect exceed one or two subjects. This greatly limits our ability to draw inferences about phylogenetic patterns.

Despite the limitations we have sketched above, it is reasonable to assert that, in studies that are conducted with appropriate measures (i.e. tests to determine that subjects are using configuration in processing images and can generalize from the original training set, accuracy and reaction time measures), we can be confident that the results are comparable to human results and thus provide evidence for specialized processing for faces.

13.2.2. Hemispheric Specialization and Face Perception

In humans, evidence from both clinical and experimental studies has shown that face processing mechanisms are lateralized predominantly (although not entirely) to the right cerebral hemisphere. Patients with brain damage in the right cerebral hemisphere are often more impaired on tasks involving face perception than patients with similar damage in the left hemisphere. Experimental studies have shown that presentation of faces to the right hemisphere (through the left visual field) leads to faster processing than presentation to the left hemisphere (Hilliard, 1973; Burton and Levy, 1991; Whitman and Keegan, 1991).

The earliest studies with primates failed to find any evidence of hemispheric asymmetries in processing faces (e.g. Overman and Doty, 1982; Hamilton and Vermeire, 1983). More recent experiments suggest, however, that primates also show a right hemisphere bias for processing faces. In a study of split-brain rhesus macaques, Hamilton and Vermeire (1983, 1988) reported a right hemisphere advantage for discriminating individual animals by face and by facial expression. The stimuli consisted of rhesus monkey faces. Specifically, subjects were required to distinguish between slides of different individuals and slides of the same individuals producing different facial expressions. Of the split-brain monkeys tested, 70% showed a right hemisphere advantage based on accuracy scores. Vermeire and Hamilton (1998) then ran this experiment again, but contrasted performance of split-brain monkeys on upright versus inverted faces. Although a right hemisphere advantage was found for upright faces, the cerebral asymmetry disappeared for inverted faces. In addition, there was enhanced right hemisphere performance on both retention and generalization tasks performed 6 months after the conclusion of the aforementioned experiment (Hamilton and Vermeire, 1991).

Morris and Hopkins (1993) used a visual half-field presentation approach to determine whether captive chimpanzees show an asymmetry in processing human chimeric facial stimuli. Three chimpanzees were trained to discriminate normal 'happy' faces from normal neutral faces. Once the criterion was reached (85% correct), subjects had to discriminate chimeric faces in which half of the facial stimulus was smiling and half was neutral. Analyses of classification accuracy indicated that the subjects demonstrated a significant left hemispatial-field advantage, and thus, a right hemisphere advantage for processing faces.

13.2.2.1. Patterns and Problems

Hamilton and Vermeire's results (1991) support the hypothesis that upright faces are processed preferentially in the right hemisphere and that this processing difference is what drives the inversion effect. One drawback of this work, however, is that no reaction time data are reported. Thus, although the accuracy scores for the inverted faces may have been similar across hemispheres, there may have been a difference in reaction time.

In summary, studies to date provide evidence of a specialized mechanism for face processing in primates. The primary evidence, mostly from studies of macaques, reveals an inversion effect in both intact (e.g. Perrett et al., 1988) and split-brain subjects (Hamilton and Vermeire, 1991). Moreover, these studies suggest that the right hemisphere plays a dominant role in face processing, and in particular, in processing upright faces (Hamilton and Vermeire, 1991).

In the future, behavioural experiments must focus on the great apes and the New World monkeys in order to determine whether the patterns obtained for rhesus macaques are consistent across the primates or differ as a function of socioecological pressures. Such comparative work, which depends critically on the use of comparable methods and standards of evaluation, is likely to shed light on the evolutionary history of the hemispheric bias for processing upright faces in the right hemisphere. For example, do more arboreal species, who may depend less on visual features for species and individual recognition, fail to show evidence of a specialization for faces? Do species that spend considerable amounts of time in inverted positions fail to show an inversion effect due to their species-typical locomotory experiences? Likewise, do some species attempt to view inverted faces in the upright position by cocking their heads? In cases where upright and inverted faces are processed equivalently (i.e. no accuracy or reaction time differences) do they nonetheless show a hemispheric bias for processing faces at the neuronal level?

13.3. The Neural Basis of Face Perception

13.3.1. Face-selective Neurones in the Temporal Cortex of Macaques

Cognitive studies, both clinical and experimental, of face perception by humans suggest that face and object recognition involve qualitatively different processes that occur in different brain regions (Damasio, Tranel and Damasio, 1990). The results of these studies suggest that face processing by humans is a behaviour that requires specialized neural mechanisms. This hypothesis has been borne out by studies using patients with lesions and functional imaging of normal subjects, which show that there are discrete regions of the brain that are essential to the recognition of facial identity and expression (Damasio, Tranel and Damasio, 1990; Kanwisher, McDermott and Chun, 1997; Puce et al., 1996, 1999; Haxby et al., 1999). Many of these studies also show a lateralization of face processing to the right hemisphere. There is also evidence suggesting that primates may have homologous neural mechanisms for face processing (reviewed in Rolls, 1999).

Bruce, Desimone and Gross (1981) and Desimone et al. (1984) published the first exhaustive descriptions of 'face' cells in the neocortex of macaque monkeys, and the existence of such cells has been confirmed by subsequent studies from several laboratories. The unique property of these neurones is that they respond selectively to the presentation of faces. That is, while they may respond to other complex visual stimuli, these neurones respond much more vigorously (at least double the magnitude) to faces or components of faces (such as eyes or mouths) than to other stimuli. Face cells are primarily found in the temporal cortex, specifically in the inferior temporal (IT) cortical area (Desimone et al., 1984; Hasselmo, Rolls and Baylis, 1989), and in the superior temporal sulcus (STS) (Bruce, Desimone and Gross, 1981; Perrett, Rolls and Caan, 1982; Hasselmo, Rolls and Baylis, 1989). These regions are homologous with regions in the human temporal lobe, which are activated by faces or facial expressions (Kanwisher, McDermott and Chun, 1997; Puce et al., 1999).

13.3.2. Gaze Direction and Face Cells

Among social species of primates (including humans), the ability to detect where another individual is looking is highly adaptive because the gaze direction of an individual can be used to predict that individual's movements or actions. In support of this, studies on rhesus monkeys have shown that, when viewing faces, conspecific or otherwise, individuals selectively attend to the eyes when compared to other features of the face (Keating and Keating, 1982; Wilson and Goldman-Rakic, 1994; Nahm et al., 1997). In these studies, selective attention was measured by tracking eye movements while the subjects freely viewed visual stimuli, and by recording the duration and location of visual fixation points. There is also evidence that primates can attend to another individual's direction of gaze, using a combination of head and eye cues (Povinelli and Eddy, 1996; Tomasello et al., 1998; Santos and Hauser, 1999).

To investigate the neural bases for the perception of gaze following, Perrett and his coworkers (1985) searched for neurones in the temporal cortex that may respond to such cues. They reasoned that since neurones in the STS are often tuned to many views of the head (face, both profiles, and the back of the head) – more views than are needed for recognition purposes – that these neurones may play a role in coding social intention. In other words, these cells may signal where another individual is attending by encoding the combination of head orientation and gaze direction. Indeed, Perrett et al. (1985) found that many of the cells responsive to head view were found to be equally (if not more) sensitive to gaze direction. The most robust responses could be elicited from these cells if head orientation and gaze direction in the stimuli were matched; that is, if the cell was selective for a head turned laterally away from the monkey (to the right, for example), then that cell gave a greater response to the stimulus if the eyes were also laterally oriented (eyes looking to the right). This finding is also supported by lesion experiments. Monkeys trained to follow eye gaze direction showed significant deficits following bilateral lesions of the STS (Campbell et al., 1990), but no deficits for face recognition in general (Heywood and Cowey, 1992).

13.3.3. Cellular Correlates of Hemispheric Specialization for Faces in Macaques

In the only study of its kind, Perrett et al. (1988) compared single unit physiology data with results from a modified version of an upright versus inverted face discrimination task to address the question of specialized processing of faces in monkeys. Reaction time measurements showed that monkeys were able to discriminate upright configurations of faces significantly faster than stimuli presented in the horizontal or inverted orientations. This result is nearly identical to those based on humans performing the same task. Moreover, the monkeys' reaction-time data corresponded nicely with the single unit data. Thus, while many face-selective cells respond with a similar magnitude to either upright or inverted faces, the onset latency (or how quickly the neurone responds) differs. A majority of face-selective neurones respond with a shorter onset latency to upright faces when compared to responses to inverted faces (Perrett et al., 1988).

To determine whether the distribution of face cells is lateralized in macaques, Perrett et al. (1988) quantified the number of face-selective neurones encountered in the STS of both the left and right hemisphere using a within-subjects design. Surprisingly, for the three monkeys tested, there was a greater probability of encountering face neurones in the *left* hemisphere than in the right.

13.3.4. Neural Encoding of Facial Identity and Expressions

For humans, there appears to be separate mechanisms for processing facial identity and facial expressions. For example, prosopagnosics have difficulty identifying individuals by their faces, but have no difficulty in categorizing facial expressions in general (Damasio, Tranel and Damasio, 1990). In normal subjects, reaction times are faster for matching identity for familiar versus unfamiliar faces, while the use of familiar versus unfamiliar has no effect on reaction times for matching facial expressions (Bruce and Young,

1986). A similar dissociation for identity and expression discrimination has not been directly observed in monkeys tested on behavioural tasks.

In contrast to studies using behavioural assays, neurophysiological experiments suggest that macaques may have different systems for processing facial identity and expression. Hasselmo, Rolls and Baylis (1989) recorded from cells in the temporal cortex to determine whether facial identity and facial expression are encoded by the same or different populations of neurones in rhesus macaques. In their experiments, the faces of three macaques with three different expressions were used as stimuli, and the neural responses measured and compared across identities and expressions. The expressions used were a calm–neutral face, a slightly open-mouth threat and a full open-mouth threat. Neurones were sampled from the IT cortex and the STS. Two significant findings emerged from this study: (1) some neurones responded selectively to different identities independent of facial expression, while other neurones responded selectively to specific facial expressions independently of facial identity; and (2) neurones sensitive to facial identity were located in the IT cortex, while neurones sensitive to facial expressions were found primarily in the STS. The difference in anatomical distribution was statistically significant.

Studies of normal humans using neuroimaging techniques, as well as studies of patient populations with focal lesions in the temporal lobe, support the notion that there are different pathways for processing facial identity and facial expression (Adolphs et al., 1996; Kanwisher, McDermott and Chun, 1997; Puce et al., 1999). Thus, taken together, these results suggest that the similarity in face-processing mechanisms between humans and macaques is a case of homology; this claim holds even if macaques, or other primates, fail to show evidence of cerebral asymmetry.

13.3.4.1. Patterns and Problems

The finding of left hemisphere bias for distribution of face-selective cells in the STS (Perrett et al., 1988) stands in contrast to the split-brain studies on macaques, the visual half-field study of chimpanzees (see Sections 13.2.1 and 13.2.2) and the human imaging data, where there is greater activation of the STS in the *right* hemisphere than in the left for face stimuli (Puce et al., 1999). Perrett and colleagues are the only group to study this issue using electrophysiological methods and, unfortunately, their subject group consisted of only three monkeys. Furthermore, there is the issue of sampling bias. As stated by the authors, in two of the monkeys, the STS of the left hemisphere was sampled more extensively than that of the right hemisphere (Perrett et al., 1988). Nevertheless, a recent human imaging study demonstrates that, when subjects are required to attend to the eye gaze of face stimuli, the *left* STS is more active than the right (Hoffman and Haxby, 2000) – a finding that lends some support to the finding of Perrett and colleagues. In the future, it will be important for studies to track the eye movements of subjects (see below). Ultimately, the question of cerebral asymmetries for face processing in primates may best be addressed by techniques with greater spatial resolution, such as functional magnetic resonance imaging (fMRI) (Logothetis et al., 1999).

Considering the neural segregation of face recognition versus expression in primates, several caveats must be considered. One potential problem concerns the limited number of 'expressive' stimuli used. Only three expressions were used, one of which was a 'calm–neutral' face. The other two were open-mouth threats of different magnitudes. Macaques have a suite of distinctive facial expressions, of which the open-mouth threat is only one exemplar among several associated with negative–withdrawal emotions (Hauser, 1993a, 1999; Hauser and Akre, 2001; Hauser et al., in preparation). Thus, these experiments may not have robustly tested the selectivity of temporal lobe neurones for facial expressions.

Many primates, including macaques, preferentially look at the eyes when viewing other monkey faces (Keating and Keating, 1982; Wilson and Goldman-Rakic, 1994; Nahm et al., 1997), and the eyes remain the primary targets for visual scanning for a range of facial expressions, including open-mouth threats, lipsmacks, yawns and fear grimaces (Nahm et al., 1997). However, for some expressions, such as lipsmacks and yawns, there were no significant differences between the average time spent looking at the eyes versus the mouth, but for other expressions, such as the open-mouth threat and the fear grimace, significantly more time was spent looking at the eyes (Nahm et al., 1997). Thus, it would be of interest to know what facial features the 'expression-selective' neurones are coding. Is it the mouth, the eyes, or a combination of both? Also, how are the responses of neurones modulated by vocal expressions in which both visual and auditory cues are present? This is an interesting problem given the fact that human infants preferentially attend to the eyes of faces, whether the face is still or talking (Haith et al., 1977), and infants are clearly sensitive to both visual and auditory information (Kuhl and Meltzoff, 1982). Future research on the neurophysiology of facial expression must explore the possibility of different responses, based in part on such factors as emotional valence (e.g. positive/approach versus negative/withdrawal), whether or not the eyes are directed at the receiver or away from him, and whether the individual giving the expression is familiar or unfamiliar.

13.4. Specialized Processing of Vocalizations: Perceptual Mechanisms

13.4.1. Specialization in Vocal Perception

The task of vocal perception involves extremely complex processing. The first challenge facing the receiver is to detect the vocalization despite environmental noise and attenuation. Once the signal is detected, the receiver must then classify the type of vocalization and the source. Successful classification requires the receiver to attend to the relevant acoustic parameters, both spectral and temporal. In humans, for example, temporal cues, such as duration, interval and order of acoustic features are important for speech perception and sound categorization (Liberman et al., 1967; Harnard, 1987). In fact, data from language-impaired children have contributed to the theory that speech perception is based on rapid processing of temporal information (Tallal, Miller and Fitch, 1993). Given that vocal classification and communication is an important component of the behavioural repertoire of most primates species, it is of interest to investigate whether other species besides *Homo sapiens* have specialized neural mechanisms for vocal perception, and whether such mechanisms are lateralized.

To determine whether primates might exhibit similar specializations for processing their own vocalizations, psychophysical experiments were conducted on Japanese macaques (*Macaca fuscata*), focusing on two functionally distinctive variants of the 'coo' call (Beecher et al., 1979; Zoloth et al., 1979). These vocalizations, referred to as the smooth early (SE) and smooth late high (SH) coos, differed in the relative temporal position of the peak frequency, as well as the social context in which they were given (Green, 1975). Experiments were designed to assess whether Japanese macaques compared to closely related species (pig-tailed macaques, bonnet macaques and vervet monkeys) have evolved specialized mechanisms for classifying conspecific vocalizations as a function of distinctive acoustic features. Two conditions were tested. In condition one, subjects were required to discriminate calls based on the position of the peak frequency of the fundamental. In condition two, discrimination was based on the initial frequency of the fundamental. While comparison species were better able to distinguish the vocalizations using the initial frequency, only Japanese macaques performed better at the task of distinguishing calls based on the peak frequency. Thus, the findings from these experiments, as well as subsequent studies (May, Moody and Stebbins, 1988), are consistent with the interpretation that Japanese macaques have evolved specialized mechanisms to distinguish between their call types. What these studies leave open is whether such specializations are largely innate as opposed to shared by experience with a particular vocal repertoire, and whether the pattern obtained for coos would be obtained for other call types within the repertoire.

13.4.2. Behavioural Asymmetries in Vocal Perception

The left hemisphere is thought to be dominant for the more formal aspects of language processing (such as syntax, semantics, etc.), while the right hemisphere appears dominant for processing the paralinguistic or prosodic cues (e.g. rhythm, melody). Evidence for cerebral asymmetries underlying language processing come from cases of brain-damaged patients, dichotic listening studies and recent neuroimaging work (e.g. Price, 1998; Mueller et al., 1999).

Petersen and colleagues (1978) conducted a series of experiments to test whether Japanese macaques, which are thought to have evolved specialized mechanisms to distinguish between their call types, exhibit behavioural asymmetries in processing conspecific vocalizations. In their initial study (Petersen et al., 1978), they trained Japanese macaques to discriminate among several natural exemplars of two functionally distinctive, tonal vocalizations (the coos mentioned in Section 13.4.1). The stimuli were presented monaurally, alternating between ears and the performance for each ear was compared. They found that the Japanese macaques performed better when the stimuli were presented to the right ear (and thus left hemisphere). Several other primate species (bonnet and pig-tailed macaques and vervets) were also tested, but only one (vervet) showed a significant ear advantage and in the same direction as Japanese macaques. Follow-up studies showed that all species attended to the same acoustic features of the call. Petersen and colleagues therefore concluded that the communicative valence of the calls for Japanese macaques was responsible for the enhanced left hemisphere performance. Consequently, the pattern observed among Japanese macaques appears to resemble the left hemisphere bias for language processing in humans.

Although the work on Japanese macaques represents a landmark in the field, several questions remain. First, because the experiments focused exclusively on coos, it is not yet known whether the enhanced right ear performance would generalize to other vocalizations within the repertoire. Second, because one vervet monkey showed the same pattern as the Japanese macaques, the extent to which the right ear bias represents a species-specific specialization remains unclear. Finally, it is unclear whether the right ear bias would be observed under more naturalistic conditions, and in young as well as older individuals.

To address some of these issues, Hauser and colleagues conducted a series of field experiments with both adult and infant (4–12 months olds) rhesus monkeys (*Macaca mulatta*) living on the island of Cayo Santiago, Puerto Rico. In the first experiment (Hauser and Andersson, 1994), a speaker was concealed in vegetation 180° behind the target subject, 10–12 m behind one side of a chow dispenser. When the subject's back completely faced the speaker, and the camera was lined up with both the speaker and subject, the playback was initiated and the subject's head-orienting response recorded on to video. Recording the response on to video allowed for an unambiguous assessment as to which direction the subject turned to listen. Three responses were possible: turn right, turn left, or no detectable response.

The experimental stimuli consisted of exemplars from 12 call types that could be separated into three broad categories: aggressive, fearful and affiliative. The underlying assumption of the experiment was that, if the subject turned its right ear toward the speaker, then the acoustic input would be biased toward the left hemisphere of the brain; although both hemispheres would receive input, the input to the left hemisphere would have greater intensity. Likewise, if the subject turned its left ear toward the speaker, then the input would be biased to the right hemisphere of the brain. As a control stimulus, Hauser and Andersson played the alarm call of the ruddy turnstone (*Arenaria intepres*), a seabird that lives on Cayo Santiago. The turnstone's alarm call is familiar to the monkeys and may have significance in alerting the monkeys to the presence of humans.

Results showed that adult subjects turned with the right ear leading (left hemisphere) in response to conspecific vocalizations, but turned with the left ear leading in response to the ruddy turnstone's call. In contrast, infants failed to show a significant head-orienting bias for either the conspecific or the ruddy turnstone calls. At present, the mechanisms underlying these age differences are not well understood. One possibility is that hemispheric differentiation develops only when the vocalizations acquire meaning (possibly occurring up to 2 years of age; see Gouzoules and Gouzoules, 1989; Cheney and Seyfarth, 1990). A second possibility is simply that complete maturation of the hemispheres requires at least 1 year to complete. As a result, the orienting response of the infants is based on general rather than selective acoustic-processing mechanisms. More research is needed to determine when rhesus monkeys acquire adult levels of comprehension and when the hemispheric asymmetries observed in adults become fully mature. Nevertheless, the overall results from this experiment provide strong evidence for the existence of a left hemisphere bias for processing conspecific vocalizations in the adult rhesus macaque. These findings parallel the results obtained for Japanese macaques as well as for humans tested in dichotic listening experiments (e.g. Bryden, 1982).

In humans, the magnitude of the lateralization effect can be altered by speeding up or slowing down formant transitions within a syllable (Schwartz and Tallal, 1980). In order to test whether rhesus macaques are similarly affected by temporal manipulations, a follow-up series of investigations have been undertaken to determine which acoustic features of the vocalizations are responsible for the head-turning preference, and thus, the presumed hemispheric bias. The first series of experiments (Hauser, Agnetta and Perez, 1998) looked at the temporal features of three different pulsatile call types: a grunt (affiliative signal involving food or conspecific), a shrill bark (alarm signal) and a copulation scream (mating signal). These calls were chosen because of the available acoustic analyses and studies on call context (e.g. Hauser, 1993b; Hauser and Marler, 1993; Bercovitch et al., 1995). The signals were presented in both manipulated and unmanipulated form. The manipulated calls consisted of (1) a reduction of the interpulse interval (IPI) to zero or the population minimum; and (2) an expansion of the interpulse interval to the population maximum or twice the population maximum. The main prediction for this experiment was that calls manipulated beyond the species-typical range should elicit a different pattern of head orientation than that which had been reported for unmanipulated conspecific calls (Hauser and Andersson, 1994). Results showed that when the IPI was eliminated from the grunts and shrill barks, there was no significant orienting bias. In contrast, the right ear bias was preserved when this manipulation was imposed on the copulation calls. When the IPI was extended to the maximum in the population, there was a tendency for subjects to orient with the left ear leading for both grunts and shrill barks, but this bias was not statistically significant; for copulation calls, the right ear bias was preserved. When IPIs were increased to twice the population maximum, subjects consistently turned left for grunts and shrill barks, but again maintained a right ear bias for copulation calls.

Overall, results from experiments on grunts and shrill barks support the hypothesis that manipulations of the IPI beyond the species-typical range of variation cause a shift from a right ear bias to either no bias (with the IPI eliminated) or to a significant left ear bias (with the IPI stretched to twice the population maximum). This pattern of response was not, however, observed for playbacks of copulation screams. The acoustic morphology of the copulation scream may provide insights into why the temporal manipulations imposed failed to elicit a change in the direction of head orientation. Grunts and shrill barks are produced with a minimum of two pulses, while copulation screams can be produced with either one or many pulses. Consequently, whereas the IPI appears relevant for classifying grunts and shrill barks, it may not be relevant for copulation screams, at least in terms of assessing whether or not it is a conspecific signal; the interpulse interval of copulation screams may, however, be relevant to male quality (e.g. Hauser, 1993b).

More recent work, using the same head-orienting procedure, has focused on a different acoustic manipulation: reversing the call (Ghazanfar and Hauser, in preparation). This type of manipulation preserves all of the spectral energy in the call while inverting the temporal relationships. For this experiment, our goal was to assess the perceptual salience of a time-reversed amplitude envelope. Specifically, we tested the hypothesis that if rhesus monkeys use the amplitude envelope of a signal to classify it as falling within or outside the category of 'conspecific', then reversing a call should eliminate the head-turning bias observed in the original experiments involving unmanipulated calls.

The calls used in this experiment included two aggressive calls (bark, pant threat), one food call (harmonic arch), and one alarm call (shrill bark); each of these calls was used in the original Hauser and Andersson (1994) study. The aggressive calls are characterized by a symmetrical amplitude envelope, and thus reversing the signal has only a minimal effect. In contrast, both the food and alarm calls have asymmetric amplitude envelopes, and thus reversing the signal should have a more noticeable effect. To human observers who have had experience with the rhesus repertoire, no perceptual differences are detected between forward and reversed aggressive calls, whereas reversed food and alarm calls sound distinctively different from forward exemplars.

Results revealed that rhesus macaques switched from a right to a left ear-orienting bias for both harmonic arches and shrill barks played backwards. However, for the aggressive calls, rhesus macaques maintained a moderate right ear bias. These data are consistent with the description of each call's characteristic amplitude envelope. Reversing a relatively symmetric call has little to no effect on the orienting response, whereas reversing an asymmetric call directly influences the direction of the orienting bias. In parallel with the manipulations of interpulse interval (Hauser et al., 1998), these results also suggest that the left hemisphere bias shown for normal, but not reversed calls, has to do with the specific, species-typical call morphology.

13.4.2.1. Patterns and Problems

Results from all of the playback experiments described above are consistent with the interpretation that the left hemisphere is more active in processing conspecific vocalizations while the right hemisphere is more active in processing sounds falling outside the species-typical repertoire. One possible explanation of these patterns is that meaningful sounds (in terms of conspecific interactions) are preferentially processed in the left hemisphere, while other acoustic signals (which may be meaningful in terms of particularly salient environmental events) are preferentially processed in the right hemisphere. Two observations provide some support for this hypothesis. The first comes from Hauser and Andersson's (1994) original study, and the use of the ruddy turnstone's alarm call as a control. It is quite possible that this call contains meaningful information that may be used to predict the presence of humans, an event that may be particularly useful in predicting the delivery of monkey chow or attempts to trap the rhesus macaques for biomedical purposes. Before this hypothesis can be evaluated, additional experiments with other potentially meaningful sounds must be conducted. For example, one could contrast the response given to turnstone alarm calls with the response given to human speech, as well as potentially meaningful, but non-biological sounds, such as the arrival of the boat (which brings the researchers to the island) or the sound of the chow placed into the dispenser. The second piece of evidence comes from studies of baboons indicating a left ear–right hemisphere bias for processing musical chords, pure tones, and human consonants and vowels (Pohl, 1983, 1984). In this case, it appears that the observed asymmetry is mediated by sounds that are unlikely to carry any meaning or significance to baboons. Additional experiments are needed in order to better determine to what extent the left hemisphere is specialized for processing conspecific signals.

Another issue that remains unresolved is the developmental course of the head-turning bias. As mentioned above, Hauser and Andersson (1994) found that infants did not display a significant head-turning bias. Future research should focus on tracking the developmental course of the head-turning preference. If the left hemisphere bias develops as the infants learn the meanings of the vocalizations, then there may be corresponding behavioural changes in vocal production. Thus we may look for a relationship between acoustic experience and the head-orienting bias. In addition, the relationship between hemispheric maturation and the head-orienting bias warrants further investigation.

To date, the manipulations used with this playback paradigm have focused on temporal features. Using sound synthesis techniques developed by Evan Balaban and Kim Beeman (Beeman, 1996), however, it is also possible to manipulate spectral parameters of rhesus calls by creating synthetic replicas; the technique exploits pitch contour algorithms as well as tools for extracting the amplitude envelopes of each harmonic. Hauser and Fitch (see Hauser, 1999) used a habituation–dishabituation paradigm to determine whether rhesus macaques classify synthetic calls as functionally similar to natural exemplars. Briefly, the experiment involved habituating the subjects to a series of natural exemplars and then playing back a synthetic replica. If subjects failed to respond (i.e. did not orient toward the speaker) to the synthetic replica, then a post-test call from a different functional category was played to ensure that the failure to respond to the test signal was not due to the subject's overall habituation to the experimental setup. Preliminary results, using tonal screams and harmonic arches, revealed that rhesus macaques transfer habituation to the synthetic exemplar, suggesting that the natural and synthetic exemplars are treated as functionally similar (for summary of the experiment, see Hauser, 1999). These results are important only in so far as they set up a methodology for systematically manipulating the spectral properties of a call and then establishing which features most significantly contribute to the head-turning bias.

Although behavioural studies of acoustic perception in macaques are relatively advanced with respect to our understanding of specialized processing, we do not know anything about the other species of primates. As we have emphasized throughout this chapter, comparative data are critical if we are to document the phylogenetic patterns associated with cerebral asymmetries underlying acoustic perception. At present, we do not know whether macaques are typical or atypical in their perceptual biases.

If we focus on the macaques, and especially rhesus macaques, several issues remain. First, the head-orienting method must be cross-validated with other species. Second, if rhesus macaques preferentially turn the right ear to listen to conspecific vocalizations, what clockwise deviation from the center line (i.e. 180° from the midpoint of the subject's back) is necessary to induce a shift from right to left ear for unmanipulated conspecific calls? This experiment would test the strength of the head-orienting effect by testing it against directional cues. Third, for rhesus macaques reared in captivity, with considerable exposure to human speech, is there a right ear–left hemisphere bias for both rhesus monkey calls and human speech? More specifically, is the orienting bias due to the frequency of exposure to rhesus calls or to an innately specified, specialized mechanism for processing conspecific calls?

The fact that infants do not develop the species-typical orienting bias until after the first year of life is not evidence against an innate bias. Fourth, when do infants develop the species-typical orienting bias? Fifth, to what extent is the orienting bias flexible? If rhesus monkeys are cross-fostered (Masataka and Fujita, 1989; Owren et al., 1993) do they show the orienting bias of their own species, or of the cross-fostered parent? Sixth, what areas of the brain are most directly involved in the orienting bias? Data reviewed in Section 13.5 provide some answers to this last question.

13.5. The Neural Bases of Vocal Perception

13.5.1. Gross Anatomical Differences between Hemispheres

The perceptual asymmetries for vocal perception are supported by both neuroanatomical and experimental lesion studies, although here the data extend beyond the macaques. For humans that have demonstrated left-hemi-sphere biases for language processing, it has been shown that the Sylvian fissure (bordering the auditory cortex) is significantly longer in the left hemi-sphere than in the right (Geschwind and Levitsky, 1968). It has been assumed that the length of the fissure corresponds to the size of auditory cortex. Based on this assumption, potential anatomical asymmetries have similarly been measured in several primate species. Left Sylvian fissure length was found to be significantly greater than right for apes (*Pan, Gorilla* and *Pongo*: LeMay and Geschwind, 1975), Old World monkeys (*M. fascicularis* and *M. mulatta*), and New World monkeys (*Saguinus oedipus* and *C. jacchus*) (Falk et al., 1986; Heilbroner and Holloway, 1998). These results support the claim that the perceptual asymmetries observed under laboratory (Petersen et al., 1984) and seminatural (Hauser and Andersson, 1994) conditions are associated with these anatomical asymmetries. Recently, more detailed analyses of potential asymmetries in the primate auditory cortex have revealed specific homologies with the language areas in humans. For instance, an auditory cortical struc-ture known in ape and human brains as the *planum temporale* is located in the posterior portion of the temporal lobe. In the majority of humans, this structure is significantly larger in the left hemisphere than the right and is considered to be a speech-processing area (Geschwind and Levitsky, 1968). A similar left–right asymmetry in the planum temporale has been shown for chimpanzees and other great apes (Gannon et al., 1998a, 1998b). In an Old World monkey species (*Macaca fascicularis*), volumetric measurements of the cytoarchitectonic area Tpt, which is the equivalent of the planum temporale in humans and pongids, revealed that the left area Tpt was significantly larger than the right (Gannon, Kheck and Hoff, 1999). Interestingly, the left Tpt appeared to have a unique neurochemical organization when compared to the right Tpt (Kheck et al., 1999). Specifically, the left Tpt had greater parvalbumin (a calcium-binding protein) and gamma-aminobutyric acid (GABA; an inhibitory neurotransmitter) immunoreactivity, while the right Tpt had greater calbindin (another type of calcium-binding protein) immunoreactivity. These data suggest that the neurochemical specialization of the language areas evolved prior to the gross anatomical asymmetries (Kheck et al., 1999).

13.5.2. Lesion Studies of the Auditory Cortex

Human patients with lesions of the left temporal lobe exhibit deficits in speech perception (Mazzocchi and Vignolo, 1979), a phenomenon known as Wernicke's aphasia. This deficit is specifically associated with lesions in the posterior portion of the left temporal gyrus, an area that contains higher-order auditory areas, such as the planum temporale (area Tpt in Old World monkeys). In rhesus macaques and squirrel monkeys, experimental lesions of auditory cortex result in a similar deficit, in that subjects become selectively impaired at discriminating species-specific vocalizations but not other types of auditory stimuli (Dewson, 1977; Heffner and Heffner, 1984, 1986; Hupfer, Jürgens and Ploog, 1977). The performance of Japanese macaques trained to discriminate between the two different coos used in the behavioural asymmetry studies (Petersen et al., 1984; see Section 13.4.2) was greatly impaired following lesions of the left superior temporal gyrus, but was unimpaired by similar lesions in the right hemisphere (Heffner and Heffner, 1984, 1986).

In squirrel monkeys, bilateral lesions of auditory cortex impaired performance of vocal discrimination (unilateral lesions did not), but this deficit was not specific to vocalizations (Hupfer, Jürgens and Ploog, 1977). The gross neuroanatomical data converge with the absence of neural lateralization in squirrel monkeys: Sylvian fissure asymmetries are not as robust in this species as in other primates (Heilbroner and Holloway, 1988). This leads to the conclusion that, in squirrel monkeys, unlike Japanese macaques, the auditory cortex processes vocal signals in the same way as it processes other complex sounds (Hupfer, Jürgens and Ploog, 1977).

13.5.2.1. Patterns and Problems

A synthesis of current results suggests that the left auditory cortex of Japanese macaques is specialized for processing vocalizations, whereas squirrel monkeys appear to lack such specialization. These findings underscore the importance of choosing an appropriate species to answer particular neuroethological questions, and caution against making generalizations from one primate species to all primates (Preuss, 1995). This is particularly important given that most neurophysiological experiments using species-specific vocalizations have used squirrel monkeys (see below). It would be of significant interest to determine whether other species of primates, which show Sylvian fissure asymmetries, demonstrate behavioural asymmetries in the processing of their conspecific vocalizations. This would directly test whether neural asymmetries (or lack thereof) mediate behavioural asymmetries, an important relationship given the findings for human speech.

Given the rather crude spatial resolution of these types of lesion experiments, it is unclear what specific auditory cortical areas are contributing to the behavioural asymmetries. In both the Japanese macaque and squirrel monkey studies, much or all of the superior temporal gyrus was removed. The temporal lobe contains many auditory and auditory-related cortical areas (Kaas and Hackett, 1998). Based on the anatomical studies conducted thus far, one would predict that area Tpt would be critically involved in macaques (Gannon, Kheck and Hof, 1999; Kheck et al., 1999).

A second difficulty associated with lesion studies is that they are confounded by the problem of behavioural recovery. For example, although the ability of Japanese macaques to discriminate conspecific calls was impaired following lesions of the left auditory cortex (Heffner and Heffner, 1986), they regained normal performance after 5–15 sessions. Thus, there is considerable, and rapid, neural plasticity following experimental lesions that cannot be controlled. Similar issues arise in studies of lesions in human patients (Naeser et al., 1987).

Again, many questions regarding neural asymmetries in primates may best be addressed by functional imaging techniques (Logothetis et al., 1999). Such an approach would offer better resolution, at the level of cortical areas, while avoiding sampling errors associated with electrophysiological methods. In comparison with lesion experiments, functional imaging would avoid non-stationary effects such as lesion-induced plasticity and problems controlling the extent of lesions. An illustrative experiment would involve imaging the temporal lobe of a rhesus macaque while presenting normal versus temporally manipulated calls (ISI manipulations or reversed calls). Given this species' differential head-orienting responses to these call categories, this would allow us to determine differences in the hemispheric distribution of active cortical areas under both conditions.

13.5.3. Call-selective Neurones in the Auditory Cortex

To date, the squirrel monkey represents the most extensively studied mammalian model system for the auditory processing of species-specific vocalizations. This is, in one sense, an unfortunate situation given the fact that most of the work was conducted before 1975; on the other hand, it speaks to the great insights of this team of researchers (reviewed in Jürgens, 1990; Hauser, 1996). Recordings of single unit activity in the superior temporal gyrus of the awake squirrel monkey revealed that more than 80–90% of the neurones in this region responded differentially to more than one of the 12 species-specific vocalizations used as stimuli (Wollberg and Newman, 1972; Newman and Wollberg, 1973). For one subpopulation of neurones, a variety of temporal firing patterns were observed across the sample of neurones for any given call type. Likewise, the firing pattern of a given neurone varied considerably as a function of call type. In another subpopulation, neurones responded selectively to one vocalization and had a relatively simple discharge pattern (Wollberg and Newman, 1972; Winter and Funkenstein, 1973). Although the relative lack of information regarding squirrel monkey architectonic boundaries in these studies limits what one can say about the functional organization of auditory cortex, they nevertheless provide substantial evidence that auditory neurones are highly responsive to species-specific vocalizations, and sometimes highly selective for vocalizations.

More recent experiments in identified subdivisions of the auditory cortex of anaesthetized rhesus and marmoset monkeys largely support the results from squirrel monkeys, demonstrating that cortical neurones, from multiple subdivisions of the auditory cortex, respond selectively to conspecific vocalizations with complex temporal patterns of firing (Rauschecker, Tian and Hauser, 1995; Wang et al., 1995; Tramo, Bellew and Hauser, 1996). Rauschecker, Tian and Hauser (1995), using vocalizations recorded from rhesus monkeys as stimuli, found several call-selective neurones localized to area Tpt of the superior temporal gyrus – an area homologous with the planum temporale (Wernicke's area) in humans.

13.5.3.1. Patterns and Problems

Given that auditory cortical neurones can be call selective, or at least call responsive, how are these complex spectrotemporal stimuli integrated by such neurones? In other words, how is selectivity achieved? One approach to answering these questions involves presenting acoustically manipulated stimuli. One can systematically alter specific features of a call and then use such perturbed signals to determine how components of the call affect neural response patterns. As has been demonstrated in the songbird (Margoliash, 1983; Margoliash and Fortune, 1992) and bat auditory systems (Suga et al., 1987; Kanwal, 1999), this approach has revealed that neurones are sensitive to particular conjunctions of acoustic features as opposed to isolated components of the vocal signal. In rhesus macaques, filtering out certain frequencies (or selectively removing harmonics) in particular species-specific vocalizations (such as the 'coo' call), results in less robust responses from call-selective neurones when compared to responses to normal intact vocalizations (Rauschecker, Tian and Hauser, 1995; Rauschecker, 1998). Similarly, in the temporal domain, it has been shown that editing out parts of, or reversing, vocalizations used as stimuli results in a decrease in neuronal responsiveness for call-selective neurones in squirrel monkeys (Wollberg and Newman, 1972), marmosets (Wang et al., 1995) and rhesus monkeys (Rauschecker, 1998). Together with studies of 'response enhancement' using tone sequences as stimuli (Brosch, Schulz and Scheich, 1999), these data suggest that neurones in the auditory cortex of primates are 'combination-sensitive' (i.e. they respond non-linearly) to conspecific vocalizations in the same way that neurones in the songbird forebrain and bat auditory cortex are combination sensitive to their own vocalizations.

Several questions remain concerning the behavioural relevance of the selectivity of primate auditory neurones. For example, how do spectrotemporal manipulations that affect responses at the *neural* level affect responses at the *perceptual* level, and vice versa. Does removing or extending particular portions of vocalizations affect how subjects respond to them? How are the temporal manipulations of vocalizations that influence behaviour processed by call-selective neurones in the auditory cortex (Hauser, Agnetta and Perez, 1998)? What is the relative importance of temporal versus spectral (frequency) cues in vocal processing. Different domains may be used to extract different categories of information. For example, temporal features may define whether a call is within the conspecific repertoire (Hauser, Agnetta and Perez, 1998), while spectral features may define the identity (or size) of the caller (Rendall, Rodman and Emond, 1996; Fitch, Miller and Tallal, 1997). Finally, what is the distribution of call selective neurones in the right and left hemispheres, and do split-brain monkeys show cerebral dominance for processing species-typical vocalizations? Future studies must not only address these questions, but extend the range of species studied beyond squirrel monkeys and macaques.

13.6. Specialized Processing of Facial Expressions and Vocalizations: Production Mechanisms

13.6.1. Production of Facial and Vocal Expressions: Behaviour and Context

Humans produce a wide range of different facial expressions (Darwin, 1872; Ekman, 1973; Fridlund, 1994). While considerable variation exists between different populations, humans around the world are able to recognize at least six basic emotions (Ekman, 1992). Such universality may indicate that humans have an innate capacity for both expressing and perceiving facial expressions. Further evidence for this claim comes from studies of human infants. For example, infants spontaneously start producing smiles within the first few weeks of life, even though there is considerable cross-cultural variation in the frequency with which parents smile at their infants (Trevarthen, 1979). Moreover, blind infants start producing smiles at the same time as infants born with normal eyesight, suggesting that visual feedback is not necessary for the production of smiles (Eibl-Eibesfeldt, 1973). Finally, smiles have a consistent, stereotyped structure and generally convey the same emotional state for humans throughout the world.

Darwin (1872) was perhaps the first to point out the continuity in the expressions of humans and non-human animals. Although he explicitly avoided providing an adaptationist's perspective, he nonetheless maintained that the observed similarity provided support for his theory of evolution by natural selection (Fridlund, 1994). Since his landmark book entitled *The Expression of Emotions in Man and Animals*, other researchers have proceeded to show quantitatively that Darwin's observations were indeed correct. Some of the most striking continuities in facial expressions occur between humans and primates. Preuschoft and van Hoof (1997) argue that the morphological similarities between the 'bared teeth' and 'open mouth' displays of many primate species, and the smile and laughter expressions produced by humans, are homologous across all primate species. They suggest that these expressions evolved to mediate social interactions in primates. The phylogenetic continuities found in studies of humans and primates suggest that some facial expressions are displays that evolved as a result of natural selection.

In contrast to the documented similarities between human and primate facial expressions, until recently, far less was known about homologues or analogues to human vocal expressions. Non-human animals, primates included, produce a range of different vocal expressions. As Darwin articulated, many of these sounds convey information about the caller's affective state (see also, Smith, 1977). Often, the connection between affective state and acoustic morphology is consistent across animals. For example, as Darwin, Morton (1977) and Hauser (1993b) have revealed vocalizations associated with aggression are often low pitched, while vocalizations associated with fear or affiliation are often high pitched. Recent studies of primates, and one avian species (i.e. the domestic chicken), also provide evidence that some vocalizations convey information about events that are external to the caller (Seyfarth, Cheney and Marler, 1980; Cheney and Seyfarth, 1982; Dittus, 1984; Gouzoules, Gouzoules and Marler, 1984; Evans, Evans and Marler, 1994; Zuberbuhler, Noe and Seyfarth, 1997; Hauser, 1998; Zuberbuhler, Cheney and Seyfarth, 1999). While these calls are not equivalent to human words, they are not simply expressions of the caller's affective or internal state. Rather, such vocalizations appear to be functionally referential, with acoustic morphology closely coupled to key objects (e.g. predators, food) and events (e.g. dominance-related social interactions) in the species-typical environment (Cheney and Seyfarth, 1990; Marler, Evans and Hauser, 1992; Hauser, 1996). Given the information conveyed, it is of interest to explore whether the patterns of hemispheric asymmetry exhibited for primate facial expressions are similar to or different from those exhibited for primate vocalizations.

13.6.2. Hemispheric Asymmetries in the Production of Human Facial and Vocal Expressions

Humans demonstrate asymmetrical use of the mouth during vocal and facial expressions. Specifically, one side of the mouth opens wider and is more expressive during the production of different facial and vocal expressions. Early work on facial expressions indicated that humans expressed emotions more strongly on the left side of the face than on the right side (Sackheim, Gur and Saucy, 1978). In the past two decades, however, a different picture has emerged. While producing expressions associated with positive/approach emotions, humans exhibit a right-side bias. In contrast, expressions mediated by negative/withdrawal emotions are associated with a left-side bias (Davidson et al., 1990; Davidson, 1995). These data suggest that the different hemispheres are responsible for opposite aspects of emotional expression in humans. Whereas the left hemisphere is dominant for positive/approach expressions, the right hemisphere is dominant for negative/withdrawal expressions. Several studies using neuroimaging techniques on brain-damaged and normal patients provide support for the relationship between emotional valence and hemispheric specialization during the production of expressions (Davidson, 1999).

Investigations of mouth asymmetries during speech production show a somewhat different pattern. Humans exhibit a strong lateral bias to the right side of the mouth while producing meaningful speech (Graves, Goodglass and Landis, 1982; Graves and Landis, 1990). This effect, however, is somewhat variable depending on the task and the sex of the subject. Graves, Goodglass and Landis (1982) conducted four separate experiments to further explore asymmetrical use of the mouth during speech production. Male and female subjects were asked to either describe a scene or generate a list of words. While males showed a significant bias for the right side of the mouth for both of these tasks, female subjects showed the right side bias only for generating the word list. Only slightly more than half of the female subjects showed a right-side bias while describing the scene (44 of 85). Since there have been no follow-up studies, we do not yet know whether certain types of words (i.e. vowel, nouns, adjectives, etc.) show greater asymmetries than others or whether different speech acts (e.g. producing nonsense words with linguistically relevant intonation or rhythm) cause other patterns of asymmetries in mouth use. More research in this area is likely to elucidate our knowledge of the asymmetrical use of the mouth during speech production.

13.6.3. Hemispheric Asymmetries in the Production of Primate Facial and Vocal Expressions

Hauser (1993a) conducted an analysis of adult rhesus monkey facial expressions to explore the possibility of hemispheric asymmetries. Video footage was taken of free-ranging rhesus monkeys producing four different facial expressions: the fear grimace, copulation grimace, open mouth threat and ear flap threat. Video records were then analysed to determine which side of the face started the expression first, and which side was more expressive. For all facial expressions, rhesus monkeys showed a left side bias. To further explore this asymmetry, chimeras of the rhesus expressions were created for the right and left side of the face. Human subjects were then asked to rate the chimeras for overall expressiveness. Forty-one of the forty-three subjects reported that the chimeras for the left side of the face were more expressive then the right-side chimeras.

All of the expressions used in Hauser's analyses were associated with negative/withdrawal emotions, with the exception of the copulation grimace. Although the copulation grimace is produced by males during copulation,

presumably an interaction associated with positive emotion, males are sometimes attacked during copulation (Hauser, 1993b); the possibility of being
attacked may cause the act of copulation to be associated with negative
emotion as well. Overall, then, the pattern of results obtained suggest that,
like humans, rhesus monkeys also exhibit right hemisphere dominance for
facial expressions associated with negative/withdrawal emotions.

In a follow-up study, Hauser (1999) found that, in contrast to other facial
expressions, adult and juvenile rhesus monkeys exhibited a right side bias
while producing the play face. The play face is the only expression in the
rhesus monkey repertoire that is clearly associated with positive/approach
emotions. In contrast to adults and juveniles, Hauser found no evidence for a
directional bias in the facial expressions of infants. The results of this study,
together with Hauser's (1993a) earlier report, suggest that the directional
asymmetries exhibited by rhesus during the production of facial expressions
are similar to those shown by humans. However, these studies were based on
relatively small sample sizes, thus making the putative similarity with human
facial expressions relatively tenuous.

To increase the sample of subjects, as well as the number of exemplars per
facial expression type, Hauser and Akre (2001) conducted a follow-up study
of rhesus monkey facial expressions. In addition to facial expressions, they
also analysed video records of vocalizations and the gestures mediating their
production. Their methodology followed that of Hauser (1993a, 1999), measuring the side of the mouth first initiating the expression. Results showed
that adults exhibited a statistically significant left side bias during the production of four different vocalizations, but no side bias for four other vocalizations. The four facial expressions first measured by Hauser (1993a) also
showed a left side bias, but with a larger sample of subjects and exemplars per
expression type. Given that the repertoire of facial expressions and vocalizations analysed include signals associated with both positive/approach and
negative/withdrawal emotions, these studies fail to support the hypothesis
that the direction of asymmetry covaries with emotional valence. Thus, and
in contrast to the earlier reports, adult rhesus monkeys show a significant
right hemisphere/left side of the mouth bias for both vocal and facial expressions, given in a variety of emotional contexts. These results stand in contrast
to the patterns obtained for humans, but support the general conclusion that
rhesus monkeys show hemispheric asymmetries for producing communicative expressions.

Analyses of infants revealed a left side bias in three of the four vocalizations, and one of the three facial expressions. Because these expressions were
associated with both positive/approach and negative/withdrawal emotions,

the pattern exhibited by infants also provides no support for the emotional valence hypothesis. That is, infants show a right hemisphere bias for producing both facial and vocal expressions, independently of emotional content.

Hauser, Akre and Goldenberg (in preparation) also analysed the extent to which expressions remained lateralized over the course of the expression, as well as the degree to which the mouth opened wider on one side than on the other. In this study, asymmetries in mouth opening (i.e. area of the left and right sides) were scored from video for each frame for four different facial expressions and six different vocalizations (Figure 13.2). Specifically, for each frame the face was bisected by dividing the face into left and right sides with a line drawn directly down the midpoint of the face. The area was then determined by measuring the area of the open mouth on each side of the midpoint line. The data were analysed for the following effects: the overall number of frames that showed a left vs. right side bias in each expression; the side of the mouth initiating the expression and completing the expression; the magnitude of the asymmetry during an expression; and the within-subject stability for producing a specific expression. Results indicated that adults show a left side bias for most vocal and facial expressions in the first three analyses. Additionally, the within-subject analyses indicated that subjects consistently showed the same lateral bias across exemplars. Infants, in contrast, showed moderately less asymmetry, but all expressions exhibiting statistically significant side biases were in the same direction as for adults: left side.

In parallel with the findings from Hauser and Akre (2001), the asymmetries revealed from the analyses of mouth opening cannot be accounted for by the emotional valence hypothesis. Specifically, although those expressions showing statistically significant side biases were all biased to the left side, some of these expressions were associated with positive/approach emotions, and some with negative/withdrawal emotions. Taken together, studies of rhesus monkeys (Hauser, 1993a, 1999; Hauser and Akre, 2001; Hauser, Akre and Goldenberg, in preparation) indicate right hemisphere dominance during the production of both facial and vocal expressions, independent of emotional content.

Asymmetrical use of the mouth during vocal production has also been shown in the common marmoset (*Callithrix jacchus*), a New World primate. Hook-Costigan and Rogers (1998) studied asymmetries in the production of the tsik and twitter vocalizations. The tsik call/expression is apparently associated with fear and is generally given when individuals come into contact with a predator. The expression can occur with the vocalization, but it is also given in isolation. The twitter is used as a social contact call (Epple, 1968). Based on contextual observations, Hook-Costigan and Rogers argue that the

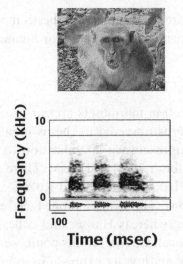

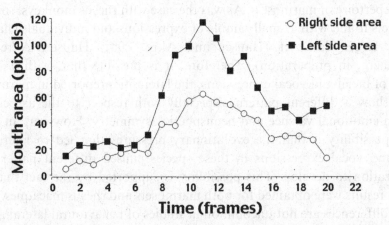

Figure 13.2. Top: A rhesus monkey producing a copulation scream. Middle: A spectrogram and smaller amplitude waveform of a copulation scream. Bottom: A graph of the area asymmetry during production of a copulation scream.

tsik call/expression is associated with negative/withdrawal emotions, while the twitter is associated with positive/approach emotion. After recording video footage of captive adult marmosets producing these facial and vocal expressions, a frame by frame analysis was conducted to assess asymmetrical use of the mouth, as well as the stability of the lateral bias over the duration of the expression. Results indicated a left side bias during production of the tsik expression, with or without the vocalization, and a right side bias when the subjects produced twitter vocalizations. In addition, these side biases were maintained over the course of the entire expression. These results stand in contrast to the patterns obtained for rhesus monkeys, but converge with the directional asymmetries documented for humans.

13.6.4. Patterns and Problems

The patterns of lateralization in marmosets appear to resemble those found in humans, while the patterns observed in rhesus monkeys expressions are different, the latter showing exclusive lateralization to the left side of the mouth, for all expressions, both vocal and facial. There are several potential reasons for this difference. Hook-Costigan and Rogers only examined asymmetries associated with two vocalizations and one facial expression unaccompanied by any vocalization, whereas Hauser and colleagues looked at four facial expressions and six vocalizations. At this point, we cannot yet ascertain the extent to which the tsik and twitter expressions are representative of the entire repertoire of marmosets. As was the case with rhesus monkeys, some of the effects found with a small sample of expressions and individuals changed with a larger data set (Hauser and Akre, 2001; Hauser, Akre and Goldenberg, in preparation). Therefore, it is possible that, with a larger sample of facial and vocal expressions, the lateralities reported in marmosets would show a different pattern, especially with respect to the association between emotional valence and hemispheric asymmetry. However, an alternative possibility is that the evolutionary pressures that led to lateralized facial and vocal expressions in these species caused different patterns of lateralization to develop. This possibility is supported by the fact that significant results were obtained for both marmosets and rhesus macaques. Such species differences are not uncommon in studies of behavioural lateralization in primates (Bradshaw and Rogers, 1994; McGrew and Marchant, 1997). However, until more data are available for comparison, any of these possibilities may prove correct.

Researchers studying cerebral asymmetries in humans have differentiated between facial and vocal expressions. Typically, the lateralization associated with speech production is separated from expressions in which the face and mouth change but no sounds are uttered. Hook-Costigan and Rogers (1998) do not, however, make this distinction. In their analysis, they apply the emotional valence hypothesis (Davidson, 1995) to both facial and vocal expression. Little is known about the relationship between affective changes and asymmetries in the use of the mouth during speech production. Therefore, it would be premature to assume that all expressions, facial and vocal, will be lateralized according to the pattern found for facial expression. Clearly, more work is needed on human and primate expressions to understand fully the relationship between asymmetries at the level of the central nervous system and asymmetries at the periphery. Furthermore, we need to extend the comparative data beyond rhesus monkeys and marmosets, and provide more careful analyses of the underlying emotions as well as the communicative content of both vocal and facial expressions.

13.7. Neural Mechanisms Underlying the Production of Facial and Vocal Expressions

13.7.1. *Neural Control of Orofacial Movements*

Surprisingly little work has been done on the neurobiology underlying the production of primate facial expressions. To our knowledge, the only study conducted to examine the contributions of the neocortex in the production of species-typical facial expressions is that of Ifune, Vermeire and Hamilton (1984). Using split-brained rhesus macaques, these investigators compared the frequency with which visual stimuli such as natural scenes (including videos of humans and primates) could elicit facial expressions. They found that visual stimulation of the right hemisphere elicited significantly more facial expressions from their subjects than the left hemisphere.

At a more basic level, electrical stimulation of motor cortical areas has been used to delineate the representations of orofacial movements. In both the owl monkey (*Aotus trivirgatus*) (Preuss, Stepniewska and Kaas, 1996) and the genus *Macaca* (Goldshalk et al., 1985; Gentilucci et al., 1988; Huang et al., 1988), it was found that electrical stimulation of the ventral regions of the primary motor cortex and the ventral premotor cortex can elicit facial and oral movements. However, the relationship between the movements evoked in these studies and species-typical expressions is ambiguous. Thus, although, these cortical areas could be involved in facial expressions, vocal production and/or food consumption, they do provide a foundation for future studies of the motor control of facial expressions.

13.7.2. *Cortical Control of Vocal Behaviour*

In humans, electrical stimulation and lesion studies have demonstrated that motor, premotor and prefrontal regions of the antero-lateral neocortex are involved in the specialized functions of speech (e.g. naming, syntax, verbal memory, etc.) (Ojemann, 1983). In the majority of right-handed subjects, these functions are lateralized to the left hemisphere. The medial part of the anterior cerebral cortex – the mesial cortex – is also involved in human speech (Sutton and Jürgens, 1988). This region includes the anterior cingulate and supplementary motor area. Based on electrical stimulation and lesion studies, the mesial cortex appears to be involved in the less cognitive aspects of speech production, such as the regulation of speech and the basic motor control of vocalizations (e.g. the initiation of a vocal response) (Sutton and Jürgens, 1988).

13.7.2.1. The Cingulate and Supplementary Motor Cortex

Using electrical stimulation and experimental lesions, at least some of the cortical regions involved in primate vocal production have been identified. Most of this work focuses on squirrel monkeys and macaques. In an intensive series of studies using electrical stimulation, Jürgens and his colleagues have mapped all the cortical and subcortical areas of the squirrel monkey brain that are involved in producing vocalizations (for a review, see Jürgens, 1992). They found that the region around the anterior cingulate sulcus, which includes the supplementary motor area and limbic cortex, is the cortical region from which vocalizations can be elicited via electrical stimulation (Sutton and Jürgens, 1988). In support of these data, bilateral lesions of these areas reduce the number of spontaneous vocalizations, although different call types are affected differentially (Kirzinger and Jürgens, 1982).

Data from macaques are nearly identical to the results found for squirrel monkeys. Electrical stimulation of the anterior cingulate cortex and the supplementary motor area elicit vocalizations, although not nearly as many call types as in squirrel monkeys (Smith, 1945; Sloan and Kaada, 1953; Robinson, 1967). Lesions of these areas, as in the squirrel monkey, reduce the rate of vocalizations (Aitken, 1981; Sutton, Larson and Lindeman, 1974; Sutton, Trachy and Lindeman, 1985). Furthermore, single unit studies have demonstrated vocalization-related neural activity in the mesial cortex of macaques (West and Larson, 1995). These results from squirrel monkeys and macaques suggest homologies with the vocalization-related areas of the mesial cortex in humans.

13.7.2.2. The Lateral Motor Cortical Areas

Studies of lateral neocortex have not established a clear-cut role for this region in primate vocal production. Electrical stimulation of the motor cortex in both squirrel monkeys and macaques elicited vocal fold movements (Hast and Milojevic, 1966; Hast et al., 1974), but few, if any, vocalizations were elicited (Green and Walker, 1938). Bilateral lesions of motor and premotor 'face' areas, presumptive homologues of Broca's area in humans, do not appear to alter the acoustic structure of calls or the rate of calling in squirrel monkeys (Kirzinger and Jürgens, 1982) and macaques (Sutton, Larson and Lindeman, 1974; Aitken, 1981). Thus, while the lateral precentral cortex may be involved in oral and laryngeal movements, its role in vocal behaviour does not appear to be critical, based on current evidence.

13.7.3. Patterns and Problems

The search for the evolutionary substrates underlying these vocalization-related areas has been only partially successful. To date, only squirrel monkeys and macaques have been studied extensively. Data from these species suggest that, like humans, the anterior cingulate cortex and supplementary motor area play an important role in vocal production. However, there is scant evidence for the participation of lateral neocortical motor areas in primate vocal behaviour, and hemispheric asymmetries at the neural level have not been investigated.

In humans, higher order vocal control of speech is generally associated with areas in the lateral neocortex. It is unclear what role, if any, the lateral neocortical motor areas play in primate vocal production (Deacon, 1997), but the manner in which their potential role has been tested has not been thorough. For example, in human studies, it is known that the application of electrical stimulation to the lateral neocortex outside of the primary motor area of awake, but quiet, human subjects *does not* elicit speech sounds (Ojemann, 1983). It is only when stimulation is applied during an ongoing language task that effects of stimulation can be identified. Thus, in the case of the primate studies, a more fruitful approach may be to apply electrical stimulation during vocalizations, especially in primates that produce long multisyllabic calls (e.g. cotton-top tamarins, capuchins, marmosets).

Another approach may be to use behavioural/lesion techniques in which the control of vocal production can be assessed by the experimental manipulation of external noise, followed by measuring the compensatory mechanisms used to transmit vocal signals. For example, voice amplitude in particular frequency bandwidths can be controlled by macaques in the presence of acoustic noise (Sinnott, Stebbins and Moody, 1975). In conjunction with experimental lesions, such studies may reveal the neural basis for such exquisite motor control. Indeed, there is evidence that voice amplitude may be affected by lesions of the motor cortical 'face' area (Green and Walker, 1938).

A major limitation of many of these neural studies is that our knowledge of the acoustic structure and variation of primate vocalizations was quite limited at the time these studies were conducted. The technology was not available to quantify carefully potential changes in the acoustic structure of vocalizations following lesions. We now have behavioural and acoustic evidence in Old World monkeys that the spectral properties of their vocalizations are in part the result of articulatory gestures, such as the movement of the lips, tongue and jaws (Hauser, Evans and Marler, 1993; Hauser and Schön Ybarra, 1994). For example, quantitative examination of the 'coo' vocalization (used in all the macaque studies cited above) indicates that the changes in the position of the jaw are reliably associated with changes in the dominant frequency (i.e. resonance frequency) but not the fundamental frequency (Hauser, Evans and Marler, 1993). Given that there are motor and premotor areas involved in voluntary oral movements in primates, including movements of the jaw, it is natural to assume that they would play an important role in the articulatory control of vocalizations. However, this remains to be tested neurophysiologically.

13.8. Conclusions

In this chapter, we have attempted to provide a summary and synthesis of current studies aimed at specialized processing of facial and vocal expressions in primates. What emerges is that primates show evidence of behavioural, neurophysiological and neuroanatomical asymmetries for the perception and production of facial and vocal expressions. What we cannot yet account for is the pattern of variation across species and, in some cases, within species but across expression types. Some of this variation is likely to be explained by differences in experimental procedure. For a better understanding of the evolution of hemispheric specialization for communicative expressions, future studies must sample a broader range of species, using comparable methodological procedures.

Acknowledgements

For research on rhesus monkeys, we thank the staff of the Caribbean Primate Research Center (P51RR00168-37). For facilitating our research on cottontop tamarins, we thank the animal care staff at Harvard University. During the writing of this paper, Asif Ghazanfar was supported by an NIH postdoctoral fellowship (F32 DC00377), and Marc Hauser was supported by grants from the National Science Foundation (SBR-9602858; SBR-9357976), and the Mind, Brain, and Behavior Program at Harvard University.

References

Adolphs, R., Damasio, H., Tranel, D. & Damasio, A.R. (1996). Cortical systems for the recognition of emotion in facial expressions. *Journal of Neuroscience*, 16, 7678–7687.

Aitken, P.G. (1981). Cortical control of conditioned and spontaneous vocal behavior in rhesus monkeys. *Brain and Language*, 13, 636–642.

Arnold, A.P. & Bottjer, S.W. (1985). Cerebral lateralization in birds. In *Cerebral Lateralization in Nonhuman Species*, ed. S.D. Glick, pp. 11–39. New York: Academic Press.

Ball, G.F. (1999). Neuroendocrine basis of seasonal changes in vocal behavior among songbirds. In *The Design of Animal Communication*, ed. M. Hauser and M. Konuishi, pp. 213–253. Cambridge, MA: MIT Press.

Beecher, M.D., Petersen, M.R., Zoloth, S.R., Moody, D.B. & Stebbins, W.C. (1979). Perception of conspecific vocalizations by Japanese macaques. *Brain, Behavior, and Evolution*, 16, 443–460.

Beeman, K. (1992). *SIGNAL User's Guide*. Belmont, MA: Engineering Design.

Bercovitch, F., Hauser, M.D. & Jones, J. (1995). The endocrine stress response and alarm vocalizations in rhesus macaques. *Animal Behaviour*, 49, 1703–1706.

Bradshaw, J. & Rogers, L. (eds) (1994). *The Evolution of Lateral Asymmetries, Language, Tool Use, and Intellect*. San Diego, CA: Academic Press.

Brosch, M., Schulz, A. & Scheich, H. (1999). Processing of sound sequences in macaque auditory cortex: Response enhancement. *Journal of Neurophysiology*, 82, 1542–1559.

Bruce, C. (1982). Face recognition by monkeys: absence of an inversion effect. *Neuropsychologia*, 20, 515–521.

Bruce, V. & Humphreys, G.W. (1994). Recognizing objects and faces. In *Object and Face Recognition. Special Issue of Visual Cognition*, Vol. 1, ed. V. Bruce and G.W. Humphreys, pp. 141–180. Hillsdale, NJ: Lawrence Erlbaum.

Bruce, V. & Young, A.W. (1986). Understanding face recognition. *British Journal of Psychology*, 77, 305–327.

Bruce, C., Desimone, R. & Gross, C.G. (1981). Visual properties of neurones in a polysensory area in superior temporal sulcus of the macaque. *Journal of Neurophysiology*, 46, 369–384.

Bryden, M.P. (1982). *Laterality: Functional Asymmetry in the Human Brain*. New York: Academic Press.

Burton, L.A. & Levy, J. (1991). Effects of processing speed on cerebral asymmetry for left- and right-oriented faces. *Brain and Cognition*, 15(1), 95–105.

Camhi, J.M. (1984). *Neuroethology. Nerve Cells and the Natural Behavior of Animals*. Sunderland, MA: Sinauer Associates Inc., Publishers.

Campbell, R., Heywood, C.A., Cowey, A., Regard, M. & Landis, T. (1990). Sensitivity to eye gaze in prosopagnosic patients and monkeys with superior temporal sulcus ablation. *Neuropsychologia*, 28, 1123–1142.

Carey, S. & Diamond, R. (1977). From piecemeal to configurational representation of faces. *Science*, 195, 312–314.

Cheney, D.L. & Seyfarth, R.M. (1982). How vervet monkeys perceive their grunts: Field playback experiments. *Animal Behaviour*, 30, 739–751.

Cheney, D.L. & Seyfarth, R.M. (1990). *How Monkeys See the World: Inside the Mind of Another Species*. Chicago, IL: Chicago University Press.

Corballis, M.C. (1998). Cerebral asymmetry: Motoring on. *Trends in Cognitive Neuroscience*, 2, 152–157.

Damasio, A.R., Tranel, D. & Damasio, H. (1990). Facial agnosia and the neural substrates of memory. *Annual Review of Neuroscience*, 13, 89–109.

Darwin, C. (1872). *The Expression of the Emotions in Man and Animals*. London: John Murray.

Davidson, R.J. (1995). Cerebral asymmetry, emotion and affective style. In *Brain Asymmetry*, eds R.J. Davidson & K. Hugdahl, pp. 361–388. Cambridge, MA: MIT Press.

Davidson, R.J. (1999). The functional neuroanatomy of emotion and affective style. *Trends in Cognitive Science*, 3, 11–21.

Davidson, R.J., Ekman, P., Saron, C.D., Senulis, J.A. & Friesen, W.V. (1990). Approach–withdrawal and cerebral asymmetry: Emotional expression and brain physiology I. *Journal of Personality and Social Psychology*, 58, 330–341.

Deacon, T.W. (1997). *The Symbolic Species: The Co-evolution of Language and the Brain*. New York: W.W. Norton.

Desimone, R., Albright, T.D., Gross, C.G. & Bruce, C. (1984). Stimulus-selective properties of inferior temporal neurones in the macaque. *Journal of Neuroscience*, 4, 2051–2062.

Dewson, J.H. (1977). Preliminary evidence of hemispheric asymmetry of auditory function in monkeys. In *Lateralization in the Nervous System*, eds S. Harnad, R.W. Doty, L. Goldstein, J. Jaynes & G. Krauthamer, pp. 63–71. New York: Academic Press.

Dittus, W.P. (1984). Toque macaque food calls: Semantic communication concerning food distribution in the environment. *Animal Behaviour*, 32, 470–477.

Doupe, A.J. & Solis, M.M. (1999). Song- and order-selective auditory responses emerge in neurones of the songbird anterior forebrain during vocal learning. In *The Design of Animal Communication*, eds M.D. Hauser & M. Konishi, pp. 343–368. Cambridge, MA: MIT Press.

Eible-Eibesfeldt, I. (1973). The expressive behavior of the deaf and blind born. In *Social Communication and Movement*, eds M. v. Cranach & I. Vine, pp. 163–194. San Diego, CA: Academic Press.

Ekman, P. (1973). *Darwin and Facial Expression*. London: Academic Press.

Ekman, P. (1992). An argument for basic emotions. *Cognition and Emotion*, 6, 169–200.

Epple, G. (1968). Comparative studies on vocalization in marmoset monkeys (*Hapalidae*). *Folia Primatologica*, 8, 1–40.

Evans, C.S., Evans, L. & Marler, P. (1994). On the meaning of alarm calls: Functional reference in an avian vocal system. *Animal Behaviour*, 45, 23–28.

Falk, D., Cheverud, J., Vannier, M.W. & Conroy, G.C. (1986). Advanced computer graphics technology reveals cortical asymmetry in endocasts of rhesus monkeys. *Folia Primatologica*, 46, 98–103.

Farah, M.J., Drain, J.W. & Maxwell, H. (1995). What causes the face inversion effect? *Journal of Experimental Psychology: Human Perception and Performance*, 21, 628–634.

Farah, M.J., Wilson, K.D., Drain, H.M. & Tanaka, J.R. (1995). The inverted face inversion effect in prosopagnosia: Evidence for mandatory, face-specific perceptual mechanisms. *Vision Research*, 35, 2089–2093.

Fitch, R.H., Miller, S. & Tallal, P. (1997). Neurobiology of speech perception. *Annual Review of Neuroscience*, 20, 331–353.

Fridlund, A.J. (1994). *Human Facial Expressions: An Evolutionary Perspective.* New York: Academic Press.

Gannon, P.J., Broadfield, D.C., Kheck, N.M., Hof, P.R., Braum, A.R., Erwin, J. M. & Holloway, R.L. (1998a). Brain language area evolution I: Anatomic expression of Heschl's gyrus and planum temporale asymmetry in great apes, lesser apes, and Old World monkeys. *Society for Neuroscience Abstracts*, 24, 160.

Gannon, P.J., Holloway, R.L., Broadfield, D.C. & Braun, A.R. (1998b). Asymmetry of chimpanzee planum temporale: Humanlike pattern of Wernicke's brain language area homolog. *Science*, 279, 220–222.

Gannon, P.J., Kheck, N.M. & Hof, P.R. (1999). Brain language area evolution III: Left hemisphere predominant hemispheric asymmetry of cytoarchitectonic, but not gross anatomic, planum temporale homolog in old world monkeys. *Society for Neuroscience Abstracts*, 25, 105.

Gentilucci, M., Fogassi, L., Luppino, G., Matelli, M., Camarda, R. & Rizzolatti, G. (1988). Functional organization of inferior area 6 in the macaque monkey: I. Somatotopy and control of proximal movements. *Experimental Brain Research*, 71, 475–490.

Geschwind, N. & Levitsky, W. (1968). Human brain: Left–right asymmetries in temporal speech region. *Science*, 161, 186–187.

Ghazanfar, A.A. & Hauser, M.D. (1999). The neuroethology of primate vocal communication: substrates for the evolution of speech. *Trends in Cognitive Sciences*, 3, 377–384.

Ghazanfar, A.A. & Hauser, M.D. (in preparation). Orienting asymmetries in rhesus macaques: Playbacks of reversed calls.

Goldshalk, M., Mitz, A.R., van Duin, B. & van der Burg, H. (1985). Somatotopy of monkey premotor cortex examined with microstimulation. *Neuroscience Research*, 23, 269–279.

Gouzoules, H. & Gouzoules, S. (1989). Design features and developmental modification in pigtail macaque (*Macaca nemestrina*) agonistic screams. *Animal Behaviour*, 37, 383–401.

Gouzoules, S., Gouzoules, H. & Marler, P. (1984). Rhesus monkey (*Macaca mulatta*) screams: Representational signalling in the recruitment of agonistic aid. *Animal Behaviour*, 32, 182–193.

Graves, R. & Landis, T. (1990). Asymmetry in mouth opening during different speech tasks. *International Journal of Psychology*, 25, 179–189.

Graves, R., Goodglass, H. & Landis, T. (1982). Mouth asymmetry during spontaneous speech. *Neuropsychologia*, 20, 371–381.

Green, H.D. & Walker, A.E. (1938). The effects of ablation of the cortical motor face area in monkeys. *Journal of Neurophysiology*, 1, 262–280.

Green, S. (1975). Communication by a graded vocal system in Japanese monkeys. In *Primate Behavior*, Vol. 4, ed. L.A. Roseenblum, pp. 1–102. New York: Academic Press.

Haith, M.M., Bergman, T. & Moore, M.J. (1977). Eye contact and face scanning in early infancy. *Science*, 198, 853–855.

Hamilton, C.R. & Vermeire, B.A. (1983). Discrimination of monkey faces by split-brain monkeys. *Behavioural Brain Research*, 9, 263–275.

Hamilton, C.R. & Vermeire, B.A. (1988). Complementary hemispheric specialization in monkeys. *Science*, 242, 1691–1694.

Hamilton, C.R. & Vermeire, B.A. (1991). Functional lateralization in monkeys. In *Cerebral Laterality*, ed. F.L. Kitterle, pp. 19–34. Cambridge: Cambridge University Press.

Harnard, S. (1987). *Categorical Perception: The Groundwork of Cognition*. Cambridge, MA: Cambridge University Press.

Hasselmo, M.E., Rolls, E.T. & Baylis, G.C. (1989). The role of expression and identity in the face-selective responses of neurones in the temporal visual cortex of the monkey. *Behavioural Brain Research*, 32, 203–218.

Hast, M.H. & Milojevic, B. (1966). The response of the vocal folds to electrical stimulation of the inferior frontal cortex of the squirrel monkey. *Acta Otolaryngolica*, 61, 196–204.

Hast, M.H., Fischer, J.M., Wetzel, A.B. & Thompson, V.E. (1974). Cortical motor representation of the laryngeal muscles in *Macaca mulatta*. *Brain Research*, 73, 229–240.

Hauser, M.D. (1993a). Right hemisphere dominance for the production of facial expressions in monkeys. *Science*, 261, 475–477.

Hauser, M.D. (1993b). Rhesus monkey (*Macaca mulatta*) copulation calls: Honest signals for female choice? *Proceedings of the Royal Society of London*, 254, 93–96.

Hauser, M.D. (1996). *The Evolution of Communication*. Cambridge, MA: MIT Press.

Hauser, M.D. (1998). Functional referents and acoustic similarity: Field playback experiments with rhesus monkeys. *Animal Behaviour*, 55, 1647–1658.

Hauser, M.D. (1999). The evolution of the lopsided brain: Asymmetries underlying facial and vocal expression in primates. In *The Design of Animal Communication*, eds M.D. Hauser & M. Konishi, pp. 597–628. Cambridge, MA: MIT Press.

Hauser, M.D. & Akre, K. (2001). Motor asymmetries in the timing of facial and vocal expressions by rhesus monkeys: implications for hemisphereic specialization. *Animal Behaviour*, 61, 391–400.

Hauser, M.D. & Andersson, K. (1994). Left hemisphere dominance for processing vocalizations in adult, but not infant rhesus monkeys: field experiments. *Proceedings of the National Academy of Sciences USA*, 91, 3946–3948.

Hauser, M.D. & Konishi, M. (eds) (1999). *The Design of Animal Communication*. Cambridge, MA: MIT Press.

Hauser, M.D. & Marler, P. (1993). Food-associated calls in rhesus macaques (*Macaca Mulatta*). I. Socioecological factors influencing call production. *Behavioral Ecology*, 4, 194–205.

Hauser, M.D. & Schön Ybarra, M. (1994). The role of lip configuration in monkey vocalizations: Experiments using xylocaine as a nerve block. *Brain and Language*, 46, 232–244.

Hauser, M.D., Evans, C.S. & Marler, P. (1993). The role of articulation in the production of rhesus monkey, *Macaca mulatta*, vocalizations. *Animal Behaviour*, 45, 423–433.

Hauser, M.D., Agnetta, B. & Perez, C. (1998). Orienting asymmetries in rhesus monkeys: The effect of time-domain changes on acoustic perception. *Animal Behaviour*, 56(1), 41–47.

Hauser, M.D., Akre, K. & Goldenberg, D. (in preparation). Asymmetries in rhesus monkey facial and vocal expressions: right hemisphere dominance across modalities and communicative content.

Haxby, J.V., Ungerleider, L.G., Clark, V.P., Schouten, J.L., Hoffman, E.A. & Martin, A. (1999). The effect of face inversion on activity in human neural systems for face and object perception. *Neuron*, 22, 189–199.

Heffner, H.E. & Heffner, R.S. (1984). Temporal lobe lesions and perception of species-specific vocalizations by macaques. *Science*, 226, 75–76.

Heffner, H.E. & Heffner, R.S. (1986). Effect of unilateral and bilateral auditory cortex lesions on the discrimination of vocalizations by Japanese macaques. *Journal of Neurophysiology*, 56, 683–701.

Heilbroner, P.L. & Holloway, R.L. (1988). Anatomical brain asymmetries in New World and Old World monkeys: Stages of temporal lobe development in primate evolution. *American Journal of Physical Anthropology*, 16, 38–48.

Heywood, C.A. & Cowey, A. (1992). The role of the 'face-cell' area in the discrimination and recognition of faces by monkeys. *Philosophical Transactions of the Royal Society of London, Series B*, 335, 31–38.

Hilliard, R.D. (1973). Hemispheric laterality effects on a facial recognition task in normal subjects. *Cortex*, 9, 246–258.

Hoffman, E.A. & Haxby, J.V. (2000). Distinct representations of eye gaze and identity in the distributed human neural system for face perception. *Nature Neuroscience*, 3, 80–84.

Hook-Costigan, M.A. & Rogers, L.J. (1998). Lateralized use of the mouth in production of vocalizations by marmosets. *Neuropsychologia*, 36, 1265–1273.

Huang, C.-S., Sirisko, M.A., Hiraba, H., Murray, G.M. & Sessle, B.J. (1988). Organization of the primate face motor cortex as revealed by intracortical microstimulation and electrophysiological identification of afferent inputs and corticobulbar projections. *Journal of Neurophysiology*, 59, 796–818.

Hupfer, K., Jürgens, U. & Ploog, D. (1977). The effect of superior temporal lesions on the recognition of species-specific calls in the squirrel monkey. *Experimental Brain Research*, 30, 75–87.

Ifune, C.K., Vermeire, B.A. & Hamilton, C.R. (1984). Hemispheric differences in split-brain monkeys viewing and responding to videotape recordings. *Behavioral and Neural Biology*, 41, 231–235.

Jürgens, U. (1990). Vocal communication in primates. In *Neurobiology of Comparative Cognition*, eds R.P. Kesner & D.S. Olton, pp. 51–76. Hillsdale, NJ: Erlbaum Associates.

Jürgens, U. (1992). Neurobiology of vocal communication. In *Nonverbal Vocal Communication: Comparative and Developmental Approaches*, eds H. Papoucek, U. Jürgens & M. Papoucek, pp. 31–42. Cambridge: Cambridge University Press.

Kaas, J.H. & Hackett, T.A. (1998). Subdivisions of auditory cortex and levels of processing in primates. *Audiology and Neuro-otology*, 3, 73–85.

Kanwal, J.S. (1999). Processing of species-specific calls by combination-sensitive neurones in an echolocating bat. In *The Design of Animal Communication*, eds M.D. Hauser & M. Konishi, pp. 159–186. Cambridge, MA: MIT Press.

Kanwisher, N., McDermott, J. & Chun, M.M. (1997). The fusiform face area: A module in human extrastriate cortex specialized for face processing. *Journal of Neuroscience*, 17, 4302–4311.

Kanwisher, N., Tong, F. & Nakayama, K. (1998). The effect of face inversion on the human fusiform face. *Cognition*, 68, B1–B11.

Keating, C.F. & Keating, E.G. (1982). Visual scan patterns of rhesus monkeys viewing faces. *Perception*, 11, 211–219.

Kheck, N.M., Hof, P. R., Deftereos, M., Lo, T. & Gannon, P.J. (1999). Brain language area evolution IV: Chemoarchitectonic interhemispheric asymmetries of the planum temporale (PT) homolog in old world monkeys. *Society for Neuroscience Abstracts*, 25, 105.

Kirzinger, A. & Jürgens, U. (1982). Cortical lesion effects and vocalization in the squirrel monkey. *Brain Research*, 233, 299–315.

Konishi, M. (1985). Birdsong: From behavior to neurone. *Annual Review of Neuroscience*, 8, 125–170.

Kuhl, P.K. & Meltzoff, A.N. (1982). The bimodal perception of speech in infancy. *Science*, 218, 1138–1141.

Kuhl, P.K. & Meltzoff, A.N. (1988). Speech as an intermodal object of perception. In *Perceptual Development in Infancy*, ed. A. Yonas, pp. 235–266. Hillsdale, NJ: Lawrence Erlbaum Associates.

Leehey, S.C., Carey, S., Diamond, R. & Cahn, A. (1978). Upright and inverted faces: The right hemisphere knows the difference. *Cortex*, 14, 411–419.

LeMay, M. & Geschwind, N. (1975). Hemispheric differences in the brains of great apes. *Brain, Behavior and Evolution*, 11, 48–52.

Liberman, A.M., Cooper, F.S., Shankweiler, D.P. & Studdert-Kennedy, M. (1967). Perception of the speech code. *Psychology Review*, 74, 431–461.

Logothetis, N.K., Guggenberger, H., Peled, S. & Pauls, J. (1999). Functional imaging of the monkey brain. *Nature Neuroscience*, 2, 555–562.

Margoliash, D. (1983). Acoustic parameters underlying the responses of song-specific neurones in the white-crowned sparrow. *Journal of Neuroscience*, 3, 1039–1057.

Margoliash, D. & Fortune, E.S. (1992). Temporal and harmonic combination-sensitive neurones in the zebra finch's HVc. *Journal of Neuroscience*, 12, 4309–4326.

Marler, P. (1970). Birdsong and speech development: could there be parallels? *American Scientist*, 58, 669–673.

Marler, P., Evans, C.S. & Hauser, M.D. (1992). Animal Signals? Reference, motivation or both? In *Nonverbal Vocal Communication: Comparative and Developmental Approaches*, eds H. Papoucek, U. Jürgens and M. Papoucek. Cambridge: Cambridge University Press.

Masataka, N. & Fujita, K. (1989). Vocal learning of Japanese and rhesus monkeys. *Behaviour*, 109, 191–199.

Matsuzawa, T. (1996). Chimpanzee intelligence in nature and in captivity: Isomorphism of symbol use and tool use. In *Great Ape Societies*, ed. W.C. McGrew & L.F. Marchant, pp. 196–209. Cambridge: Cambridge University Press.

May, B., Moody, D.B. & Stebbins, W.C. (1988). The significant features of Japanese macaque coo sounds: A psychophysical study. *Animal Behaviour*, 36, 1432–1444.

Mazzocchi, D. & Vignolo, L.A. (1979). Localisation of lesions in aphasias: Clinical-CT scan correlations in stroke patients. *Cortex*, 15, 627–654.

McGrew, W.C. & Marchant, L.F. (1997). On the other hand: Current issues in and meta analysis of behavioral laterality of hand function in nonhuman primates. *Yearbook of Physical Anthropology*, 40, 201–232.

Morris, R.D. & Hopkins, W.D. (1993). Perception of human chimeric faces by chimpanzees: Evidence for a right hemisphere advantage. *Brain and Cognition*, 21, 111–122.

Morton, E.S. (1977). On the occurrence and significance of motivation-structural rules in some bird and mammal sounds. *American Naturalist*, 111, 855–869.

Mueller, R.-A., Rothermel, R.D., Behen, M.E., Muzik, O., Chakraborty, P.K. & Chugani, H.T. (1999). Language organization in patients with early and late left-hemisphere lesion: A PET study. *Neuropsychologia*, 37, 545–557.

Naeser, M.A., Helm-Estabrooks, N., Haas, G., Auerbach, S. & Srinivasan, M. (1987). Relationship between lesion extent in 'Wernicke's area' on computed tomographic scan and predicting recovery of comprehension in Wernicke's aphasia. *Archives of Neurology*, 44, 73–82.

Nahm, F.K.D., Perret, A., Amaral, D.G. & Albright, T.D. (1997). How do monkeys look at faces? *Journal of Cognitive Neuroscience*, 9, 611–623.

Newman, J.D. & Wollberg, Z. (1973). Multiple coding of species-specific vocalizations in the auditory cortex of squirrel monkeys. *Brain Research*, 54, 287–304.

Nottebohm, F. (1999). The anatomy and timing of vocal learning in birds. In *The Design of Animal Communication*, eds M.D. Hauser & M. Konishi, pp. 37–62. Cambridge, MA: MIT Press.

Nottebohm, F., Stokes, T.M. & Leonard, C.M. (1976). Central control of song in the canary, *Serinus canarius*. *Journal of Comparative Neurology*, 165, 457–486.

Ojemann, G.A. (1983). The intrahemispheric organization of human language, derived with electrical stimulation techniques. *Trends in Neurosciences*, 6, 184–189.

Overman, W.H. & Doty, R.W. (1982). Hemispheric specialization displayed by man but not macaques for analysis of faces. *Neuropsychologia*, 20, 113–128.

Owren, M.J., Dieter, J.A., Seyfarth, R.M. & Cheney, D.L. (1993). Vocalizations of rhesus (*Macaca mulatta*) and Japanese (*M. fuscata*) macaques cross-fostered between species show evidence of only limited modification. *Developmental Psychobiology*, 26, 389–406.

Parr, L.A., Hopkins, W.D. & de Waal, F.B.M. (1996). A preliminary report on face recognition in chimpanzees (*Pan troglodytes*). Paper presented at the *ASP/IPS Joint Congress*, Madison, WI.

Parr, L.A., Dove, T. & Hopkins, W.D. (1998). Why faces may be special: Evidence of the inversion effect in Chimpanzees. *Journal of Cognitive Neuroscience*, 10, 615–622.

Parr, L.A., Winslow, J.T. & Hopkins, W.D. (1999). Is the inversion effect in rhesus monkeys face-specific? *Animal Cognition*, 2, 123–129.

Perrett, D.I., Rolls, E.T. & Caan, W. (1982). Visual neurones responsive to faces in the monkey temporal cortex. *Experimental Brain Research*, 47, 329–342.

Perrett, D.I., Smith, PA.J., Potter, D.D., Mistlin, A.J., Head, A.S., Milner, A.D. & Jeeves, M.A. (1985). Visual cells in the temporal cortex sensitive to face view and gaze direction. *Proceedings of the Royal Society of London, Series B*, 223, 293–317.

Perrett, D.I., Mistlin, A.J., Chitty, A.J., Smith, P.A., Potter, D.D., Broennimann, R. & Harries, M. (1988). Specialized face processing and hemispheric asymmetry in man and monkey: evidence from single unit and reaction time studies. *Behavioral Brain Research*, 29, 245–258.

Petersen, M.R., Beecher, M.D., Zoloth, S.R., Moody, D.B. & Stebbins, W.C. (1978). Neural lateralization of species-specific vocalizations by Japanese macaques. *Science*, 202, 324–326.

Petersen, M.R., Beecher, M.D., Zoloth, S.R., Green, S., Marler, P.R., Moody, D.B. & Stebbins, W.C. (1984). Neural lateralization of vocalizations by Japanese macaques: Communicative significance is more important than acoustic structure. *Behavioral Neuroscience*, 98, 779–790.

Phelps, M.T. & Roberts, W.A. (1994). Memory for pictures of upright and inverted faces in humans (*Homo sapiens*), squirrel monkey (*Saimiri sciureus*) and pigeons (*Columba livia*). *Journal of Comparative Psychology*, 198, 114–125.

Pohl, P. (1983). Central auditory processing V: Ear advantage for acoustic stimuli in baboons. *Brain and Language*, 20, 44–53.

Pohl, D. (1984). Ear advantages for temporal resolution in baboons. *Brain and Cognition*, 3, 383–401.

Povinelli, D.J. & Eddy, T.J. (1996). What young chimpanzees know about seeing. *Monographs of the Society for Research in Child Development*, 247.

Preuschoft, S. (1995). *'Laughter' and 'Smiling' in Macaques – An Evolutionary Approach*. Utrecht: University of Utrecht Press.

Preuschoft, S. & van Hooff, J.A.R.A.M. (1997). The social functioning of 'smiling' and 'laughter'. Variations across primate species and societies. In *Nonverbal Communication: Where Nature Meets Culture*, ed. U. Segersträle & P. Molná, pp. 171–191. Mahwah, NJ: Lawrence Erlbaum.

Preuss, T.M. (1995). The argument from animals to humans in cognitive neuroscience. In *The Cognitive Neurosciences*, ed. M.S. Gazzaniga, pp. 1227–1242. Cambridge, MA: MIT Press.

Preuss, T.M., Stepniewska, I. & Kaas, J.H. (1996). Movement representation in the dorsal and ventral premotor areas of owl monkeys: a microstimulation study. *Journal of Comparative Neurology*, 371, 649–676.

Price, C.J. (1998). The functional anatomy of word comprehension and production. *Trends in Cognitive Sciences*, 2, 281–288.

Puce, A., Allison, T., Asgari, M., Gore, J.C. & McCarthy, G. (1996). Differential sensitivity of human visual cortex to faces, letterstrings, and textures: A functional magnetic resonance imaging study. *Journal of Neuroscience*, 16, 5205–5215.

Puce, A., Allison, T., Bentin, S., Gore, J.C. & McCarthy, G. (1999). Temporal cortex activation in humans viewing eye and mouth movements. *Journal of Neuroscience*, 18, 2188–2199.

Rauschecker, J.P. (1998). Cortical processing of complex sounds. *Current Opinion in Neurobiology*, 8, 516–521.

Rauschecker, J.P., Tian, B. & Hauser, M. (1995). Processing of complex sounds in the macaque nonprimary auditory cortex. *Science*, 268, 111–114.

Rendall, D., Rodman, P.S. & Emond, R.E. (1996). Vocal recognition of individuals and kin in free-ranging rhesus monkeys. *Animal Behaviour*, 51, 1007–1015.

Robinson, B. (1967). Vocalizations evoked from the forebrain in *Macaca mulatta*. *Physiology and Behavior*, 2, 345–354.

Rolls, E. (1999). *Brain and Emotion*. Oxford: Oxford University Press.

Ryan, M.J. & Rand, A.S. (1995). Female responses to ancestral advertisement calls in tungara frogs. *Science*, 269, 390–392.

Ryan, M.J. & Rand, A.S. (1999). Phylogenetic influence on mating call preferences in female tungara frogs, *Physalaemus pustulosus*. *Animal Behaviour*, 57, 945–956.

Ryan, M.J., Fox, J.H., Wilczynski, W. & Rand, A.S. (1990). Sexual selection for sensory exploitation in the frog, *Physalaemus pustulosus*. *Nature*, 343, 66–67.

Sackheim, H.A., Gur, R.C. & Saucy, M.C. (1978). Emotions are expressed more intensely on the left side of the face. *Science*, 202, 434–436.

Santos, L.R. & Hauser, M.D. (1999). How monkeys see the eyes: Cotton-top tamarins' reaction to changes in visual attention and action. *Animal Cognition*, 2, 131–139.

Schwartz, J. & Tallal, P. (1980). Rate of acoustic change may underlie hemispheric specialization for speech perception. *Science*, 207, 1380–1381.

Seyfarth, R.M., Cheney, D.L. & Marler, P. (1980). Monkey responses to three different alarm calls: Evidence of predator classification and semantic communication. *Science*, 210, 801–803.

Simmons, J.A. (1971). Echolocation in bats: Signal processing of echoes for target range. *Science*, 171, 925–928.

Sinnott, J.M., Stebbins, W.C. & Moody, D.B. (1975). Regulation of voice amplitude by the monkey. *Journal of the Acoustical Society of America*, 58, 412–414.

Sloan, N. & Kaada, B. (1953). Effects of anterior limbic stimulatin on somato-motor and electrocortical activity. *Journal of Neurophysiology*, 16, 203–220.

Smith, W.J. (1977). *The Behavior of Communicating*. Cambridge, MA: Harvard University Press.

Smith, W.K. (1945). The functional significance of the rostral cingular cortex as revealed by its responses to electrical stimulation. *Journal of Neurophysiology*, 8, 241–255.

Suga, N. (1989). Principles of auditory information-processing derived from neuroethology. *Journal of Experimental Biology*, 146, 277–286.

Suga, N., Niwa, H., Taniguchi, I. & Margoliash, D. (1987). The personalized auditory cortex of the mustached bat: Adaptation for echolocation. *Journal of Neurophysiology*, 58, 643–654.

Suthers, R.A. (1997). Peripheral control and lateralization of birdsong. *Journal of Neurobiology*, 33, 632–652.

Sutton, D. & Jürgens, U. (1988). Neural control of vocalization. *Comparative Primate Biology*, 4, 625–647.

Sutton, D., Larson, C. & Lindeman, R.C. (1974). Neocortical and limbic lesion effects on primate phonation. *Brain Research*, 71, 61–75.

Sutton, D., Trachy, R.E. & Lindeman, R.C. (1985). Discriminative phonation in macaques: Effects of anterior mesial cortex damage. *Experimental Brain Research*, 59, 410–413.

Tallal, P., Miller, S. & Fitch, R.H. (1993). Neurobiological basis of speech: A case for the preeminence of temporal processing. In *Temporal Information Processing in the Nervous System: Special reference to dyslexia and disphasia*, eds P. Tallal & A.M. Galaburda, *Annals of the New York Academy of Sciences*, Vol. 682, pp. 27–47. New York: New York Academy of Sciences.

Tomasello, M., Call, J. & Hare, B. (1998). Five primate species follow the visual gaze of conspecifics. *Animal Behaviour*, 55, 1063–1069.

Tomonaga, M. (1994). How laboratory-raised Japanese monkeys (*Macaca fuscata*) perceive rotated photographs of monkeys: Evidence for an inversion effect in face perception. *Primates*, 35, 155–165.

Tomonaga, M., Itakura, S. & Matsuzawa, T. (1993). Superiority of conspecific faces and reduced inversion effect in face perception by a chimpanzee. *Folia Primatologia*, 61, 110–114.

Tramo, M.J., Bellew, B.F. & Hauser, M.D. (1996). Discharge patterns of auditory cortical neurones evoked by species-specific vocalizations and synthetic complex signals in alert *Macaca mulatta*. Paper presented at the *Society for Neuroscience Abstracts*.

Trevarthen, C. (1979). Instincts for human understanding and for cultural cooperation: Their development in infancy. In *Human Ethology: Claims and Limits of a New Discipline*, eds M. v. Cranach, K. Foppa, W. Lepenies & D. Ploog. Cambridge: Cambridge University Press.

Valentine, T. & Bruce, V. (1986). The effect of race, inversion and encoding activity upon face recognition. *Acta Psychologica*, 61, 259–273.

Vermeire, B.A. & Hamilton, C.R. (1998). Inversion effect for faces in split-brain monkeys. *Neuropsychologia*, 36, 1003–1014.

Wang, X., Merzenich, M.M., Beitel, R. & Schreiner, C.E. (1995). Representation of a species-specific vocalization in the primary auditory cortex of the common marmoset: Temporal and spectral characteristics. *Journal of Neurophysiology*, 74, 2685–2706.

Weiss, D.J., Kralik, J.D. and Hauser, M.D. (2001). Face processing in cotton-top tamarins (*Saguinus oedipus oedipus*). *Animal Cognition*, 4, 191–205.

West, R.A. & Larson, C.R. (1995). Neurones of the anterior mesial cortex related to faciovocal activity in the awake monkey. *Journal of Neurophysiology*, 74, 1856–1869.

Whitman, R.D. & Keegan, J.F. (1991). Lateralization of facial processing: A spatial frequency model. *International Journal of Neuroscience*, 60, 177–185.

Wilson, F.A.W. & Goldman-Rakic, P.S. (1994). Viewing preferences of rhesus monkeys related to memory for complex pictures, colours, and faces. *Behavioural Brain Research*, 60, 79–89.

Winter, P. & Funkenstein, H.H. (1973). The effect of species-specific vocalizations on the discharge of auditory cortical cells in the awake squirrel monkey (*Saimiri sciureus*). *Experimental Brain Research*, 18, 489–504.

Wollberg, Z. & Newman, J.D. (1972). Auditory cortex of squirrel monkeys: Response properties of single cells to species-specific vocalizations. *Science*, 175, 212–214.

Wright, A.A. & Roberts, W.A. (1996). Monkey and human face perception: Inversion effects for human faces but not for monkey faces or scenes. *Journal of Cognitive Neuroscience*, 8, 278–290.

Yin, R.K. (1969). Looking at upside-down faces. *Journal of Experimental Psychology*, 81, 141–145.

Yin, R.K. (1970). Face recognition by brain-injured patients: A dissociable ability. *Neuropsychologia*, 8, 395–402.

Zoloth, S.R., Petersen, M.R., Beecher, M.D., Green, S., Marler, P., Moody, D.B. & Stebbins, W.C. (1979). Species-specific perceptual processing of vocal sounds by monkeys. *Science*, 204, 870–873.

Zuberbuhler, K., Noe, R. & Seyfarth, R.M. (1997). Diana monkey long distance calls: Messages for conspecifics and predators. *Animal Behaviour*, 53, 589–604.

Zuberbuhler, K., Cheney, D.L. & Seyfarth, R.M. (1999). Conceptual semantics in a nonhuman primate. *Journal of Comparative Psychology*, 113, 33–42.

Part Four

Lateralization and Memory

14
Memory and Lateralized Recall

AMY N.B. JOHNSTON AND STEVEN P.R. ROSE

14.1. Introduction

This chapter reviews the evidence for asymmetrical involvement of different forebrain structures at different times during learning and memory consolidation in the chick, focusing primarily on imprinting and one-trial passive avoidance. In doing so it covers the lateralized behavioural, electrophysiological, biochemical and structural processes that occur during memory consolidation. It concludes with a discussion of the functional relevance of the observed lateralization to theories of memory formation and storage.

Despite their overall gross anatomical similarities, the two cerebral hemispheres of many of the avian species studied to date show a remarkable degree of structural and functional lateralization. The directionality of lateralization in, for instance, chicks, zebra finches and canaries is the same in almost all individuals. Indeed, some of the earliest demonstrations of functionally lateralized brains in non-human species were in avian species (Nottebohm, 1971; Rogers and Anson, 1979; Scharff and Nottebohm, 1991; Cynx, Williams and Nottebohm, 1992).

This chapter is primarily concerned with the evidence for functional lateralization in one of the most thoroughly investigated avian models of learning and memory, the domestic chick (*Gallus gallus domesticus*). As we will describe, not only do chicks show brain asymmetries in processing information, akin to the lateralized functions of the mammalian brain (Bisazza, Rogers and Vallortigara, 1998) but the processes of memory consolidation and storage engage structures in left and right hemispheres differentially. These asymmetries may in part be the consequence of developmental constraints, such as the experience-dependent early maturation of the visual projections to the left compared to the right hemisphere (Chapter 6 by Deng and Rogers), or they may be ontogenetically specified differences in cerebral organization that are relatively unmodifiable by epigenetic experiential factors. Whatever the origins, evolutionarily or developmentally, of these differences, in this chapter we limit ourselves to describing the evidence for lateralization of the processes engaged in memory consolidation and expression. This evidence, derived from a number of – primarily visual – training paradigms, also casts light on the important and general neurobiological question of sequential versus parallel consolidation of memory in the minutes to hours following a learning experience.

14.2. Relevant Aspects of Visual Processing

Most avian species studied to date use their two eyes quite differently when observing stimuli, responding to different features with each eye. For example, chicks show perceptual asymmetries for many characteristics, such as recognition of conspecifics (Vallortigara et al., 1996; Bradshaw and Rogers, 1993). Visual scanning is often performed independently with each eye (Wallman and Pettigrew, 1985) and, as the eyes scan separately, the information received by the two visual systems can differ markedly (Bradshaw and Rogers, 1993; Rogers, 1995). In many species with laterally placed eyes, such as the chick and the pigeon (*Columba livia*), the binocular field of view is around only 30° (Turkel and Wallman, 1977; Bloch, Lemeignan and Martinoya., 1987; Bischof, 1988) and each eye can focus independently in the lateral field (Schaeffel, Howland and Farkas, 1986). The two visual systems and the two forebrain hemispheres can attend to different visual stimuli or to quite different aspects of the same stimulus. In the chick there is almost complete decussation of the optic fibres (Karten and Hodos, 1970; Cohen and Karten, 1974; described in Chapter 6 by Deng and Rogers) and there are relatively minor interhemispheric connections (Pearson, 1972; Cuenod, 1973; Ehrlich and Saleh, 1982; Ehrlich, Zappia and Saleh, 1988). Communication between the two eye systems in young chicks is thus likely to be minimal, as the supraoptic decussation (DSO) does not appear to be functional at this early age (at least up to 3 days post-hatching; Rogers and Ehrlich, 1983; Saleh and Ehrlich, 1984). While the tectal and posterior commissures appear to be functional (Parsons and Rogers, 1993), they do not contain many fibres directly from the visual pathways that cross over (Phillips, 1966; Pearson, 1972; Ngo et al., 1994; but see Chapter 7 by Güntürkün).

This relative independence of each hemisphere's visual system, which does not appear to be the case for other modalities, such as acoustic inputs (Bock and Braun, 1998; Braun et al., 1999) also lends itself to experimental manipulation. Thus, it is possible to examine lateralized functioning by assessing the performance of chicks in specific tasks when they have one or the other eye covered by a patch. Covering the right eye of a chick forces it to use the left eye to view stimuli and, therefore, its right hemisphere is primarily used to perceive and respond to them. A chick that has a patch over the right eye is, therefore, mostly using its left eye system (LES) and will be referred to in this chapter as a LES chick. Chicks with their left eye covered, using the right eye (and primarily the left hemisphere), will be referred to as RES chicks (Andrew, 1988). It is also possible to sever the supraoptic decussation and the commissures, thus producing a split-brain chick in which the two visual systems process information independently (Horn, Bradley and Bateson, 1979).

Such experiments permit the interpretation that the LES processes information about the details of the object, including novelty, spatial position, colour and shape (Andrew and Brennan, 1983, 1984; Andrew, 1991b; Dharmaretnam and Andrew, 1994; reviewed in Rogers, 1995). These characteristics are important in learning about specific stimuli and also about the organization of the environment (Andrew, 1991b; Dharmaretnam and Andrew, 1994). In contrast to this, the RES appears to categorize objects based on generalized characteristics rather than topographical, spatial cues (Andrew, 1991b; van Kampen and de Vos, 1992). A chick using the RES is then able to initiate an appropriate response to a 'class' of cues without reference to slight differences between different objects within the class (Andrew, 1991b); see also Chapter 3 by Andrew and Rogers.

14.3. Paradigms for the Study of Memory Formation

The merits of the young chick as 'God's organism' for the study of learning and memory have been discussed elsewhere (Rose, 2000). The learning tasks available include imprinting, passive avoidance learning and visual categorization (or pebble/bead-floor learning), though young chicks will also rapidly learn their way through simple mazes (Regolin and Rose, 1999). For all of these tasks, the brain processes engaged can be studied both correlatively and interventively. Correlative approaches involve training the bird and observing consequent and subsequent brain changes using morphological, electrophysiological or biochemical techniques. Interventive approaches involve disrupting memory consolidation by selective (where appropriate, unilateral) lesions or drug injections. The soft, incompletely ossified skull of the newly hatched chick enables rapid and accurate intracerebral injection of various chemical agents, or placement of lesioning probes or electrodes into one or the other hemisphere without the requirement for anaesthetic (which in itself has been shown to affect consolidation of some forms of learning; see Bolhuis and Horn, 1997; Parsons and Rogers, 1997) and with a minimal level of discomfort to the chick (Andrew, 1991a).

14.3.1. Imprinting

Filial imprinting is the process whereby a young chick learns to restrict its social preference to a single stimulus: in normal development, the chick imprints on its mother (sexual imprinting, normally to conspecifics, occurs when the birds are a few days older). The age of the chick when it is exposed to the imprinting stimulus is critical, as filial imprinting occurs during a relatively short sensitive period during the first few days post-hatching (Lorenz, 1935; Immelmann and Suomi, 1981; Bateson, 1985; Parsons and Rogers, 1997). Imprinting engages a number of modalities, primarily visual and auditory, but also olfactory (Burne and Rogers, 1996; Vallortigara and Andrew, 1994b). Irrespective of the stimulus used, imprinting is clearly a biologically important learning experience for the chicks. Chicks may run several hundred metres in as little as 5 min in an endeavour to approach the imprinting stimulus during re-exposure. In the laboratory, visually and/or auditorily naïve chicks are exposed to a visually conspicuous moving object and/or the sounds of a clucking hen or a tone. A stuffed hen or rotating patterned box are routinely used as the conspicuous visual object (Bateson, 1991; Horn, 1999), although now more sophisticated techniques using back-projected moving slides are also common (Griffiths and Bateson, 2000). Purely auditory imprinting, the brain correlates of which have been studied extensively by Scheich, Braun and their collaborators (Braun, Bock and Wolf, 1992; Heil and Scheich, 1992; Bock and Braun, 1998; Braun et al., 1999), does not seem to involve lateralized responses within the forebrain and will not be considered further here. For imprinting experiments eggs are almost always incubated and chicks hatched in the dark. Incubation in the light has been shown to alter at least some lateralized measures associated with imprinting. For example, glutamate subtype receptor binding in the brain region known as intermediate medial hyperstriatum ventrale (IMHV), which is lateralized in dark-incubated chicks, is not lateralized in chicks incubated in the light (Johnston et al., 1997).

In visual imprinting, the chick is exposed to the stimulus for a period of up to 2 h, during which time it learns the characteristics of the object and endeavours to move towards it. The strength of the preference can then be tested by placing the chick in a running wheel or runway and monitoring its activity when offered a choice between the imprinted and a novel stimulus

(Bateson and Wainwright, 1972; Bateson, 1991; Johnston, Rogers and Johnston, 1993). This makes possible the calculation of a preference score, a measure of how much the chick prefers to move towards the familiar or imprinting object compared to any other object. This serves as a measure of recall.

14.3.2. Passive Avoidance Learning

One-trial passive avoidance learning exploits the young chick's propensity to peck at small bright objects, a behaviour also shown by other gallinaceous birds and waterfowl (Schaller and Emlen, 1962). The task involves offering the day-old chick reared on a normal light dark cycle a small bead (attached to a thin rod), which the chick views binocularly (Cherkin, 1969; Watts and Mark, 1971; Mark, 1974; Gibbs and Ng, 1979, 1984; Rose, 1995a; Rose and Stewart, 1999). If the bead is coated with, or smells of, an aversive substance such as methylanthranilate (MeA) or euganol (Burne and Rogers, 1997), the chick will peck, show a characteristic disgust response (head shaking, bill wiping, etc.) and avoid pecking at a similar (but dry) bead presented subsequently. As the training experience involves only a single, sharply timed trial, usually not exceeding 30 s in duration, the brain processes involved in registering and learning the experience can be separated from those involved in subsequent memory consolidation. Brain processes consequent on training can be compared between birds that have pecked the aversive bead and those that have pecked a water-coated or dry bead; the latter continue to peck on test. However, pecking at a water-coated bead is itself an appetitive experience and has brain sequelae that differ from those in quiet control animals held in equivalent conditions (Johnston, 1999; Salinska, Chaudhury, Bourne and Rose, 1999).

Retention of memory for the avoidance of an MeA-coated bead persists for 48 h or longer, and can be calculated as a population measure of percent avoidance. If the strength of the stimulus is reduced, for example, by using a 10% solution of MeA, or replacing MeA with quinine, chicks evince a similar disgust reaction, but will remember to avoid the previously weakly aversive bead for only some 6–8 h (Sandi and Rose, 1994; Clements and Bourne, 1996). This weak training version of the task lends itself to a variant approach to studying its biochemical correlates, by exploring the possibility of pharmacologically enhancing the salience of the experience and hence extending the period of memory retention (Sandi and Rose, 1994). One disadvantage of both strong and weak versions of the task is that they are based on an all-or-none response in each bird, which is trained and tested only once, although additional behavioural measures such as latency to peck or numbers of pecks in a given time can also be recorded. A further measure of retention involves a discrimination test in which the chick must avoid the previously aversive bead but continue to peck at a neutral one of a different colour. This enables a discrimination ratio to be calculated (Gibbs, 1991). While the details of training procedures vary from laboratory to laboratory and, indeed, small variations in procedure have been shown to have profound effects on the duration of memory retention (Burne and Rose, 1997), the elements of the task are the same. One interesting variant derives from the fact that the chicks are usually held in pairs in pens during training, in order to reduce isolation stress. Johnston, Burne and Rose (1998) have shown that simply watching another chick peck at a bitter-tasting bead is itself sufficient to evoke a long-term (> 24 h) avoidance response. It should be noted that all chicks included in the standard form of the passive avoidance task must meet the minimum behavioural criteria of pecking at the pre-training bead in at least two out of three pre-training trials, pecking at the training bead and evincing a disgust response (headshaking/beak wiping) following pecking at the bitter-tasting training bead.

14.3.3. Pebble/Bead-floor Learning

Imprinting is sometimes argued to be a unique form of learning with little relation to other more familiar forms of association or instrumental learning (but see Hess, 1964; Bateson, 1966). Passive avoidance training is aversive. The third, the bead or pebble-floor task, like the first two involves the visual modality, but is appetitive and requires visual discrimination or categorization. The pebble-floor design was originally introduced by Rogers and Andrew (1982) for chicks between approximately 3 and 14 days of age (Rogers, 1990), at a time when they are no longer supplied with nutrients from the yolk sac (Romanoff, 1944; Hogan, 1971) and are therefore hungry (Hogan, 1984; Hale and Green, 1988). More recently the task has been modified, as a bead floor, by Tiunova and Anokhin (Tiunova et al., 1998). Food-deprived chicks are placed in pens on a floor in which food grains are scattered amongst either coloured beads or small 'food-like' pebbles stuck to the floor of the cage. The chicks first peck indiscriminately at the objects on the floor but soon, usually within about 20 pecks, learn to peck selectively at grains. This categorization memory is usually retained by the chick for at least 48 h, so that when the chick is put back onto the pebble-floor it confines its pecks almost entirely to the grains. Memory is defined in terms of savings – pecks at grain as opposed to beads or pebbles on test. Pebble-floor learning is intriguing in that the hemisphere responsible for acquisition of the task depends on the incubation conditions during the sensitive period for visual system lateralization (see Chapter 6 by Deng and Rogers).

Some of the earliest evidence for asymmetries in learning and memory formation came from work using the pebble-floor technique, although they have not been followed up in any molecular detail. Therefore, before discussing the other two tasks, we will review these data briefly. The first demonstration of lateralized forebrain use during learning in the chick was revealed by injecting the protein synthesis inhibitor cycloheximide (CXM) or the excitotoxin glutamate prior to training on the pebble-floor (Rogers and Anson, 1979; Rogers, 1990, 1995). These experiments were based on the premise that, by injecting glutamate into one forebrain hemisphere, the usual functioning of that hemisphere becomes disrupted, forcing the other hemisphere to become engaged. Light-incubated chicks (incubated so that the right eye is exposed to light), irrespective of their sex, were not able to learn the pebble-floor task effectively subsequent to an injection of cycloheximide or glutamate into the left hemisphere, but similar treatment of the right hemisphere appeared to have no effect (Hambley and Rogers, 1979; Howard, Rogers and Boura, 1980; Rogers and Hambley, 1982; Rogers, 1990). The reverse was true when chicks were incubated with the left eye exposed to light. In this condition, injection of the left hemisphere had no effect, while injection of the right hemisphere prevented chicks from acquiring the discrimination (see also Chapter 6 by Deng and Rogers).

Although there is much debate about how glutamate exerts these effects, they are profound, despite the fact that very low doses of glutamate which do not cause lesions are effective. The effects of glutamate show both age and time dependency. The retarding effects of an intrahemispheric injection of glutamate at day 2 post-hatching on pebble-floor learning continued for at least 8 days (cf. long effects of CXM; Rogers, Drennen and Mark, 1974). However, injections later than day 3 post-hatching, at a stage when the blood–brain barrier in the chick appears to be fully functional and so is able to regulate intracranial free amino acid levels (Purdy and Bondy, 1976; Hambley and Rogers, 1979; Ribatti, Nico and Bertossi, 1993), had no effect. By contrast, although chicks injected with glutamate into the left forebrain on day 2 post-hatch did not demonstrate effective pebble-floor learning on days 6, 7 or 9 after hatching, they were able to perform the task by day 12 (Johnston and Rogers, 1993a, 1993b). Thus, the effects of a glutamate injection on visual discrimination learning appear to be transient (see Chapter 6 by Deng and Rogers for a discussion of the effects of light).

14.4. Imprinting

In the initial studies on the cellular correlates of imprinting, Bateson, Horn and Rose (1969; 1971) found that chicks exposed to a visual imprinting stimulus showed enhanced incorporation of radiolabelled precursors of RNA and protein into the upper part of the forebrain (the 'roof'). Moreover, the amount of incorporation was correlated with the preference chicks showed for the imprinting object, strongly suggesting that it was learning related (Bateson, Horn and Rose, 1975). Horn, Bradley and Bateson (1979) went on, using an autoradiographic technique, to localize these imprinting-related changes in RNA synthesis to a specific brain area, now known as the intermediate medial hyperstriatum ventrale (IMHV). The involvement of the IMHV was soon corroborated by Kohsaka et al. (1979) using a 2-deoxyglucose (2-DG) technique, and it is on this region in particular that most subsequent work has concentrated.

The IMHV is not a primary sensory area, but receives secondary inputs from many other sensory regions, functioning perhaps as the avian equivalent of an association cortex. Horseradish peroxidase, fast blue and nuclear yellow tracers injected into the IMHV all reveal monosynaptic ipsilateral and contralateral connections to the hippocampus, the palaeostriatum augmentatum and the dorsal part of the archistriatum (see Horn, 1985a). The IMHV also has direct connections to the caudal part of the hyperstriatum ventrale, all of the laminae of the visual Wulst, and the intermediate and caudal portions of the neostriatum (Bradley, Davies and Horn, 1985). It receives what is presumably visual input from the stratum album centrale of the ipsilateral (and to a lesser degree, the contralateral) optic tectum and from parts of the hyperstriatum (Bradley and Horn, 1978; Bradley, Horn and Bateson, 1981; Davies et al., 1991) and auditory and somatosensory input through its connections to the neostriatum (Karten, 1969; Bradley and Horn, 1978; Miceli and Repérant, 1985). There are no direct connections between left and right IMHV (Bradley and Horn, 1978, 1979; Bradley et al., 1985) but the left and right IMHV may interact through the left and right hippocampi, which are linked by the hippocampal commissure, and through the left and right archistriatum and neostriatum, which are linked by the anterior and posterior commissures (Phillips, 1966; Pearson, 1972). In a manner somewhat akin to the mammalian hippocampus, the IMHV may act as a coordination region, given that it has connections to regions that are associated with the generation of emotional behaviours (such as the archistriatum; Martin, Delanerolle and Phillips, 1979; Phillips and Youngren, 1986) and the modification of motor output [such as via the paleostriatum augmentatum (PA); Zeier and Karten, 1971].

14.4.1. Monocular Testing

The initial imprinting studies were not designed to reveal lateralization but, as the work proceeded, it soon became apparent that left and right IMHV were not responding identically to the imprinting experience. Horn, Rose and Bateson (1973) noted that, even when chicks were imprinted using only the LES or only the RES, they showed increased RNA synthesis compared with controls in both the left and right forebrain roof. This suggested that both hemispheres were involved in the consolidation of imprinting memory, possibly involving a form of interhemispheric transfer of information (Horn, Rose and Bateson, 1973). Moreover, chicks trained with one eye occluded showed a preference for the imprinting object irrespective of whether they were tested with the 'trained' or 'non-trained' eye (Bateson, Horn and Rose, 1975; Johnston and Rogers, 1993b). However, if the supraoptic decussation (probably along with the smaller intrahippocampal commissures and the anterior commissure) was cut before the birds were trained, enhanced synthesis was found only in the roof of the forebrain hemisphere contralateral to the eye exposed to the imprinting stimulus (Horn et al., 1971). Under these conditions, the chick preferentially approached the imprinting stimulus at test only when using the 'trained' eye.

It is also possible to alter, at least in part, the hemisphere responsible for imprinting memory consolidation by altering the incubation conditions. Irrespective of sex, chicks incubated with their right eye exposed to light during the latter few days of incubation do not show recall of imprinting memory 8 h after training following pharmacological disruption of the right hemisphere (by intrahemispheric injection of 500 nmol glutamate) but do show recall following a similar injection of the left hemisphere (see also Johnston and Rogers, 1998). In the reverse condition (left eye exposed to light during the latter part of incubation), the opposite hemisphere is affected, so that recall is disrupted by injection of the left rather than the right hemisphere (Johnston and Rogers, 1999).

One possible explanation for this functional effect of lateralized stimulation light is that it differentially alters the cues to which the chicks attend. The two eye systems of the chick respond differently to imprinting stimuli, be they inanimate objects such as a table tennis ball or animate objects such as another chick (Vallortigara and Andrew, 1991). When chicks were placed in a 45 cm long runway with the imprinting, or familiar, object at one end and an unfamiliar object (an 'imprinting' object that had been subjected to transformation of one of its features) at the other end, the object they moved towards depended on which eye system was exposed during the test (Vallortigara and Andrew, 1991). The chicks chose to move towards a familiar chick when they were tested binocularly or using the LES. Thus the LES notices the difference between the familiar and unfamiliar chick. By contrast, RES chicks approached both stimuli equally (Vallortigara and Andrew, 1991).

Differential use of the eye systems has also been revealed using monocular testing of chicks imprinted on table tennis balls (Vallortigara and Andrew, 1991). Chicks were raised with a red ball, with a white horizontal stripe on it, and tested for preference on day 3 posthatching. When the angle of the stripe on the ball was changed from a familiar horizontal line to an unfamiliar, vertical stripe, binocularly tested chicks and chicks using the LES preferentially approached the familiar imprinting stimulus, whereas RES chicks chose at random, spending equal amounts of time at both ends of the testing laneway (Vallortigara and Andrew, 1991). The same pattern of eye preference was revealed in chicks that were required to choose between an imprinting stimulus and one transformed with oblique stripes (45° or 135° from horizontal) (Vallortigara and Andrew, 1991). However, chicks tested using the RES discriminated when the changed visual appearance of the imprinting stimulus was relatively large (Vallortigara and Andrew, 1991). In situations in which the transformation was large, chicks using the RES, as well as those using the LES, preferentially approached the familiar compared to the unfamiliar stimulus. Vallortigara and Andrew (1991) concluded that the LES attends to details of the stimulus, whereas the RES attends to features that categorize the object into 'imprinting stimulus' or 'not imprinting stimulus'.

14.4.2. Lesioning and Imprinting

In an early study, Salzen, Parker and Williamson (1975) found that two regions in the lateral forebrain, the lateral neostriatum and archistriatum, were necessary for the preferential selection of the imprinting stimulus. They found that chicks preferentially approached the imprinting stimulus (in these experiments a red ball or a green cube) 3 days after bilateral lesions of the IMHV. In the absence of the archistriatum and lateral neostriatum, chicks would approach objects, but did not appear to show any preference for the familiar (Salzen, Parker and Williamson, 1975, 1978; Salzen, Williamson and Parker, 1979), suggesting that these regions were essential for recall of visual imprinting (Salzen, Bell and Parker, 1983). This result appeared to be confirmed by 2-DG uptake and glutamate binding studies, showing an increase in uptake and binding in some of these regions following imprinting (Takamatsu and Tsukada, 1985a).

More recent lesion studies have directly implicated the IMHV in learning the characteristics of the imprinting stimulus (Bateson, Horn and McCabe, 1978; McCabe, Horn and Bateson, 1981; Cipolla-Neto, Horn and McCabe, 1982; McCabe et al., 1982; Horn, McCabe and Cipolla-Neto, 1983; Johnson and Horn, 1984, 1986; Horn, 1985a, 1991; Bolhuis et al., 1989; McCabe, 1991). Bilateral lesions of the IMHV prior to training prevented imprinting on a visual stimulus (McCabe, Horn and Bateson, 1981; McCabe et al., 1982; Takamatsu and Tsukada, 1985a, 1985b), while bilateral lesions made up to 6 h after training prevented recall. Thus, the IMHV is involved in both the consolidation and recall of information associated with imprinting to a visual stimulus (McCabe, Horn and Bateson, 1981; Davey, McCabe and Horn, 1987). Lesioning studies also suggested that the IMHV is involved in the acoustic elements of imprinting (Salzen, Bell and Parker, 1983).

However, left and right IMHV are differentially engaged in these tasks. Experiments in which the left and right IMHV were destroyed sequentially by radiofrequency coagulation after imprinting showed that imprinted chicks that were tested with a single functional IMHV region were still able to recall the stimulus (Cipolla-Neto, Horn and McCabe, 1982; Horn, McCabe and Cipolla-Neto, 1983; Davey, McCabe and Horn, 1987). However, when a second lesion was placed in the left IMHV approximately 26 h (23–28 h) after the end of training of right IMHV-lesioned chicks, they no longer showed a preference for the imprinting stimulus. By contrast, when a second lesion was placed in the right IMHV in left IMHV-lesioned birds, they were still able to recall the stimulus (Cipolla-Neto, Horn and McCabe, 1982). Thus, the two IMHVs do not function in the same way in the consolidation and recall of imprinting in the chick.

This sequence of experiments led Horn and coworkers to suggest that an unknown region, designated S′, present in the right hemisphere (but distinct from the right IMHV), stored a 'longer term' form of imprinting memory. This store can be accessed in the absence of a right IMHV as long as the right IMHV had been present for the first 26 h after training; chicks given bilateral IMHV lesions 26 h after training showed significant imprinting preferences (Cipolla-Neto, Horn and McCabe, 1982). The involvement of the region S′ did not depend on the absence of the left IMHV after training, and memory may move to S′ in an intact chick as rapidly as 6 h after training (Davey, McCabe and Horn, 1987). Chicks in which both IMHV regions were lesioned 6 h after training demonstrated a preference for the imprinting stimulus, but not if the IMHV regions were lesioned 1.5, 3 or 5 h after training (Davey, McCabe and Horn, 1987).

Chicks imprinted with only one IMHV intact could recall the memory. However, if the other IMHV was then removed and the chicks were then retested, they did not show recall, irrespective of which IMHV was present at training (Horn, McCabe and Cipolla-Neto, 1983). In the absence of the left IMHV, chicks with the right IMHV intact were able to acquire and retain imprinting memory, but not to initiate the formation of S'. Thus the differential use of either hemisphere appears to depend, at least in part, on the integrity of both hemispheres and possibly on some form of interhemispheric communication between left and right IMHV (Horn, 1985b; McCabe, 1991).

As always, one must be cautious when drawing conclusions from lesioning data, not least because the young chick brain seems to be extremely plastic. The fact that a process occurs in a particular region in a lesioned brain does not imply that it also involves the same region and process in the intact brain. Examples of just this are the demonstrations that the formation of the S' store of imprinting memory depends on the presence of both IMHVs, and the possibility that chicks with bilateral lesions of the IMHV can demonstrate recall 3 days after training (Salzen, Parker and Williamson, 1975; Salzen, 1992). Thus, lesioning may not be the most appropriate method for examining the differential consolidation of imprinting memory (Horn, McCabe and Cipolla-Neto, 1983; McCabe, 1991).

A further problem might arise from the anaesthetic agents used when the chicks are lesioned. A number of anaesthetic agents are known to alter the sensitive period for visual imprinting (Bolhuis and Horn, 1997; Parsons and Rogers, 1997). The anaesthetic agents used in the lesioning studies include nembutal, ether, halothane, pentobarbital and ketamine/xylazine, which have been shown to inhibit the formation and/or retention of imprinting memory (Hess, 1957; Guntekunst and Youniss, 1963; Bradford and MacDonald, 1969; Lecanuet, 1984; Parsons and Rogers, 1997), presumably by interacting directly with some of the neurochemical processes [N-methyl-D-aspartate (NMDA)-type glutamatergic or noradrenergic changes] responsible for some part of acquisition of imprinting memory.

14.4.3. Electrophysiology

Some of the problems of interventive studies can be avoided by looking instead for the direct neural correlates of memory formation. Horn and his colleagues have focused on the electrophysiological sequelae of imprinting in the IMHV recorded from trained and anaesthetized animals. Asymmetrical changes in neuronal activity occur in the left and right IMHV during the consolidation and recall of imprinting memory (Payne and Horn, 1984; Davey and Horn, 1991; McLennan and Horn, 1992; Bradford and McCabe, 1994; McCabe and Nicol, 1999). 'Spontaneous' multiple-unit activity recorded in the left IMHV is affected by exposure to a visual or auditory imprinting stimulus (Payne and Horn, 1982, 1984; Brown and Horn, 1992). The responses recorded in the left and right IMHV vary depending on the interval between training and recording. Consistent with the notion that the right IMHV may also be involved with the consolidation of imprinting memory, spontaneous activity in the right IMHV at 6 h, but not immediately after training, was shown to be significantly different from that in the left IMHV (Davey and Horn, 1991). This may be due to a decrease in firing rate in the right IMHV or an increase in the left IMHV, or both. This effect may not, however, have been due solely to consolidation of imprinting memory, as a similar pattern of activity change was recorded in the hyperstriatum accessorium (HA) 6 h after training (Davey and Horn, 1991) and the HA does not appear specifically to support imprinting memory (Horn, 1991). More recent studies have indicated that imprinted chicks have more than twice the proportion of sites in the left IMHV activated by a presentation of an imprinting stimulus than do untrained chicks (Brown and Horn, 1992, 1993, 1994; see also below).

Recordings from the left IMHV during the presentation of similarly shaped, but differently coloured, boxes suggest that the differential responses from the left and right IMHV regions may be due to the different forms of processing occurring in these regions (McLennan and Horn, 1992). Recordings from the left IMHV of trained chicks were similar (either excitatory or inhibitory) when the chick was presented with the imprinting object or with an object of the same shape as the imprinting object but of a different colour (McLennan and Horn, 1992). This contrasted with recordings from the left IMHV of visually naïve chicks, which depended on the colour of the object; excitatory responses were evoked by objects of one colour while inhibitory responses were evoked by a differently coloured object (McLennan and Horn, 1992). Recordings from awake chicks showed that neurones in the left and right IMHV of untrained chicks responded equally to two novel stimuli (a rotating red box and a rotating blue cylinder). However, imprinting chicks on one or the other stimulus reduced the proportion of neurones in the right IMHV that responded to the non-imprinted stimulus (Nicol et al., 1995; McCabe and Nicol, 1999). Thus, a measure of selectivity is present in the right IMHV that does not appear to occur in the left IMHV. These results support the view that the left IMHV categorizes the stimulus as 'the imprinting object' according to broad characteristics and may not attend to small differences (Vallortigara and Andrew, 1991, 1994a; McLennan and Horn, 1992).

Perhaps the most exciting of all of the electrophysiological evidence acquired to date with chicks is the finding that individual neurones in both the left and right IMHV respond differentially to different elements of the imprinting stimulus. In a series of elegant dissociation experiments using electrophysiological recordings from awake chicks imprinted on a red box or a blue cylinder and then exposed to a red box, a blue cylinder, a red cylinder or a blue box, Nicol, Brown and Horn (1995, 1998a, 1998b) found neurones that responded to these specific characteristics. While a high proportion of neurones (30%) showed selective responses to the training stimulus, more neurones (40%) responded particularly to one feature only of the stimulus (colour or shape). Nicol et al. also found IMHV neurones sensitive to the distance of the imprinting stimulus. (In addition there are hippocampal neurones that show the same sorts of selectivity to colour or shape as IMHV neurones, although there do not appear to be the 'trained' neurones in the hippocampus.) Thus far, however, there is little understanding of the temporal dynamics of neuronal responses to imprinting objects. Given the time-dependent development of receptor binding changes (see later) and the time-dependent development of the S′ memory store, such investigations are essential.

Thus, in the intact chick, both the left and the right IMHV play a role in stimulus recognition and hence imprinting preference – results that run counter to the lesioning studies that suggested the right IMHV was not so important in longer term recall of imprinting memory.

14.4.4. Biochemistry

Coupled with the electrophysiological results, lateralized changes in a variety of correlative biochemical measures have also been found following exposure to an imprinting stimulus. These include changes in glutamate receptor binding, which increases in the IMHV at certain post-training times. Intriguingly, there is a direct relationship between the strength of memory and NMDA-displaceable glutamate receptor binding in the left IMHV region (McCabe and Horn, 1988, 1991; Johnston, Dodd and Rogers, 1995). Preference scores of 2-day-old chicks trained on a visual imprinting task were found to correlate with the increases in NMDA binding in the left, but not the right, IMHV and in glutamate binding in the left, but not the right, archistriatal/paleostriatal complex 8–9 h following the end of exposure to the imprinting stimulus (Johnston et al., 1993). There were no learning-related left–right differences at earlier times up to 6 h post-training. Moreover, it is unlikely that light exposure alone accounts for the lateralization found in imprinted chicks, as visual experience alone does not cause an elevation of NMDA-type binding levels in the left IMHV of chicks (McCabe and Horn, 1988; Johnston, Dodd and Rogers, 1995).

However, some asymmetries are present prior to exposure to an imprinting stimulus. [^{3}H]MK-801 binding studies show an asymmetry in NMDA-type glutamate receptor binding in the IMHV of dark-reared, visually naïve chicks (MK-801 is an NMDA receptor ligand). This asymmetry may depend on the absence of light exposure prior to hatching. In chicks incubated and hatched in the light, there was no asymmetry in MK-801 binding (Johnston, Dodd and Rogers, 1995). Exposure to the imprinting stimulus resulted in significant changes to MK-801 binding levels in the left IMHV, while receptor binding in the right IMHV remained relatively constant, thus reversing the lateralization found in the dark-reared birds (Johnston, Dodd and Rogers, 1995). Although there does not seem to be an imprinting-induced change in NMDA receptor binding in the right IMHV, injection of glutamate 1, 3 or 6 h, but not 9 h, post-training into the right but not the left forebrain resulted in transient amnesia for the imprinting in chicks tested 8 h later, but recall had reappeared by 48 h (Johnston and Rogers, 1998).

Transient changes in receptor binding, if they are to be translated into long-term changes in synaptic connectivity, must result in due course in gene activation, the synthesis and insertion into the synaptic membranes of structural proteins, and consequent morphological change. Increases in *fos*-like immunoreactivity in dorsal hippocampus, hyperstriatum accessorium and IMHV have been described following imprinting (Ambalavanar, McCabe and Horn, 1993; McCabe and Horn, 1994). The authors noted that almost all *fos*-expressing cells stain for gamma-aminobutyric acid (GABA), perhaps indicating an inhibitory function. The expression of *fos*, one of a family of immediate early genes, correlated with the degree of memory shown, but there was no evidence of lateralization.

14.4.5. Structural Changes

Changes in electrical activity, receptor binding and gene expression are generally thought to underlie the longer term encoding of memory. This long-term storage has, following a Hebbian-type model of memory formation, been assumed to involve changes in structural connectivity in associated brain regions. Early electron microscopic studies showed that there was a transient increase in the length of the synaptic apposition zone in the right, but not the left, IMHV of chicks trained for as little as 20 min on the imprinting task (Bradley, Horn and Bateson, 1981). This disappeared by 140 min (Payne and Horn, 1984). This could reflect a developmental change (Rostas et al., 1983, 1992; Rostas, Brent and Güldner, 1984; Horn, Bradley and McCabe, 1985). Subsequently Horn, Bradley and McCabe (1985) reported a positive relationship between the overall length of the postsynaptic density (PSD) of axospinous synapses in the left, but not the right, IMHV of chicks trained for 140 min and the level of recall of imprinting memory (Horn, Bradley and McCabe, 1985). A similar increase in the mean length of the PSD, and in levels of recall, was observed in the left IMHV, but not the right IMHV or the left or right hyperstriatum accessorium, 3 h after a sequential training session (chicks were trained for 1 h at 22, 24 and 42 h posthatching; Horn, Bradley and McCabe, 1985).

However, Horn, Bradley and McCabe (1985) indicated that the length of the PSD may also change in the right IMHV concurrent with the consolidation of imprinting memory. An initially higher mean PSD length in the right IMHV (approximately 9% higher) appears to decrease following imprinting (by approximately 5%). In contrast, the mean length of the PSD in the left IMHV increased by 10% (Horn, Bradley and McCabe, 1985). Thus, the PSD of these synapses in both the left and right IMHV may change during consolidation, but do so in opposite ways. Alternatively it may be that the right IMHV is involved at an early stage and the left IMHV 'catches up' later on.

An overall model reconciling these findings in a theory of the cellular and molecular processes underlying imprinting memory has yet to be constructed. Clearly, any such model must include reference to the role of brain regions both in the left and the right forebrain, the temporal as well as regional (and hemispheric) dependence of memory consolidation and the range of possible mechanisms by which memory might be maintained. Some of these elements are discussed in more detail in the final part of this chapter.

14.5. Passive Avoidance

Initial studies using the passive avoidance protocol followed something of the same route as the imprinting work. Using the 2-DG technique, Kossut and Rose (1984) found that chicks that had pecked an MeA-coated bead showed increased activity in the IMHV and the lobus parolfactorius (LPO; also known as rostral striatum). The LPO and IMHV are connected by a reciprocal disynaptic pathway passing through the archistriatum and palaeostriatum augmentatum (caudal striatum) (Bradley, Davies and Horn, 1985; Davies et al., 1991; Lowndes and Stewart, 1994). Although there was also increased uptake into the paleostriatum augmentatum in trained compared to quiet control chicks, this seemed to be related to the motor components of the task, as it did not differ between MeA and water-trained chicks. Later, more detailed analysis (Rose and Csillag, 1985) of 2-DG uptake into chicks injected 5 min before, 10 min or 30 min after training and killed 40 min later showed that, whilst there was raised uptake into the left IMHV of MeA-trained chicks at all times, increased uptake into the right IMHV was limited to the 10 min post-training time of injection. Uptake into LPO showed a similar temporal as well as lateral distribution. There was higher uptake into the LPO in the left hemisphere at both the 5 min pre-training and the 10 min post-training injection time but not at the 30 min post-training time. Thus, both hemispheres and regions seem to be engaged in response to the experience, but at different times.

14.5.1. Monocular Effects

Differential temporal involvement of the two hemispheres in passive avoidance memory is also revealed using monocular testing. When intact chicks are tested monocularly, two stages of enhanced recall occur; one associated with the LES at 25 min after training and the other associated with the RES at 30–32 min after training (Andrew and Brennan, 1985). Trained chicks avoided fixating on a bead the same colour as the previously bitter-tasting bead significantly more often using their LES at 25 min after training and with their RES at 32 min after training (Andrew and Brennan, 1985). In addition, chicks tested for avoidance responses using their RES demonstrated recall when injected with a disruptive agent (sotalol) into the right hemisphere between 5 and 15, and between 30 and 80 min after training, but not when injected 25 min after training. By contrast, sotalol disrupted recall from the left hemisphere when it was administered 15–25 min after training, but not prior to 15 min or after 25 min (Andrew and Brennan, 1985). It has been suggested that these differential results are due either to interhemispheric interaction at specific stages after training (Andrew and Brennan, 1985; Andrew, 1991b) or, more probably, differential accessing of information stores in the left and right hemispheres (see Chapter 15 by Andrew).

For example, when small coloured beads were presented repeatedly to chicks previously habituated to bead presentation, the chicks' responses varied according to which eye system was used to observe the bead. LES chicks responded to a change in a single feature (e.g. red compared to violet beads) by increasing levels of pecking (dishabituation; Andrew, 1983). By contrast, RES chicks did not exhibit dishabituation following a change in the colour of the bead or when the position of the bead during presentation was varied. Both LES and RES chicks respond fearfully to red or violet beads, especially during the first 5 days post-hatch; however, LES chicks respond more fearfully to novel objects than chicks tested using the RES. LES chicks continue to have elevated fear responses until 9 days post-hatching (Andrew and Brennan, 1983). LES viewing of novel objects is known to evoke much more rapid distress calling than RES viewing of the same novel objects (Andrew, 1983, 1991b).

14.5.2. Lesioning

These differences between the hemispheres have been followed using the entire armamentarium of techniques described in the context of imprinting. We turn first to the lesioning studies (Benowitz and Lee-Teng, 1973; Davis, Taylor and Johnson, 1988; Patterson, Gilbert and Rose, 1990; Gilbert, Patterson and Rose, 1991; Rose, 1991a, 1991b; Sandi, Patterson and Rose, 1993). Bilateral lesions of the IMHV prior to training did not prevent acquisition of the task, but disrupted memory for it, just as in the imprinting case. However, the effect of unilateral lesions was subtly different. Lesioning the left, but not the right, IMHV prior to training resulted in chicks showing amnesia for the task when tested 3 h later (Patterson et al., 1990), suggesting an essential role of the *left* IMHV in the early stages of memory consolidation. However, the IMHV seemed to be required only for the first hour following training, as bilateral lesions either 1 or 6 h after training did not disrupt recall. Thus, it appeared that the IMHV is not the site of longer term memory storage, but rather that it is involved only in shorter term processes (Patterson, Gilbert and Rose, 1990; Rose, 1991a). However, a more complex series of lesioning experiments showed that this was not always the case. When left or right IMHV was lesioned pre-training and the chicks were trained and tested monocularly, a different pattern emerged (Sandi, Patterson and Rose, 1993). Chicks using the LES during training and testing showed recall if the left IMHV had been lesioned prior to training, but not if the right IMHV had been lesioned. By contrast, chicks using the RES during training and testing showed recall irrespective of which IMHV had been lesioned. Thus, when the *right* IMHV is forced to take on the 'memory-processing' role (because the left IMHV has been lesioned) it is able to do so in the absence of the *left* IMHV but, when the *left* IMHV is forced to take on the memory-processing role, chicks using the left eye are unable to access the memory.

As the 2-DG studies had found increased activity in the LPO as well as the IMHV, another set in this series of experiments explored the effect of bilateral LPO lesions in bilaterally trained and tested chicks (unilateral lesions were without effect). The results were the inverse of those with the IMHV. Pre-training LPO lesions were without effect on either acquisition or recall, but bilateral lesions made 1 h post-training resulted in amnesia in chicks tested 24 h later (Gilbert, Patterson and Rose, 1991). Unilateral LPO lesions made 1 h after training did not effect recall 24 h after training (Gilbert, Patterson and Rose, 1991). If both LPOs were lesioned before training and the *left* IMHV was lesioned after training, chicks still showed recall. However, if the LPOs were lesioned prior to training and then the *right* IMHV was lesioned, chicks were amnesic. On the basis of these results, one of us (Rose, 1991b) proposed a 'linear flow' model of memory consolidation in which a hypothetical memory trace moved from the left to the right IMHV and then to both LPOs. Others have represented the same model as an 'interocular transfer' of memory stores, suggesting that such a transfer occurs more readily from the right eye to the left than vice versa (Rogers, 1995). Both depend on a memory store 'moving' from one hemisphere to the other. A prediction from this model is that a pre-training lesion of the right IMHV (which is not usually amnestic) followed by post-training lesions of the LPO would result in amnesia (as the memory 'flow' from left IMHV to right IMHV and on to LPO regions would be disrupted). Indeed, this was the case. However, the reverse scenario, in which the right IMHV was lesioned pre-training followed by a post-training lesion of the left IMHV, did not lead to amnesia. This is contrary to the logic of the memory 'flow' model previously suggested by Rose and coworkers (see Figure 14.2). The model suggested that memory 'moves' from the right IMHV to the left IMHV, and thence on to the left and right LPO regions. Also according to the model, if the right IMHV is not present at training, the 'flow' is disrupted, the 'memory' is stranded in the left IMHV and a post-training lesion of the left IMHV should be amnestic. Thus, in the absence of the IMHV memory store and when the LPO regions are not present, the memory seemed to be retained elsewhere. The sequence of these lesioning experiments is shown schematically in Figures 14.1 and 14.2.

However, this 'hydraulic-flow' model is simplistic, ignoring as it does both plasticity and compensatory effects available to the young developing brain. Moreover, it assumes that 'memory', even for so simple an object as a bright bead, is unitary. A later experiment unpicked this complexity. In this, the

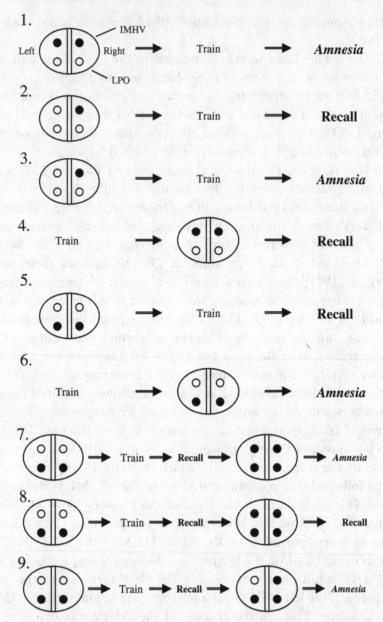

Figure 14.1.   Effects of intermediate medial hyperstratum ventrale (IMHV) and lobus parolfactorius (LPO) lesions on recall for passive avoidance learning are represented schematically 1–9 (unilateral LPO lesions are without effect; redrawn from Rose, 1991a). The solid circles represent lesioned brain regions while open circles represent undamaged brain regions. The outcomes of each passive avoidance test are indicated.

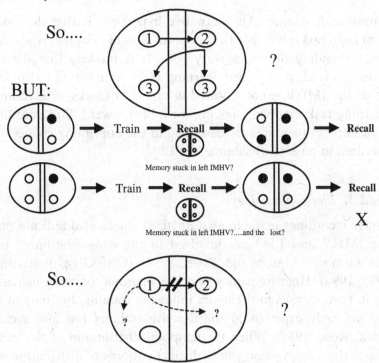

Figure 14.2. The upper part symbolizes the possible model of 'memory flow'. The lower parts show a schematic representation of subsequent modification of the 'memory flow' model, based on further lesioning studies. Based on Gilbert, Patterson and Rose (1991); redrawn from Rose (1991a).

trained and lesioned chick was challenged not by a simple avoidance test, but by a colour discrimination between the previously aversive and a novel neutral bead (Patterson and Rose, 1992). Chicks with bilateral lesions of the IMHV given 1 h after training failed to discriminate between a yellow 'aversive' (training) bead and a neutral (blue) bead, but instead avoided both. By contrast, lesions of the LPO had no effect on colour discrimination. This led to the conclusion that memory for the bead was distributed, and that memory for colour discrimination, in particular, required an intact left IMHV.

Comparing this lesion data with that from the imprinting studies, it is apparent that the left IMHV is not simply a site that is involved in all forms of early visual learning in the chick. Lesioning of the left IMHV before training induces amnesia for passive avoidance learning but does not disrupt the development of imprinting preferences (for discussion, see also McCabe, 1991). McCabe suggests that perhaps the right IMHV requires activation before it can support 'acquisition' and so can support learning tasks with extended training, such as imprinting, but not tasks performed more rapidly, such as passive avoidance. An alternative hypothesis is that the IMHV is involved in both tasks, but that the temporal dynamics of its involvement in each vary, depending on the sensory cues. It is unlikely that the IMHV regions are involved in any early learning task. Johnson and Horn (1986) showed that the IMHV is not required in order for chicks to learn an associative learning task involving moving for a heat reward. Still unresolved is the relationship between area 'S' described in the imprinting studies and an analogous area in passive avoidance, the LPO.

14.5.3. Electrophysiology

Extracellular recordings made in anaesthetized chicks also indicate that left and right IMHV and LPO are involved in the consolidation of passive avoidance memory (Mason and Rose, 1987, 1988; Gigg, Patterson and Rose, 1993, 1994). High-frequency bouts of neuronal bursting increase significantly in both regions in the hours following training, but only in chicks that have not only experienced the aversive training but also recalled it (Mason and Rose, 1988). When the temporal distribution of the increased bursting activity is analysed, it shows distinct patterns of distribution in time and space. Increased activity was found in both left and right IMHV 3–4 h post-training, while by 6–7 h the increase was largely in the right IMHV (Gigg, Patterson and Rose, 1994). Increases in mean bursts were observed in both the left and right LPO 4–7 h post-training (Gigg, Patterson and Rose, 1994). A correlation between bursting activity recorded in the LPO and the IMHV (Gigg, Patterson and Rose, 1993, 1994) suggests that these regions may interact. Thus the learning-related changes in electrical activity in the right IMHV support the evidence of its importance for memory consolidation derived from the lesion studies.

14.5.4. Biochemistry

In the minutes to hours following training on the passive avoidance task, the left IMHV is the site of a cascade of molecular processes, which begin with synaptic transients, are mediated by the activation of intracellular transcription factors and immediate early genes, and culminate in the insertion into presynaptic and postsynaptic membranes of members of a family of glycoprotein cell adhesion molecules. These serve to affect synaptic connectivity, in accord with the view that memory requires Hebbian-type alterations in neuronal pathways. That such changes in local circuits are not in themselves sufficient to account for memory should be apparent from the discussion of the fluidity of the putative memory trace indicated by the evidence cited in the previous sections. Nonetheless, these molecular processes, which have been reviewed extensively elsewhere and will only be summarized here (e.g. Rose and Stewart, 1999; Rose, 2000) are clearly both necessary for and dependent upon memory formation. These events have been followed using both interventive and correlative procedures. As indicated earlier, the sharply timed nature of the training experience, and the possibility of injecting putative disruptive agents directly into the IMHV without the need for anaesthetic, has made it possible not merely to identify components of the cascade but the approximate time at which they become engaged. The sequence is shown in Figure 14.3, and the relevant agents affecting it and the times at which they are effective are shown in Table 14.1.

Unfortunately, in the context of the focus of the present chapter, although the processes described in Figure 14.3 have all been shown to occur in the left IMHV, whether they also occur in the right IMHV (and/or LPO) is not always known. Not all studies have examined left and right IMHV independently. In some cases, putative interventive agents have been injected unilaterally as well as bilaterally, and in some but not all of these, left and right hemisphere injections have differing effects, or there are indications that the

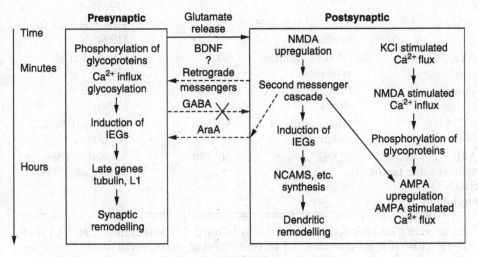

Figure 14.3. Cellular cascade during memory formation for passive avoidance. A schematic representation of the time-based presynaptic and postsynaptic changes associated with consolidation of passive avoidance memory in the chick. AMPA, α-amino-3-hydroxy-5-methylisoxazole-4-proprionic acid; AraA, arachadonic acid; BDNF, brain-derived neurotrophic factor; GABA, gamma-aminobutyric acid; IEGs, immediate early genes; NCAM, neural cell adhesion molecule; NMDA, *N*-methyl-D-aspartate.

Table 14.1. *Time of onset of amnesia for the passive avoidance task following injection of blocking agents just before or after training*

Agent	Process blocked	Onset of amnesia (min)	Reference
Muscimol	GABA agonist	10–15	Clements and Bourne (1996)
Nitroarginine	Nitric oxide synthesis	15–30	Holscher and Rose (1993)
TFP: W13; A3W9; HA1004	Ca^{2+}/CAM kinase/B50?	15–30	Serrano et al. (1994)
Ω-Conotoxin GVIA	N-type calcium channels	15–30	Clements, Rose and Tiunova (1995)
2-D-galactose	Glycosylation	30–40	Crowe et al. (1994)
Anti-L1 antibodies	Signal transduction	30–40	Scholey et al. (1995)
H7; HA156; H8; H9; ML9	PKA, PKC, PKG, B50	60	Serrano et al. (1994)
c-*fos* antisense	c-*fos* expression	60	Mileusnic et al. (1997)
β-Apolipoprotein (APP) antisense	β-APP expression	60	Mileusnic et al. (2000)
Anisomycin[a]	Protein synthesis	60	Freeman and Rose (1995)
NDGA	Arachadonic acid synthesis	75	Holscher and Rose (1994)
AP5, MK-801	NMDA receptor activation	90	Rickard et al. (1994)
Dantrolene	Intracellular Ca^{2+} stores	60–180	Salinska et al. (1999)
RU38486	Glucocorticoid receptors	60–180	Sandi and Rose (1994)
Anti-brain-derived neurotrophic factor (BDNF) antibodies	BDNF actions	60–180	Johnston, Clements and Rose (1999)

[a]Anisomycin also produces amnesia if it is injected 4 h post-training; anti-L1 and 2-D-galactose if it is injected at 5.5 hr post-training. Agents effective only during the 'second wave' are not listed here. GABA, gamma-aminobutyric acid; PKA, protein kinase A; PKC, protein kinase C; PKG, protein kinase G; Ca^{2+}, calcium ions.

two hemispheres are engaged differentially or at different post-training times. We will, therefore, discuss in detail here only those aspects of the cascade that have been shown to occur differentially in the two hemispheres.

14.5.5. *Glutamatergic Effects*

One of the first detectable biochemical events occurring after training is the enhanced release of glutamate in the left but not right IMHV, occurring within 10 min after training and no longer present by 20 min after training (Daisley et al., 1998). This increased release of glutamate was, interestingly, recapitulated when a dry, but previously aversive, bead was presented at test. It was matched by an upregulation of NMDA receptor binding in the left IMHV and LPO 30 min post-training. This surge, determined using *in vivo* microdialysis in awake chicks, was also present, but to a lesser extent, in chicks trained with a water-coated bead. (In untrained chicks, there is a higher level of NMDA binding in the right than the left IMHV, and training effectively abolishes this asymmetry; Johnston, Dodd and Rogers, 1995, 1997.) A 67% increase in NMDA-sensitive glutamate binding in the left IMHV and an 80% increase in the left LPO compared to water-trained chicks occurs in what appears to be a learning-related manner (Steele, Stewart and Rose, 1995). MK801-binding studies showed a similar, if less dramatic, increase in binding in the left IMHV and LPO (Stewart, Bourne and Steele, 1992) together with a small increase (13%) in binding in the right IMHV and a decrease in the right LPO (15%). Rendering the chick amnesic by subconvulsive transcranial electroshock immediately after training abolished these changes, indicating that they are specifically related to memory formation rather than other effects of the experience of pecking the bitter bead.

The specific involvement of the left IMHV is also suggested by the findings that blockade of the glycine binding site of the NMDA receptor in the left but not the right IMHV with 7-chlorokynurenane is amnestic at 30 min and, conversely, injection of the partial agonist cycloserine, to this site, is said to enhance retention (tested at 24 h) for the weak version of the task (bead coated in 10% MeA, usually only recalled for 6–8 h; Steele and Stewart, 1994). The AMPA receptor site is not involved at all at these earlier post-training times, but does become engaged, seemingly bilaterally, later in the cascade, at 5 h post-training during the so-called 'second wave' of protein synthesis (Stewart, Bourne and Steele, 1992; Salinska et al., 1999).

As also might be expected from the engagement of glutamate receptors in the cascade (Rickard, Poot, Gibbs and Ng, 1994), injection of glutamate itself can be amnestic; when 500 nmol glutamate was injected into the left, but not the right hemisphere 5 min before training, amnesia was apparent in chicks tested at 24 h (Patterson et al., 1986). Finally, in this context, it should be noted that injections of glutamate 5 min pre-training into the right, but not the left, lateral neostriatum were also amnestic (Patterson et al., 1986). However, the neurotoxicity of these high doses and large volumes of glutamate make interpretation of such results, in the absence of morphological controls, uncertain.

14.5.6. Calcium Flux

A key consequence of the activation of NMDA channels is the entry of calcium ions into the postsynaptic cell. Calcium flux is necessary for memory consolidation in this task: blockade of (N-type) calcium channels by bilateral injections of Ω-conotoxin GVIA just prior to training results in amnesia within 30 min post-training. By contrast and to our surprise, blockade of L-type Ca^{2+} channels was without effect, despite previous reports to the contrary (Clements, Rose and Tiunova, 1995; Deyo, 1996). In adult mammals, the N-type channels are largely presynaptic, whereas L-type channels are postsynaptic. However, this may not be the case either in young animals in general or in birds, and the possibility that this represents a postsynaptic effect is still open. That training results in changes in calcium flux was established directly by measuring intracellular calcium concentrations in synaptoneurosomes (i.e. preparations containing pinched off and resealed presynaptic and postsynaptic elements) prepared from left or right IMHV at various post-training times. K^+-stimulated Ca^{2+} uptake was enhanced in left IMHV samples prepared between 1 and 30 min, but not 180 or 360 min post-training (smaller elevations were found in samples from water-trained compared with 'quiet' non-trained birds). The increase in NMDA-stimulated Ca^{2+} flux occurred a little downstream, in samples of either left *or* right IMHV taken 10 or 30 min post-training, but not either earlier or later. In accord with the data on the later involvement of AMPA receptors, described above, AMPA-stimulated Ca^{2+} flux was found only in samples prepared at the later post-training times of 180 and 360 min. Two caveats are required against overinterpreting these data. The first is that, as the chicks were tested prior to preparing the synaptoneurosomes, this in itself could affect ion flux, as shown from the microdialysis data described above. The second, perhaps more important, is the problem of extrapolating from the *in vitro* results to the *in vivo* situation.

The cascade of events indicated in Figure 14.3 suggests that, following enhanced glutamate release, NMDA receptor upregulation and Ca^{2+} flux, the putative retrograde messenger nitric oxide (NO) is released, inferred from studies showing that blocking NO synthase with nitroarginine results in amnesia setting in for the passive avoidance some 30 min post-training (Holscher and Rose, 1993). This effect, however, is not lateralized: as with MK-801, injection of nitroarginine into either hemisphere resulted in amnesia with an identical time of onset; nor is it known whether changes in the phosphorylation state of the synaptic membrane protein GAP43/B50 or the translocation of its phosphorylating enzyme protein kinase C to the membrane are lateralized. Therefore, it is the data on glutamate neurotransmission and Ca^{2+} flux on which we must, so far, rely for insight into the temporal and lateralized dynamics of memory consolidation. In part, these do confirm the implications of the lesioning studies. That is, training with MeA results in early synaptic events in the left IMHV, followed later by a less marked and more ambiguous process in the right IMHV. Ca^{2+} flux in the LPO or other brain regions has yet to be studied.

14.5.7. Other Transmitter Systems

Just a 30-min delay following training with the MeA-coated bead was sufficient time to enable significant increases in the delta opioid (but not the mu or kappa) receptors in both left and right LPO and also the right palaeostriatum augmentatum (Csillag et al., 1993). These increases were accompanied by significant decreases in the delta, but not other types of opioid receptors in the right lateral hyperstriatum ventrale. As with glutamate binding, these training-related changes are superimposed on an initial right–left asymmetry in receptor binding levels at least in dark-incubated chicks (Johnston, Dodd and Rogers, 1995). Almost as rapid were changes in α-noradrenergic receptors following training. Stamatakis, Stewart and Dermon (1998) found significant and lateralized increases in receptor binding (Bmax) in LPO, archistriatum, and both the dorsal and ventral portions of the hyperstriatum ventrale, with an initially higher level in right than left regions being reversed. A reciprocal pattern was evident in the archistriatum, the initially higher binding in the left compared to the right archistriatum being abolished by training with MeA. However, in the absence of interventive control experiments, it is unclear whether these changes in receptor binding are specifically required for memory formation or are aspects of more general forebrain responses to the experience of pecking the bead. What remains is the evidence that rapid upregulation or downregulation of a number of receptor systems can occur, in widely distributed brain regions, as a consequence of the training experience. These, as with the upregulation of the NMDA-receptor, may be the result either of the recruiting of a pool of preformed receptors into the synaptic membrane, or the unmasking of existing membrane receptor sites; they are unlikely to involve *de novo* synthesis of receptors, given the time scales involved.

14.5.8. BDNF

As well as the more obvious synaptic transients involved in memory consolidation for the passive avoidance task, we have recently also been able to show a requirement for one of the neurotrophins, BDNF (brain-derived neurotrophic factor). BDNF has a number of actions at the synapse, mediated via its specific tyrosine kinase receptor, and possibly involving the GABAergic system. Very recently it has also been shown to have a direct depolarizing effect on neurones even stronger than that of glutamate, suggesting a direct neurotransmitter function (Kafitz et al., 1999). Suspecting a possible involvement in memory consolidation, we were able to show that antibody to BDNF, injected 5 min pre-training, resulted in amnesia that was apparent 3 h and 24 h later (Johnston, Clements and Rose, 1999). This amnestic effect was lateralized, being significantly greater in chicks injected unilaterally into the left IMHV than into the right (Johnston and Rose, 1998). With both bilateral and unilateral injections, BDNF appears to alter the later stages of memory consolidation, as recall is evident up to an hour after training.

14.5.9. Gene Activation and Protein Synthesis

As the flow diagram of Figure 14.3 shows, within an hour of training chicks on the passive avoidance task, there is enhanced expression of the immediate early gene c-*fos* and c-*jun* (Anokhin et al., 1991; Freeman, 1994). As such enhanced expression is known to occur as a result of many forms of stimulation, irrespective of whether they are associated with memory formation, it was important to show that blocking c-*fos* expression, for instance with antisense (Mileusnic, Anokhin and Rose, 1997; Mileusnic, 2000) was amnestic, and indeed this turned out to be so. The amnestic effect of protein synthesis inhibitors such as anisomycin administered around the time of training is also attributable to their effect on expression of these early gene products (Freeman, 1994). However, there is no clear-cut evidence on the lateralization of either the enhanced expression or the effect of blocking it in these experiments.

In order to control for possible c-*fos* expression related to non-specific effects of training, a further experiment in which chicks were trained on the passive avoidance task monocularly was conducted. Using this it was possible to show an almost complete suppression of c-*fos* in the hemisphere ipsilateral to the eye in use (Anokhin et al., 1991). Chicks using their left eye during training showed c-*fos* expression only in the right IMHV, LPO and primarily in the right ectostriatum, although there was some expression in the ipsilateral or left ectostriatum, consistent with the reverse crossover (i.e. to ipsilateral) visual pathways proposed in the chick, whereas chicks trained using their right eye showed expression in the left IMHV, left ectostriatum and left LPO, although low level expression was also noted in the right LPO consistent with the notion of concurrent processing in the left and right hemisphere. Moreover, there was almost no expression in the right ectostriatum of birds looking with their right eye.

The activation of immediate genes such as c-*fos*, like that of the cascade of transcription factors (notably CREB, cAMP response element binding protein) has engaged the interest of other groups working on memory consolidation processes. Such activation forms part of the molecular biological housekeeping mechanisms that must lead, ultimately, to the synthesis of the structural proteins that are embedded into presynaptic and postsynaptic membranes. The latter affect adherence and hence connectivity. The observation that not only does blocking protein synthesis with anisomycin around the time of training result in amnesia, but that anisomycin is also amnestic if injected 4 h post-training, pointed to the occurrence of a 'second wave' of protein synthesis as being necessary for long-term memory (Freeman, 1994; Freeman and Rose, 1995). These, we argue, constitute the relevant structural proteins. Amongst them are the microtubular protein tubulin (Scholey et al., 1995) and members of the family of glycosylated cell adhesion molecules, notably NCAM (neural cell adhesion molecule) and L1 (aka NgCAM) (Mileusnic et al., 1995; see also Mileusnic et al., 2000). The synthesis of this latter group is blocked by the metabolic inhibitor 2-deoxygalactose (2-D-gal), which competes with galactose for insertion into the growing glycoprotein chain. The molecular structure of 2-D-gal prevents it forming glycosidic bonds with the terminal fucose, and hence the chain terminates prematurely. Bilateral injections of 2-D-gal 5–8 h post-training are amnestic (Bullock, Potter and Rose, 1990; Crowe et al., 1994). As discussed in other reviews (e.g. Rose, 1995a), the time displacement between the amnestic effects of blocking protein and glycoprotein synthesis relates to the sequence of post-translational modifications that are required to convert the former into the latter. There is some evidence that, using a slightly modified version of the passive avoidance task (with green and yellow diodes rather than beads), perhaps recall might be affected to a greater degree by injection of 2-D-Gal into the right hemisphere (Barber and Rose, 1991). However, the volumes used for intracerebral injection (10 μl per hemisphere) were such that the effects of interhemispheric transfer cannot be excluded. In the present context, it is unfortunate that there is no unequivocal evidence for the lateralized effect of either of these inhibitors, as detailed investigation may well provide important evidence regarding the significance of changes in cell adhesion for memory processes in either/both forebrain hemispheres.

The correlative approach has been a little more revealing so far as the glycoproteins are concerned. Glycoprotein synthesis can be monitored by measuring the rate of incorporation of fucose into the macromolecules, and fucose synthesis is indeed enhanced in the hours following training. There was a specific increase in fucose incorporation in tissue slices prepared from a region containing the right LPO of trained compared to control chicks 30 min, 3 h and as long as 24 h after training (McCabe and Rose, 1985). Two of the enzymes involved in fucosylation of glycoproteins are fucokinase and fucosyltransferase. The activity of the former was increased in an area including the right but not the left LPO, 1 h post-training (Lossner and Rose, 1983).

These studies were performed prior to the identification of the LPO as a specific region of interest. However, when considered in the light of the later lesion studies, it seemed that they could be interpreted as indicating that the longer term molecular processes leading to changes in synaptic connectivity were LPO located (Rose, 1991a) and attempts to identify the glycoproteins concerned focused on changes in the LPO (Rose, 1995a, 1995b). However, the more precise molecular work that has followed, and has led to the identification of NCAM, Li and other cell adhesion molecules, has shown that the synthesis of these is required in the IMHV as well as the LPO. However, we have no evidence of lateralization (for a complete review, see Rose, 1995a, 1995b, 2000); of course, lack of evidence should not be interpreted as evidence of lack.

14.5.10. Morphological Changes

Memory consolidation is supposed to culminate in synaptic remodelling, which in principle should be detectable morphologically. Such synaptic modulation has indeed been found, in both hemispheres, well downstream of the training experience, and has been studied at both light and electron microscopic levels – the latter requiring careful three-dimensional reconstruction and statistical methodology (Rose and Stewart, 1999).

Some changes are rapid. As little as 30 min is required for there to be a detectable increase in the number of synaptic vesicles around the active zone of synapses in the IMHV (Rusakov et al., 1993). Within 1 h there is a 30% increase in the number of synapses in the right IMHV accompanied by a significant decrease in synaptic size (Doubell and Stewart, 1993). This bimodal response is transient (possibly an artefact of synapse splitting) and is not protein synthesis dependent (Rose and Stewart, 1999) and it is no longer present by 24 h. However, by this time there are significant alterations in a number of morphological measures in IMHV. These fall into two categories, those in which an initial asymmetry is abolished by MeA training, and those in which training imposes an asymmetry not previously there. The former includes an increase in the length of the post-synaptic thickening in axospinous synapses in the left IMHV, eliminating a previous asymmetry in favour of right IMHV (Bradley and Galal, 1988). The latter includes an increase in the volume of tissue occupied by synapses in the left IMHV (Stewart et al., 1984). The most notable finding of this study was the 61% increase in the number of vesicles per synapse in the left compared to the right IMHV of MeA-trained chicks. Later studies showed that memory formation is also accompanied by decreases in PSD length in the right IMHV (Stewart et al., 1984; Doubell and Stewart, 1993) and also by a concurrent increase in the number of synaptic vesicles per synapse in the right IMHV by 24 h after training (Stewart et al., 1984; Stewart and Rusakov, 1995). The increase in volume density of the larger presynaptic boutons in the left IMHV of MeA-trained chicks might reflect changes in membrane vesicle recycling rate (Stewart et al., 1984; Johnston, Clements and Rose, 1999).

The time-course of the changes in synaptic density in the LPO following training were mapped by Hunter and Stewart (1993). Increases first became apparent 12 h after training in both right and left LPO. By 24 h there was a significantly greater synaptic density in the left LPO of MeA-trained chicks and by 48 h the increase was evident in both left and right LPO. These changes, unlike the transient changes noted in the right IMHV, were protein synthesis dependent (see Rose and Stewart, 1999).

Postsynaptic measures have focused on dendritic spine density (i.e. spines per micrometre of dendritic branch), observed in Golgi-stained preparations. Density is greater, by 47%, in right than left IMHV in water-trained chicks (Patel and Stewart, 1988). This asymmetry was abolished by training; significantly greater (89–113%) spine density was observed in the left IMHV of

MeA-trained than water-trained chicks 24 h post-training. The smaller increase (37–69%) in the right IMHV was barely significant. These increases did not occur in chicks trained but rendered amnesic by electroshock (Patel, Rose and Stewart, 1988). Suggestions of changes in the dendritic branching pattern in these experiments have been carefully examined using 3-dimensional microscopy reconstruction techniques (see Lowndes and Stewart, 1994). These studies suggest that greater dendritic branching (accompanied by a 9% increase in dendritic length) occurs in the left IMHV 24 h after training. An increase of 35% in spine density was found in the left LPO, accompanied by a barely detectable change in this measure in the right LPO (Lowndes and Stewart, 1994).

One problem in interpreting such longer-lasting changes, either in morphology or biochemistry, is that any effects of training are superimposed on rapid developmental processes occurring in these young animals. Even though synaptogenesis peaks prior to hatching in these precocial birds, it continues well into the posthatching period (Rostas, 1991). Although control chicks can be drawn from the batch, it can always be argued that the training experience serves merely to accelerate a developmental sequence that would have occurred anyway. In so far as many, especially the morphological studies, find asymmetries in the 'quiet' control (untrained) or water-trained birds that are abolished or reversed by training, then this argument needs to be taken seriously. In this context, the morphological studies made by Scheich and his collaborators on chicks exposed to acoustic imprinting must be taken into account. Using similar Golgi methods to those in the Patel et al. (1988) studies described above, they found a marked *decrease* in spine density in the medial neastriatum hyperstriatum (MNH), a region identified as necessary for acoustic imprinting (see Braun et al., 1999). They argue that this dendritic pruning represents an increasing specification of connectivity, for example, to ensure responses only to appropriate, previously experienced stimuli. Such dendritic and synaptic pruning is, of course, a well-established means of ensuring specificity during development (Edelman, 1991), although others (e.g. Purves, 1988) have argued the contrary case: that developmental specificity requires growth, not loss, of connectivity. The results of the two chick studies can, of course, be reconciled in a number of ways: the two forms of learning are different; the group examining auditory imprinting (see Braun et al., 1999) has not studied events close to the time of training, but only from 2 days later, and it could be that both forms of training result in an initial efflorescence followed by a slower pruning to leave intact only those connections essential for retention of the altered behavioural responses (Bock and Braun, 1998; Braun et al., 1999).

14.6. Imprinting and Passive Avoidance Compared

It is useful to compare and contrast the patterns of lateralization shown in imprinting and passive avoidance learning studies. Both tasks utilize both left and right IMHV, as well as other brain regions (LPO and S'), but seemingly with different temporal dynamics. However, the different temporal parameters of the training regimes involved in the two tasks make direct comparisons problematic. Passive avoidance training involves a single bead presentation, which takes at the most 30 s, whereas imprinting involves at least 30 min exposure to an object, and more commonly around 1.5 h (often broken by several periods of up to 30 min). Thus, it is not possible to simply match the spatiotemporal dynamics of the two tasks. As noted earlier, McCabe (1991) has suggested, on the basis of pre-training lesioning studies of the IMHV, that the right IMHV may require several minutes of activation before it can support 'acquisition'. Therefore, it may be able to be involved in tasks with extended training such as imprinting, but not more rapid tasks such as passive avoidance learning. An alternative hypothesis is that it is equally involved in both tasks, but that the temporal dynamics of its involvement in each varies depending on the relevant sensory cues. The left IMHV appears to be crucial for the very early stages of memory consolidation for passive avoidance, with measures such as increased glutamate release and Ca^{2+} flux recorded in the early minutes following training. While the left IMHV is also significant for consolidation of imprinting memory, its role appears to be more delayed. Thus significant glutamate binding changes do not appear in either IMHV or the archistriatum/LPO regions until at least 8 h after imprinting. Indeed, some behavioural pharmacological studies as well as the morphological studies cited earlier suggest that it is regions in the right hemisphere (perhaps the right IMHV) that are crucial in the early stages of imprinting recall (Bradley, Horn and Bateson, 1981; Johnston and Rogers, 1998). The results of lesioning and electrophysiological experiments suggest that the right IMHV becomes important for passive avoidance memory some time downstream of the initial experience, but in both imprinting and passive avoidance memory some other region seems to contribute to 'holding' longer term forms of memory.

The S' region, as yet unidentified, seems to hold imprinting memory by 6 h after the end of training in an intact chick. The left and right LPO are essential for the longer term store of passive avoidance memory. In both models, some sort of system of memory 'flow' has been proposed. The system proposed a decade ago for passive avoidance involves (as shown in Figure 14.2) the 'movement' of a putative memory trace from the left to the right

IMHV, and then on to the left and right LPO regions. That for imprinting memory is less clear, involving the (possibly) sequential flow of memory from the left IMHV through the right IMHV and back to the left, as well as a shunt to S′. Both of these systems were proposed some time ago and the proponents readily concede that they are overly simplistic. One confounding factor negating such a simple 'linear flow' model of memory formation is the experimental finding that different brain regions play different roles in memory consolidation. Bateson and Horn (1994) have proposed what is possibly a more useful model system, which breaks down a specific memory system into its various elements (analysis, execution, etc.) in a manner reminiscent of the breakdown of passive avoidance memory into various functional (and regionally located) elements, such as recall of colour and shape (Patterson and Rose, 1992). In their model, each feature of a stimulus activates a specific 'feature detector', enabling it to be encoded together with information about the intensity with which that detector was activated (the relative importance of that feature to the whole). If, as behavioural evidence suggests, chicks can distinguish and classify items on the basis of colour, shape, size, patterns and spatial location, then recalling and avoiding the 'aversive' bead or recalling and approaching the imprinting object may require the chick to access many pieces of information relating to any or all of these features. It may be that, drawing on the systems level ('medial temporal lobe') model of declarative memory (Squire, 1992), these separate elements of 'memory' (colour, shape, size, location, etc.) are consolidated in different brain regions and different brain hemispheres, depending on the sort of information they contain. These individual 'bits' of information might equate with the 'feature detectors' described in the Bateson and Horn (1994) model of memory consolidation. Recognition would require reassembly of the key components of memory 'activators', drawing upon all of the previously activated feature detectors. How the different and dispersed components of the memory are bound together to provide some sort of coherent unity remains, of course, a central problem for current memory theory.

The primary confounding factor in resolving such models in the chick is that evidence for the involvement of various regions in each hemisphere for each task is, at best, partially understood. The majority of passive avoidance studies have focused on the cellular–biochemical mechanisms of memory formation almost to the exclusion of the role of distinct regions in either hemisphere. One research group that has provided some valuable insights into the mechanisms of passive avoidance memory consolidation has carried out the studies almost exclusively bilaterally; e.g. Ng et al., 1992; but see Chapter 15 by Andrew). Studies examining imprinting memory consolidation have either focused on behavioural measures (from which the separate use of the hemisphere can, at best, only be inferred) or exclusively on the mechanisms active in the left IMHV.

It is important to remember the ethological or functional basis of these tasks. In the first few days after hatching, a chick has to both imprint on the hen for protection and guidance, *and* learn to feed and avoid non-food or aversive items. Thus, perhaps the early glutamatergic and other receptor changes that occur following passive avoidance training reflect a rapid response to an immediately potentially life-threatening situation (the need to not eat noxious things), while the more delayed receptor changes following imprinting reflect the more gradual acquisition and implementation of a preference. Moreover, the majority of the biochemical evidence for memory consolidation suggests quite clearly that memory consolidation is not a unitary event, or even a series of linear stages, each of which proceeds from the former, but rather a complex of parallel but time-shifted events in more than one brain region (see Rose, 2000; McGaugh, 2000). Indeed, given the complex and ostensibly contrary patterns of brain structures and processes involved in memory formation, it seems likely that we are dealing with multiple memory holding mechanisms of which only a shifting subset are actually necessary at any specific time point for gross forms of consolidation or memory expression.

14.7. What is the Significance of Lateralization?

Irrespective of the mechanism(s) of memory consolidation, we are still left with the question of whether lateralization of memory processes is important. There is little evidence for such dynamic and seemingly significant lateralization occurring in mammalian memory models, whether rodent or primate, although imaging studies reveal some asymmetries in humans. Here we must again emphasize that lack of evidence (such as in mammals) cannot necessarily be taken as evidence of a lack. One of the reasons for the importance of studies such as those conducted using the chick models is that they lend themselves particularly successfully to just such experimental dissection of memory processes and may indeed lay significant guidelines for future studies in mammals.

However, it may also be the case that the lateralization of memory processes described in chicks is a phenomenon unique to chicks (even birds). One speculative proposition is that brains that lack the ordered layering of cortical structures characteristic of mammals, compensate functionally by increasing the packing density of neurones and functional lateralization, confining processes to one rather than two hemispheres, to increase processing capacity at the expense of redundancy. The rapid and profound lateralization, neurogenesis and subsequent neuronal loss in song nuclei described in songbirds would lend itself to just such a proposal. (Again, however, it should be noted that studies of neurogenesis in birds led the field of neurogenesis in adult animals, and provided the impetus to examine and find just such neurogenesis in adult mammals.)

It is also currently an open question as to whether the lateralization found in the young bird fully persists in the adult or is a transient consequence of differential relative rates of maturation of the two hemispheres. While there is some evidence that lateralization in (adult) hens raised from chicks produced commercially (and presumably incubated in at least partially illuminated incubators) show similar patterns of functional lateralization to those of light-incubated chicks, this is limited (McKenzie, Andrew and Jones, 1998). Thus, some types of lateralization in *Gallus gallus domesticus* could depend in part on the presumably contingent fact of the differential light exposure of the RES and LES *in ovo*, as described by Deng and Rogers (Chapter 6), rather than having a profound functional significance. Such differential maturation would then simply reflect the relative ability of each hemisphere to deal with incoming information. It is certainly the case that the direction of lateralization is modifiable in the chick, possibly as a result of differential rates of stimulation-driven maturation of either hemisphere. Experience such as asymmetrical exposure to light during incubation can alter the direction, and even the presence of some forms of lateralization as shown by the findings of Rogers and coworkers with the pebble-floor visual discrimination task (see earlier chapters). This could certainly alter the direction of lateralization in chicks trained on the passive avoidance task, most of which are exposed to light during incubation (see also Sui and Rose, 1997). However, even dark-incubated chicks show lateralization for imprinting memory: lateralization that is only partially altered by asymmetrical exposure to light (Johnston and Rogers, 1999). The lateralization demonstrated by chicks is modality dependent, and both hemispheres are used, but differently. Moreover, the experiments involving training chicks using one of the eye systems show only that it remains possible, within limits, to drive the chick to learn the task with the normally non-preferred hemisphere. Thus, the possibility of increased plasticity, while certainly not constituting evidence against the importance of lateralization in non-manipulated animals, certainly cautions against proposing universal mechanisms of memory lateralization based solely on chick work. Further investigation in both chick and other models is required.

While the notion of confining processes to one rather than two hemispheres to increase processing capacity may well help explain the importance of lateralization in chicks and other avian species, it remains intriguing that so far, in this respect at least, birds resemble humans more closely than do the usual mammalian models, such as rodents. Given the closer genetic relationship between humans and chicks than between humans and rats (genes shared between humans and chicks are more highly conserved than those shared between humans and rats; Burt et al., 1999), perhaps the chick is the best species for the study of learning and memory (Rose, 2000).

References

Ambalavanar, R., McCabe, B.J. & Horn, G. (1993). *Fos*-like immunoreactivity in γ-aminobutyric acid (GABA)-containing neurones in a forebrain region of the domestic chick. *Journal of Physiology*, 467, 350P.

Andrew, R.J. (1983). Lateralization of emotive and cognitive function in higher vertebrates, with special reference to the domestic chick. In *Advances in Vertebrate Neuroethology*, eds J.P. Ewert, R.P. Caprinica & D.J. Ingle, pp. 477–509. New York: Plenum Press.

Andrew, R.J. (1988). The development of visual lateralization in the domestic chick. *Behavioural Brain Research,* 29, 201–209.

Andrew, R.J. (1991a). The chick in experiments: techniques and tests. In *Neural and Behavioural Plasticity: The Use of the Domestic Chick as a Model*, ed. R.J. Andrew, pp. 5–57. Oxford: Oxford University Press.

Andrew, R.J. (1991b). The nature of behavioural lateralization in the chick. In *Neural and Behavioural Plasticity: The Use of the Domestic Chick as a Model*, ed. R.J. Andrew, pp. 536–554. Oxford: Oxford University Press.

Andrew, R.J. & Brennan, A. (1983). The lateralization of fear behaviour in the male domestic chick: A developmental study. *Animal Behaviour*, 31, 1166–1176.

Andrew, R.J. & Brennan, A. (1984). Sex differences in lateralization in the domestic chick. *Neuropsychologia*, 22, 503–509.

Andrew, R.J. & Brennan, A. (1985). Sharply timed and lateralized events at time of establishment of long-term memory. *Physiology and Behavior*, 34, 547–556.

Anokhin, K.V., Mileusnic, R., Shamakina, I.Y. & Rose, S.P.R. (1991). Effects of early experience on c-*fos* gene expression in the chick forebrain. *Brain Research*, 544, 101–107.

Barber, A.J. & Rose, S.P.R. (1991). Amnesia induced by 2-deoxygalactose in the day-old chick: Lateralization of effects in two different one-trial learning tasks. *Behavioral and Neural Biology*, 56, 77–88.

Bateson, P.P.G. (1966). The characteristics and context of imprinting. *Biological Reviews*, 41, 177–220.

Bateson, P. (1985). Imprinting as a process of competitive exclusion. In *Brain Plasticity, Learning and Memory*, eds B.E. Will, P. Schmitt & J.C. Dalrymple-Alford, pp. 151–168. New York: Plenum Press.

Bateson, P P.G. (1991). The nature of behavioural lateralization in the chick. In *Neural and Behavioural Plasticity: The Use of the Domestic Chick as a Model*, ed. R.J. Andrew, pp. 12–15. Oxford: Oxford University Press.

Bateson, P.P.G. & Horn, G. (1994). Imprinting recognition memory: A neural net model. *Animal Behaviour*, 48, 695–715.

Bateson, P.P.G. & Wainwright, A.A.P. (1972). The effects of prior exposure to light on the imprinting process in domestic chicks. *Behaviour*, 42, 279–290.

Bateson, P.P.G., Horn, G. & McCabe, B.J. (1978). Imprinting: The effect of partial ablation of the medial hyperstriatum ventrale of the chick. *Journal of Physiology*, 285, 23P.

Bateson, P.P.G., Horn, G. & Rose, S.P.R. (1969). Effects of an imprinting procedure on regional incorporation of tritiated lysine into protein in the chick brain. *Nature*, 223, 534–535.

Bateson, P.P.G., Horn, G. & Rose, S.P.R. (1971). Effects of early experience on regional incorporation of precursors into RNA and protein in the chick brain. *Brain Research*, 39, 449–465.

Bateson, P.P.G., Horn, G. & Rose, S.P.R. (1975). Imprinting: Correlations between behaviour and incorporation of [^{14}C]uracil into chick brain. *Brain Research*, 84, 207–220.

Benowitz, L. & Lee-Teng, E. (1973). Contrasting effects of three forebrain ablations on discrimination learning and reversal learning in chicks. *Journal of Comparative and Physiological Psychology*, 84, 391–397.

Bisazza, A., Rogers, L.J. & Vallortigara, G. (1998). The origins of cerebral asymmetry: A review of evidence of behavioural and brain lateralization in fishes, reptiles and amphibians. *Neuroscience and Biobehavioral Reviews*, 22, 411–426.

Bischof, H.-J. (1988). The visual field and visually guided behaviour in the zebra finch (*Taeniopygia guttata*). *Journal of Comparative Physiology*, 163, 329–337.

Bloch, S., Lemeignan, M. & Martinoya, C. (1987). Coordinated vergence for frontal viewing, but independent eye movements for lateral viewing, in the pigeon. In *Eye Movements*, eds J.K. O'Regan & A. Levy-Schoen, pp. 47–56. New York: Elsevier.

Bock, J. & Braun, K. (1998). Synaptic reorganisation induced by a juvenile learning process and the role of the NMDA-receptor. *European Journal of Neuroscience*, 10, (S10), 5661.

Bolhuis, J.J. & Horn, G. (1997). Delayed induction of a filial predisposition in the chick after anaesthesia. *Physiology and Behavior*, 62, 1235–1239.

Bolhuis, J.J., Johnson, M.H., Horn, G. & Bateson, P. (1989). Long-lasting effects of IMHV lesions on social preferences in domestic fowl. *Behavioral Neuroscience*, 103, 438–441.

Bradford, C.M. & McCabe, B.J. (1994). Neuronal activity related to memory formation in the intermediate medial part of the hyperstriatum ventrale of the chick brain. *Brain Research*, 640, 11–16.

Bradford, J.P. & MacDonald, G.E. (1969). Imprinting: Pre- and posttrial administration of pentobarbital and the approach response. *Journal of Comparative and Physiological Psychology*, 68, 50–55.

Bradley, P.M. & Horn, G. (1978). Afferent connections of the hyperstriatum ventrale in the chick brain. *Journal of Physiology*, 128, 46P.

Bradley, P.M. & Horn, G. (1979). Efferent connections of the hyperstriatum ventrale in the chick brain. *Journal of Anatomy*, 128, 414.

Bradley, P.M., Davies, D.C. & Horn, G. (1985). Connections of the hyperstriatum ventrale of the domestic chick (*Gallus domesticus*). *Journal of Anatomy*, 140, 577–589.

Bradley, P.M., Horn, G. & Bateson, P. (1981). Imprinting: An electron microscopic study of the chick hyperstriatum ventrale. *Experimental Brain Research*, 41, 115–120.

Bradley, P.M. & Galal, K.M. (1988). State-dependent recall can be induced by protein synthesis inhibition: Behavioural and morphological observations. *Brain Research*, 468, 243–251.

Bradshaw, J.L. & Rogers, L.J. (1993). *The Evolution of Lateral Asymmetries, Language, Tool Use, and Intellect*, pp. 37–97. San Diego: Academic Press.

Braun, K., Bock, J. & Wolf, A. (1992). NMDA-mediated mechanisms in auditory filial imprinting in chicks. *Proceedings of the Fifth Conference on the Neurobiology of Learning and Memory*, Irvine, CA.: George Thieme Verlag, no. 113.

Braun, K., Bock, J., Metzger, M., Jiang, S. & Schnabel, R. (1999). The dorsal neostriatum of the domestic chick: A structure serving higher associative functions. *Behavioural Brain Research*, 98, 211–218.

Brown, M.W. & Horn, G. (1992). Neurones in the intermediate and medial part of the hyperstriatum ventrale (IMHV) of freely moving chicks respond to visual and/or auditory stimuli. *Journal of Physiology*, 452, 237P.

Brown, M.W. & Horn, G. (1993). The influence of learning (imprinting) on the visual responsiveness of neurones of the intermediate and medial part of the hyperstriatum ventrale (IMHV) of freely moving chicks. *Journal of Physiology*, 459, 161P.

Brown, M.W. & Horn, G. (1994). Learning-related alterations in the visual responsiveness of neurones in a memory system of the chick brain. *European Journal of Neuroscience*, 6, 1479–1490.

Bullock, S., Potter, J. & Rose, S.P.R. (1990). Effects of the amnesic agent 2-deoxygalactose on incorporation of fucose into chick brain glycoproteins. *Journal of Neurochemistry*, 54, 135–142.

Burne, T.H.J. & Rogers, L.J. (1996). Responses to odorants by the domestic chick. *Physiology and Behavior*, 60, 1441–1447.

Burne, T.H.J. & Rogers, L.J. (1997). Relative importance of odour and taste in the one-trial passive avoidance learning paradigm. *Physiology and Behavior*, 62, 1299–1302.

Burne, T.H.J. & Rose, S.P.R. (1997). Effects of training procedure on memory formation using a weak passive avoidance learning paradigm. *Neurobiology of Learning and Memory*, 68, 133–139.

Burt, D.W., Bruley, C., Dunn, I.C., Jones, C.T., Ramage, A., Law, A.S., Morrice, D.R., Paton, I.R., Smith, J., Windsor, D., Sazanov, A., Fries, R. & Waddington, D. (1999). The dynamics of chromosome evolution in birds and mammals. *Nature*, 402, 411–413.

Cherkin, A. (1969). Kinetics of memory consolidation: Role of amnestic treatment of parameters. *Proceedings of the National Academy of Sciences USA*, 63, 1094–1101.

Cipolla-Neto, J., Horn, G. & McCabe, B.J. (1982). Hemispheric asymmetry and imprinting: The effect of sequential lesions to the hyperstriatum ventrale. *Experimental Brain Research*, 48, 22–27.

Clements, M.P. & Bourne, R.C. (1996). Passive avoidance learning in the day-old chick is modulated by GABAergic agents. *Pharmacology, Biochemistry and Behavior*, 53, 629–634.

Clements, M.P., Rose, S.P.R. & Tiunova, A. (1995). Ω-Conotoxin GVIA disrupts memory formation in the day-old chick. *Neurobiology of Learning and Memory*, 64, 276–284.

Cohen, D.H. & Karten, H.J. (1974). The structural organization of the avian brain: An overview. In *Birds, Brain and Behaviour*, eds I.J. Goodman and M.W. Schein, pp. 29–73. New York: Academic Press.

Crowe, S.F., Zhao,W.-Q., Sedman, G.L. & Ng, K.T. (1994). 2-Deoxygalactose interferes with an intermediate processing stage of memory. *Behavioral and Neural Biology*, 61, 206–213.

Csillag, A., Stewart, M.G., Szekely, A.D., Magloczky, Z., Bourne, R.C. & Steele, R.J. (1993). Quantitative autoradiographic demonstration of changes in binding to delta opioid, but not mu or kappa receptors, in chick forebrain 30 minutes after passive avoidance training. *Brain Research*, 613, 96–105.

Cuenod, M. (1973). Commissural pathways in interhemispheric transfer of visual information in the pigeon. In *The Neurosciences, Third Study Program*, ed. F.O. Schmidt and F.G. Worden, pp. 21–29. Cambridge, MA: MIT Press.

Cynx, J., Williams, H. & Nottebohm, F. (1992). Hemispheric differences in avian song discrimination. *Proceedings of the National Academy of Sciences USA*, 89, 1372–1375.

Daisley, J.N., Gruss, M., Rose, S.P.R. & Braun, K. (1998). Passive avoidance training and recall are associated with increased glutamate levels in the intermediate medial hyperstriatum ventrale of the day-old chick. *Neural Plasticity*, 6, 53–61.

Davey, J.E. & Horn, G. (1991). The development of hemispheric asymmetries in neuronal activity in the domestic chick after visual experience. *Behavioural Brain Research*, 45, 81–86.

Davey, J.E., McCabe, B.J. & Horn, G. (1987). Mechanisms of information storage after imprinting in the domestic chick. *Behavioural Brain Research*, 26, 209–210.

Davies, D.C., Csillag, A., Székely, A.D. & Kabai, P. (1991). The efferent connections of the domestic chick archistriatum. *European Journal of Neuroscience Supplement*, 4, 21.

Davis, C.D., Taylor, D.A. & Johnson, M.H. (1988). The effects of hyperstriatal lesions on one-trial passive-avoidance learning in the chick. *Journal of Neuroscience*, 8, 4662–4666.

Deyo, R.A. (1996). Technical note on Ω-conotoxin GVIA disrupts memory formation in the day-old chick. *Neurobiology of Learning and Memory*, 66, 89–90.

Dharmaretnam, M. & Andrew, R.J. (1994). Age- and stimulus-specific effects on the use of right and left eyes by the domestic chick. *Animal Behaviour*, 48, 1395–1406.

Doubell, T.P. & Stewart, M.G. (1993). Short-term changes in the numerical density of synapses in the intermediate and medial hyperstriatum ventrale following one-trial passive avoidance training in the chick. *Journal of Neuroscience*, 13, 2230–2236.

Edelman, G.M. (1991). Cell adhesion molecules: Implications for a molecular biology. *Annual Review of Biochemistry*, 60, 155–190.

Ehrlich, D. & Saleh, C.N. (1982). Composition of the tectal and posterior commissures of the chick (*Gallus domesticus*). *Neuroscience Letters*, 33, 115–121.

Ehrlich, D., Zappia, J.V. & Saleh, C.N. (1988). Development of the supraoptic decussation in the chick (*Gallus gallus*). *Anatomy and Embryology*, 177, 361–370.

Freeman, F.M. (1994). Protein synthesis and long-term memory formation in the day-old chick. Ph.D. thesis, The Open University, Milton Keynes.

Freeman, F.M. & Rose, S.P.R. (1995). Two time windows of anisomycin-induced amnesia for passive avoidance training in the day old chick. *Neurobiology of Learning and Memory*, 63, 291–295.

Gibbs, M.E. (1991). Behavioural and pharmacological unravelling of memory formation. *Neurochemical Research*, 16, 715–726.

Gibbs, M.E. & Ng, K.T. (1979). Behavioural stages in memory formation. *Neuroscience Letters*, 13, 279–283.

Gibbs, M.E. & Ng, K.T. (1984). Dual action of cycloheximide on memory formation in day-old chicks. *Behavioural Brain Research*, 12, 21–27.

Gigg, J., Patterson, T.A. & Rose, S.P.R. (1993). Training-induced increases in neuronal activity recorded from the forebrain of the day-old chick are time dependent. *Neuroscience*, 56, 771–776.

Gigg, J., Patterson, T.A. & Rose, S.P.R. (1994). Increases in neuronal bursting recorded from the chick lobus parolfactorius after training are both time-dependent and memory-specific. *European Journal of Neuroscience*, 6, 313–319.

Gilbert, D.B., Patterson, T.A. & Rose, S.P.R. (1991). Dissociation of brain sites necessary for registration and storage of memory for a one-trial passive avoidance task in the chick. *Behavioral Neuroscience*, 105, 553–561.

Griffiths, D. & Bateson, P.P.G. (2000). Reversal of an imprinted preference: predictions from the Bateson-Horn neural net model. In *The Evolution of Cognition*, ed. C. Heyes and L. Huber, pp. 85–102. Cambridge, MA: MIT Press.

Guntekunst, R. & Youniss, J. (1963). Interruption of imprinting following anesthesia. *Perceptual and Motor Skills*, 16, 340.

Hale, C. & Green, L. (1988). Effects of early ingestional experiences on the acquisition of appropriate food selection by young chicks. *Animal Behaviour*, 36, 211–224.

Hambley, J.W. & Rogers, L.J. (1979). Retarded learning induced by intracerebral administration of amino acids in the neonatal chick. *Neuroscience*, 4, 677–684.

Heil, P. & Scheich, H. (1992). Spatial representation of frequency-modulated signals in the tomotopically organized auditory cortex analogue of the chick. *Journal of Comparative Neurology*, 322, 548–565.

Hess, E.H. (1957). Effects of metaprobate on imprinting in waterfowl. *Annals of the New York Academy of Sciences*, 67, 724–733.

Hess, E.H. (1964). Imprinting in birds. *Science*, 146, 1128–1139.

Hogan, J.A. (1971). The development of a hunger system in young chicks. *Behaviour*, 39, 129–201.

Hogan, J.A. (1984). Pecking and feeding in chicks. *Learning and Motivation*, 15, 360–376.

Holscher, C. & Rose, S.P.R. (1993). Inhibiting synthesis of the putative retrograde messenger nitric oxide results in amnesia in a passive avoidance learning task in the chick. *Brain Research*, 619, 189–94.

Horn, G. (1985a). *Memory, Imprinting, and the Brain*. Oxford: Clarendon Press.

Horn, G. (1985b). Imprinting and the neural basis of memory. In *Brain Plasticity, Learning and Memory*, eds B.E. Will, P. Schmitt & J.C. Dalrymple-Alford, pp. 138–150. New York: Plenum Press.

Horn, G. (1991). *Cerebral Function and Behaviour Investigated Through a Study of Filial Imprinting*. Cambridge: Cambridge University Press.

Horn, G. (1999). Visual imprinting and the neural mechansims of recognition memory. *Trends in Neuroscience*, 21, 300–305.

Horn, G., Bradley, P. & Bateson, P.P.G. (1979). An autoradiographic study of the chick brain after imprinting. *Brain Research*, 168, 361–373.

Horn, G., Bradley, P. & McCabe, B.J. (1985). Changes in the structure of synapses associated with learning. *Journal of Neuroscience*, 5, 3161–3168.

Horn, G., McCabe, B.J. & Cipolla-Neto, J. (1983). Imprinting in the domestic chick: The role of each side of the hyperstriatum ventrale in acquisition and retention. *Experimental Brain Research*, 53, 91–98.

Horn, G., Rose, S.P.R. & Bateson, P.P.G. (1973). Monocular imprinting and regional incorporation of tritiated uracil into the brains of intact and 'split-brain' chicks. *Brain Research*, 56, 227–237.

Horn, G., Horn, A.L.D., Rose, S.P.R. & Bateson, P.P.G. (1971). Effects of imprinting on uracil incorporation into brain RNA in the 'split-brain' chick. *Nature*, 229, 131–132.

Howard, K.J., Rogers, L.J. & Boura, A.L.A. (1980). Functional lateralization of the chicken forebrain revealed by use of intracranial glutamate. *Brain Research*, 188, 369–382.

Hunter, A. & Stewart, M.G. (1993). Long-term increases in the numerical density of synapses in the chick lobus parolfactorius after passive avoidance training. *Brain Research*, 605, 251–255.

Immelmann, K. & Suomi, S.J. (1981). Sensitive phases in development. In *Behavioural Development: The Bielefeld Interdisciplinary Project*, eds K. Immelmann, G. Barlow, L. Petrinovich & M. Main, pp. 395–431. Cambridge: Cambridge University Press.

Johnson, M.H. & Horn, G. (1984). Differential effects of brain lesions on imprinting and an associative learning task. *Neuroscience Letters Supplement*, 18, S131.

Johnson, M.H. & Horn, G. (1986). Dissociation of recognition memory and associative learning by a restricted lesion of the chick forebrain. *Neuropsychologia*, 24, 329–340.

Johnston, A.N.B. (1999). Appetitive and aversive forms of the same one-trial learning task may be consolidated via different mechanisms. *British Neuroscience Association Meeting*, April, Harrogate, UK.

Johnston, A.N. & Rogers, L.J. (1993a). A new look at the potential of the brain to change as a result of early experience: The possible role of glutamate in early learning and retardation. *Proceedings of the Australian Psychological Society*, 27, 130.

Johnston, A.N. & Rogers, L.J. (1993b). Glutamate mechanisms in early learning: Hemispheric differences. *Proceedings of the Australian Neuroscience Society*, 4, 53.

Johnston, A.N.B. & Rogers, L.J. (1998). Right hemisphere involvement in imprinting memory revealed by glutamate treatment. *Pharmacology, Biochemistry and Behavior*, 60, 863–871.

Johnston, A.N.B. & Rogers, L.J. (1999). Light exposure of chick embryo influences lateralized recall of imprinting memory. *Behavioral Neuroscience*, 113, 1267–1273.

Johnston, A.N.B. & Rose, S.P.R. (1998). Role of BDNF, but not NGF, in memory formation. *European Journal of Neuroscience*, 10, S211.

Johnston, A.N.B., Burne, T.H.J. & Rose, S.P.R. (1998). Observation learning in day-old chicks using a one-trial passive avoidance learning paradigm. *Animal Behaviour*, 56, 1347–1353.

Johnston, A.N.B., Clements, M. P. & Rose, S.P.R. (1999). Role of BDNF and presynaptic proteins in passive avoidance learning in day-old domestic chicks. *Neuroscience*, 468, 1033–1042.

Johnston, A.N., Dodd, P. & Rogers, L.J. (1995). [^{3}H]MK-801 binding asymmetry in the IMHV region of the forebrain of dark-reared chicks is reversed by imprinting. *Brain Research Bulletin*, 37, 5–8.

Johnston, A.N., Rogers, L.J. & Johnston, G.A.R. (1993). Glutamate and imprinting memory: The role of glutamate receptors in the encoding of imprinting memory. *Behavioural Brain Research*, 54, 137–143.

Johnston, A.N.B., Bourne, R.C., Stewart, M.G., Rogers, L.J. & Rose, S.P.R. (1997). Exposure to light prior to hatching induces asymmetry of receptor binding in specific regions of the chick forebrain. *Developmental Brain Research*, 103, 83–90.

Kafitz, K.W., Rose, C.R., Thoenen, H. & Konnerth, A. (1999). Neurotrophin-evoked rapid excitation through TrkB receptors. *Nature*, 401, 918–921.

Kampen, H. S. van & Vos, G. J. de (1992). Memory for the spatial position of an imprinting object in junglefowl chicks. *Behaviour*, 122, 26–40.

Karten, H.J. (1969). The organization of the avian telencephalon and some speculations on the phylogeny of the amniote telencephalon. *Annals of the New York Academy of Science*, 167, 164–179.

Karten, H.J. & Hodos, W. (1970). Telencephalic projections of the nucleus rotundus in the pigeon (*Columba livia*). *Journal of Comparative Neurology*, 140, 35–52.

Kohsaka, S.-I., Takamatsu, K., Aoki, E. & Tsukada, Y. (1979). Metabolic mapping of chick brain after imprinting using [^{14}C]2-deoxyglucose. *Brain Research*, 172, 539–544.

Kossut, M. & Rose, S.P.R. (1984). Differential 2-deoxyglucose uptake into chick brain structures during passive avoidance training. *Neuroscience*, 12, 971–977.

Lecanuet, J.P. (1984). Selective effects of post-imprinting anaesthesia on choice-test behaviour. *Behavioural Processes*, 9, 191–203.

Lorenz, K. (1935). Der kumpan in der umwelt des vogels. *Journal für Ornithology*, 83, 137–214, 289–413. [Translated by R. Martin (1970), as Companions as factors in the bird's environment. In *Studies in Animal and Human Behaviour*, Vol. 1, pp. 101–258. Cambridge, MA: Harvard University Press.]

Lossner, B. & Rose, S.P.R. (1983). Passive avoidance training increases fucokinase activity in right forebrain base of day-old chicks. *Journal of Neurochemistry*, 41, 1357–1363.

Lowndes, M. & Stewart, M.G. (1994). Dendritic spine density in the lobus parolfactorius of the domestic chick is increased 24 h after one-trial passive avoidance training. *Brain Research*, 654, 129–136.

Mark, R.F. (1974). Pharmacology of short-term memory. *International Congress of Physiological Sciences*, 23, 153.

Martin, J.T., DeLanerolle, N. & Phillips, R.E. (1979). Avian archistriatal control of fear-motivated behaviour and adrenocortical function. *Behavioural Processes*, 4, 283–293.

Mason, R.J. & Rose, S.P.R. (1987). Lasting changes in spontaneous multi-unit activity in the chick brain following passive avoidance training. *Neuroscience*, 21, 931–941.

Mason, R.J. & Rose, S.P.R. (1988). Passive avoidance learning produces focal elevation of bursting activity in the chick brain: Amnesia abolishes the increase. *Behavioral and Neural Biology*, 49, 280–292.

McCabe, B.J. (1991). Hemispheric asymmetry of learning-induced changes. In *Neural and Behavioural Plasticity: The Use of the Domestic Chick as a Model*, ed. R.J. Andrew, pp. 262–276. Oxford: Oxford University Press.

McCabe, B.J. & Horn, G. (1988). Learning and memory: Regional changes in *N*-methyl-D-aspartate receptors in the chick brain after imprinting. *Proceedings of the National Academy of Sciences USA*, 85, 2849–2853.

McCabe, B.J. & Horn, G. (1991). Synaptic transmission and recognition memory: Time course of changes in *N*-methyl-D-aspartate receptors after imprinting. *Behavioral Neuroscience*, 2, 289–294.

McCabe, B.J. & Horn, G. (1994). Learning-related changes in Fos-like immunoreactivity in the chick forebrain after imprinting. *Proceedings of the National Academy of Sciences USA*, 91, 11417–11421.

McCabe, B.J. & Nicol, A.U. (1999). The recognition memory of imprinting: Biochemistry and electrophysiology. *Behavioural Brain Research*, 98, 253–260.

McCabe, B.J. & Rose, S.P.R. (1985). Passive avoidance training increases fucose incorporation into glycoproteins in chick forebrain slices *in vitro*. *Neurochemical Research*, 10, 1083–1095.

McCabe, B.J., Horn, G. & Bateson, P.P.G. (1981). Effects of restricted lesions of the chick forebrain on the acquisition of filial preferences during imprinting. *Brain Research*, 205, 29–37.

McCabe, B.J., Cipolla-Neto, J., Horn, G. & Bateson, P.P.G. (1982). Amnesic effects of bilateral lesions placed in the hyperstriatum ventrale of the chick after imprinting. *Experimental Brain Research*, 48, 13–21.

McGaugh, J.L. (2000). Memory – a century of consolidation. *Science*, 287, 248–251.

McKenzie, R., Andrew, R.J. & Jones, R.B. (1998). Lateralization in chicks and hens: New evidence for control of response by the right eye system. *Neuropsychologia*, 36, 51–58.

McLennan, J.G. & Horn, G. (1992). Learning-dependent changes in the responses to visual stimuli of neurones in a recognition memory system. *European Journal of Neuroscience*, 4, 1112–1122.

Miceli, D. & Repérant, J. (1985). Telencephalic afferent projections from the diencephalon and brainstem in the pigeon. A retrograde multiple-label fluorescent study. *Experimental Biology*, 44, 71–99.

Mileusnic, R. (2000). Use of antisense. *Methods in Enzymology*, 314, 213–223.

Mileusnic, R., Anokhin, K. & Rose, S.P.R. (1997). Antisense oligodeoxynucleotides to c-fos are amnestic for passive avoidance in the chick *NeuroReport*, 7, 1269–1272.

Mileusnic, R., Lancashire, C., Johnston, A.N.B. & Rose, S.P.R. (2000). APP is required during an early phase of memory formation. *European Journal of Neuroscience*, 12, 4487–4495.

Mileusnic, R., Rose, S.P.R., Lancashire, C. & Bullock, S. (1995). Characterization of antibodies specific for chick brain NCAM which cause amnesia for a passive avoidance task. *Journal of Neurochemistry*, 64, 2598–2605.

Ng, K.T., Gibbs, M.E., Crowe, S.F., Sedman, G.L., Hua, F., Zhao, W., O'Dowd, B., Rickard, N., Gibbs, C.L., Syková, E., Svoboda, J. & Jendelová, P. (1992). Molecular mechanisms of memory formation. *Molecular Neurobiology*, 5, 333–350.

Ngo, T.D., Davies, D.C., Egedi, G.Y. & Tömböl, T. (1994). A phaseolus lectin anterograde tracing study of the tectorotundal projections in the domestic chick. *Journal of Anatomy*, 184, 129–136.

Nicol, A.U., Brown, M.W. & Horn, G. (1995). Neurophysiological investigations of a recognition memory system for imprinting in the domestic chick. *European Journal of Neuroscience*, 7, 766–776.

Nicol, A.U., Brown, M.W. & Horn, G. (1998a). Hippocampal neuronal activity and imprinting in the behaving domestic chick. *European Journal of Neuroscience*, 10, 2738–2741.

Nicol, A.U., Brown, M.W. & Horn, G. (1998b). Neural encoding of subject-object distance in a visual recognition system of the domestic chick. *European Journal of Neuroscience*, 10, 2738–2741.

Nottebohm, F. (1971). Neural lateralization of vocal control in a passerine bird. 1. Song. *Journal of Experimental Zoology*, 177, 229–261.

Parsons, C.H. & Rogers, L.J. (1993). Role of the tectal and posterior commissures in lateralization of the avian brain. *Behavioural Brain Research*, 54, 153–164.

Parsons, C.H. & Rogers, L.J. (1997). Pharmacological extension of the sensitive period for imprinting in *Gallus domesticus*. *Physiology and Behavior*, 62, 1303–1310.

Patel, S.N. & Stewart, M.G. (1988). Changes in the number and structure of dendritic spines 25 hours after passive avoidance training in the domestic chick, Gallus domesticus. *Brain Research*, 449, 34–46.

Patel, S.N., Rose, S.P.R. & Stewart, M.G. (1988). Training induced dendritic spine density changes are specifically related to memory formation processes in the chick, *Gallus domesticus*. *Brain Research*, 463, 168–173.

Patterson, T.A. & Rose, S.P.R. (1992). Memory in the chick: Multiple cues, distinct brain locations. *Behavioral Neuroscience*, 106, 465–470.

Patterson, T.A., Gilbert, D.B. & Rose, S.P.R. (1990). Pre- and post-training lesions of the intermediate medial hyperstriatum ventrale and passive avoidance learning in the chick. *Experimental Brain Research*, 80, 189–195.

Patterson, T.A., Alvarado, M.C., Warner, I.T., Bennett, E.L. & Rosenzweig, M.R. (1986). Memory stages and brain asymmetry in chick learning. *Behavioral Neuroscience*, 100, 856–865.

Payne, J.K. & Horn, G. (1982). Differential effects of exposure to an imprinting stimulus on 'spontaneous' impulse activity in two regions of the chick brain. *Brain Research*, 232, 191–193.

Payne, J.K. & Horn, G. (1984). Long-term consequences of exposure to an imprinting stimulus on 'spontaneous' impulse activity in the chick brain. *Behavioural Brain Research*, 13, 155–162.

Pearson, R. (1972). *The Avian Brain.* London: Academic Press.

Phillips, R.E. (1966). Evoked potential study of the connections of the avian archistriatum and neostriatum. *Journal of Comparative Neurology*, 127, 89–100.

Phillips, R.E. & Youngren, O.M. (1986). Unilateral kainic acid lesions reveal dominance of right archistriatum in avian fear behaviour. *Brain Research*, 377, 216–220.

Purdy, J.L. & Bondy, S.C. (1976). Blood–brain barrier and selective changes during maturation. *Neuroscience*, 1, 125–129.

Purves, D. (1988). *Body and Brain: A Trophic Theory of Neural Connections.* Harvard: Harvard University Press.

Regolin, L. & Rose, S.P.R. (1999). Long-term memory for a spatial task in young chicks. *Animal Behaviour*, 57, 1185–1191.

Ribatti, D., Nico, B. & Bertossi, M. (1993). The development of the blood–brain barrier in the chick. Studies with Evans blue and horseradish peroxidase. *Annals of Anatomy*, 175, 85–88.

Rickard, N.S., Poot, A.C., Gibbs, M.E. & Ng, K.T. (1994). Antagonism of either NMDA or non-NMDA glutamate receptors impairs long-term memory consolidation in the day-old chick. *Behavioral and Neural Biology*, 62, 33–40.

Rogers, L.J. (1990). Light input and the reversal of functional lateralization in the chicken brain. *Behavioural Brain Research*, 38, 211–221.

Rogers, L.J. (1995). *The Development of Brain and Behaviour in the Chicken*, Wallingford: CAB International.

Rogers, L.J. & Andrew, R.J. (1982). Light experience and asymmetry of brain function in chickens. *Nature*, 297, 223–225.

Rogers, L.J. & Anson, J.M. (1979). Lateralization of function in the chicken forebrain. *Pharmacology, Biochemistry and Behavior*, 10, 679–686.

Rogers, L.J. & Ehrlich, D. (1983). Asymmetry in the chicken forebrain during development and a possible involvement of the supraoptic decussation. *Neuroscience Letters*, 37, 123–127.

Rogers, L.J. & Hambley, J.W. (1982). Specific and non-specific effects on neuro-excitatory amino acids on learning and other behaviours in the chicken. *Behavioural Brain Research*, 4, 1–18.

Rogers, L.J., Drennen, H.D. & Mark, R.F. (1974). Inhibition of memory formation in the imprinting period: Irreversible action of cycloheximide in young chickens. *Brain Research*, 79, 213–233.

Romanoff, A.L. (1944). Avian spare yolk and its assimilation. *Auk*, 61, 235–241.

Rose, S.P.R. (1991a). What the chick can tell us about the process and structure of memory. In *Memory: Organisation and Locus of Change*, eds L.R. Squire, N.M. Weinberger, G. Lynch & J.L. McGaugh, pp. 382–412. Oxford: Oxford University Press.

Rose, S.P.R. (1991b). How chicks make memories: The cellular cascade from c-*fos* to dendritic remodelling. *Trends in Neuroscience*, 14, 390–397.

Rose, S.P.R. (1995a). Glycoproteins and memory formation. *Behavioural Brain Research*, 66, 73–78.

Rose, S.P.R. (1995b). Cell adhesion molecules, glucocorticoids and long-term memory formation. *Trends in Neuroscience*, 18, 502–506.

Rose S.P.R. (2000). God's organism? The chick as a model system for the study of learning and memory. *Learning and Memory*, 7, 1–17.

Rose, S.P.R. & Csillag, A. (1985). Passive avoidance training results in lasting changes in deoxyglucose metabolism in left hemisphere regions of the chick brain. *Behavioral and Neural Biology*, 44, 315–324.

Rose, S.P.R. & Stewart, M.G. (1999). Cellular correlates of stages of memory formation in the chick following passive avoidance training. *Behavioural Brain Research*, 98, 237–243.

Rostas, J.A.P. (1991). Molecular mechanisms of neuronal maturation: A model for synaptic plasticity. In *Neural and Behavioural Plasticity: The Use of the Domestic Chick as a Model*, ed. R.J. Andrew, pp. 177–211. Oxford: Oxford University Press.

Rostas, J.A.P., Brent, V.A. & Güldner, F.H. (1984). The maturation of post-synaptic densities in chicken forebrain. *Neuroscience Letters*, 45, 297–304.

Rostas, J.A.P., Brent, V., Güldner, F.H. & Dunkley, P.R. (1983). Maturation of post-synaptic densities in chicken forebrain. In *Molecular Pathology of Nerve and Muscle: Noxious Agents and Genetic Lesions*, eds A.D. Kidman, J.K. Tomkins, N.A. Cooper & C. Morris, pp. 67–79. New Jersey: Humana Press.

Rostas, J.A.P., Kavanagh, J.M., Dodd, P.R., Heath, J.W. & Powis, D.A. (1992). Mechanisms of synaptic plasticity. *Molecular Neurobiology*, 5, 203–216.

Rusakov, D.A., Stewart, M.G., Davies, H.A. & Harrison, E. (1993). Spatial re-arrangement of the vesicle apparatus in forebrain synapses of chicks 30 min after passive avoidance training. *Neuroscience Letters*, 154, 13–16.

Saleh, C.N. & Ehrlich, D. (1984). Composition of the supraoptic decussation in the chick (*Gallus gallus*). A possible fact limiting interhemispheric transfer of visual information. *Cell and Tissue Research*, 236, 601–609.

Salinska, E., Chaudhury, D., Bourne, R.C. & Rose, S.P.R. (1999). Passive avoidance testing results in increased responsiveness of synaptosomal voltage in ligand gated channels in chick brain. *Neuroscience*, 93, 1507–1514.

Salzen, E.A. (1992). Parallel processing and two brain regions for learning in chicks. *European Journal of Neuroscience, Supplement*, 5, 149.

Salzen, E., Bell, G. & Parker, D. (1983). The neural site of imprinting – lateral neostriatum of medial hyperstriatum ventrale. *London Conference of the British Psychological Society, London*, 37–39.

Salzen, E.A., Parker, D.M. & Williamson, A.J. (1975). A forebrain lesion preventing imprinting in domestic chicks. *Experimental Brain Research*, 24, 145–157.

Salzen, E.A., Parker, D.M. & Williamson, A.J. (1978). Forebrain lesions and retention of imprinting in domestic chicks. *Experimental Brain Research*, 31, 107–116.

Salzen, E.A., Williamson, A.J. & Parker, D.M. (1979). The effects of forebrain lesions on innate and imprinted colour, brightness and shape preferences in domestic chicks. *Behavioural Processes*, 4, 295–313.

Sandi, C. & Rose, S.P.R. (1994). Corticosterone enhances longterm retention in one day old chicks trained in a weak passive avoidance learning paradigm. *Brain Research*, 6, 106–112.

Sandi, C., Patterson, T.A. & Rose, S.P.R. (1993). Visual input and lateralization of brain function in learning in the chick. *Neuroscience*, 52, 393–401.

Schaeffel, F., Howland, H.C. & Farkas, L. (1986). Natural accommodation in the growing chicken. *Vision Research*, 26, 1977–1993.

Schaller, G.B. & Emlen, J.T. (1962). The ontogeny of avoidance behaviour in some precocial birds. *Animal Behaviour*, 10, 370–381.

Scharff, C. & Nottebohm, F. (1991). A comparative study of the behavioural deficits following lesions of various parts of the zebra finch song system: Implications for vocal learning. *Journal of Neuroscience*, 11, 2896–2913.

Scholey, A.B., Mileusnic, R., Schachner, M. & Rose, S.P.R. (1995). A role for a chicken homologue of the neural cell adhesion molecule L1 in consolidation of memory for a passive avoidance task. *Learning and Memory*, 2, 17–25.

Serrano, P.E., Beniston, D.S., Oxonian, M.G., Rodriguez, W.A., Rosenzweig, M.R. and Bennett, E.L. (1994). Differential effects of protein kinase inhibitors and activators on memory in the 2-day-old chick. *Behavioral and Neural Biology*, 61, 60–72.

Squire, L.R. (1992). Declarative and nondeclarative memory – Multiple brain systems supporting learning and memory. *Journal of Cognitive Neuroscience*, 4, 232–243.

Stamatakis, A., Stewart, M.G. & Dermon, C.R. (1998). Passive avoidance learning involves α-noradrenergic receptors in a day old chick. *NeuroReport*, 9, 1679–1683.

Steele, R.J. & Stewart, M.G. (1994). 7-Chlorokynurenate, an antagonist of the glycine binding site on the NMDA receptor, inhibits memory formation in day old chicks (*Gallus domesticus*). *Behavioral and Neural Biology*, 60, 89–92.

Steele, R.J., Stewart, M.G. & Rose, S.P.R. (1995). Increases in NMDA receptor binding are specifically related to memory formation for a passive avoidance task in the chick: A quantitative autoradiographic study. *Brain Research*, 674, 352–356.

Stewart, M.G. & Rusakov, D.A. (1995). Morphological changes associated with stages of memory formation in the chicks following passive avoidance training. *Behavioural Brain Research*, 66, 21–28.

Stewart, M.G., Bourne, R.C. & Steele, R.J. (1992). Quantitative autoradiographic demonstration of changes in binding to NMDA-sensitive [^{3}H]glutamate and [^{3}H]MK801, but not [^{3}H]AMPA receptors in chick forebrain after passive avoidance training. *European Journal of Neuroscience*, 4, 936–943.

Stewart, M.G., Rose, S.P.R., King, T.S., Gabbott, P.L.A. & Bourne, R.C. (1984). Hemispheric asymmetry of synapses in chick medial hyperstriatum ventrale following passive avoidance training: A stereological investigation. *Developmental Brain Research*, 12, 261–269.

Sui, N. & Rose, S.P.R. (1997). Effects of dark rearing and light exposure on memory for a passive avoidance task in day-old chicks. *Neurobiology of Learning and Memory*, 68, 230–238.

Takamatsu, K. & Tsukada, Y. (1985a). Neurological basis of imprinting in chick and duckling. In *Perspectives on Neuroscience From Molecule to Mind*, ed. Y. Tsukada, pp. 187–206. Berlin: Springer-Verlag.

Takamatsu, K. & Tsukada, Y. (1985b). Neurochemical studies on imprinting behaviour in chick and duckling. *Neurochemical Research*, 10, 1371–1391.

Tiunova, A., Anokhin, K.V., Schachner, M. & Rose, S.P.R. (1998). Three time windows for amnestic effect of antibodies to cell adhesion molecule L1 in chicks. *Neuroreport*, 9, 1645–1648.

Turkel, J. & Wallman, J. (1977). Oscillatory eye movements with possible visual function in birds. *Neuroscience*, 3, 158.

Vallortigara, G. & Andrew, R.J. (1991). Lateralization of response by chicks to a change in a model partner. *Animal Behaviour*, 41, 187–194.

Vallortigara, G. & Andrew, R.J. (1994a) Differential involvement of the right and left hemisphere in individual recognition in the domestic chick. *Behavioural Processes*, 33, 41–58.

Vallortigara, G. & Andrew, R.J. (1994b). Olfactory lateralization in the chick. *Neuropsychologia*, 32, 417–423.

Vallortigara, G., Regolin, L., Bortolomiol, G. & Tommasi, L. (1996). Lateral asymmetries due to preferences in eye use during visual discrimination learning in chicks. *Behavioural Brain Research*, 74, 135–143.

Wallman, J. & Pettigrew, J.D. (1985). Conjugate and disjunctive saccades in two avian species with contrasting oculomotor strategies. *Journal of Neuroscience*, 5, 1418–1428.

Watts, M.E. & Mark, R.F. (1971). Separate actions of ouabain and cycloheximide on memory. *Brain Research*, 25, 420–423.

Zeier, H. & Karten, H.J. (1971). The archistriatum of the pigeon: Organization of afferent and efferent connections. *Brain Research*, 31, 313–326.

15

Memory Formation and Brain Lateralization
RICHARD J. ANDREW

15.1. Introduction

Johnston and Rose (Chapter 14) review extensive evidence that processes of memory formation are markedly asymmetrical in the chick. Asymmetrical involvement of brain structures in memory formation and recall is now well established for humans, as will be shown in Section 15.11 of this chapter.

I argue here that in birds (and probably other non-mammalian vertebrates) left–right differences in brain function, coupled with the common occurrence of independent viewing by the two eyes, result in the initial establishment of right and left traces. These have different content, even though they derive from the same experience. The initial differences are altered during memory formation, as is revealed, for example, by what is available to recall. In the chick, at least, much of this change is caused by processes initiated at 'retrieval events'. These are cyclically recurring points of trace reactivation; the cycles have different periods in right and left hemispheres.

The terms right or left eye systems (RES, LES) will be used, when evidence for left–right differences consists of differences in spontaneous eye use or differences between performance with right or left eye (RE, LE). This avoids the implication that structures which are exclusively contralateral to the eye in use are the only ones involved. In other cases, direct application of amnestic agents to the right or left cerebral hemisphere (RHem, LHem) reveals lateralization: here functions will be ascribed to RHem or LHem for the sake of brevity, although again other structures are no doubt involved to some extent (see Chapter 6 by Deng and Rogers). In most studies that are cited here, agents were applied in the neighbourhood of a structure [the intermediate medial hyperstriatum ventrale (IMHV), see below], that is important in memory formation; it will be made clear when other sites were used.

Early evidence for independent recording of information in the two eye systems was provided by Friedman (1975), who showed that foraging doves record only the type of seed, which they have just taken, with the eye system by which it was detected.

Differences in what each eye sees are clearly central to the way in which different traces are set up in the two eye systems. Two major strategies ensure that the full visual field of each eye, frontal as well as lateral, sees quite different things. These are the assumption of the 'principal position of gaze' in which the eyes are held diverged, or the use of independent scanning by each eye (Wallman and Pettigrew, 1985; Wallman and Velez, 1985). In addition, Martin and Katzir (1995) pointed out that depth estimation depends in birds on the accommodative state of each eye, rather than calculation of binocular disparities. The independent accommodation that results is likely to increase the differences between what is seen by each of the two eye systems for much of the time. Such independence presumably disappears during convergence for binocular fixation and pecking.

The differing properties of the two eye systems may be summarized as follows (see also Chapter 3 by Andrew and Rogers). The RES/LHem tends to control response, once the target of the response is visible: thus during approach to a manipulandum, this is fixated by the RE. During approach under identical conditions, except that there is no visible target for the bill to peck or grasp, the LE is used instead (Andrew, Tommasi and Ford, 2000; see also McKenzie, Andrew and Jones, 1998). This means that the LES is likely to see, and so to record, cues relating to the position and spatial context of the area that is being approached, while the RES sees and so records properties of the object that is to be manipulated.

The spontaneous use of the RE (McKenzie, Andrew and Jones, 1998), when a naïve chick is deciding whether to approach a potential social partner (an imprinting object) may also reflect use of the RES when decisions have to be taken, with accompanying preparations for response. Similarly, naive chicks turn the right ear towards source of clucks (Miklósi, Andrew and Dharmaretnam, 1996), when the decision to approach or not has to be taken.

Chicks using the RE also perform learned strategies of response in a way which LE chicks do not. (1) Chicks are trained, with both eyes in use, where to find two different types of food, both of which are acceptable, and then one food is devalued by an extensive meal. Following this, chicks will choose only the food that has not been devalued when the RE is in use (Cozzutti and Vallortigara, 2001). Choice is random when the LE is used. (2) In Section 15.8, it is shown that RES/LHem control may be accompanied by the use of a shift strategy in food search, which is absent when LES/RHem is controlling. (3) Recording of consequences of response by the RES is also strongly suggested by the fact that the RE shows marked advantage in learning to choose food grains from amongst inedible novel targets (pebble-floor test; Rogers and Anson, 1979).

The properties of the LES/RHem are different and complementary. As already noted, the LE is used to monitor familiar objects and surroundings. It is likely that this is a default condition that is used when there is no strong reason to use the RE. Monocular experiments show that the LES is more likely to detect (or at least respond to) novelty (e.g. in an imprinting object; Vallortigara and Andrew, 1991; Vallortigara, 1992). The use of the LE in monitoring familiar environments and objects is thus adaptive, since it facilitates the detection of unexpected but potentially important novelty.

The LES also has marked advantage in topographical (Rashid and Andrew, 1989), and spatial tasks (e.g. learning to choose the centre of an arena and to continue to do this after changes in arena shape; Tommasi, Vallortigara and Zanforlin, 1997). Further, the LES records and responds to change in the contextual cues associated with the type of stimulus that is used in the standard passive avoidance task (see below and also Chapter 14 by Johnston and Rose). In tests that are based on dishabituation of pecking by change in appearance of beads, the LE shows clear dishabituation to change in the angle of the rod on which the bead is presented; the RE is quite unaffected by such change (Andrew, 1983).

In studies of hoarding (Section 15.9), titmice use position in the environment, when retrieving food with the LE, but local cues when using the RE. Note that local cues are associated with response performance: they identify the hole into which the bill was inserted in order to store the item of food.

15.2. The Standard Passive Avoidance Task

Most of the data to be discussed were obtained using variants of a standard passive avoidance task ('bead task'): a bead is presented to a chick, which pecks it spontaneously; the chick then experiences an unpleasant taste and odour, associated with the bead. As a result, pecks at beads of the same appearance are inhibited at subsequent tests.

The task begins with pre-training, which was initially intended to measure naïve responsiveness to the two colours of beads that are commonly used in the test, and to accustom the chick to bead-like objects. However, the precise conditions of pre-training have proved to affect the outcome of learning in marked and revealing ways. In pre-training, the chick is presented with beads, at least one of which is the same in appearance as the bead that will be used at training, but (of course) lacks any unpleasant taste. It has long been known that prior experience with a non-aversive bead, similar in appearance to the bead that will be used at training, can compete with subsequent training. The effectiveness of competition can be increased by the action of hormones (e.g. testosterone) at the time of pre-training (Andrew, Clifton and Gibbs, 1981; Andrew, 1991). Early mammalian studies that used comparable competition procedures are reviewed by Spear, Gordon and Chiszar (1972).

Recently, it has been shown that the wetting of the bead with water at pre-training powerfully increases competition (Burne and Rose, 1997). Gibbs (1983) also gives evidence that wetting with water directly affects what is learned: the whole topic is discussed further in Section 15.4.8. In general, the use of beads at pre-training that are wet with water has been confined to the work of Gibbs, Ng and their associates; this is extensively cited in later sections. The procedure was intended to increase the resemblance to the appearance of the bead to that used at training, which is also wet, but with the distasteful substance methyl anthranilate.

All variants of pre-training are liable to result in records of the consequences of pecking, which differ to a greater or lesser degree from what is experienced as a result of training. As a result, the chick has to reconcile different information, derived from the two experiences (at the simplest: no taste or unpleasant taste). It will be argued here that this reconciliation occurs to some extent during memory formation itself. If discrepancy remains at recall, then loss of inhibition of pecking may reflect confusion, just as well as complete loss of the training trace.

15.3. Cellular Mechanisms of Consolidation

These are briefly considered here as an introduction to the evidence from the effects of amnestic agents and promotors of consolidation in the chick, which is considered in the sections that follow. The consolidatory cascade is discussed at much greater length by Johnston and Rose (Chapter 14).

Early involvement of N-methyl-D-aspartate (NMDA)-sensitive glutamate channels is likely: Steele and Stewart (1993) found that an antagonist of the glycine-binding site of the NMDA receptor produced amnesia, when given before training. Burchuladze and Rose (1992) report the same, and also that non-NMDA receptors are not involved. However, Rickard et al. (1994) found that both are important. For the present purposes, the main point is that changes in glutamate channels are important early in memory formation. The very early action of glutamate as an amnestic agent (see below) is entirely consistent with this. The consequent entry, and rise in intracellular levels of calcium ions is likely to be a key early integrating signal in the cascade in the chick (Gibbs, Gibbs and Ng, 1979; Serrano et al., 1994).

Calcium levels, together with noradrenergic (NE) effects (Cirelli, Pompeiano and Tononi, 1996; Section 15.5), probably converge to bring about increase in cyclic adenosine monophosphate (cAMP). Stanton and Sarvey (1985a, 1985b) showed for that repetitive electrical stimulation had to be combined in the dentate gyrus with adequate NE to produce the cAMP rise necessary for long-term potentiation. The manipulation of $\beta2$ and $\beta3$ action of NE has been very important in the recent study of memory formation in the chick (Gibbs and Summers, 2000).

Subsequent steps are, in the main, consequences of the phosphorylation of key molecules (Kaang, Kandel and Grant, 1993), which affect the current responsiveness of the cell, and initiate transcription of selected genes (e.g. ones sensitive to phosphorylation of cyclic adenosine monophosphate response element-binding protein, CREB). In the chick, c-*fos* shows elevation as a result of learning (Anokhin et al., 1991).

The final steps are the result of consequent protein synthesis. For chick memory formation, there is a considerable amount of evidence confirming this, which is derived from general block of protein synthesis by agents like anisomycin, as well as from more specific interference (Chapter 14 by Johnston and Rose).

15.4. The Temporal Framework of Memory Formation: 0–100 minutes

The *specializations of LHem and RHem* suggest that left and right traces will differ, following training on the bead task. The left trace (set up by the RES) is likely initially to hold key cues possessed by the stimulus object to which response was directed (e.g. bead), and information relating to the character of the response and its outcome (no taste, water, unpleasant taste). Such a record requires that perceptual inputs, which were experienced sequentially, should be linked appropriately.

The right trace (set up by the LES) is likely initially to hold fuller descriptions of the bead (e.g. spatial context, appearance). It is here suggested that, while consequences of response may well be included, particularly when they involve an unpleasant taste, this information will not be associated as clearly with a record of the response as in the left hemisphere trace.

A series of endogenous reactivations of the traces appear to allow further processing, involving interaction between the right and left records of the training experience. These have been termed retrieval events, since they are marked by brief enhancement of recall. The evidence for the existence of retrieval events has been reviewed by Andrew (1991, 1997). A more recent treatment (Andrew, 1999) showed that such events occurred over a quite extended period (to at least 7 h after training), and were paralleled in other birds. Subsequently, work by Crowe, Gibbs, Ng, MJ Summers and RJ Summers (see below) has greatly extended the evidential base both for the events, and for the processes that they initiate.

I begin with *the first 100 min* after learning, and sketch the temporal framework of memory formation in this period. Over this period, retrieval events recur with a period of about 16 min for the LHem trace and about 25 min for the RHem trace. These episodes of enhanced recall were initially demonstrated (Andrew and Brennan, 1985) for tasks that involved no pre-training, and so were not complicated by two overlapping processes of consolidation. Chicks had only one eye in use during learning and recall tests, so that one or other eye system had main control of both learning and recall.

With the LE in use, and following habituation of response to a coloured bead, points of good recall were revealed by maximum retention of habituation at recall at 25, 50, 75 and 100 min (Andrew, 1991). Training consisted of a brief period of repeated presentation of a clean red bead; recall was measured by the behaviour of turning or backing away, which developed by the end of exposure. This showed peak recall at 25 min in LE birds (i.e. tested monocularly using the left eye), and at 16 and 32 min in RE birds (i.e. using the right eye). An early peak also occurred in RE birds at 3 min (Andrew, 1991).

The best estimate of the duration of retrieval events comes from a study (Andrew and Brennan, 1985) in which the temporal resolution of the time-course was 1 minute. After training with an ill-tasting red bead, with both eyes in use, generalization of inhibition of pecking to a blue bead showed very clear minima, centred on 25 and 32 min. Note that very good recall allows the chick not only to withhold pecks at the training bead (red), but also to discriminate a bead of a different colour (blue) with sufficient confidence as to fail to generalize any inhibition of pecking to it. Poorer recall results in some generalization of inhibition of pecking to the blue bead, while failure of recall results in pecking at both beads. The durations of very good recall were 3 and 4 min, at 25 and 30 min events, respectively. Durations of 3–4 min agree with evidence from other time-courses for the duration of retrieval events.

The terminology used here is to identify LHem retrieval events as L1, L2 and so on, and RHem as R1, R2 and so on; since the events at 48 and 50 min have now been resolved, they are included as L3 and R2. Note that this estimate of event duration implies overlap (coincidence) of L3 and R2, but not of R1 and L2 (Figure 15.1). This will be·shown to have functional importance. L3–R2 will be used to indicate this probable overlap, and R1–L2 will be used to refer to the period from 25 to 32 min, as an indication that there is here probably no overlap.

The event at 3 min is outside the periodic structure, and is unusual in its brevity (being present at 3 but not 2 nor 4 min) and so will be termed L0.

Evidence that the RES/LHem series of events continue at least to 96 min (L4, L5, and L6) comes from the effects of unilateral administration of the $\beta 1$ and $\beta 2$ antagonist sotalol, which are considered later.

Competition experiments have been of considerable use in establishing the nature of the processes that go on at L0, L1 and L2. The design of such experiments at Sussex University has been: pre-training (red clean bead) – varying interval – training (red distasteful bead). When the experience with the distasteful bead coincided with L1 or L2 in memory formation based on pre-training (Andrew, 1991), there was competition, opposing the inhibition of pecks at subsequent recall tests.

The pattern of competition was markedly enhanced by the action of testosterone at the time of pre-training (Clifton, Andrew and Gibbs, 1982). Following such action, the character of competition was strongly dependent on the interval between pre-training and training. When intervals were varied from very short (10 s) up to 3 min (the time at which the pre-training trace should undergo L0), in all cases competition prevented the establishment of the new and discrepant information from training.

This type of competition probably depends on stabilization of attention. Testosterone is known to stabilize attention, so that, for example, search for

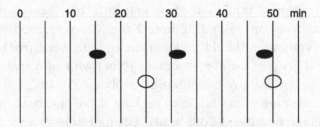

Figure 15.1. Retrieval events in the early phases of memory formation. The distribution of retrieval events is shown for the left hemisphere by filled ellipses and for the right hemisphere by open ellipses. The approximate duration of each is shown by the length of each ellipse. Note that the first overlap of events in the two hemispheres occurs just before 50 min after learning.

a particular type of visual target persists, despite the presence of alternative and distracting targets, which in untreated chicks cause repeated shifts in choice (Andrew and Rogers, 1972; Andrew, 1972, 1976). It is likely that, here, the stabilization in use of the record of pre-training, in the evaluation of perceptual input, interferes with the association of ill-taste with the bead that normally results from training.

At L0, competition ceases. The importance of the RES/LHem in the control of response in general, and selection of targets in search in particular (Section 15.1), suggests a possible explanation for this. The onset of L0 may clear the trace derived from pre-training, which is up to that time held activated in the LHem, and used in evaluation of perceptual inputs. While the pre-training trace is held activated, it determines what is learned about bead-like objects.

There is also a striking transition between R1 and L2 during memory formation based on pre-training. At L2, competition reappears and thereafter is present at all intervals between pre-training and training that have been examined. The reasons for this change are considered below.

The period 25–30 min (R1, L2) is also picked out by offsets of sensitivity to certain amnestic agents: this was first described for the $\beta 1 + 2$ antagonist sotalol (Stephenson and Andrew, 1981). The metabolic inhibitor 2,4-dinitrophenol also shows the same timing of offset of sensitivity (Gibbs and Ng, 1984). A second key point occurs immediately after L3–R2, when there can be either delayed loss or delayed return of memory, according to what amnestic agent was used. As a result, these two points have been described as transitions between memory phases: intermediate memory A and B (ITM.A/ITM.B) in the first case, and ITM.B and long-term memory (LTM) in the second (for a review, see Ng and Gibbs, 1991).

I now discuss in more detail the retrieval events that occur in the first 100 min.

15.4.1. L0: 3 Minutes

The enhanced recall shown by chicks using the RE at this time suggests endogenous reactivation of the trace, such as is characteristic of retrieval events. Evidence that is consistent with clearance of any currently active LHem trace at this time has already been noted. One possible explanation for such clearance is that reactivation brings about change in the cellular processes responsible for holding the trace.

L0 is also revealed by the offset of sensitivity to the bilateral administration of agents like 4 mM glutamate, and KCl at specific molarities (Gibbs and Ng,

1977; for a review, see Ng and Gibbs, 1991). These agents are effective at 2.5, but not 5 min after training. Recently, action of glutamate at this time has been tied to the LHem (Gibbs and Ng, in preparation): it is effective in the LHem, but not the RHem, at 0 and 3 min, but not 5 min.

Action confined in this way to the LHem is also exemplified by the finding (Steele and Stewart, 1993) that an antagonist of the glycine binding site of the NMDA receptor, given 30 min before training, is amnestic in the left IMHV, but not in the right IMHV.

15.4.2. R0: 5 Minutes

The existence of an event at this time has only recently been demonstrated, and it is not yet certain that it has all the properties of a retrieval event. However, on balance it seems likely to be the RHem counterpart of L0.

The key evidence is a brief window of sensitivity to the specific $\beta 3$ antagonist CL 316243 in the RHem at 5, but not 0 or 7.5 min; over the same period the antagonist was without effect in the LHem (Gibbs, in preparation). Another amnestic agent, hypoxia, acts in the same window of time: it is effective at 2.5 and 5 min, but not at 0 or 10 min (Allweis et al., 1984). Finally, ouabain is effective only up to 5 min when given bilaterally (Gibbs and Ng, 1977); when given unilaterally, ouabain also acts later (see below). The offset of sensitivity at 5 min for bilateral administration could well be due to action at R0, as will be shown.

In view of this evidence for RHem processes centred at 5 min, it seemed worthwhile examining other existing evidence for possible changes in recall in LE chicks at this time. Two time-courses are available that cover the relevant period. The first is a time-course for recall following persistent presentation of a clean red bead (Andrew, 1991). The response measured was turning the head away, which appears once pecking has habituated. At the test, a red bead and then a blue bead were presented. In the case of RE birds, it was clear that the two beads were treated identically up to L0 and after R1 (i.e. beginning at L2), but that, in between, response to the two beads varied quite independently. LE chicks responded much more strongly to the blue bead at almost all points. There were only two points at which response levels were identical: these were at R1 (25 and 27.5 min) and at 5 min. It is thus possible that there are indeed changes in recall at R0 that resemble those at R1.

Secondly, Gibbs (1991) described a sharp 'dip' in recall at 5 min (but not at 4, nor 6 min). Chicks were pre-trained with beads wet with water and trained with dilute methyl anthranilate. In two out of three time-courses there was a clear dip at 5 min. In the third there was no clear dip at all. In Section 15.4.8

dips are discussed further, and shown to be associated with retrieval events, following pre-training and training of this kind.

15.4.3. L1: 16 Minutes

From this point on, the model predicts the timings of events, and these timings have been confirmed. The clean red bead task revealed L1 as a point of enhanced recall; there is also enhanced competition at the time of L1 (Andrew, 1991). L1 is also revealed by extensive data from the laboratory of Gibbs and Ng (Gibbs and Ng, 1979; Ng and Gibbs, 1991), but as a dip in retrieval (again, see Section 15.4.8 for a discussion of dips). Here timing is the issue: L1 is unambiguously a point when retrieval is unusual.

There is also action of ouabain at the time of L1. This agent blocks the sodium–potassium transporter; the reader is referred to Ng and Gibbs (1991) for further discussion of possible routes of action on consolidation. As already noted, ouabain is effective, when given bilaterally, only up to R0. However, when given unilaterally, it is effective in the RHem at 10 and 15, but not 5 and 20 min (Gibbs and Ng, in preparation); over the same period it is without effect on the LHem. Note that the action of ouabain persists at the cellular level for somewhat over 5 min (Ng and Gibbs, 1988). Application at 10 min means that it should still be acting at L1 (15 min).

This odd pattern of action is best explained by supposing that the induction of amnesia induced by ouabain administered at the time of L1 requires that there should be interaction between the two hemispheres. This interaction must involve the trace reactivated in the LHem at the time of L1, and a RHem trace that is disturbed by the action of ouabain. Action of ouabain on the LHem blocks trace reactivation in that hemisphere and, as a result, there are no processes open to disturbance. Action on the RHem allows L1 to occur, while disturbing the RHem trace in a way that affects subsequent recall. Most probably this is because the LHem trace has been linked to a degraded RHem trace; comparable cases are considered below (e.g. Section 15.4.4).

Some amnestic agents cause delayed loss of memory at about the time of L1; it is not possible to establish whether this loss coincides precisely with L1 or not, since the resolution of the available time courses is in general only 5 min.

Both ethacrynic acid and ouabain, acting bilaterally at the time of training, cause loss of recall somewhere between 10 and 15 min after training. This can also be explained by interaction between left and right records of the experience, one of which has been disturbed by the amnestic agent; there is insufficient evidence to decide which hemisphere is disturbed in this case. Effects of this sort at L3/R2 are discussed below (Section 15.4.5).

15.4.4. R1: 25 Minutes

This event is marked by enhanced recall in LE, and by a sharp offset of sensitivity to the $\beta 1 + 2$ antagonist sotalol. Recently, a similar $\beta 1 + 2$ antagonist, propanolol, has been shown to be effective in the LHem, but not the RHem, up to 25, but not at 30 min (Gibbs, in preparation). This confirms the time of action but unexpectedly implicates the LHem, rather than the RHem, as the hemisphere that is affected.

The existence of R1 is well established (see above); further evidence for this is given in Section 15.5. This LHem action of propanolol thus seems to be another case where amnestic action requires the normal occurrence of the retrieval event (here R1), coupled with disturbance of the partner hemisphere (here the LHem).

15.4.5. L2: 32 Minutes

L2 is revealed by (1) a period of enhanced recall extending over a period including 30 and 32.5 min after the clean red bead test (see above; Andrew, 1991), and (2) by the reappearance of competition following pre-training, which was carried out while testosterone was acting.

L2 is also associated with the offset of sensitivity to anisomycin (ANI), which acts on ribosomes to inhibit protein synthesis, at least in part by inhibiting the binding of tRNA (Gale et al., 1981; Schumacher and Hall, 1982). ANI is effective in the LHem, but not the RHem, at 10 and 20, but not 30 min (Gibbs and Ng, in preparation). In view of probable delays in the onset of action of ANI at the cellular level, it seems likely that its effect is here at about the time of L2, and so is on processes initiated within the LHem by L2.

It thus seems that tRNA must begin to act within ribosomes close to, or during L2, or the consolidatory step cannot occur. Other coincidences of key events in cellular consolidation and retrieval events exist; possible explanations are discussed later.

The absence of RHem sensitivity to ANI at this time does not mean that protein synthesis is not important in the RHem. It will be shown later (Section 15.10) that the RHem trace may depend on structures other than the IMHV in a way that is not true for the LHem trace.

15.4.6. L3–R2: 48–50 Minutes

The left and right retrieval events should overlap (Figure 15.1), if the pre-dicted timings are correct. These timings have now been confirmed by two different procedures (Andrew, 1997). In the first, chicks were trained using a modification of the standard passive avoidance bead task, in which pre-train-ing (with a dry red bead) preceded training by only 10 s, so that there was no ambiguity about the time from which timings should be taken. At the test, chicks were presented with the bead, but at a little distance, so that they tended to fixate it with one or other eye. At 49 min there was strong and significant bias to LE use, which was less marked, but still significant at 50 min, whilst at 47 and 48 min there was no bias; significant differences between the times of L3 and R2 resulted. Away from L3–R2, there was slight (and non-significant) bias to LE use. Here, then, the LE was used during R2, thus allowing a direct view of the bead by the LES. Note that this pattern of eye use promotes the use of the reactivated trace in perceptual assessment of the bead.

The second procedure involved a delayed match-to-sample procedure, or rather a procedure that was intended to cause match-to-sample. Test involved choice between a pattern, which had been seen once before in asso-ciation with a highly preferred food item, and a pattern not seen before. At the times of L1 and L2, there was clear choice of the novel pattern. The presence of such choice differed markedly and significantly from the absence of clear choice at the time of R1. In the light of new findings about the control of response by RES (Section 15.1), it is not surprising to find the RES operating a foraging strategy of shift, since it was clear to the chick at training that the food item had been consumed, leaving nothing behind.

In the period of L3–R2, there was extreme choice of the new pattern at 47, and absence of choice at 49 and 50 min, resulting in a significant difference between L3 and R2 (Andrew, 1997). Taken together the two studies show that L3 begins before R2, and that the two are sufficiently close as to be able to affect hemispheric interaction through overlap of trace reactivation in the LHem and the RHem.

Another striking property of L3–R2 is that a range of agents, which inter-fere with protein synthesis, cause delayed loss of memory immediately after L3–R2. ANI given 5 min before training is followed by good recall up to, and including, 50 min, but not at 55 min or thereafter (Bernard et al., 1983). Cycloheximide given after training causes delayed loss with the same timing (Gibbs and Ng, 1984). The same is true for bilateral administration of anti-chick anti-Thy-1 antibodies (Lapukke et al., 1987). This suggests that the glycoprotein Thy-1 is important amongst the proteins whose failure to be available at this time interferes with memory formation.

The occurrence of loss immediately after L3–R2 suggests that the final step leading to loss in these cases is interaction between the two activated traces. The evidence (Section 15.4.5) for a LHem site of action of ANI (specifically on processes associated with L2) suggests that it is the LHem trace, which is affected at L2 by ANI. Despite this, the LHem trace, remains accessible to recall and adequate to support inhibition of pecking up to L3–R2. At this time, it is proposed, two factors combine to disturb the LHem trace and make it unsuitable subsequently to support inhibition of pecking at recall. Firstly, there is interaction with the RHem trace, which records the consequences of pecking less clearly than the LHem trace and, secondly, the earlier action of ANI has left the LHem trace in a decaying state. As a result it is the LHem trace that is lost in the course of the interaction. A fuller discussion is given in Section 15.7.

After hypoxia, there is delayed return of memory at what may be exactly the same time: memory was absent at 50 min and had returned at the next time point in the time-course, which was 60 min (Allweis, 1991). It has been argued (above) that hypoxia acts on processes associated with R0. This suggests that, in this instance also, it may be the RHem trace that is disturbed. If so, the effects of hypoxia are the reverse of the effects of protein synthesis inhibition. Hypoxia clearly has effects unlike those of other agents. It is here suggested that its action at the time of R0 is to disturb the RHem trace. The changes in recall following hypoxia would then be due, firstly, to the establishment of persisting access by the LHem to the RHem trace at L1, so that subsequent recall is disturbed. Secondly, at L3–R2 (Section 15.7), changes occur that eliminate the disturbed material held by the RHem, and so allow recall once again to be effective.

15.4.7. Events Between L3–R2 and 100 Minutes

Retrieval events continue during this period with the predicted timings. RHem events at 75 (R3) and 100 (R4) min were revealed in LE chicks by points of good recall, in a test that measured the persisting degree of habituation of pecking at a clean violet bead. The bead was presented for long enough at training to produce substantial habituation of the initially high rates of pecking (Andrew, 1991).

LHem events were demonstrated by a study (McKenzie, 1991; McKenzie and Andrew, 1996; Andrew, 1991) in which chicks were trained and tested with the LE in use. The $\beta 1 + 2$ blocker sotalol was injected into the LHem (i.e. the hemisphere not receiving the main visual input) at a range of intervals following training. Injection at the time of a LHem event, but not at other times, caused subsequent disturbance of recall, either by loss of inhibition of pecking at the training bead, or by increased generalization of inhibition to a bead of another colour. This was true for all LHem events from L1 to L6; there were no data for the times of possible subsequent such events.

In an earlier such study (Andrew and Brennan, 1985), a similar effect was found in RE birds injected in the RHem at 25 min (R1). In addition to confirming timings of events, these findings suggest disturbance of memory in the hemisphere predominantly responsible for memory formation and for recall (namely, that receiving the main direct visual input). This is brought about by depression of noradrenergic activity in the other less involved hemisphere, while it is undergoing a retrieval event. These two studies thus provide yet further examples of disturbance of memory by interaction between the hemispheres during retrieval events, in which the functioning of only one hemisphere is disturbed. An important feature of these findings is that the disturbance was due to depression of β function at any one of a series of retrieval events. At each reactivation, the trace becomes vulnerable again to such depression.

15.4.8. 'Dips' in Recall

A most important index of change in memory formation, revealed by studies from the laboratories of Gibbs and Ng, is the occurrence of points when inhibition of pecking due to training is much reduced. These have (reasonably enough) been characterized as impairment of, or 'dips' in recall. In the absence of special effects of hormones or other agents, dips occur at 15 and 55 min, and are said to mark shift from short-term to intermediate memory, and from intermediate to long-term memory, respectively (e.g. Ng and Gibbs, 1991). There have been few reports of such dips from laboratories using different procedures (but see Rosenzweig et al., 1991), and an attempt to demonstrate them failed in chicks using both eyes (McKenzie and Andrew, 1996). However, dips with the predicted timings were demonstrated by this latter study in monocular chicks (see below). The phenomenon thus was confirmed but with puzzling discrepancies.

The one experimental condition that was not replicated in the Sussex study was the use of water to wet the red bead used at pre-training. Precisely this procedure has recently been shown to have profound effects on the outcome of training. Burne and Rose (1997) compared the use of wet and dry beads in pre-training, in a procedure that results in delayed loss of memory, as a result of use of dilute, rather than full-strength aversant. Here, the use of a wet bead in pre-training caused early loss of memory. Much the same was earlier reported for the same pre-training procedure by Crowe, Ng and Gibbs (1989, 1990): the time of loss varied, but always followed L2 and preceded L3–R2. In the study by Burne and Rose, matched groups receiving pre-training with a dry bead showed a much later time of loss, similar to that already reported by Sandi and Rose (1994).

With hindsight, it is easy to see that pre-training with a bead wet with water introduces strong discrepancy between the two bead experiences in the key feature of after-taste. Water is certainly noticed by the animal in a way that dry beads are not: drinking movements can sometimes be seen. Indeed, there is direct evidence that water makes pecking a bead reinforcing in a way in which pecking a dry bead is not.

Gibbs (1983) showed that, after training with full-strength methyl anthranilate, inhibition of pecking was resistant to repeated tests with a dry red bead (which might have been expected to be equivalent to extinction trials). Specifically, with a test repeated every 5 min starting at 15 min (at which time the usual retention 'dip' was observed), there was strong inhibition of pecking up to 45 min. This inhibition suddenly disappeared at 50 min, and pecking continued to be shown up to 120 min (tests at 90 and 120 min).

For the present purposes, the key finding is that a single trial with a bead wet with water at 15 min at once abolished inhibition of pecking at subsequent trials, while experience with a dry bead had no such effect.

McKenzie and Andrew (1996), using a dry bead at pre-training, followed by training with full-strength aversant, demonstrated a 'dip' at 55 min (i.e. a return of pecking at the aversive bead). However, this dip was present only in chicks that used the RE at all trials. One consequence of monocular pre-training and training is that the traces held by the hemisphere fed by the covered eye are likely to be very incomplete visually. The extent to which this occurs is discussed in Section 15.9.

Nevertheless, it is likely that a trace is set up in the 'non-seeing' eye system: Gaston (1980) produced aversion in chicks to water which was novel both in colour and taste. Subsequent recall, using the eye that saw at training, gave avoidance of the colour, while recall using the other eye gave avoidance of the taste (which would have been a cue freely available even to the 'non-seeing' hemisphere). Monocular experience thus probably gives rise to traces in both hemispheres, so that interaction between them can still occur when reactivations coincide.

The loss of inhibition of pecking at 'dips', after such monocular pre-training and training, is best explained by the existence of a very imperfect version of the trace in one hemisphere (i.e. that ipsilateral to the eye in use). This imperfect version affects recall only when there is unusually good access to it. In the RE condition, this occurs immediately after L3–R2, as a result of overlapping activation of RHem and LHem traces.

The pattern of dips that was shown by the LE chicks in the same study (McKenzie and Andrew, 1996) was very different from that shown by RE chicks. Dips followed immediately after L4, L5 and L6, but not after L1 or L2. This suggests that access by one hemisphere to material held by the other is improved by changes caused by L3–R2. A possible explanation is that the two traces become linked at this time. Thereafter, the occurrence of a LHem event affects recall by the LE in a way that does not occur before L3–R2. As a result, after L3–R2 the (imperfect) LHem trace is strongly involved when reactivated in recall based on LE input, and so disturbs such recall.

It is now possible to interpret the dip at L1, following pre-training with a wet red bead. This is likely to be due to interaction between pre-training and training traces with very different content: one trace associates water with the bead, and the other unpleasant taste. At L1, it is proposed, the pre-training trace is temporarily strongly involved in recall. Given that the key discrepancy between the two traces is in the consequence of response (water or methyl anthranilate in the mouth), it is likely that the main interaction is between two different LHem traces.

15.4.9. Summary

This section has shown that the timings of left and right retrieval events are indeed as predicted up to 100 min. In later sections this will be shown to be true for predicted right–left coincidences up to 400 min.

The other points of especial importance are as follows.

1. Reactivation at the time of a retrieval event not only gives improved access at recall, but also reinstates sensitivity to some amnestic agents.
2. Access to the trace held by the partner hemisphere is important in producing changes during memory formation. Such access becomes easier after changes associated with the overlap of activation at L3–R2, suggesting the establishment of links between the right and left traces at this time.
3. Interactions between traces with discrepant information content (or in one of which consolidation has been disturbed) can produce changes in what is subsequently available to recall. Such interactions depend upon trace reactivation.

15.5. Modulation of Memory Formation by Noradrenergic Activity and by Hormones

Noradrenergic activity has long been known to be in some instances permissive to (and even necessary for) neural plasticity. Thus, the establishment of binocular units in the visual cortex during development is blocked when NE levels are chronically lowered due to locus coeruleus lesions and is restored by local perfusion of NE (Kasamatsu, Pettigrew and Ary, 1979). The general involvement of noradrenergic mechanisms in plasticity in the forebrain is confirmed by Cirelli et al. (1996). Facilitation of plasticity by noradrenergic mechanisms may underlie the return of sensitivity to an amnestic agent (electrical stimulation of the cortex), which is produced by the administration of adrenalin at a time when such sensitivity is normally long gone (Gold and Riegel, 1980).

NE promotes ongoing memory formation over a strikingly wide range of times. Action of this sort at or immediately after learning has been demonstrated by the following two approaches.

1. In the chick, Crowe, Ng and Gibbs (1990) showed that NE opposed the delayed loss of memory (which would otherwise have occurred), when given bilaterally and intracranially 10 s after training with diluted aversant. Stephenson (1981) found that NE, when given bilaterally 10 s before pre-training with a dry red bead, greatly enhanced subsequent competition with training.
2. Noradrenergic action of this sort during perception, as well as learning, is suggested by the fact that activation of the locus coeruleus occurs in monkeys during search for, and detection of, positive conditioned stimuli (Foote and Aston-Jones, 1995). Activity in the locus coeruleus should cause forebrain release of NE. Direct modulation of perception by NE also occurs: microinfusion of NE enhances the selectivity of visual units in cat primary visual cortex (McLean and Waterhouse, 1994).

It is likely that β action plays some part at later LHem events from L3 to (at least) L6, in view of the effects of sotalol given at the times of these events (see Section 15.4.7), but nearly all the information so far available relates to the period up to L3.

In chick memory formation, NE action on $\beta 2$ and $\beta 3$ receptors seems to be of great importance (Gibbs and Summers, 2000). $\beta 3$ activation in the RHem at the time of R0 promotes memory formation (see Section 15.4.2). Both $\beta 2$ and $\beta 3$ action are important in the period R1–L2 (25–30 min). It is thus striking that NE levels peak in the forebrain at about this time following training, and appear to be higher after normal training than after training with dilute aversant (Crowe, Ng and Gibbs, 1991). Promotion of memory formation by this natural rise in NE, acting on β receptors, is likely. The delayed loss of memory during the period from just after L2 to L3–R2 that normally follows the dilute aversant procedure (Crowe, Ng and Gibbs, 1989) can be prevented by β agonists given in the period R1–L2 (Gibbs, in preparation).

Processes in the two hemispheres appear to be differentially sensitive to $\beta2$ and $\beta3$ action at this time. The memory loss that normally ensues after dilute aversant is opposed by β agonists. This is true for a specific $\beta3$ agonist in the RHem but not the LHem, and for $\beta2$ agonists in the LHem but not the RHem; in both cases, the time-course showed the agonist to be effective at 20 min, but not 30 min (Gibbs, in preparation). However, fuller time-courses for bilateral administration showed the $\beta3$ agonist to be effective at 25 (R1), but not 30 min, and the $\beta2$ agonist to be effective at 30 (L2), but not 35 min. Taken together, these findings strongly suggest the need for $\beta3$ action at R1 in the RHem and for $\beta2$ action at L2 in the LHem, at least under these specific pre-training and training conditions.

It is unlikely that this striking asymmetry is due solely (or perhaps at all) to an asymmetry in the distribution of β receptor types: thus the $\beta1 + 2$ antagonist propanolol (acting here almost certainly as a $\beta2$ blocker) is effective in the RHem at 5 min (R0) (Gibbs, in preparation).

There is other evidence that $\beta2$ and $\beta3$ actions affect memory formation in different ways. The most extensive such evidence comes from a variety of studies that suggest that $\beta2$ (but not $\beta3$) agonists, like hormones such as testosterone, stabilize activated traces.

Both $\beta2$ agonists and some hormones cause the postponement of the 'dips' that normally occur at 15 and at 55 min. This is true of pituitary hormones (Gibbs and Ng, 1984) and of testosterone (Gibbs, Ng and Andrew, 1986), as well as of the $\beta2$ agonist zinterol (Gibbs and Summers, 2000). In both cases, the postponement is similar in that it reaches a limit at the time of a later retrieval event (with postponement being greater for larger doses, and for administration nearer the normal time of the dip). In the case of the 15-min dip, this limit is usually 25 min (R1; testosterone and zinterol). In the case of the 55-min dip, the limit is 95 or 100 min (L6R4) when postponement is due to testosterone. When zinterol is used, postponement of the 55-min dip to 85 min has been recorded; it is not clear whether further postponement to the 95–100 min limit is possible.

Although testosterone and $\beta2$ agonists postpone dips in very similar ways, the ways in which they promote competition may well differ substantially. One problem is that the available data are for the administration of NE, so that several types of receptor may have been affected; clearer resemblance to testosterone might be found if data were available for $\beta 2$ agonists.

Testosterone promotes competition due to pre-training with a clean dry red bead, when the interval between pre-training and training is so short that the pre-training trace has not yet passed L0 (see Section 15.4). Here, competition appears to be caused by stabilization of the record of the recent bead experience, which is used to assess the training experience, so that the establishment of a record including unpleasant taste is opposed. Interestingly, with such short intervals it is not necessary for the two beads to be of the same colour (Clifton, Andrew and Gibbs, 1982): the general resemblance between the bead presentations is enough for the use of the record of pre-training to interfere with the association of bead with unpleasant taste. It is not known whether NE has action of this sort.

NE, given 10 s before pre-training (dry red bead), causes pre-training to compete with training up to 20, but not at 30 min after pre-training (Stephenson, 1981). The timing of change in competition is thus similar, perhaps even identical to that found for stabilization of pre-training due to testosterone. However, although the timing of change is the same, competition ceases in the case of NE at the time at which it reappears in the case of testosterone. Possible explanations are considered in Section 15.7.

For the moment, the points to stress are, firstly, there are marked resemblances between effects on attention and on the postponement of markers like dips; these suggest that reactivated traces may be in a similar state to traces, which have just been established based on perceptual input. Secondly, it is possible that trace stabilization interferes, while it lasts, with some processes in memory formation, including ones that generate dips. This would provide an explanation for the postponement of dips up to, but not beyond, a subsequent retrieval event, given that (as the evidence from the end of competition at L0 suggests) trace stabilization may be terminated by the onset of a retrieval event. This possibility becomes important in later discussion (see Section 15.7).

15.6. Effects of Reminders on Memory Formation

Important new insights into the processes underlying memory formation in the chick have recently been provided by reminder procedures (Summers, Crowe and Ng, 1995, 1996; Litvin and Anokhin, 1999). A reminder involves the presentation of the context of training, without presentation of the reinforcer, and without the opportunity to perform the response that brought about reinforcement. There has been some controversy as to whether such procedures allow further learning (for a review, see Lewis, 1976). It is difficult to see why they should not, and the chick results suggest that further material is indeed added to existing traces. Two striking effects of reminders have been described for mammals (Lewis, 1979): they can oppose the development of amnesias due to prior application of amnestic agents (such as protein synthesis inhibitors) and they can restore sensitivity to subsequently applied amnestic agents. Both effects have been demonstrated in the chick (Radushkin and Anokhin, 1997; Litvin and Anokhin, 1999).

The reminder procedure that was used was the presentation of a red bead like that used at training, without the possibility of pecking. Following such a reminder, memory became once more sensitive to glutamate; this sensitivity was extraordinarily persistent, being still present at least 2 h after the reminder, when the latter followed 120 min after training (Summers, Crowe and Ng, 1995).

The timing of amnesia onset, when glutamate was given immediately after the reminder, suggested that memory formation is restarted as a result of the reminder. When the reminder was given 7.5 min after training, loss was rapid (perhaps immediate). A reminder 40 or 120 min after training resulted in good recall up to 10 min after the reminder, with loss of inhibition of pecking at 15 min (i.e. at the time of L1 in memory formation based on the reminder). A reminder after 24 h gave recall up to 20 min, with return of pecking at 25 min (i.e. at the time of R1 in memory formation based on the reminder). It thus seems that memory loss occurs at the times at which retrieval events would be expected were a reminder to initiate a fresh process of memory formation.

Such fresh initiation of memory formation is supported by the fact that, when a reminder is immediately followed by a protein synthesis inhibitor (cycloheximide; CXM), amnesia appears with a longer delay than when glutamate is used (Litvin and Anokhin, 1999). Recall is normal at 30, but not at 60 min after the reminder. This is clearly different from timings obtained with glutamate (above), and is consistent with interference by CXM with memory formation based on the reminder.

It seems almost certain (and is assumed to be true by all discussions of reminders) that the reminder causes recall and reactivation of the training trace. Any new learning based on the reminder is likely to be markedly similar in content to that which is available from recall of training. In the chick reminder procedure, the new information might be that the distasteful bead stayed at a distance, rather than approaching at once into pecking range. I propose that the outcome of a reminder is the formation of linked traces, with the overlap between training and the new reminder material being sufficient to cause them to be subsequently associated in both recall and reactivation.

1. Litvin and Anokhin (1999) have already argued that, as a result of a reminder, 'already consolidated memory' 'reconsolidates afresh' in the chick. Evidence has just been reviewed, which suggests that events in memory formation initiated by a reminder have the usual standard timings. The return of sensitivity to early- and late-acting amnestic agents, which is commonly caused by reminders (in mammals as well as the chick), can also be explained if such sensitivity were to be associated with consolidation of new material derived from the reminder experience.
2. In the case of glutamate at least, it is clear that the prolonged sensitivity, which follows a reminder, requires some part of the new material to be in a quite abnormal state. It seems that consolidation is unable to proceed to a point at which insensitivity to glutamate develops. This prolonged sensitivity is evidently due to effects of recall of the established training trace at the time of the reminder on the learning that is then initiated. Note that no amnestic agent is involved at the time of the reminder. Such interaction must be common in normal learning, in which few experiences will be without earlier parallels.

It is possible that the linkage between the more fully established training trace and the reminder material opposes the consolidation step that would normally occur at L0. One way in which this might happen is that linkage to the training trace helps to stabilize the new reminder material, so that it fails to be cleared at L0.

The fact that a reminder 40 or 120 min after training (followed by glutamate) gives a loss at L1 (timings here and in the rest of this section are in relation to the reminder), whereas a reminder at 24 h gives a loss at R1, may indicate that the LHem trace (of training) becomes resistant to linkage with reminder material earlier than does the RHem trace.

The final feature, which is clear both for studies involving glutamate and for ones involving CXM, is that amnesia is not permanent. In the case of CXM, the less advanced the memory formation of the training trace, the longer amnesia lasts. Thus a reminder given at 120 min after training is followed by amnesia up to 24 h, after the reminder, and has gone at 48 h. A reminder at 24 h is followed by amnesia at 1 h after the reminder, and has gone at 24 h (with recall at intermediate levels at 3 h). A reminder at 48 h is followed by amnesia at 1 h, which has gone at 3 h. The evidence for glutamate suggests a similar pattern of change. A reminder at 120 min gives amnesia at 3 h, which has gone at 24 h. A reminder at 24 h gives amnesia at 1 h, which has gone at 3 h. A reminder at 48 h gives only a hint of amnesia at 35 min, and none at all at 40 or 60 min. Glutamate is thus probably followed by slightly shorter periods of amnesia than is true of CXM. However, for both agents, the timing of the return of memory is largely dependent on the stage reached in memory formation based on training, at the time of the reminder.

The most likely cause of the return of memory is that the disturbed material due to a reminder followed by an amnestic agent is eventually removed. The times of recall tests are too widely spaced to give any clue as to whether this is gradual, or associated with an event or events in memory formation.

15.7. Retrieval Events and Consolidation

The period of memory formation so far discussed (namely the first 100 min) is that to which models that postulate 'phases' or 'stages of memory' have chiefly been applied. Such models (e.g. Ng and Gibbs, 1991) have been very useful tools, and they remain useful for some applications. However, for present purposes, they have serious drawbacks. The most fundamental is that they confound consequences of trace reactivations (whether endogenous, or consequences of recall at reminders) with steps in the consolidatory cascade at cellular level.

An associated problem is that phases become closely linked in discussion with particular consolidatory steps; indeed, they are commonly defined in this way. This becomes especially confusing when there is evidence that consolidation may be restarted by an appropriate experience, so that amnestic agents, which had long ceased to be effective, are once again effective. Further, if the consequences of linking new traces with traces advanced in consolidation are as powerful (and common) as the evidence just cited suggests, then further sources of confusion will arise in applying phase models.

However, this said, it is clear that issues, which were central in the development of phase models, remain as important as ever. These include the delayed development of effects of amnestic agents, and of behavioural manipulations such as competition and reminders, and evidence that some steps in consolidation occur with remarkably sharp and standard timings. These issues are examined here from the point of view of the possible interactions between retrieval events and steps in consolidation.

A specific hypothesis will be presented, namely that trace reactivation allows a process of testing by which the plastic connections within the trace that are fully active, are retained (or strengthened), and ones that are not fully active, are weakened or lost. This happens at more than one step in consolidation. In view of the intervention of noradrenergic mechanisms at a number of such points, it is possible that the same integrating signal plays a part at both early and late consolidatory steps. As has already been noted, rises in cAMP as a result of both NE and of neuronal activation is one (but only one) such possible signal.

It is assumed in the argument that follows that the information content of traces, and the relative timing of events in RHem and LHem will strongly affect the outcome of such reactivation.

1. When two versions of the same experience are involved, then the main effect is expected to be on material that is represented only in one version; the outcome would be reconciliation of the two versions, perhaps usually by elimination of material not present in both versions ('pruning'). The inter-action of right and left versions provides almost all the evidence that is discussed here. However, differences between what is held in different struc-tures of one hemisphere could also play a part. Thus Patterson, Gilbert and Rose (1990) showed that, early in memory formation based on the bead task, the lateral neostriatum plays a more important part in the RHem, and the IMHV in the LHem.

2. The route by which activation occurs may be important: thus interaction between right and left traces, which begins with a retrieval event in one hemisphere, may access only some of the information held in the other. Overlap or coincidence of activation may cause more complete comparison, and be more likely to lead to pruning, when there are differences.

3. When two traces relating to the same type of stimulus or situation are set up at different times, the outcome of subsequent memory formation depends both on the information held in each, and on the stage of consolidation reached by the first trace. The information content of the two traces may be nearly identical (e.g. following a reminder) or strongly discrepant (e.g. in the case of competition). Most evidence is available for processes that occur at 25–32 min (R1–L2) and at the first overlap (48–50 min: L3–R2).

15.7.1. R1–L2

When pretraining involves a red bead that is wet with water, and training involves a red bead wet with dilute aversant, processes occur at R1–L2, which lead to memory loss soon after (Crowe, Ng and Gibbs, 1989). This is clear from the fact that β agonists given at the time of R1–L2, but not later, can oppose such subsequent loss (Gibbs and Summers, 2000). It can also be opposed by a reminder (the sight, but no more, of a red bead) at 25, but not 40 min after training (Summers, Crowe and Ng, 2000). The most likely way in which this might come about is that recall, as a result of the reminder, activates the training trace at the crucial period of R1–L2. This in turn implies that the loss, which would occur in the absence of the reminder, results in some way from inadequate activation.

A second clue is given by the fact (Gibbs, in preparation) that a $\beta3$ agonist can also prevent loss caused in this way. However, this only occurs if the agonist is given in the RHem, and then only if it is given before a point 25 min after training. It has already been argued (see Section 15.5) that this is probably caused by agonist action at 25 min. There is another example of $\beta3$ action that is confined to a brief window of time and to the RHem (action at R0; see above). It is thus likely that $\beta3$ action in the RHem tends to be within a brief period that coincides with a retrieval event. One possibility is that $\beta3$ action promotes full reactivation of the trace (as does a reminder at this time; see above), which would not otherwise occur at R1, after training with weak aversant. The fact that weak aversant results in lower than normal NE levels in the forebrain at about the time of R1–L2 provides a possible causal mechanism for inadequate activation (Crowe, Ng and Gibbs, 1991).

Adequate activation of the RHem trace at R1 could prevent loss either by effects confined to the RHem trace or by effects that depend on interaction between the two traces.

As has just been noted, the processes that lead to loss can also be opposed at L2. $\beta2$ agonists act in the LHem, but not the RHem, and up to 30 min, but not later, to oppose loss (Gibbs, in preparation). Since $\beta2$ action is known to differ from $\beta3$ action (which is here ineffective in the LHem) in that it causes trace stabilization (see Section 15.5), it is possible that loss is opposed by prolonging the activation of the LHem trace of training. This, in turn, presumably acts to promote the initiation of translation in the LHem that has been shown to occur at this time (see Section 15.4.5). $\beta2$ agonists may not be effective in opposing loss when given in the RHem at R1, because stabilization of the RHem trace brings about overlap with L2, which, in turn, interferes with processes at L2 that lead to translation, rather than promoting them.

The resemblances between the effects of testosterone and of $\beta2$ agonists justifies the use here of evidence from work with testosterone. Competition due to testosterone action on the pre-training trace (Clifton, Andrew and Gibbs, 1982; Andrew, 1991) is present briefly at L1 (15 min after pre-training) and is then absent again until L2 (absent at 25, but present at 30 min). However, once it returns at L2, competition continues to be effective, rather than disappearing again. Thus the trace of pre-training becomes persistently more effective in competition, once some key step is taken at L2. Most probably initiation of translation is made more effective by more sustained reactivation of the LHem trace at L2 due to testosterone.

There are sharply timed offsets of sensitivity to $\beta 1 + 2$ antagonists and to the metabolic inhibitor 2,4-dinitrophenol (Gibbs and Ng, 1984) between 25 and 30 min. This suggests that these agents act by opposing the initiation of translation at 25–30 min, the former agent via the β mechanisms just discussed, and the latter perhaps via direct effects on cAMP levels.

15.7.2. L3–R2

The processes that cause delayed loss or return of memory just after L3–R2 are of central importance to the understanding of interaction between left and right traces. Two types of processes are suggested by the data: (1) one or both versions (right and left) are changed by loss of discrepant material ('pruning'); and (2) linkages between the two versions become more effective. These will be considered in turn.

1. Delayed loss or return of memory are more probably caused by pruning than by changes in linkages. Here, interaction at L3–R2 involves one trace (the left or right), which has been disturbed earlier in memory formation, while the other is unaffected. A simple mechanism that could explain what happens at interaction is that each trace provides neural input to the other. If activated sites of plasticity fail to receive inputs from the partner hemisphere, then they tend to be lost; if such inputs are received, then the changes at the sites of plasticity are made more durable.

The delayed loss of memory just after R2, which results from the application of protein synthesis inhibitors soon after learning, has been demonstrated repeatedly (e.g. Ng and Gibbs, 1991). It can be explained as follows: for ANI, it is now known that the action, which brings this loss about, is in the LHem in the period R1–L2 (and probably specifically on the L2 event; see Section 15.4.5). As a result of such action at L2, the LHem trace is in a state by the time of L3–R2 in which it is not capable of full and sustained reactivation, although it is still accessible to recall up to this point, and therefore supports inhibition of pecking. Trace interaction due to L3–R2 causes sufficient pruning in the LHem trace to prevent the inhibition of pecking at subsequent recall. Note that this explains the sharp and standard timing of loss.

Similar processes can explain the delayed return of memory at the time of L3–R2, as a result of hypoxia at 5 min. Hypoxia here probably acts on the RHem at R0 (Section 15.4.6). Ability to recall is disturbed when the LHem trace becomes more closely linked to the disturbed RHem material at L1. The disturbed RHem material tends to be lost at interaction associated with L3–R2, so that effective recall returns. The main difference from the delayed loss due to ANI is that hypoxia disturbs the RHem trace in such a way as to degrade its information content, whereas ANI has no such immediate effect.

2. The strengthening of linkages between left and right traces at L3–R2 would, on an extension of the same hypothesis, occur between parts of the two traces that were both persistently activated. Such strengthening is clear at L3–R2 following monocular training and testing (McKenzie and Andrew, 1996; see also Section 15.4.7).

At this point it is necessary to consider the exact timing of changes associated with L3–R2. For both delayed loss and delayed return of memory, it is clear that change follows the end of R2 (50 min). The dip (return of pecking), which occurs under conditions that have already been specified, is precisely timed: it occurs at 55 min. It thus seems that important changes in what can be recalled occur after the end of the RHem event that is involved in L3–R2.

One procedure produces memory loss that unambiguously occurs at R2. When training is followed by a long series of tests with a dry red bead (at 5-min intervals), these 'extinction' trials have no effect, until 50 min, when there is a sudden disappearance of inhibition of pecking (Gibbs, 1983). This suggests that experience with tasteless red beads is largely recorded by the RHem, and therefore suddenly affects behaviour at the reactivation of the RHem trace at 50 min. Before then, performance at recall depends on the content of the LHem trace.

It is possible that interaction begins as the endogenous reactivation of the RHem trace is withdrawn, and it is this that is revealed by the various changes at 55 min.

Finally, in an earlier discussion, based on far less adequate evidence (Andrew, 1991), delayed return or loss of memory were explained differently. It was suggested that delayed return occurred when the LHem trace, but not the RHem trace was disturbed. This caused immediate disturbance of recall. At L3–R2, there was establishment of linkage to an adequate RHem trace; this allowed subsequent recall to support inhibition of pecking.

Delayed loss was explained as being due to disturbance of the RHem trace, which came to affect recall only when linked to the LHem trace at L3–R2. The evidence (see Section 15.4.6) showing that action of amnestic agents on the LHem causes delayed loss, and suggesting that action on the RHem causes delayed return, argues strongly against both suggestions.

15.7.3. L0

Effects that occur at L0 provide a final example of effects of trace reactivation on steps in consolidation. The evidence is most striking for reminder procedures, in which association of new material, set up at the reminder, with the existing trace, causes sustained sensitivity to glutamate. I argued in Section 15.6 that it is linkage of the new material to the existing trace, which is here responsible for the failure to initiate the next step in consolidation: the step that gives invulnerability to glutamate. This step normally coincides with L0, so that it is possible that it is a failure of L0 to occur in the new material that is responsible for the disturbance of consolidation.

15.8. Later Exact Coincidences between LHem and RHem Retrieval Events

If the cycles of retrieval events were to continue after the first overlap at L3–R2, then further coincidences should occur, culminating in an exact coincidence at 400 min. It has already been shown that retrieval events do continue up to about 100 min.

A task that was intended to be a delayed match to sample task was used to explore the (possible) further events in memory formation up to points after 400 min (Andrew, 1997, 1999). Behaviour at points before, during and after three predicted close coincidences was measured: the coincidences were 125/ 128 (R5–L8), 175/176 (R7–L11) and 400 min (R16–L25).

In the training trial, the already experienced chick found a mealworm (a much prized reward), at the base of a complex pattern, never seen before, the whole being placed at the end of the test arena. At the test, two patterns, one that was seen at training and another complex pattern, matched for general appearance, but in detail never seen before, were placed, one on each side of the central position that had been occupied by the first pattern at training. Positions were counterbalanced across chicks and across successive tests of individual chicks. No mealworms were present, but this was not immediately apparent, since sawdust was present at the base of each pattern.

Two measures were taken: the relative extent of search at each pattern, and the first choice. The former was (potentially) largely independent of the latter, since chicks moved freely between patterns during the test. In each experiment a particular stretch of time-course was examined and, over a series of tests, all chicks were tested at all points within this stretch. Within-individual differences were used in further analysis.

With intervals between training and test of up to 50 min, behaviour was clearly different at times of LHem and of RHem retrieval events. Direction of initial turn at L1 was strongly to the right, differing significantly in this from turn at R0, which was predominantly to the left (Andrew, unpublished observations). It is likely that both these findings reflect hemispheric control at the start of the test, with greater attention being directed to objects in the hemi-space contralateral to the controlling hemisphere. In addition, at the times of L1 and L3, there was marked and significant choice of the novel pattern. This resulted in significant differences between choice at the times of LHem, and of RHem events: that between L3 and R2 has already been described (see Section 15.4.6). At RHem events there was usually a trend towards choice of the pattern seen at training, but this was not marked (with the exception of R0): the results are compatible either with random choice, or choice of the training pattern by only some chicks.

Choice of the novel object at the times of LHem events probably reflects a strategy of shift away from a 'foraging patch', where the chick knows that all food was consumed at training. The disappearance of such choice at RHem events could reflect either absence of such a strategy, or inability to recall adequate information.

After about 100 min, choice of the novel object was usual, and was more marked the further the time of test was from the time of a coincidence (Andrew, 1999). In contrast, at each of the coincidences that were studied, choice was random. It is likely that this was due to greater involvement of the RHem at these times, since there was then also significant turning to the left, which was absent at neighbouring points in the time-courses (Andrew, unpublished observations). The overall pattern suggests that LHem control of choice had become usual, with the exception of the times when, thanks to a coincidence, there was enhanced access to the RHem trace.

The evidence for a coincidence at 400 min (6.7 h) is of particular impor-tance, since there is other evidence for processes of memory formation in birds at about this time.

In the chick, the standard passive avoidance bead task is followed by a late episode of glycoprotein synthesis, which results in a window of sensitivity to anti-NCAM (anti-neural cell adhesion molecule). Sensitivity is present at 6, but not 8 h (Scholey et al., 1993).

Evidence for an event in memory formation almost exactly at 400 min (6.7 h) in the zebra finch is provided by a remarkable study (Chew, Vicario and Nottebohm, 1996) that demonstrated a series of sharply timed points, at which sounds were particularly likely to be forgotten. The measure of for-getting that was used was the loss of rapid habituation of the response of acoustic neurones, during exposure to the sound to which the neurone was specifically responsive.

Memory of sounds (human or canary) that were likely to have little sig-nificance for the bird was lost at about 180 min. Memory of sounds, which had some species-specific properties (e.g. reversed zebra finch vocalization), was lost at 6.5–7 h. This timing was independently confirmed (Vicario and Nottebohm, 1998). Protein synthesis inhibitors caused loss of memory for sounds (such as were normally remembered for longer), when the inhibitors were given in a window that extended from 6 to 7 h (ANI, 6–6.5 h; cyclohex-imide, 6.5–7 h). Such windows of sensitivity, in the absence of any recall, suggest that reactivation of the trace(s) occurs, somewhat as at retrieval events in the chick.

Later points of precisely timed loss occurred for sounds such as conspecific song. These were again accompanied by sensitivity to protein synthesis inhi-bitors. Vicario and Nottebohm (1998) report greater variability of timing of these later points of loss in a study which used only conspecific song (and so did not examine possible points of loss as early as 6.7 h). However, at least one later point at 47–48 h was confirmed by this second study.

In addition to the point at 6.7 h, there is one at about 3 h, which agrees well with the chick coincidence at 175–176 min. The sort of processes that might generate the later zebrafinch points remain obscure. The point at 47.5 h could be generated as the result of seven 400-min cycles; equally it could be pro-duced by two 24-h cycles (although endogenous circadian cycles are rarely precisely 24 h in period). However, at least one reported point of forgetting, around 18 h (18.5 h for forgetting male conspecific long calls; 17–18 h for conspecific song, learned under conditions that should weaken learning) fits neither of these suggestions.

It is striking that forgetting should occur at a point so similar in timing to the first exact coincidence (400 min) in the chick. The evidence of unusual RHem involvement at coincidences in the chick suggests that LHem interac-tion with a changing RHem trace may be important in forgetting, an issue to which I return at the end of the next section.

15.9. Interocular Transfer

In the procedure of 'interocular transfer' animals are trained with one eye in use; at recall, either the same, or the other eye is used. Poor performance, when the eye in use is changed, has been ascribed to failure to transfer the trace from one side of the brain to the other. The hypothesis of trace transfer is particularly convincing when there is a shift from competence with the eye used in training to competence confined to use of the other (Clayton, 1993). However, bulk transfer of a trace is a surprising phenomenon (see Section 15.10). It is clearly worth examining alternatives, as such transfer is an issue important to the arguments of this chapter.

15.9.1. Visual Mechanisms

The interpretation of results from monocular training or testing in birds (and so the interpretation of transfer experiments) depends very much on what function is ascribed to visual projections from the ipsilateral eye to forebrain visual areas: the main supply to such areas is, of course, from the contralateral eye. Specifically, do ipsilateral projections allow any visual memory to be established, during monocular experience, in the eye system ipsilateral to the uncovered eye?

Two major visual projections exist: one ('thalamofugal') by a direct thalamic relay to the visual Wulst (roughly equivalent to structures recipient of projections from the lateral geniculate in mammals), and one ('tectofugal') via the optic tectum to the ectostriatum (see Chapter 6 by Deng and Rogers). In the pigeon, the thalamofugal route is predominantly fed by the lateral visual field and the frontal field, used in close examination, chiefly feeds the tectofugal route (Güntürkün and Hahmann, 1999; but see Chapter 6 by Deng and Rogers for differences in the chick). The bead task largely depends on the use of frontal fixation, and so I will concentrate on the tectofugal system.

It is important, when considering results from monocular tasks, to note the frequent adoption by birds of the principal position of gaze, in which the eyes are held diverged (Wallman and Pettigrew, 1985; Wallman and Velez, 1985). While the eyes are in this position, the frontal field of each eye sees quite different things. The same is true of search in which the eyes are moved independently.

It is thus not possible for an ipsilateral supply to be used continuously (as in mammals with obligate yoking of eye movements) to provide inputs from a stimulus seen with both eyes to binocular visual units. There is also indirect evidence confirming that binocularity is not the main role of ipsilateral supply: in the tectofugal route, that supply consists mainly of collaterals from tectal units (Deng and Rogers, 1998). These have been shown to be mainly unmyelinated, unlike the myelinated fibres of the contralateral supply (Ehrlich and Stuchbery, 1986). The slower ipsilateral conduction, which this implies, has been directly confirmed for the zebrafinch (Engelage and Bischof, 1988).

An alternative function is thus needed for at least much of the ipsilateral supply. The most likely is that it allows targets viewed within the visual field of one eye to be more readily identified within the other field, when a head movement brings the other field to bear. Such a function would make it probable that ipsilateral routes would also be functional in experiments using monocularity.

Güntürkün and Hahmann (1999) found left–right asymmetry in tectofugal functioning in a test based on visual acuity in the frontal field (see also Chapter 7, by Güntürkün). Pigeons were trained binocularly and then tested monocularly after lesions of the thalamic relay nuclei in the tectofugal pathway. These lesions cut all supply in this pathway to the hemisphere supplied by the nuclei. As a result, lesion of the right relay nuclei meant that input from the LE reached the RHem by the contralateral route, and input from the RE (potentially) reached the RHem by the ipsilateral route; the LHem was cut off. The reverse pattern held for left-side lesions.

In monocular tests, following unilateral lesions, performance was good in birds that were using the RE, only if the supply to the LHem was intact: left lesions greatly depressed performance, while right lesions had little or no effect. In contrast, LE birds showed some depression with both right and left, which was, if anything, more marked with left lesions; this is surprising, since left lesions left intact the main contralateral route from the LE to the RHem.

Bilateral lesions depressed performance as profoundly in LE as in RE birds, so that LE birds were indeed relying on the tectofugal route. It thus seems that, following binocular training (in this specific task), the LHem plays a disproportionately large role in setting up memory and, in consequence, in processing visual input at subsequent test.

Further, a key conclusion is that the evidence from LE birds shows that both the ipsilateral and the contralateral component of the tectofugal route separately allow memory to be established and then accessed to allow successful performance. It is likely therefore that traces can be set up bilaterally as a result of monocular training, under some circumstances.

15.9.2. Transfer

Two striking examples of asymmetries of 'transfer' have recently been described. In the chick, marked differences between chicks given left to right, or right to left reversal of eye use appear late in memory formation.

Anokhin and Tiunova (1996) used a procedure in which chicks fed on familiar food scattered over a floor covered with (firmly fixed) beads of a variety of colours ('bead floor'). Learning was measured by the extent to which pecks at beads came to be inhibited. Monocular training and test resulted in good recall at 3 h, only when the same eye was in use at both times. However, at 24 h there was good recall when LE training was followed by RE test. Recall remained poor for the other reversal, namely RE training and LE test. Use of the same eye as at training still resulted in good recall at this time, whether it was RE or LE.

This asymmetry is present in more dramatic form in the case of memory of hoarding in the marsh tit. Here the two eye systems are known to show the same specializations as in the chick. The test was 5 min after an item had been hoarded, using both eyes. At the test, only one eye was in use. RE birds used local cues in food retrieval (i.e. the appearance of the local surround of the hole into which the food was placed), while LE used position in the overall environment (Clayton and Krebs, 1994). The results for monocular training were as follows: at 3 h after training, use of the same eye (RE or LE) at the test gave good recall, while change to the other gave poor recall. At 7 h there was the first change: performance with the LE in use became poor, even when the LE had been used at training. By 24 h a second change had occurred. There was good performance at test with the RE, whichever eye had been used at training. Performance with the LE was still poor (Clayton, 1993).

The most obvious interpretation, which is that advanced by Clayton (1993), is that a trace initially set up in the RHem is transferred to the left at some point before 24 h, while a trace already resident in the LHem remains where it is.

The evidence that was given in the preceding section suggests an alternative explanation. Ipsilateral input to the LHem from the LE (in monocular training) sets up a trace there, in addition to that set up in the RHem by the contralateral route. The LHem trace is not initially of a character to allow effective recall at test with the RE but, after changes during memory formation, this becomes possible. RE training results in a LHem trace, which is available for effective recall using the RE from the start; however, it does not set up a second trace in the RHem by the ipsilateral route.

It was suggested above that the immediate function of the ipsilateral input might be to provide information allowing a target identified from the input from one eye to be more readily acquired by the other eye system after a head movement. Such information could be quite sketchy (position in the environment; key cues). It is probably provided by linkage to the full trace held by the other hemisphere.

Following LE training in both chick and marsh tit, it is proposed that, at some point in memory formation after 3 h, and before 24 h, a LHem record of an ipsilateral input of this sort becomes linked to the RHem trace (Figure 15.2a). As a result, recall using the RE allows access to both traces, and recall allowing good performance.

Neither study provides a time-course with sufficient resolution to establish whether this linkage occurs in a particular time window. The special state at 7

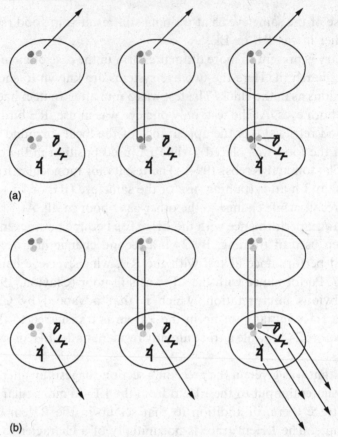

(a)

(b)

Figure 15.2. Recall, linkages and forgetting. (a) The initial trace is shown as similar in both hemispheres, and consisting in cartoon style of groups of dots. Recall is shown as from the left hemisphere (LHem), which is the upper of the pair of hemispheres in each case. In fact, recall would normally be from both hemispheres, when both eyes are in use. If, as is commonly the case, the eyes are used differently, the content of right and left traces would differ. This is here disregarded, and the two are shown as identical. Another oversimplification is that no links are shown between trace fragments or between right and left traces at the beginning of the process of memory formation.

The first pair represents a recall trial before the establishment of new linkage between the left and right traces at coincidences of reactivation (see text). Recall using the right eye makes available the LHem trace in its initial form. After linkage (second pair) recall, which uses the right eye, makes material from the right hemisphere (RHem) trace available more effectively.

As memory formation proceeds, other experiences result in the establishment (or more probably the extension and strengthening) of links with related material in the RHem. This material is shown as established memory, so that links at learning with the consolidating trace could have already formed at learning and affected it (as occurs with reminder procedures; see text). Nevertheless, recall through the LHem is still effective, since it directly accesses only the initial RHem trace.

h, described for memory based on hoarding a food item, establishes only that change occurs in two steps. However, the finding is consistent with the hypothesis that drop in performance of birds, trained and tested with the LE, occurs at a transition at about the 6.7 h point, and so coincident with the 400 min coincidence.

A further assumption is required to explain the findings for hoarding: it is necessary to suppose that the character of the information held by the RHem makes the RHem trace more vulnerable to failures of recall. This is consistent with RHem recording of the layout of the environment; information of this sort must make up part of all the memories of visits to the hoarding area. The evidence from reminder studies (see Section 15.6) shows that interaction with established memories, with which the new material overlaps, can reduce the duration of availability of that material. However, recall through the RE/ LHem still allows access to the relevant RHem material: the LHem trace itself, it has been proposed, records the outcome of response to a specific local cue. Not only would such a trace be more resistant to the development of recall failures but, once activated, its links with the RHem trace would be specific, and might allow access, even when this failed through the LE (Figure 15.2).

There is further evidence for more effective recall via the LHem. Pigeons were trained binocularly in an extensive series of discriminations (725 pairs of discriminations; von Fersen and Güntürkün, 1990). They showed consistently better performance at recall when the RE, rather than the LE, was used at test. Interference from past learning must be powerful during recall with such large numbers of discriminations, much as in human learning of word lists. LHem recording of pairing of certain patterns with reinforcement would explain the finding.

The continuing good recall at 24 h shown in the bead-floor task by chicks trained and tested with the LE is not unexpected, even though it differs from

Caption for Fig. 15.2. (*contd.*)
   (b) Recall through the RHem initially accesses the trace effectively (first pair). Linkage with the LHem trace, once established (second pair), allows its content to affect recall. This is true for both directions of interaction between the hemispheres, and is responsible for the delayed effects on recall (see text) of a variety of experimental manipulations, which are effective only unilaterally.
   As further experiences link the RHem trace more and more extensively with other related material, recall through the RHem becomes less and less effective in allowing the material from the specific learning experience to be accessed in usable form. Instead, a complex of material is accessed (third pair) and acts as general knowledge.

the results for hoarding in tits. Learning in the bead-floor task is based on very many comparable experiences in a short period of time. At each choice, recall might be expected to be dominated by immediately preceding experiences (rather than well-established ones). Further, the beads were very novel, so that the resulting memories are much less likely to have had extensive overlap with existing ones, even in the RHem.

It remains to briefly consider the *thalamofugal route* to the Wulst. Remy and Watanabe (1993) argued that successful 'interocular transfer' depended on use of the tectal supply to the forebrain, rather than the thalamofugal route. This is consistent with the evidence (Goodale and Graves, 1982; Mallin and Delius, 1983) that successful transfer is associated with the use of frontal, rather than lateral fixation in training and test.

It may thus be possible to disregard the thalamofugal route when discussing evidence from tasks that are largely dependent on frontal input like the bead task. Whether this is also true of tasks such as hoarding can only be determined by more detailed studies of eye use during the task.

15.9.3. Forgetting

Evidence has just been reviewed, which suggests that forgetting due to failure of recall may develop sooner in the RHem. In Section 15.8, coincidences of trace reactivation in LHem and RHem were shown to be points of unusually good access to the RHem trace, and it was suggested that interaction at these points might generate forgetting. Forgetting of acoustic memories does indeed occur at the time of the first exact coincidence (6.7 h). If this finding is to be used to flesh out the hints of RHem vulnerability to forgetting, it is necessary to show that appropriate hemispheric specialization for the analysis of sounds is present in the zebrafinch.

This may well be the case. When acoustic input to the forebrain was interrupted unilaterally at the thalamic level, the LHem proved to be better at learning to distinguish two familiar songs, used as conditioned stimuli. In contrast, the RHem was better at detecting a single subtle difference between syllables in similar training (Cynx, Williams and Nottebohm, 1992). It is thus possible that, for sounds too, the RHem has to store many detailed records, so that new records will be exposed to extensive overlap with earlier records. In consequence, difficulties are likely eventually to develop for recall via the RHem. If this were indeed to happen, forgetting at the time of the 6.7 h exact coincidence would be due to effects of linkage of the LHem trace to the corresponding RHem material (above), which is difficult to recall effectively via the RHem (Figure 15.2).

15.10. Trace Transfer Theories and Lesion Studies

There is a growing consensus, based largely on primate and human studies, that memories are distributed as 'fragments', resident in the structures in which perceptual, and further, analysis occurred at the time of learning. Damasio (1989) has even stressed that no structure, which is a potential candidate as a general store, remains unexamined and acceptable in this capacity in the vertebrate brain. There has been movement towards this position in chick studies. Thus Bateson and Horn (1994) argue that memory of the imprinting object appears to be broken down into different properties. Johnston and Rose (Chapter 14) point out that, if these are resident in different structures, then we are probably dealing with a distributed memory.

Nevertheless, there are striking findings from lesion studies, where bulk transfer of the trace seems the obvious explanation. Bulk transfer of traces is not necessarily incompatible with the effects of interactions between right and left versions of the trace of the sort that have been discussed here. Disturbance of recall, as a result of such interaction, for example, could be due to the transfer of degraded material from one hemisphere to the other. However, transfer (or, for that matter, copying) blocks of material from one structure to another seems a hazardous process, likely to introduce errors and open to failure.

It is almost impossible to develop models of memory formation without taking a position on the issue. I here examine explanations alternative to trace transfer for findings for lesion studies in the chick.

A number of points, which need to be made in introduction, are listed below.

1. The trace transfer models are based on simplifying assumptions, which do not necessarily hold, namely that there is a single trace (in each hemisphere), which is resident in one structure at a time. Further, all or nearly all effects of lesions are held to be due to destruction of the trace.

2. In fact, in the case of the bead task, it is clear that more than one structure within each hemisphere is involved in memory formation (Chapter 14 by Johnston and Rose) and that this is true from very early in memory formation (Patterson et al., 1986).

3. Most (or all) lesion studies in the chick have been carried out during the course of memory formation. Note that the evidence already reviewed suggests that, in birds, changes continue in memory formation for at least 24 h, and probably longer. A lesion induced before or during memory formation could have a variety of actions. (a) It might remove the trace or part of it. However, it is worth remembering that the removal of distributed traces

might require lesions so extensive as to interfere with any normal brain functioning. It is possible therefore that lesions capable of removing the entire memory trace have never been studied in chicks. (b) It might abolish modulatory inputs (e.g. noradrenergic) to structures elsewhere, which are essential for normal memory formation. (c) It might produce disturbing inputs (e.g. due to abnormal firing of injured neurones) to neurones elsewhere, involved in memory. (d) It might cut fibres of passage with a variety of functions in memory formation.

4. Differences between the hemispheres in the role played by different structures in memory formation depends in some cases on the normal involvement of both eye systems in learning.

In the case of the bead task, Patterson et al. (1988) showed that amnestic agents, which are effective at a wide range of points in memory formation, acted in the region of the IMHV in the LHem but in the lateral neostriatum in the RHem. Here learning was binocular and the two hemispheres evidently played different parts in the learning process.

However, when learning was monocular, Bell and Gibbs (1977) found the region of IMHV to be sensitive to the full series of agents that have been normally employed. Such sensitivity was present in the hemisphere contralateral to the eye in use, and not the other. It was normal in that each agent showed the normal timing of offset of sensitivity. Crucially, there were now no differences between RE and LE chicks in the consequences of disturbances of IMHV function.

It thus seems that the hemisphere that receives the main contralateral input in the monocular condition is the one that makes important use of the IMHV. This conclusion has already been reached in studies of memory formation following imprinting (for a review, see McCabe, 1991; and below). As has been shown, in the binocular condition, it is predominantly the LHem that does this. This suggests that the IMHV is important in memory formation based on response and the outcome of response; in other words, the IMHV is important in memory based on the sort of processing that engages the LHem.

Unilateral manipulations, including brain lesions, if they depressed general hemispheric function, might shift control to the intact hemisphere in binocular chicks. Evidence that left IMHV lesions, induced before learning, force the RHem into a pattern of function like that normally shown by the LHem has indeed been provided by lesion studies (see below).

5. An important constraint on the interpretation of lesion studies has been that learning causes structural and biochemical changes in the IMHV (McCabe, 1991; Stewart, 1991), which have reasonably been identified as

the basis of (part of) the trace. However, it must also be remembered that comparable changes have been identified in other structures (e.g. for the bead task, lobus parolfactorius (LPO) and palaeostriatum augmentatum: Stewart, 1991; Chapter 14 by Rose and Johnston). Other sites may remain to be discovered. The real issue may be whether, in any particular instance, the kind of information that may be held as a result of changes in the IMHV is essential for adequate performance at recall. Note also that a role in storage does not exclude, but rather makes more likely interaction with other structures during memory formation.

15.10.1. Bead Task

With these points in mind, I turn to a particular study (Gilbert, Patterson and Rose, 1991), the results of which are remarkable, however interpreted. The hypothesis advanced at the time of publication (but see Chapter 14 by Johnston and Rose) is also a convenient introduction to the results. After training, memory is 'collated' in the left IMHV, then transferred briefly to the right IMHV, before finding a perhaps permanent home bilaterally in the lobus parolfactorius. Alternative explanations for the key findings on which this model is based are listed below.

1. Unilateral lesions of the left IMHV 24 h before training cause the loss of memory after training, but right IMHV lesions do not. The explanation that is suggested here is that the nature of the task provides so strong a bias to LHem control that such control occurs despite prior left IMHV lesion. As a result, the initial course of memory formation is so disturbed by the left IMHV lesion that memory loss results. Right IMHV lesions, in contrast, act on a hemisphere with lesser involvement in the task, and reduce that involvement to a point at which any disturbances of function due to the right IMHV lesion are without serious consequences for memory formation.

2. Bilateral IMHV lesions 60 min after training are ineffective, although such lesions induced before training did produce amnesia (Patterson, Gilbert and Rose, 1990), at least by the time of test (180 min). This is consistent with the hypothesis that insult to the IMHV has an effect on memory formation only during an early window of sensitivity.

3. Lesions of the LPO have effects different from those of IMHV lesions: they are effective 1 h after training, but not 24 h before. One possible explanation is that, for some time after such lesions, abnormal discharge from neurones, which have been damaged or deafferented, reaches structures involved in memory formation and disturbs their normal functioning. Evidence has been given that important steps in chick memory formation

continue until 400 min (at least). It is also necessary to assume that, 24 h after the lesion, this sort of discharge has reduced or disappeared, so that the lesions no longer disturb memory formation by this mechanism.

4. This hypothesis (3, above) requires that the LPO is widely and effectively connected with structures crucial to memory formation. It is therefore not surprising to find that LPO lesions 24 h before training, which must prevent the LPO from playing its proper role in memory formation, do have effects that can be revealed by further insult. Although IMHV lesions 1 h after training are normally ineffective, bilateral LPO lesions 24 h before training make right, but not left, IMHV lesions effective, when induced 1 h after training.

Under these conditions both hemispheres take part normally in learning, and the normal asymmetries of function should hold. However, in normal memory formation, input from the LPO (or from systems passing through the LPO) promotes successful consolidation. This may be compared with the effects of noradrenergic activity. After LPO lesions such an effect is lacking and traces are in consequence vulnerable at L3–R2. The fact that only right IMHV lesions are effective at 1 h can be explained if it is assumed that memory loss requires interaction at LHem events (for which a normally functioning LHem is required) with a disturbed RHem. Examples of just such asymmetries have already been reviewed (see Sections 15.4.3, 15.4.4, 15.7 and 15.8).

15.10.2 Imprinting

Trace transfer models were first developed in the chick to explain the effects of lesions in studies using imprinting as the learning task (e.g. Cipolla-Neto, Horn and McCabe, 1982).

Imprinting requires long exposure (e.g. Bolhuis and Johnson, 1988), making it difficult to decide when the processes of memory formation might begin. In general, therefore, it is difficult to use data from imprinting studies in a discussion of sharply timed events in memory formation and this has meant that their coverage here is brief. Lateralization of imprinting processes is reviewed more extensively by Johnston and Rose (Chapter 14). However, the possibility of trace transfer, following imprinting, does require discussion here.

A first point is that it is likely that the two hemispheres are differently, but both extensively, involved in imprinting. McKenzie, Andrew and Jones (1998) found that chicks view an imprinting object with the RE when naïve, but shift to the LE after exposure of 1 h or more. Biochemical and structural changes, consequent on imprinting, are more marked in the LHem. This has indicated that the LHem is of particular importance in the learning involved in imprinting (e.g. McCabe and Horn, 1988). However, the memory of imprinting is persistently sensitive to glutamate given to the RHem (in the region of the IMHV), but not when the glutamate is given to the LHem (Johnston and Rogers, 1998).

The experimental findings (for a review, see McCabe, 1991) may be considered as follows. When lesions follow imprinting, so that both hemispheres take part normally in initial memory formation, there is right–left asymmetry. When IMHV lesions are induced sequentially, the first at 3 h, and the second at 26 h after the end of imprinting, then right first gives amnesia but left first does not. The trace transfer model explains this by supposing that, in the RHem (but not the LHem), the trace that is resident initially in the IMHV moves out bodily to another store at some transition. This occurs later than 3 h, but before 24 h. As a result, when two IMHV lesions are induced, the earlier lesion has to be in the RHem, if it is to catch the trace before it moves. The alternative explanation suggested here is that this is yet another instance where amnesia results from interaction in which a normally functioning LHem accesses a disturbed RHem. This interaction occurs after 3 h, and is prevented by depression of LHem function, if it is the LHem that is lesioned first.

However, when one lesion precedes imprinting and the other follows it (20 h), there is no such asymmetry: amnesia results in both cases. McCabe (1991, p. 271) notes that this suggests that, when the left IMHV is absent during imprinting, this causes the right IMHV 'to take on the putative storage function of the left IMHV'. This may be reformulated in more general terms, as has already been discussed: a left IMHV lesion makes the RHem play the major role in memory formation during imprinting, so that there is now strong RHem involvement in learning and early memory formation.

When both hemispheres are normal during imprinting, recall is sensitive to bilateral lesions of the IMHV up to the time of a transition beginning after 3 h and complete by 6 h (Davey, McCabe and Horn, 1987). The hypothesis of trace transfer explains this by transfer (between 3 and 6 h) of the trace resident in the right IMHV out into some other store, where it is invulnerable to the removal of its former residence. The alternative 'trace interaction' hypothesis is that lesioning the IMHV before a major event in memory formation has occurred causes overall disturbance of the trace. This results from interaction between distributed versions of the memory for which normal IMHV function is necessary. Once that event (or events) is over, memory formation becomes invulnerable to such lesions. The timing of the offset of sensitivity is consistent with the last opportunity for interaction being at the 6.7 h coincidence.

15.11. Comparison with Mammals

Human evidence shows unambiguously that there is asymmetrical involvement of brain structures in both memory formation and in recall. Left prefrontal activation is greater than right during the encoding of episodic memory, even when the material is non-verbal (Tulving et al., 1994). Recall, especially that involving the use of multiple cues in parallel search of memory, followed by verification of match (Shallice and Burgers, 1996), is associated with right prefrontal activation. Further, it seems certain that the specialization of the human right and left hemispheres for different tasks will lead to their differential involvement in the setting up of traces.

However, this does not prove the existence of a temporal structure of events in memory formation like that found in birds. The lack of independent use of the two eyes in mammals must reduce the possibility of the use of independent strategies of search and scanning, at least for vision. Without this independence, left and right traces might not differ in the same way as in birds. The independent scanning by right and left ears, which is shown by some mammals (e.g. *Galago*, Artiodactyla, Perissodactyla), is a hint that this may not be equally true of all mammals, and of all sensory systems.

Consolidatory processes do show some resemblances of timing to those in birds. Thus Gold and Van Buskirk (1975) note that, in the rat, the promotion of consolidation by adrenalin is possible only up to about 30 min after learning. Allweis (1991) compared chick and rat studies, which were carried out under comparable conditions. Hypoxia caused memory loss in the rat after 15 min, much as in the chick; however, memory returned at a much later time, namely by about 180 min (rather than 60 min, as in the chick). In the same study, steps preceding RNA synthesis were deduced to end at about 30 min after learning in the rat, another possible resemblance to the chick. Finally, both chick and rat show a window of sensitivity to anti-NCAM at 6–8 h (Scholey et al., 1993).

Phases of memory formation have been distinguished in the rat on behavioural evidence from a passive avoidance task, as well as by the use of amnestic agents (Izquierdo et al., 1999). A transition at about 6 h (between 6 and 9 h) is suggested by the fact that repeated unreinforced trials do not lead to extinction before 6 h, but do so thereafter. An earlier transition is suggested by the fact that exposure to a novel environment 1 or 2 h after training, but not at 6 h, is amnestic.

Zornetzer, Abraham and Appleton (1978) found that, in mice, unilateral lesions of the locus coeruleus that were induced following (but not before) training made memory susceptible to electroconvulsive shock long after this would normally have ceased to be effective. Zornetzer et al. explained the effectiveness of unilateral, but not bilateral lesions by supposing that 'the conversion of labile to stable memory may normally require a coordinated harmonious processing of information between the two cerebral hemispheres'. There are clearly parallels with the examples of memory loss associated with asymmetrical disturbance in the chick.

None of the above discussion bears on the question of periodic variations in recall. There is some suggestion of the latter in mammals: Holloway (1978) described good recall in rats at 1.5, 12 and 24 h after learning, and poor recall at 6 and 18 h. This was the full set of times studied, so any shorter periodicity would not have been revealed.

Periodic alternation of right and left hemisphere activity is also present, but of a character very different to the events that are timed in relation to a particular learning experience in the chick. Klein and Armitage (1979) described in humans a cyclic alternation in the ability to perform tasks in which either right or left hemisphere involvement gave advantage, which had a mean period of 90–110 min. They linked this to the alternation of rapid eye movement sleep (REM) with the phase of sleep in which REM is absent (non-REM; NREM): awakening in REM gives better performance in spatial, and awakening in NREM gives better performance in verbal tasks. This association has been confirmed subsequently (Gordon, Frooman and Lavie, 1982; for a review, see Shannahoff-Khalsa, 1993).

Broughton (1985) argued that the human 90-min cycle was part of a harmonic sequence running up through 3, 6 and 12 h to 24 h. He confirmed that there was a highly significant cycle with a period of about 90 min. A linkage to processes of memory formation is provided by the finding that deprivation of REM (rather than NREM) tends to oppose consolidation, making memory sensitive to electroconvulsive shock, for example (Fishbein, McGaugh and Swarz, 1971).

It remains to be seen whether the alternation of substantial blocks of time in which left or right control predominates is a peculiarly mammalian condition. If anything similar proves to be present in birds, it presumably is superimposed in some way on the timed processes initiated by each learning experience.

15.12. Physiological Mechanisms

Novel mechanisms are required to explain trace reactivation, timed action of noradrenalin during memory formation and the timing of retrieval events.

A solution to the first problem may be provided by recent evidence that neurones, which are involved in the analysis of a particular stimulus, show associated firing, because all receive facilitation at the same phase of cyclic changes in potential. These changes are simultaneous in large distributed populations of neurones (for a review, see Bressler, 1990; for mechanisms, see Engel et al., 1991; Stern, Jaeger and Wilson, 1998). They may provide a key part of the 'binding codes' postulated by Damasio (1989) as necessary for the simultaneous and linked neural activity, distributed across a number of brain structures, which is required for recall of a memory. Buzaki and Chrobak (1995) have suggested that neuronal activity linked in this way is used to allow 'fast replay' in CA3 during 'sharp wave bursts'. The associations between neurones, that were studied, were set up in rats during exploration. They could be identified, both after acquisition and during replay by a characteristic sequence of discharge amongst the linked units. Replay resembles trace reactivation at a retrieval event, in that it is endogenous (rather than dependent on recall initiated by a perceptual input), and involves associated activation of a population of neurones.

The noradrenergic supply to the forebrain that modulates plasticity derives from the locus coeruleus, which in turn is driven predominantly by the nucleus paragigantocellularis (Aston-Jones et al., 1986). This nucleus, in turn, is driven by startling and painful stimuli, and activates the locus coeruleus in parallel with sympathetic outflow. At the other end of the scale, locus coeruleus activity ceases almost entirely during slow-wave sleep (Gervasoni et al., 1998). Clearly, noradrenergic activity during any particular course of memory formation can vary substantially and randomly due to events unrelated to the experience that gave rise to the consolidating trace. A solution may be provided by specificity conferred by special effects of NE on active neurones, which are active as part of reactivation at a retrieval event.

However, a second problem is that, if NE modulation during memory formation is to be appropriately related to the conditions at learning, it must somehow vary at the right times and places to affect one trace specifically. One possible mechanism is that the linkages that are established at trace formation include an appropriate drive on units in the locus coeruleus. If the learning experience is assessed as important as a result of analysis, and noradrenergic activity is commanded, then this could be made part of the recorded trace. Reactivation would then initiate noradrenergic activity again, at exactly the right times to have effects specific to consolidation of the trace.

Finally, there is the question of timing mechanisms. The changes in potential that may serve to bind trace fragments vary at such high frequencies (10 and 40 Hz) that they seem unlikely to generate the 16- and 25-min cycles present in the chick. Clock genes are a more promising possibility. In *Drosophila, per* affects behavioural cycles that are about a minute in period, as well as circadian rhythms (Kyriacou and Hall, 1980; Kyriacou, van den Berg and Hall, 1990).

15.13. Conclusions

I have argued in this chapter that the specializations of the LHem and RHem, together with the processing made possible by retrieval events, serve a number of functions in memory formation.

1. The fact that the LHem is responsible for visual guidance of response ensures that salient cues, which identify the category of the object and which are used in guidance of response, are selected for recording. These are associated in memory with the character of the response and of its outcome. The recording by the RHem of the detailed properties of the object and of its context means that the two types of information can be efficiently gathered, and stored with the minimum of interference. I assume that the two traces will be to some extent linked at the time of learning.

2. The initial version of a LHem record will be based on a period of attention in which a particular type of target is sought, found after a number of trials, and then a response is planned, carried out and has consequences. The production of such a record is likely to require further processing. The evidence that has been summarized here provides a mechanism by which this might happen. At L0, the first reactivation of the LHem trace gives the opportunity for setting up links that are not solely constrained by order of establishment of records in time. These might, for example, stress relation of response and its outcome. R0 may do the same job for the RHem; since it follows L0, it may allow the record of context to be so organized, that it is appropriately related to the LHem trace as already set up.

The existence of traces in LHem and RHem that are different in content provides the basis for effective processing during memory formation.

1. Perceptual input that is discrepant with the record that is controlling attention may be excluded. The process of comparison between perceptual input and recalled record, which is involved, resembles the interactions

between RHem and LHem versions that occur during memory formation (see below).

2. Discrepant material may be pruned at one of a series of points of interaction between RHem and LHem traces (as exemplified by competition studies). The likelihood of this is modulated by noradrenergic mechanisms, and this in turn depends on conditions at learning (e.g. degree of reinforcement). As a result of pruning, recently acquired material may dominate recall only for a period, before it is lost. This provides a mechanism by which a single experience, which is not associated with effective reinforcement, might affect behaviour only temporarily.

3. When new material is very like established material (e.g. as a result of reminder procedures), association with the established material results, with the new material surviving for some time but in a state such that it is eventually lost. This is more important in the RHem, where it may provide a means of allowing recall of recently acquired material of a kind that overlaps very extensively with established records, without eventually leading to the degradation of established general knowledge.

4. New material, which overlaps with existing material, may remain open to recall through the LHem and its records of specific responses and their outcomes, after recall through the RHem becomes impossible.

Acknowledgement

Most of the work carried out at Sussex University that is discussed here was supported by the Biotechnology and Biological Research Council of the UK, to which, and to the Leverhulme Trust for its support of recent work on chick lateralization, I am very grateful.

References

Allweis, C. (1991). The congruity of rat and chick multiphasic memory-formation models. In *Neural and Behavioural Plasticity; The Use of the Domestic Chick as a Model*, ed. R.J. Andrew, pp. 370–393. Oxford: Oxford University Press.

Allweis, C., Gibbs, M.E., Ng, K.T. & Hodge, R.J. (1984). Effects of hypoxia on memory consolidation: Implications for a multistage model of memory. *Behavioral Brain Research*, 11, 117–121.

Andrew, R.J. (1972). Changes in search behaviour in male and female chicks, following different doses of testosterone. *Animal Behaviour*, 20, 741–750.

Andrew, R.J. (1976). Attentional processes and animal behaviour. In *Growing Points in Ethology*, eds P.P.G. Bateson & R.A. Hinde, pp. 95–133. Cambridge: Cambridge University Press.

Andrew, R.J. (1983). Lateralization of emotional and cognitive function in higher vertebrates, with special reference to the domestic chick. In *Advances in Vertebrate Ethology*, eds J.-P. Ewert, R.R. Capranica, & D.J. Ingle, pp. 477–507. New York: Plenum.

Andrew, R.J. (1991). Cyclicity in memory formation. In *Neural and Behavioural Plasticity: the Use of the Domestic Chick as a Model*, ed. R.J. Andrew, pp. 476–502. Oxford: Oxford University Press.

Andrew, R.J. (1997). Left and right hemisphere traces: Their formation and fate. Evidence from events during memory formation in the chick. *Laterality*, 2, 179–198.

Andrew, R.J. (1999). The differential roles of right and left sides of the brain in memory formation. *Behavioural Brain Research*, 97, 289–295.

Andrew, R.J. & Brennan, A. (1985). Sharply timed and lateralized events at the time of establishment of long term memory. *Physiology and Behavior*, 34, 547–556.

Andrew, R.J. & Rogers, L.J. (1972). Testosterone, search behaviour and persistence. *Nature*, 237, 343–346.

Andrew, R.J., Clifton, P.G. & Gibbs, M.E. (1981). Enhancement of effectiveness of learning by testosterone in domestic chicks. *Journal of Comparative and Physiological Psychology*, 95, 406–417.

Andrew, R.J., Tommasi, L. & Ford, N. (2000). Motor control by vision and the evolution of cerebral lateralization. *Brain and Language*, 73, 220–235.

Anokhin, K.V. & Tiunova, A. (1996). Interocular transfer of learning associated with the formation of categoric memory in chicks. *Proceedings of the Russian Academy of Sciences*, 348, 564–566.

Anokhin, K.V., Mileusnic, R., Shamakina, I.Y. & Rose, S.P.R. (1991). Effects of early experience on c-fos expression in the chick forebrain. *Brain Research*, 544, 101–107.

Aston-Jones, G., Ennis, M., Pieribone, V.A., Nickell, W.T. & Shipley, M.T. (1986). The brain nucleus *locus coeruleus*: Restricted afferent control of a broad efferent network. *Science*, 234, 734–737.

Bateson, P.P.G. & Horn, G. (1994). Imprinting recognition memory. A neural net model. *Animal Behaviour*, 48, 695–715.

Bell, G.A. & Gibbs, M.E. (1977). Unilateral storage of monocular engram in day old chick. *Brain Research*, 245, 263–270.

Bernard, C.C.A., Gibbs, M.E., Hodge, R.J. & Ng, K.T. (1983). Inhibition of long-term memory in the chicken by anti chick Thy-1 antibody. *Brain Research Bulletin*, 11, 111–116.

Bolhuis, J.J. & Johnson, M.H. (1988). Effects of response-contingency and stimulus presentation schedule on imprinting in the chick (*Gallus gallus domesticus*). *Journal of Comparative Psychology*, 102, 61–65.

Bressler, S.L. (1990). The gamma wave: a cortical information carrier? *Trends in the Neurosciences*, 13, 161–162.

Broughton, R.J. (1985). Three central issues concerning ultradian rhythms. In *Ultradian Rhythms in Physiology and Behaviour*, eds H. Schulz & P. Lavie, pp. 217–233. Berlin: Springer Verlag.

Burchuladze, R. & Rose, S.P.R. (1992). Memory formation in day-old chicks requires NMDA but not non-NMDA glutamate receptors. *European Journal of Neuroscience*, 4, 533–538.

Burne, T.H.J. & Rose, S.P.R. (1997). Effects of training procedure on memory formation using a weak passive avoidance learning paradigm. *Neurobiology of Learning and Memory*, 68, 133–139.

Buzaki, G. & Chrobak, J.J. (1995). Temporal structure in spatially organised neuronal assemblies: a role for interneuronal networks. *Current Opinion in Neurobiology*, 5, 504–510.

Chew, S.J., Vicario, D.S. & Nottebohm, F. (1996). Quantal duration of auditory memories. *Science,* 274, 1909–1914.

Cipolla-Neto, J., Horn, G. & McCabe, B.J. (1982). Hemispheric asymmetry and imprinting: The effect of sequential lesions to the hyperstriatum ventrale. *Experimental Brain Research*, 48, 22–27.

Cirelli, C., Pompeiano, M. & Tononi, G. (1996). Neuronal gene expression in the waking state: A role for the locus coeruleus. *Science*, 274, 1211–1215.

Clayton, N. (1993). Lateralization and unilateral transfer of spatial memory in marsh tits. *Journal of Comparative Physiology A*, 171, 799–806.

Clayton, N. & Krebs, J.R. (1994). Memory for spatial and object specific cues in food-storing and non-storing birds. *Journal of Comparative Physiology A*, 174, 371–379.

Clifton, P.G., Andrew, R.J. & Gibbs, M.E. (1982). Limited period of action of testosterone on memory formation in the chick. *Journal of Comparative and Physiological Psychology*, 96, 212–222.

Cozzutti, C. & Vallortigara, G. (2001). Hemispheric memories for the content and position of food caches in the domestic chick. *Behavioral Neuroscience*, 115, 305–313.

Crowe, S.F., Ng, K.T. & Gibbs, M.E. (1989). Memory formation processes in weakly reinforced learning. *Pharmacology, Biochemistry and Behavior*, 33, 881–887.

Crowe, S.F., Ng, K.T. & Gibbs, M.E. (1990). Memory consolidation of weak training experiences by hormonal treatments. *Pharmacology, Biochemistry and Behaviour*, 37, 729–734.

Crowe, S.F., Ng, K.T. & Gibbs, M.E. (1991). Forebrain noradrenaline concentration following weakly reinforced training. *Pharmacology, Biochemistry and Behavior*, 40, 173–176.

Cynx, J., Williams, H. & Nottebohm, F. (1992). Hemispheric differences in avian song discrimination. *Proceedings of the National Academy of Sciences USA*, 89, 1372–1375.

Damasio, A.R. (1989). Time-locked multiregional retroactivation: A systems-level proposal for the neural substrate of recall and recognition. *Cognition*, 33, 25–62.

Davey, J.E., McCabe, B.J. & Horn, G. (1987). Mechanisms of information storage after imprinting in the domestic chick. *Behavioural Brain Research*, 26, 209–210.

Deng, C. & Rogers, L.J. (1998). Bilaterally projecting neurones in the two visual pathways of chicks. *Brain Research*, 794, 281–290.

Ehrlich, D. & Stuchbery, J. (1986). A note on the projection from the rostral thalamus to the visual hyperstriatum of the chicken (*Gallus gallus*). *Experimental Brain Research*, 62, 207–211.

Engel, A.K., König, P., Kreiter, A.K. & Singer, W. (1991). Interhemispheric synchronisation of oscillatory neuronal responses in cat visual cortex. *Science*, 252, 1177–1179.

Engelage, J. & Bischof, H.-J. (1988). Enucleation enhances ipsilateral flash evoked responses in the ectostriatum of the zebra finch (*Taeniopygia guttata castanotis* Gould). *Experimental Brain Research*, 70, 79–89.

Fersen, L. von & Güntürkün, O. (1990). Visual memory lateralization in pigeons. *Neuropsychologia*, 28, 1–7.

Fishbein, W., McGaugh, J.L. & Swarz, J.R. (1971). Retrograde amnesia: Electroconvulsive shock effects after termination of rapid eye movement sleep deprivation. *Science*, 172, 80–82.

Foote, S.L. & Aston-Jones, G.S. (1995). Pharmacology and physiology of central noradrenergic systems. In *Psychopharmacology: The Fourth Generation of Progress*, eds F.E. Bloom & D.J. Kupfer, pp. 335–345. New York: Raven Press.

Friedman, M.B. (1975). How birds use their eyes. In *Neural and Endocrine Aspects of Behaviour in Birds*, eds P Wright, P.G. Caryl & D.M. Vowles, pp. 181–204. Amsterdam: Elsevier.

Gale, E.F., Cundliffe, E., Reynolds, P.E., Richmond, M.H. & Waring, M.J. (1981). *The Molecular Basis of Antibiotic Action*. London: John Wiley.

Gaston, K.E. (1980). Evidence for separate and concurrent avoidance learning in the two hemispheres of the normal chick brain. *Behavioral and Neural Biology*, 28, 129–137.

Gervasoni, D., Darracq, L., Fort, P., Soulière, F., Chouvet, G. & Luppi, P.H. (1998). Electrophysiological evidence that noradrenergic neurones of the rat *locus coeruleus* are tonically inhibited by GABA during sleep. *European Journal of Neuroscience*, 10, 904–970.

Gibbs, M.E. (1983). Memory and behaviour; birds and their memories. *Bird Behaviour*, 4, 93–107.

Gibbs, M.E. (1991). Behavioural and pharmacological unravelling of memory formation. *Neurochemical Research*, 16, 715–726.

Gibbs, M.E. & Ng, K.T. (1977). Psychobiology of memory: Towards a model of memory formation. *Biobehavioural Reviews*, 1, 113–136.

Gibbs, M.E. & Ng, K.T. (1979). Neuronal depolarisation and the inhibition of short-term memory formation. *Physiology and Behavior*, 23, 369–375.

Gibbs, M.E. & Ng, K.T. (1984). Dual action of cycloheximide on memory formation in day old chicks. *Behavioural Brain Research*, 12, 21–27.

Gibbs, M.E. & Summers, R.J. (2000). Separate roles for $\beta2$- and $\beta3$-adrenoceptors in memory consolidation. *Neuroscience*, 95, 913–922.

Gibbs, M.E., Gibbs, C.L. & Ng, K.T. (1979). The influence of calcium on short-term memory. *Neuroscience Letters*, 14, 355–360.

Gibbs, M.E., Ng, K.T. & Andrew, R.J. (1986). Effect of testosterone on intermediate memory in day-old chicks. *Pharmacology, Biochemistry and Behavior*, 25, 823–826.

Gilbert, D.B., Patterson, T.A. & Rose, S.P.R. (1991). Dissociation of brain sites responsible for registration and storage of memory for a one-trial passive avoidance task in the chick. *Behavioural Neuroscience*, 105, 553–561.

Gold, P.E. & Riegel, J.A. (1980). Extended retrograde amnesia gradients: Peripheral epinephrine and frontal cortex stimulation. *Physiology and Behavior*, 24, 1101–1106.

Gold, P.E. & Van Buskirk, R.B. (1975). Enhancement of time dependent memory processes with posttrial epinephrine injections. *Behavioural Biology*, 13, 145–153.

Goodale, M.A. & Graves, J.A. (1982). Interocular transfer in the pigeon: retinal focus as a factor. In *Analysis of Visual Behaviour*, eds D. Ingle, M.A. Goodale & R. Mansfield, pp. 211–240. Cambridge, MA: MIT Press.

Gordon, H.W., Frooman, B. & Lavie, P. (1982). Shift in cognitive asymmetries between wakings from REM and NREM sleep. *Neuropsychologia*, 20, 99–103.

Güntürkün, O & Hahmann, U. (1999). Functional subdivisions of the ascending visual pathways in the pigeon. *Behavioural Brain Research*, 98, 193–201.

Holloway, F.A. (1978). State dependent learning based on time of day. In *Drug Discrimination and State Dependent Learning*, eds B.T. Ho, D.W. Richards & D.L. Chute, pp. 319–343. New York: Academic Press.

Izquierdo, I., Medina, J.H., Vianna, M.R.M., Izquierdo, L.A. & Barros, D.M. (1999). Separate mechanisms for short- and long-term memory. *Behavioural Brain Research*, 103, 1–11.

Johnston, A.N.B. & Rogers L.J. (1998). Right hemisphere involvement in imprinting memory revealed by glutamate treatment. *Pharmacology, Biochemistry and Behavior*, 60, 863–871.

Kaang, B.-K., Kandel, E.R. & Grant, S.G.N. (1993). Activation of cAMP-responsive genes by stimuli that produce longterm facilitation in *Aplysia* sensory neurones. *Neuron*, 10, 427–435.

Kasamatsu, T., Pettigrew, J.D. & Ary, M. (1979). Restoration of visual cortical plasticity by local microperfusion of norepinephrine. *Journal of Comparative Neurology*, 185, 163–187.

Klein, R. & Armitage, R. (1979). Rhythms in human performance: 1.5 hour oscillations in cognitive style. *Science*, 204, 1326–1328.

Kyriacou, C.P. & Hall, J.C. (1980). Circadian rhythm mutations in *Drosophila melanogaster* affect short-term fluctuations in the male's courtship song. *Proceedings of National Academy of Sciences USA*, 77, 6729–6733.

Kyriacou, C.P., van den Berg, M.J. & Hall, J.C. (1990). *Drosophila* courtship song cycles in normal and *period* mutant males revisited. *Behavioural Genetics*, 20, 617–644.

Lappuke, R., Bernard, C.C.A., Gibbs, M.E., Ng, K.T. & Bartlett, P.F. (1987). Inhibition of memory in the chick using a monoclonal anti-Thy-1 antibody. *Journal of Neuroimmunology*, 14, 317–324.

Lewis, D.J. (1976). A cognitive approach to experimental amnesia. *American Journal of Psychology*, 89, 51–80.

Lewis, D.J. (1979). Psychobiology of active and inactive memory. *Psychological Bulletin*, 86, 1054–1083.

Litvin, O.O. & Anokhin, K.V. (1999). Mechanisms of memory reorganisation during retrieval of acquired behavioural experience in chicks: Effects of protein synthesis blockade in the brain. *Journal of Higher Nervous Activity*, 49, 554–565.

Mallin, H.D. & Delius, J.D. (1983). Inter- and intraocular transfer of colour discriminations with mandibulation as an operant in the head-fixed pigeon. *Behavioural Analysis Letters*, 3, 297–309.

Martin, G.R. & Katzir, G. (1995). Visual fields in ostriches. *Nature*, 374, 19–20.

McCabe, B.J. (1991). Hemispheric asymmetry of learning-induced changes. In *Neural and Behavioural Plasticity: The Use of the Domestic Chick as a Model*, ed. R.J. Andrew, pp. 262–276. Oxford: Oxford University Press.

McCabe, B.J. & Horn, G. (1988). Learning and memory: Regional changes in *N*-methyl-D-aspartate receptors in the chick brain after imprinting. *Proceedings of the National Academy of Sciences USA*, 85, 2849–2853.

McLean, J. & Waterhouse, B.D. (1994). Noradrenergic modulation of cat area 17 neuronal response to moving visual stimuli. *Brain Research*, 667, 83–97.

McKenzie, R. (1991). Lateralization of auditory and visual learning in the domestic chick. D.Phil. thesis, University of Sussex.

McKenzie, R. & Andrew, R.J. (1996). Brief retention events associated with cyclically recurring left hemisphere events in the domestic chick. *Physiology and Behavior*, 60, 1323–1329.

McKenzie, R., Andrew, R.J. & Jones, R.B. (1998). Lateralization in chicks and hens: New evidence for control of response by the right eye system. *Neuropsychologia*, 36, 51–58.

Miklósi, A., Andrew, R.J. & Dharmaretnam, M. (1996). Auditory lateralization: Shifts in ear use during attachment in the domestic chick. *Laterality*, 1, 215–224.

Ng, K.T. & Gibbs, M.E. (1988). A biological model for memory formation. In *Information Processing by the Brain*, ed. H. Markowitsch, pp. 151–178. Toronto: Hans Huber.

Ng, K.T. & Gibbs, M.E. (1991). Stages in memory formation: a review. In *Neural and Behavioural Plasticity: The Use of the Domestic Chick as a Model*, ed. R.J. Andrew, pp. 351–369. Oxford: Oxford University Press.

Patterson, T.A., Gilbert, D.B. & Rose, S.P.R. (1990). Pre- and post-training lesions of the intermediate medial hyperstriatum ventrale and passive avoidance learning in the chick. *Experimental Brain Research*, 80, 189–195.

Patterson, T.A., Alvarado, M.C., Warner, I.T., Bennett, E.L. & Rosenzweig, M.R. (1986). Memory stages and asymmetry in chick learning. *Behavioural Neuroscience*, 100, 850–859.

Patterson, T.A., Alvarado, M.C., Rosenzweig, M.R. & Bennett, E.L. (1988). Time courses of amnesia, development in two areas of the chick forebrain. *Neurochemistry Research*, 13, 643–647.

Radushkin, K.A. & Anokhin, K.B. (1997). Recovery of memory disrupted during learning: Reversal of the effect of agents inhibiting the protein synthesis. *Russian Journal of Physiology*, 83, 11–18.

Rashid, N. & Andrew, R.J. (1989). Right hemisphere advantage for topographical orientation in domestic chick. *Neuropsychologia*, 27, 937–948.

Remy, M. & Watanabe, S. (1993). Two eyes and one world: Studies of interocular and intraocular transfer in birds. In *Vision, Brain and Behaviour in Birds*, eds H.P. Ziegler & H.-J. Bischof, pp. 137–158. Cambridge, MA: MIT Press.

Rickard, N.S., Poot, A.C., Gibbs, M.E. & Ng, K.T. (1994). Both non-NMDA and NMDA glutamate receptors are necessary for memory consolidation in the day-old chick. *Behavioral and Neural Biology*, 62, 33–40.

Rogers, L.J. & Anson, J.M. (1979). Lateralization of function in the chicken forebrain. *Pharmacology, Biochemistry and Behavior*, 10, 678–686.

Rosenzweig, M.R., Bennett, E.L., Martinez, J.L., Beniston, D., Colombo, P.J., Lee, D.W., Patterson, T.A., Schulteis, G. & Serrano, P.A. (1991). Stages of memory formation in the chick: Findings and problems. In *Neural and Behavioural Plasticity: The Use of the Domestic Chick as a Model*, ed. R.J. Andrew, pp. 394–418. Oxford: Oxford University Press.

632   R. J. Andrew

Sandi, C. & Rose, S.P.R. (1994). Corticosterone enhances long-term retention in one day-old chicks trained in a weak passive avoidance learning paradigm. *Brain Research*, 647, 106–112.

Scholey, A.B., Rose, S.P.R., Zamani, M.R., Bock, E. & Schachner, M. (1993). A role for the neural cell adhesion molecule in a late, consolidating phase of glycoprotein synthesis six hours following passive avoidance training of the young chick. *Neuroscience*, 55, 499–509.

Schumacher, D.P. & Hall, S.S. (1982). An efficient stereospecific total synthesis of (±)-Anisomycin and related new synthetic antibiotics. *Journal of the American Chemical Society*, 104, 6076–6080.

Serrano, P.E., Beniston, D.S., Oxonian, M.G., Rodriguez, W.A., Rosenzweig, M.R. & Bennett, E.L. (1994). Differential effects of protein kinase inhibitors and activators on memory formation in the 2-day old chick. *Behavioral and Neural Biology*, 61, 60–72.

Shallice, T. & Burgers, P. (1996). The domain of supervisory processes and temporal organisation of behaviour. *Philosophical Transactions of the Royal Society of London Series B*, 351, 1405–1412.

Shannahoff-Khalsa, D. (1993). The ultradian rhythm of alternating cerebral hemispheric activity. *International Journal of Neuroscience*, 70, 285–298.

Spear, N.E., Gordon, W.C. & Chiszar, D.A. (1972). Interaction between memories in the rat: Effect of prior learning on forgetting after short intervals. *Journal of Comparative and Physiological Psychology*, 78, 471–477.

Stanton, P.K. & Sarvey, J.M. (1985a). Depletion of norepinephrine, but not serotonin, reduces long-term potentiation in the dentate gyrus of rat hippocampal slices. *Journal of Neuroscience*, 5, 2169–2176.

Stanton, P.K. & Sarvey, J.M. (1985b). The effect of high-frequency electrical stimulation on cyclic AMP levels in normal versus norepinephrine-depleted rat hippocampal slices. *Brain Research*, 358, 343–348.

Steele, R.J. & Stewart, M.G. (1993). 7-Chlorokynurenate, an antagonist of the glycine binding site on the NMDA receptor, inhibits memory formation in day-old chicks (*Gallus domesticus*). *Behavioral Neural Biology*, 60, 89–92.

Stephenson, R.M. (1981). Memory processing in the domestic chick (*Gallus gallus*): A psychopharmacological investigation. D.Phil. thesis, University of Sussex.

Stephenson, R.M. & Andrew, R.J. (1981). Amnesia due to β-antagonists in a passive avoidance task in the chick. *Pharmacology and Biochemistry of Behavior*, 15, 597–604.

Stern, E.A., Jaeger, D. & Wilson, C.J. (1998). Membrane potential synchrony of simultaneously recorded striatal spiny neurones *in vivo*. *Nature*, 394, 475–478.

Stewart, M.G. (1991). Changes in dendritic and synaptic structure in chick forebrain consequent on passive avoidance learning. In *Neural and Behavioural Plasticity: The Use of the Domestic Chick as a Model*, ed. R.J. Andrew, pp. 305–328. Oxford: Oxford University Press.

Summers, M.J., Crowe, S.F. & Ng, K.T. (1995). Administration of glutamate following a reminder induces transient memory loss in day-old chicks. *Cognitive Brain Research*, 3, 1–8.

Summers, M.J., Crowe, S.F. & Ng, K.T. (1996). Administration of lanthanum chloride following a reminder induces transient loss of memory retrieval in the day-old chick. *Cognitive Brain Research*, 4, 109–119.

Summers, M.J., Crowe, S.F. & Ng, K.T. (2000). Modification of a weak learning experience by memory retrieval in the day-old chick. *Behavioral Neuroscience*, 114, 713–719.

Tommasi, L., Vallortigara, G. & Zanforlin, M. (1997). Young chickens learn to localise the centre of a spatial environment. *Journal of Comparative Physiology A*, 180, 567–572.

Tulving, E., Kapur, S., Craik, F.I.M., Moscovitch, M. & Houle, S. (1994). Hemispheric encoding/retrieval asymmetry in episodic memory; positron emission tomography findings. *Proceedings of the National Academy of Sciences USA*, 91, 2016–2020.

Vallortigara, G. (1992). Right hemisphere advantage for social recognition in the chick. *Neuropsychologia*, 30, 761–768.

Vallortigara, G. & Andrew, R.J. (1991). Lateralization of response to change in social partner in chick. *Animal Behaviour*, 41, 187–194.

Vicario, D.S. & Nottebohm, F. (1998). Quantal memory durations: Observations not reproduced. *Science*, 279, 1437.

Wallman, J. & Pettigrew, J.D. (1985). Conjugate and disjunctive saccades in two avian species with contrasting oculomotor strategies. *Journal of Neuroscience*, 5, 1418–1428.

Wallman, J. & Velez, J. (1985). Directional asymmetries of optokinetic nystagmus: Developmental changes and relation to the accessory optic system and to the vestibular system. *Journal of Neuroscience*, 5, 319–329.

Zornetzer, S.F., Abraham, W.C. & Appleton, R. (1978). Locus coeruleus and labile memory. *Pharmacology and Biochemistry of Behavior*, 9, 227–234.

Epilogue

RICHARD J. ANDREW AND LESLEY J. ROGERS

Recent progress in our understanding of cerebral lateralization in vertebrates, as a whole, has been striking. The field has moved from studies designed only to show whether asymmetry of some sort is present in a range of different species to research extending knowledge of the most basic features of lateralization and the conditions under which they probably evolved and in which they develop. The flowering of studies on fish has brought better understanding of the conditions under which lateralization may have first evolved. Major advances (and surprises) are bound to come from a full elucidation of the asymmetries of the brain of Amphioxus, and perhaps the central nervous system of other primitive chordates. Although functional (and behavioural) studies remain to be done, the mere existence in Amphioxus of a direct input from the homologue of the paired eyes of vertebrates to motor mechanisms, which exists only on the left side of the brain, is enough to push our evolutionary horizon back to beyond the origin of the vertebrates.

At present, the clearest hypothesis, which might guide new research, is the proposition that the basic properties of cerebral lateralization are present and similar throughout the vertebrates. This now calls for rigorous testing. Past experience shows that it is not enough to fail to find any evidence of asymmetry in a single type of test (particularly behavioural). Indeed, in the case of the rhesus monkey, extensive and varied testing failed to find lateralization, which subsequent work (largely in the same laboratory) has nevertheless revealed to be extensive and in line with that shown by other primates. The best strategy is likely to be to concentrate effort on a few widely separated species. Aside from the primates, rat, chick and zebra fish are obvious candidates, on the basis of existing knowledge both of lateralization and of supporting fields (e.g. genetic control of central nervous system development in the zebra fish and knowledge of developmental events in the chick).

It is important to note that the hypothesis that basic properties of lateralization are similar throughout the vertebrates does not imply that all forms of lateralization are generated by the direct consequences of gene expression within the developing central nervous system (CNS). Extensive evidence, discussed in this book, shows how consistent asymmetries outside the CNS can affect the development of asymmetric sensory input to the CNS. The effects of head posture in developing bird embryos is a striking example of this. Lateralization of learning dependent on such asymmetries in early life is, perhaps, an equivalent effect.

At the same time, it would be surprising if no changes in lateralization had appeared in some species or populations, and it is likely to be rewarding to examine candidate species for which preliminary evidence suggests that lateralization is unusual (or even absent). There are hints that some fish may show marked heterozygosity of genetic control of asymmetries in visual structures, which may in turn imply differences from the usual pattern of lateralization in tetrapods. No doubt other differences between systematic groups will be found.

However, the interpretation of such differences awaits a better understanding of exactly what features of lateralization vary between individuals in those species for which we already have adequate knowledge of individual variation in lateralization (i.e. humans and rats). Two categories of individual differences should be distinguished. The first category includes those variations that are sufficiently rare to be explicable by mutation pressure: complete reversal of the left–right axis would be the obvious example. It is certainly rare and should not markedly affect viability, since all asymmetries remain congruent. The second category includes those variations between individuals that are so common that it is likely that they are maintained within the population by some selective advantage. The latter category may be generated by the interaction of genetic, hormonal and experiential factors, the relative contributions of each to the outcome varying between individuals (e.g. in the chick, by the interaction of light experience and varying steroid hormone levels). If the thesis that the basic properties of lateralization remain constant is correct for species with such variability, it is necessary to distinguish properties that are always associated from ones for which left–right assignment varies.

In rats and humans, a major source of individual variation is known to be asymmetry in the effectiveness of dopaminergic inputs to the cerebral hemispheres. A simple interpretation is that variation in this asymmetry affects which side of the brain is likely to take control of response. In the chick (and probably some other vertebrates) there are shifts in which hemisphere is controlling behaviour during postembryonic development; the mechanisms responsible are unknown, but they are too fast to be explicable by changes in neural connectivity. Instead, they could depend on changes in the balance of monoaminergic inputs to the hemispheres.

If bias in monoaminergic input to the forebrain is a main source of individual variation in humans and rats, then the resulting differences would be very similar to those postulated by theories of 'hemisphericity' in humans, with individuals having a dominant hemisphere that sets their cognitive style. Despite the mixed fortunes of this approach in human work, it deserves testing in the animal models that are now available.

Animal studies, which use tests based on perceptual processing, attentional strategies and visual control of response, have in general suggested that reversals of allocation of lateralized functions to the hemispheres are rare in normal development. What this evidence amounts to is that significant right–left differences have been obtained with modest group sizes.

Another related theme is the surprising extent to which environmental inputs during development affect brain asymmetries. The most dramatic examples are in the visual system of the chick, in which the exposure of the right eye to light entering through the shell generates asymmetries in the thalamofugal projections to the forebrain, and in the pigeon, in which light exposure generates asymmetries in the tectofugal projections. The head position that generates the exposure of the right eye is almost or quite invariable: no data on the existence of variations appear to exist. This mechanism of generating central nervous system asymmetries thus could have evolved as the most effective way of generating widespread and functional left–right differences. There has been some research on reversed individuals in the chick, achieved by experimentally exposing the left eye of the embryo to light, instead of the usual right eye exposure. Clearly more work of this kind is needed to investigate the associations between different types of lateralization, as well as the effects of experience on the development of lateralization.

Although the normal exposure of the avian embryo's right eye to light can hardly be a source of reversal of normal lateralization, natural variation in the degree of such exposure does seem likely to generate variation in the degree of consequent asymmetries. Such variation in the pattern of lateralization is also likely to result from the effects of hormones, which modify the sensitivity of the visual pathways to the light exposure. In rodents, steroid hormones (from intrauterine partners) are also likely to cause individual variations in lateralization in rodents; the intensity of the effect depends on the sex of the partners. These shared hormones affect motor bias (which also is related to dopaminergic asymmetries) as well as species-specific patterns like aggressive and sexual behaviour. The importance of such variation between individuals deserves examination.

A variety of factors affect lateralization: in fact, lateralization is a classic example of interaction between genetic, hormonal and experiential factors. A potentially unifying hypothesis is as follows: patterns of lateralization function well only if a number of functional asymmetries are consistently organized (i.e. associated with one hemisphere or the other). Behavioural lateralization is subject to intense selection pressure, since it affects not only cognitive ability, but also the likelihood and intensity of performance of, for example, avoidance, and aggressive, sexual and parental behaviour. Disturbance of the harmonious allocation of function may be avoided by using modulation later in development as the means of generating the necessary range of phenotypes.

Modulation of lateralization during development may generate behaviour appropriate to the particular conditions that the individual is likely to encounter in later life. For this to be possible, the external factor that brings about modulation must correlate with likely future conditions. Thus stress during uterine life in rodents like mice and rats may predict continuing crowded conditions in which the litter will have to survive. Alternatively modulation may act at random and serve to generate variability in patterns of behaviour. In rats and mice, modulation dependent on the sex of uterine partners produces phenotypic variation in behaviour by routes that include effects on patterns of lateralization. This may ensure that some offspring at least will be well suited to whatever conditions are encountered in the future. Such an evolutionary strategy is probably available only to species that produce large numbers of offspring during a lifetime.

Most of the available evidence on which the arguments advanced in this book are based derives from vision. It has, indeed, been argued in this book that lateralization of the central nervous system arose in association with the evolution of vision in microphagous chordate ancestors of the vertebrates (with strongly asymmetric bodies): visual lateralization provided a means of timing and directing the responses involved in the capture of prey. Further, frequent independent movement, and use, of the eyes in non-mammalian vertebrates seems to be ideally suited to sustaining the initial asymmetry and driving its further evolution.

Despite the importance of visual lateralization, hearing, touch and olfaction certainly all show lateralization. Other senses may well do so as well: the electric sense and the lateral line are obvious candidates for study. It is inevitable that the senses would evolve or develop associated patterns of lateralization, since they are used together in the tasks that animals perform. If one sense, say vision, is being used by the left hemisphere to control a response and, at the same time, the right hemisphere is being used in diffuse attention to the environment, then other senses are likely to show similar left–right differences.

This is not to say that senses other than vision have not shaped some steps in evolution. A particularly interesting possibility is that, during the reduction in the use of vision by mammalian ancestors that led to the subsequent reshaping of visual mechanisms and crucially the evolution of obligate yoking of eye movements, hearing may have dominated lateralized mechanisms. Sounds are not routed to one or other side of the brain; instead, it is their perceived position to the left or right that determines which hemisphere undertakes their analysis. Reliance on hearing in mammals may have maintained cerebral lateralization, and in later evolution emphasized the role of perceived position for vision as well.

Even in lower vertebrates (toads), visual lateralization sometimes depends on perceived position, in the left or the right visual hemifield, based on the animal's midline, rather than on whether the left or right eye is being used in viewing. There is evidence that something seen by the LE only just over the midline in the right hemifield is treated as appropriate to LHem processing. A map of the body relative to surrounding space may thus be essential for both visual as well as auditory lateralization.

It is worth commenting on the richness of the behaviour by which lateralization is revealed (and can be studied). Animals bring the right or left eye or ear to bear according to the task in hand. Asymmetries of facial expression reveal lateralized control; perhaps they reveal this to social fellows as well as experimenters. Complexities that at first seemed certain to be exclusively human or primate turn out to be present in anuran amphibians: the need for postural support had to be balanced against choice of forelimb for use in response as soon as tetrapods evolved. The handedness of this may have built on lateralization of visual responding retained from teleost ancestors.

Most striking of all, the availability of two processing systems, each with differing specializations, has evidently been fundamental to learning and memory formation for a very long time in vertebrates. The evidence from memory formation in the chick indicates the elaboration of two records of a single experience, one stressing response and consequence of response (LHem), and one forming a relatively complete record including contextual information (RHem). The subsequent interaction of these records appears to provide a robust and simple way in which information can be held available for appropriate periods, without causing the degradation of established bodies of information, and with account being taken of the likely importance of the new information.

The study of lateralization is likely to become more and more important in research on animal cognition. Until very recently, the main thrust of research had been to characterize structures of right and left each as the mediator of particular (readily definable) functions. Emphasis is likely to shift to understanding the way in which right and left structures interact. It is already clear that it is not enough to characterize RHem as spatial and so responsible by itself for tasks with a large spatial component. Sustained approach to a selected distant target is likely to require LHem control, perhaps making use of memory records in the right hemisphere.

Although little has been said here of genetic control of lateralization, it is from such studies that many novel findings concerning the nature of lateralization, at least in lower vertebrates, are to be expected in the next decades. For example, genes that are involved in the generation of visceral asymmetries are known to be expressed asymmetrically in the brain of the zebrafish. Studies of zebrafish will identify the brain structures in which asymmetries originate. Mutants may (depending on viability) provide means of studying particular asymmetries.

Despite the new promise of studying genetic determinants of lateralization, it remains important to keep in mind the importance of experience (including learning) in the development of lateralization. Only in this way will we fully understand the interaction of genetic, hormonal and experiential factors. This, in turn, is necessary for adequate study of the function of lateralization in individuals and in populations.

Author Index

Subject Index

Almost all entries are related to lateralization or, more generally, to asymmetries of anatomy or behaviour. 'Lateralization' and 'asymmetry' are therefore in general not included in entries.

Richard Andrew

Printed in the United States
By Bookmasters